BING C. MEI
bmei@ecs.umass.edu

Partially Ordered Systems

Springer

New York
Berlin
Heidelberg
Hong Kong
London
Milan
Paris
Tokyo

Partially Ordered Systems

Editorial Board: L. Lam • E. Guyon • D. Langevin • H.E. Stanley

Maurice Kleman   Oleg D. Lavrentovich

Soft Matter Physics

An Introduction

Foreword by J. Friedel

With 256 Illustrations, 4 in Full Color

 Springer

Maurice Kleman
Laboratoire de Minéralogie-Cristallographie
  de Paris
Université Pierre et Marie Curie
4 Place Jussieu, Case 115
F-75005 Paris
France
maurice.kleman@mines.org

Oleg D. Lavrentovich
Chemical Physics Interdisciplinary
  Program
Liquid Crystal Institute
Kent State University
Kent, OH 44242
USA
odl@lci.kent.edu

Cover illustration: Painting by Jackson Pollock: *Shimmering Substance* from the Sounds in the Grass Series. (1946) Oil on canvas, 30 1/8 x 24 1/4" (76.3 x 61.6 cm). The Museum of Modern Art, New York. Mr. and Mrs. Albert Lewin and Mrs. Sam A. Lewisohn Funds. Photograph © 2001 The Museum of Modern Art, New York.

Library of Congress Cataloging-in-Publication Data
Kleman, Maurice.
    Soft matter physics: an introduction / Maurice Kleman, Oleg D. Lavrentovich.
        p.   cm.—(Partially ordered systems)
    Includes bibliographical references and index.
    ISBN 0-387-95267-5 (alk. paper)
    1. Soft condensed matter.   2. Order-disorder models.   I. Lavrentovich, Oleg D.   II. Title
III. Series.
QC173.458.S62 K55 2001
530.4'13—dc21   2001020449

ISBN 0-387-95267-5   Printed on acid-free paper.

Printed in the United States of America.

9 8 7 6 5 4 3 2 1   SPIN 10556875

www.springer-ny.com

Springer-Verlag   New York Berlin Heidelberg
A member of BertelsmannSpringer Science+Business Media GmbH

Foreword

Introductions to solid state physics have, ever since the initial book by F. Seitz in 1940, concentrated on simple crystals, with few atoms per cell, bonded together by strong ionic, covalent, or metallic bonds. References to weaker bonds, such as van der Waals forces in rare gases, or to geometric or chemical disorder (e.g., alloys or glasses) have been limited.

The physical understanding of this field started well before Seitz's book and led to a number of Nobel prizes after the last war. Applications cover classical metallurgy, electronics, geology and building materials, as well as electrical and ionic transport, chemical reactivity, ferroelectricity and magnetism.

But in parallel with this general and well publicized trend, and sometimes earlier as far as physical concepts were concerned, an exploration and increasingly systematic study of softer matter has developed through the twentieth century. More often in the hands of physical chemists and crystallographers than those of pure physicists, the field had for a long time a reputation of complexity. If progress in polymers was steady but slow, interest in liquid crystals had lain dormant for forty years, after a bright start lasting through 1925, to be revived in the late 1960s based on their possible use in imaging techniques. The optoelectronic properties of the field in general are even more recent.

Maurice Kleman's initial research interests have been in the study of magnetoelastic effects in ferromagnetic crystals and films, a field where he was able to apply Kröner's techniques of infinitesimal dislocations to the study of inhomogeneous magnetism in walls, lines, and points, as initiated by P. Weiss and developed notably by L. Néel. When P.G. de Gennes started developing an interest in liquid crystals in Orsay, it was natural for Kleman to turn his own attention to this field, where many mesoscopic phenomena have to be attacked from somewhat similar points of view. True to his initial interests, Kleman kept as a main objective the understanding of the possible defects of such structures, a line of attack first explored by my grandfather G. Friedel and revived in the 1950s by Kleman's friend C.F. Frank. Kleman is now well known for his work in the general field of defects in soft matter, summarized in part in a little book on points, lines, and walls. The other author, Oleg D. Lavrentovich, studied in Kiev the structures of liquid crystal droplets, which he produced by a new method, with controlled surface anchoring conditions, which allowed him to recognize rather early the presence of a TGB phase as foreseen by analogy with superconductors of the second kind. After a long stay as a visiting scientist in Orsay with

Kleman, where he studied, with P. Boltenhagen, the splitting of oily streaks into focal conic domains, he has taken a position at Kent State University. Maurice Kleman and Oleg D. Lavrentovich now present us the results of their researches and teachings at the graduate and postgraduate levels in soft matter physics. Structure properties only are considered here.

As made clear in Chapters 1 to 4, soft matter is here mostly made of molecules weakly bound by intermolecular forces of various origins, excluding metallic, covalent, and ionic bonds. Thus soft matter can encompass biological matter, although not much of this is treated in this book, except for some properties of assembled DNA molecules. Because of the often complex and flexible forms of the molecules involved, entropy plays a leading role at the origin of various possible forms of "mesomorphic" phases of the liquid crystals, as well as of many easily produced distortions or defects. Extensions of similar concepts to liquid, amorphous, or quasicrystalline phases met in "strong" solids are stressed. To study simply the phase changes involved, one has first to define an order parameter with its characteristic phase and amplitude, smoothly varying in space and time. This approach, popularized by Landau, effectively neglects local atomic or thermal inhomogeneities but provides a general framework applicable to similar mesoscopic problems in superfluidity, magnetism, or phase changes in strong solids. The general concepts derived for analyzing phase changes in the main types of liquid crystals are thus compared with similar approaches for strong superconductivity or very anisotropic magnetism.

The two following chapters relate to possible static or dynamic distortions in liquid crystals: Chapter 5, on elasticity, gives a particularly clear presentation of a field where boundary conditions play a leading role; Chapter 6 presents many aspects of dynamics and viscosity.

Chapter 7, on fractals and growth phenomena, introduces the subject of surface effects, which is pursued further in Chapter 13. Here again, the two aspects of static and dynamical properties are clearly distinguished, even if the field treated could justify longer developments.

Finally, a large part of the book covers the problems of line and point defects in the various liquid crystal phases, as compared with classical strong crystals. This rich field, which covers much original work by the authors, is presented in a rather complete and original way. The two last chapters cover colloids and polymers in a clear, albeit summary, way.

Taken as a whole, this book provides a good introduction to the general background in the study of soft matter. The main concepts involved are presented in a clear and simple way. Short exercises at the end of each chapter together with a short bibliography help readers to broaden their knowledge. The core of the book concentrates on liquid crystals, their numerous phases and their possible static and dynamic conformations, with an emphasis on the role played by boundary conditions and especially free surfaces. But soft matter is not restricted to liquid crystals: Polymers and colloids are also considered, if more briefly. And the various concepts developed for the study of liquid crystals find their equivalents in some problems of "strong" matter: the role of lines and points defects in magnetism, various types of dislocations in liquids, amorphous or quasicrystalline phases; these various aspects are properly mentioned, though not treated to the same depth.

As rightly emphasized by the authors, a striking feature of soft matter is the specific role played by the *mesoscopic range*. In most cases, the direct molecular interactions are of short range. But the way to pave space with such molecules, of various forms, flexibilities, and viscosities, keeping a reasonably compact and stable arrangement, can lead to a variety of different solutions that might differ at long range only. In the search for such solutions, the concept of *coherence length* was first developed in similar problems of magnetism and superfluidity: This is a measure of the size of the mesoscopic range where a type of arrangement imposed on the border of a range is transmitted, with decreasing strength, to the other border. But from the study of soft matter, a new concept has emerged, that of *range of frustration*; in many cases, a local or especially stable arrangement of atoms or molecules cannot be extended far, because it creates too large intermolecular tensions. Examples referred to in this book are the double twist in cholesterics and the icosahedral packing of atoms in liquid, amorphous, or quasicrystalline phases. In such cases, the frustration range is limited by the development and suitable folding of disclination lines, as for instance in blue phases. Thus the nature and symmetry of short-range intermolecular forces can dominate not only the size of the network of dislocations but the long-range structure of the network. The authors have contributed greatly to the emergence of such a concept, and if anything, it could have been developed even further in this book.

Like most of their predecessors, the authors introduce the *line singularities* (dislocations, disclinations) by the Volterra process, then classify possible singularities (points, lines, walls) by the topological approach introduced by G. Toulouse and M. Kleman. The subject is treated in a progressive way, first in solids, then in smectics A, with their dislocations, disclinations, and specific focal conics, where the authors have recently added to our knowledge. Cholesterics and nematics are then treated in depth, with a discussion of liquid relaxation and the general importance of topological classification.

It is indeed rightly stressed that the Volterra process starts with a solid medium, while the topological approach assumes complete viscous *relaxation* of stresses on an at least partly liquid medium. As this relaxation increases from smectics to cholesterics and nematics, the passage through a Volterra process might look more and more artificial, and indeed it is a pity that no more has been done on the physical properties of disclinations in nematics, related to the noncommutativity of their topology: What is the equivalent of F.C. Frank's "kinks," produced by the crossing of two dislocations in a crystal, when two disclinations cross in a nematic?

However, liquid relaxation after a Volterra process in a frozen medium helps us to understand a number of characteristic features of the singular lines in liquid crystals: It reduces the stored energy of dislocations and disclinations and allows all these singular lines to be flexible and mobile; for disclinations, it fixes the orientation of the cut surface of the initial Volterra process, so as to minimize the energy; it allows the topological elimination of some lines by an escape into the third dimension, thus creating pairs of singular points; it also allows the characteristic rotation to be tangent to the disclination line, even a curved one. The Volterra process can finally explain in a natural way why a number of dislocation configurations can be maintained, although not predicted by topological arguments: This can refer, for instance, to slip at low temperatures in amorphous solids or in quasicrystals; it can also refer to boundary or initial conditions, which can maintain lines with a relaxed and continuous core, such as cracks in motion and disclinations in a tube of nematics with molecules perpendicular to the surface of the tube.

These remarks justify, I think, the plan followed by the authors, although liquid relaxation could have been introduced earlier in the book, for dislocations as well as for disclinations or focal conics. The continuous distribution of infinitesimal dislocations produced by such relaxations is precisely that first imagined by Volterra in continuous solids and which J.F. Nye, B.A. Bilby, and E. Kröner developed later in the context of flexion and torsion of solids, as recalled earlier in the book.

Some other comments could be made in the presentation of the book, if not in its substance

- The long-range Landau approach to phase changes neglects *short-range order* effects, which can be significant even in first-order transitions. Thus, as already pointed out in 1930 by G. Foëx, the short-range effects observed in the magnetic properties on both sides of a nematic isotropic transition do not necessarily imply a second-order phase transition at the equivalent of a Curie point in a ferromagnet; but their effects, known in the nematic phase since before the First World War as "swarms" (responsible for turbidity and correctly analyzed in the long-wave limit by P.G. de Gennes), as well as the equivalent effects observed by light scattering in G. Durand's group in the early 1970s in the isotropic phase, strongly reduce the latent heat and increase the temperature variations of the effective Landau parameters. Indeed, the large (optical) range of these fluctuations poses the question of the convergence of a Landau development, which in fact limits itself to very small groups of molecules. Similarly, short-range orientational order can exist in polymers without them showing a transition to a nematic phase, as shown, for example, by B. Deloche in molten polymers as well as in polymeric membranes, using resonance techniques. By skipping rather quickly over such effects, the authors might give too rigid a picture of a field where fluctuations are all important.

- Some "historical" references could have been usefully more fully developed. Thus to say that dislocations in solids began to be studied just before the Second World War probably refers to the fundamental work produced by J.W. Burgers and by R. Peierls in 1939; but the concept of dislocation and disclination lines in continuous solids dates from Volterra, before the First World War; and its transfer to crystals dates from the early 1920s, together with many applications to crystal plasticity. On the other hand, the "Cano" geometry of a tilt boundary of a smectic or cholesteric in a wedge is due to Grandjean using a mica crack and was most probably understood as presented in Figure 8.22 by G. Friedel in the early 1930s. Cano only added a specific way of aligning molecules in a definite direction along glass plates.

- Finally, research in soft matter has been helped by a transfer of concepts developed in strong solids, and this is made very clear in this book. Conversely, concepts developed in soft matter have been transferred to the study of strong solids or of biological materials. This is mentioned here in a number of cases, but it is not the main subject of this book. It can be hoped that another publication will cover in depth recent progress in these fields, where the authors have been active.

In conclusion, I am very happy to introduce a book that presents in a condensed but clear way many facets of a very rich and fascinating field.

Jacques Friedel
Paris, France
January 2001

Series Preface

Partially Ordered Systems

Many familiar materials have neither the precise order of crystalline solids nor the completely random structure of liquids and gases. Colloids such as milk, soap, and detergent solutions; liquid crystals, well known from flat electronic displays; gels; and many kinds of ultrastrong fibers are all partially ordered systems. Such systems have emerged as an important field of study not only from the point of view of basic physics, but also with practical applications in materials science and other disciplines.

This series includes research monographs and graduate-level texts that deal with condensed systems at microscopic, mesoscopic, or macroscopic scales that do not have full long-range spatial and orientational orders. These systems—some of which have also been called soft matter, complex fluids, or supermolecular fluids—include complex liquids with molecules or aggregates of molecules organized on long scales; liquid crystals, composed of monomers or polymers; colloids; molecular crystals; quasicrystals; granular materials; disordered systems; and aggregates. Books in the series cover all aspects of the materials, their structures, properties, and formation, as well as percolation and the formation of fractals and spatiotemporal patterns.

Lui Lam
San Jose, California

xi

Preface

What Is Soft Matter? Scope of This Book

What we call "soft matter" covers a large variety of systems, from polymers to colloids, from liquid crystals to surfactants, and from soap bubbles to solutions of macromolecules. All of these materials are of increasing industrial importance. Although they have long been an eminent domain of research for chemists, physicists are now taking a keen interest in them. Soft matter systems indeed raise problems of physics of completely new types. What makes their unity is difficult to formulate precisely (one speaks of "complex systems," a qualification that at least does justice to their structural properties). We try to distinguish some characteristic proper to them all.

All systems that fall under the name of soft matter belong, with very few exceptions, to organic chemistry. In fact, when one speaks of colloidal gold, or of colloidal silica, reference is made more to a material texture than to the material itself, whereas *colloid science*, in the general understanding, addresses organic solutions characterized by dispersion or solution of one phase in another, such that interface phenomena are of great relevance. The term *colloid* is widely accepted, and even favored, but no clear unified definition of the concept has yet emerged.

This digression being made, let us note that the building blocks of soft matter are organic molecules with often complicated architectures, anisometric in shape, and bound by *weak interactions*. The stability range of these phases is, therefore, close to room temperature, and small changes in temperature are enough to induce phase transitions accompanied most usually by small latent heat or sometimes by chemical decomposition. This is to say that entropy, rather than enthalpy, is a quantity to be considered first and foremost. *Biological matter* (proteins, membranes, DNA and their associates, like viruses or microtubules) enter into the class of materials under this heading of soft matter when studied by physicists. An important characteristic of these materials is that they are not in equilibrium *in vivo*, but this fundamental property of living matter is completely outside the scope of this volume.

Soft matter, in particular *liquid crystals*, display phenomena of *order* of a very original nature, intermediary between those of crystalline solids and those of disordered phases. A considerable outgrowth of the theory of phase transitions, and of the theory of the order parameter singularities, has followed their discovery, with some remarkable features like the "phases with defects" (frustrated phases). The concepts that have been developed in this area have found applications in other parts of physics (quasicrystals, amorphous media, superfluids).

The specific nature of disorder in *polymers* has also required development of a completely new type of description of random media, the physics of *scaling laws*. Examples of new types of phase transitions (e.g., the sol–gel transition) have also encouraged new insights in the physics of tenuous media, using the notion of fractals.

Finally, let us stress the importance, at a fundamental as well as at an applied level, of their transport properties (diffusivity, viscosity), of their viscoelastic properties (e.g., flow under shear of liquid crystals, phase transitions under shear, plastic deformation of polymers), and of surface and interface phenomena (wetting, role of long-range forces), which define a large domain of renewal of *mesoscopic physics*, where we find the interplay of molecular and macroscopic concepts (hydrodynamics, rheology, capillarity). Not all of these topics will be covered in this textbook, which is essentially an *introduction* to the physics of soft matter.

Soft matter physics is *condensed* matter physics, and it goes hand in hand with solid state physics. One expects that a certain number of phenomena display neighboring, if not similar, aspects in both disciplines, to their mutual enrichment. We have therefore included in the introductory chapters a number of developments common to the whole of condensed matter physics, relating to stucture (atomic and molecular arrangements), cohesion (chemical bond), defects, and phase transitions. However, the reader should not expect that we have put the bases of both disciplines on an equal footing. For example, in our discussion of the interactions between atoms or molecules, we favor an exposition that emphasizes the chemical bond picture, not the infinite body electrons spatial configuration, which would fit better the description of phenomena in solid crystals. Therefore, a traditional classification like conductors versus insulators is hardly mentioned. The reader will also notice that we have put stress on the question of defects, extending it largely, this time, to the case of solids, in order to place it in a general perspective. The subject dates back to the beginning of the last century—liquid crystals are indeed the material for which the concept of defects was first developed—and has always looked particularly difficult to many students. Personal interest led us to develop this topic, but this is not the only reason. The general theory of defects has benefited from the discovery of many liquid crystalline phases; on the other hand, a new interest in the rheological properties of complex materials can benefit from our knowledge of the plasticity of solids. A number of concepts well investigated in this field for solids take their place in liquid crystals, with obvious differences: For example, the viscous relaxation of defects in a nematic yields situations that present less hysteresis than does solid friction in solids. Apart from this particular emphasis, we believe that all essentials are treated in a sufficiently detailed way to offer access to the whole subject of soft matter. The chapter on the hydrodynamics of nematics is introduced by a reminder of the standard hydrodynamics of isotropic fluids.

****

Some technicalities: The list of references has been restricted on purpose to textbooks, review papers, and articles, when the subject they treat is not accessible in a review. Each chapter is accompanied by a few problems, generally with solutions, which either permit readers to test their understanding of the concepts developed in the chapter, or to extend some special points not treated in the body of the text. This is also the role of some appendices. We have also added a table of conversions of units.

Note added in proof:

In the color insert, please note that the scale bar in Fig. 3.14 should read "100 μm" instead of "100 ∞ m."

Acknowledgments

This textbook results from a collaboration between the two authors over many years, and from our individual experiences of teaching students at the graduate and upper undergraduate levels (MK, at École Polytechnique and DEA of Physique des Solides at Orsay and of Physique des Liquides in Paris; OL, at Chemical Physics Interdisciplinary Program, Kent State University, Ohio). MK wishes to thank Albert Libchaber, Rockefeller University, for welcoming him in his laboratory, where he found a pleasant and quiet atmosphere to work on this book. Thanks are also due to Loïc Auvray, Pascale Fabre, Jean-Baptiste Fournier, Paul Sotta, and André Thiaville, for discussion and help as members at different times of the teaching team of MK. OL is grateful to Sergiy Shiyanovskii, Victor Pergamenshchik, and Tomohiro Ishikawa, who have helped to teach his course. Claudine Fradin (of the Laboratoire de Physique des Solides, in Orsay) typed a large part of the initial draft: The authors want to thank her for her unfailing help, in spite of the distance.

We have greatly benefited from the critical reading and constructive comments of Vladimir Dmitrienko, Jacques Friedel, Efim Kats, Randall Kamien, Grégoire Porte, Harald Pleiner, Charles Rosenblatt, Vladimir Shalaev, Tim Sluckin, and Madeleine Veyssié. The authors thank them heartily.

We also acknowledge the support of CNRS and NSF for making possible the collaboration between the two authors.

Maurice Kleman
Paris, France

Oleg D. Lavrentovich
Kent, Ohio

Contents

Condensed Matter: General Characteristics, the Chemical Bond, and Particle Interactions

It is classic to distinguish three states of matter: gas, liquid, and solid. In the first approximation, the gas state is characterized by the absence of interactions between atoms or molecules, which therefore display statistical disorder. The requirement of maximum of entropy controls this state. Gases are outside the scope of this textbook. In contrast, the more condensed liquid and solid phases are controlled not only by entropy, but also by interparticle interactions. These states are stabilized by a complex interplay between attractive interactions, which are responsible for the condensation of chemical species, and repulsive interactions. The balance yields a local order, defined on some characteristic length. In crystals, this local order reproduces on distances much larger than the interparticle distances, so that one can speak of long-range order. Long-range order of crystalline materials is described in terms of the elements of the group of symmetry under which this order is invariant. Independently of the type of arrangement, all condensed matter media in liquid or solid states present some common features as follows:

1. The energies of interactions that stabilize the local order vary in the range $0.1\,\mathrm{eV} \div 10\,\mathrm{eV}$ per atom or molecule. The order of magnitude that is frequently met is 1 eV.

2. The molar volumes are all of the same order of magnitude; typical distances between atoms are of the order of 1 Å.

3. Vibration frequencies of atomic bonds are of the order of 10^{13}–10^{14} Hz.

   Differences are to be attributed to the chemical nature of the bonds, and to molecular organization.

1.1. Entropy in Disordered Systems

The number η_1 of ways to introduce one atom, one molecule, one object, in brief, what we shall call one *particle*, in a volume V of *solvent* much larger than the volume displaced by the particle, is proportional to V; i.e., $\eta_1 = AV$, where A is some constant of proportionality. Let us introduce n identical solute particles and make a hypothesis that A does not depend on the number of particles already present; this is more true the smaller the volume occupied by n particles is compared with V. Then, the number of ways to introduce these n particles is

$$\eta_n = \frac{(AV)^n}{n!}, \tag{1.1}$$

where the number of equivalent permutations $n!$ takes into account that the particles are identical. The associated entropy $S = k_B \ln \eta_n$ is given by

$$S \cong k_B n \ln(AVe/n), \tag{1.2}$$

if one uses the approximate formula $\ln n! = n \ln \frac{n}{e}$, valid when n is large. The free energy of the solution of noninteracting particles reads as

$$F = U - TS = n[f_0(T) - k_B T \ln(V/n)], \tag{1.3}$$

where the A and the e terms have been lumped into the internal energy $nf_0(T)$. Let us introduce, in accordance with a standard thermodynamic definition, a "partial pressure" related to the solute particles

$$p \equiv -(\partial F/\partial V)_{T,n} = nk_B T/V = ck_B T; \tag{1.4}$$

here, $c = n/V$ is the concentration of particles. The physical meaning of p, the partial derivative of the free energy at constant n, can be understood as follows. Imagine that the solution of particles and a pure solvent are in contact through a membrane that is permeable for the solvent but not for the particles. The quantity $-p\,dV$ is the free energy variation of the system in a reversible process in which the volume of the solution is varied by an infinitesimal amount dV. Therefore, p is the normal force per unit area exerted by the solution on the membrane, measured along the outer normal; p is called the *osmotic pressure* (see also Chapter 4). Note that with the present free energy (1.3), equation (1.4) can be rewritten as $pV = RT$, i.e., as the equation of state of an ideal gas, if V is understood as the volume of solution that contains 1 mole of the solute; $R = 8.31 \text{ J}/(\text{mole} \cdot \text{K})$ is the gas constant.

Equation (1.4) is modified when one takes into account interactions between particles, hence, the use of the so-called "virial" expansion

$$p = ck_B T(1 + cv + \cdots), \tag{1.5}$$

where the second virial coefficient v has the dimension of volume and comes from pair interactions between particles. Terms of higher order in (1.5) would correspond to interactions involving three particles, four particles, and so on. If no attractive interactions exist and the particles are approximated by hard spheres of radius r_0, then it can be shown that v is an *excluded volume*:

$$v = \frac{1}{2} \times \frac{4}{3}\pi(2r_0)^3 = 4v_p, \tag{1.6}$$

where v_p is the volume of one particle (see Chapter 4). Generally, v accounts for both repulsions and attractions and is calculated from the pair interaction potential $w_{12}(r)$ function of the distance, assuming central forces,

$$v = \frac{1}{2}\int\limits_{r=0}^{\infty}[1 - \exp(-w_{12}/k_BT)]4\pi r^2\,dr. \tag{1.7}$$

Note that in the presence of solvent, the potential $w_{12}(r)$ depends not only on the direct particle-particle interactions, but also on particle-solvent and solvent-solvent interactions. Thus, the last expression is justified only when the solvent can be treated as a continuum medium whose presence can be accounted for in $w_{12}(r)$.

The virial expansion is used to describe dilute solutions of micelles, macromolecules, proteins, and so on. Consider a solution made of two constituents, solvent A and solute B, say. Let w_{AA}, w_{BB}, and w_{AB} be the interactions energies of the pairs AA, BB, and AB at their distance of closest approach a, which we take as the same for all pairs (the volume of a molecule of constituent A or B is a^3): the quantity $w = w_{AB} - (w_{AA} + w_{BB})/2$ measures the effective interaction between two particles A and B (see Chapter 15, where it is also shown that $w > 0$ if the forces of interaction are van der Waals forces). In such a case, the molecules A and B separate at low temperature. One can formally distinguish two types of situations: (1) weak attractive interactions, $w \ll k_BT$; (2) strong attractive interactions. These two situations are often referred to as the "good" and the "poor" solvents, respectively. The related excluded volume is $v = (1 - 2w/k_BT)a^3$ (see Chapter 15). In poor solvents, a temperature exists at which attractive and repulsive forces compensate each other, $v = 0$, and the system behaves as an ideal solution with noninteracting particles of vanishing volumes. Conditions under which the second virial coefficient vanishes are often called "theta-conditions."

1.2. Central Forces and Directional Forces Between Atoms

We give here a brief summary. For details, refer to the textbooks cited at the end of this chapter.

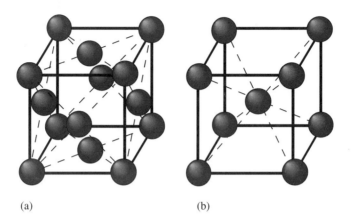

(a)   (b)

Figure 1.1. Face-centered (a) and body-centered (b) cubic lattices.

1.2.1. Metallic Bond

The valence electrons circulate freely in a metallic crystal, making it a unique, gigantic molecule. The globally attractive bonding effect of this sea of valence electrons is nondirectional, and, hence, responsible for *close-packed structures*, i.e., where each atom, considered as a sphere, is surrounded by as many atoms as possible. The interatomic distances are determined by the repulsive steric forces. Close packing is often an excellent first-order approximation in the study of local ordering. Face-centered cubic (FCC) and hexagonal close-packed lattice (HCP) crystals are the most packed structures of atoms of equal sizes (Fig. 1.1a). The body-centered cubic structure (BCC) is less compact (Fig. 1.1b) and is favored at higher temperatures in the same systems, due to the larger entropy of vibration.

1.2.2. Bond Formed by Fluctuating Dipoles

Electrically neutral species such as rare gas atoms or nonionic organic molecules are held together by *van der Waals attraction forces* (also known as *dispersion forces*). For rare gas atoms (e.g., argon), in which the averaged positions of negative and positive charges coincide, these forces were first analyzed by London: Fluctuating electric dipoles of one neutral particle induce a dipole moment in the other; see Section 1.3.3. The resulting crystalline structures are close-packed, also in relation to the nondirectional character of this bond. The London energies as well as other van der Waals energies in general are weak, often less than $k_B T$ per atom at room temperatures. The corresponding crystalline arrangements are stable only at low temperatures; they melt with a low latent heat.

Table 1.1. Energies of typical atomic bonds.

Bond	Structure	Energy per bond, kJ/mole	Energy per bond, eV/molecule	$k_B T$
Ionic	Na^+Cl^-	750	7.77	310
Covalent	$C \equiv N$	870	9.02	360
	(example: HCN)			
	$C = O$ (H_2CO)	690	7.15	285
	$C = C$ (C_2H_4)	600	6.22	
	O–H (H_2O)	460	4.77	
	C–H (CH_4)	430	4.46	
	C–C (C_2H_6)	360	3.73	
	F–F (F_2)	150	1.55	
Metallic	Na	109	1.13	
	Al	311	3.22	
Hydrogen bond	N–H...O = C;	10–50	0.10–0.51	2–20
	$C = O$...H–O–C			

1.2.3. Covalent Bond

In a pair of atoms connected by a covalent bond, each atom contributes one electron to form an electron pair shared by the two atoms. A covalent structure is characterized by the number z of bonds per atom: $z = 4$ for C, Si, and Ge; and $z = 2$ for S, Se, and Te. In the first case, one expects three-dimensional (3D) atomic arrangement (e.g., two interpenetrating FCC lattices for C in diamond). In the second case, one gets more or less coiled polymeric chains, which can arrange in 3D strongly anisotropic crystals, where the chains are interstabilized by dispersion forces. Covalent bonds are strongly directional. Their energy is high, of the order of $100 \, \text{kcal/mole} \sim 10^2 - 10^3 \, \text{kJ/mole} \sim 1 - 10 \, \text{eV/molecule}$ (Table 1.1).

1.2.4. Ionic Bond

Ionic bonding occurs because of complete electron transfer from one atom to another. This transfer of one or more electrons converts the neutral atoms into ions of an opposite charge with a strong electrostatic interaction between them. The ionic bonds are strong, yielding crystals that are stable at room temperature. A classic example is a sodium chloride (NaCl) crystal (melting point 801°C) with a cubic structure in which each negatively charged chloride is surrounded by six positively charged sodium ions (and each sodium is surrounded by six chloride ions) (Fig. 1.2).

The main contribution to the energy of interaction of two ions with charges z_1e and z_2e is the Coulomb potential $w_{att} = \frac{z_1 z_2 e^2}{4\pi \varepsilon \varepsilon_0 r}$, which is attractive when the ionic valencies

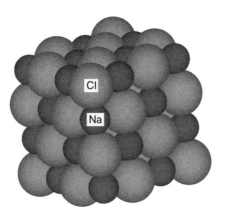

Figure 1.2. Cubic lattice of NaCl.

z_1 and z_2 are of opposite signs; here, ε is the relative dielectric permittivity of the medium in which the ions are located (also known as the dielectric constant), ε_0 is the permittivity of free space, and r is the distance between the atoms. When the two atoms come close to each other, repulsion becomes important.

To calculate the energy of cohesion of an ionic crystal such as NaCl, one has to sum up all Coulomb forces between the atoms. The energy of cohesion carried by an ion ze is the sum of the pair potentials (changed sign) of interactions of that ion with all other ions $z_i e$; i.e.,

$$\mu = -\sum_i w(r_i) = -\frac{ze^2}{4\pi\varepsilon\varepsilon_0}\sum_i \frac{z_i}{r_i}. \tag{1.8}$$

Let $\mu = M\frac{e^2}{4\pi\varepsilon\varepsilon_0 a}$, where M is the Madelung constant defined by the geometry of packing and a is the distance between two opposite charges. For the cubic NaCl lattice, $M = 1.748$ and $a = 0.276$ nm. With $\varepsilon = 1$, $\varepsilon_0 = 8.854 \times 10^{-12}\,\mathrm{F\,m^{-1}}$, $e = -1.602 \times 10^{-19}$C, one gets $\mu \approx 1.5 \times 10^{-18}$ J, which is substantially larger than the thermal energy $k_B T \approx 4 \times 10^{-21}$ J at room temperature. The energy needed to dissociate 1 mole of the crystal into a gas of ions is then estimated as $U = \mu N_{AV} = 890\,\mathrm{kJ/mole}$ (≈ 8 eV/atom), where $N_{AV} = 6.022 \times 10^{23}\,\mathrm{mole^{-1}}$ is the Avogadro number. In reality, repulsion forces make U somewhat smaller; however, the correction is small ($\approx 15\%$). Thus, the ionic bonds are very strong.

Intermediate situations, in which the same atoms can be bound by forces of mixed type (e.g., ionocovalent) can exist.

1.2.5. From Ionic Bond to Covalent Bond in Crystals

Alkali halides (I-VII crystals) are the standard examples of nearly ideal ionic crystals. They all crystallize in the NaCl FCC structure, except Cs halides. Doubly ionized II-VI crystals like CaSe are weakly covalent, to the extent that the Ca^{2+} ion is partly shielded by a fraction of electron in its neighborhood, electron fraction that is not transferred to Se^{2-}, and that the electronic cloud is slightly distorted along the directions joining Ca^{2+} to Se^{2-}. These characters are accentuated in III-V crystals like AsGa, where the ionic transfer is small and more electrons shared among atoms. AsGa crystallizes in the zincblende structure, which is characteristic of tetravalent crystals. Like diamond, the zincblende structure is made of two interpenetrating FCC lattice, but each of them is occupied by a different type of atom.

1.3. Forces Between Molecules

As soon as the predominant forces are not purely electrostatic or covalent, the bonding energies between particles are weak and we enter the domain of "soft matter." Predominant interactions are now screened ionic, dipolar, and van der Waals interactions.

1.3.1. Electrostatic Bond in a Dielectric Medium

We often deal with atoms or molecules soluble in water, so that even when the ions are present, their interactions are screened by water with a large dielectric constant, $\varepsilon \approx 78$. *Salts* such as NaCl are easily dissolved in water, because the electric fields are strongly reduced (by a factor of $1/\varepsilon$), so that μ is of the order of $k_B T$ or even smaller. Moreover, the entropy of the statistical mixture of ions also contributes to decreasing the free energy. *Ionic surfactants* such as sodium stearate, $NaC_{18}H_{35}O_2$ (Fig. 1.3), also easily ionize in water according to this principle.

Note that each hydrophilic anion of sodium stearate carries a "hydrophobic" aliphatic moiety $C_n H_{2n+1}$ that shows no affinity to the aqueous environment. When the concentration of such amphiphilic molecules is higher than a so-called *critical micellar concentration* (CMC), they aggregate in the form of roughly spherical micelles. The hydrocarbon tails are hidden in the core of the micelle, whereas the polar heads are located on the sur-

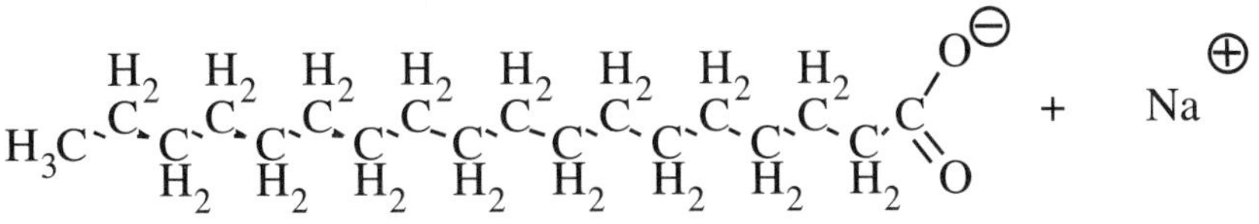

Figure 1.3. Ionic surfactant: molecule of sodium stearate dissolves in water.

face of the micelle where they can interact with the surrounding water and counterions (such as Na^+). As the concentration increases, the micelles can transform into bilayers that extend and organize in space in various fascinating geometries to be described later.

1.3.2. Electric Dipoles

Molecules of "weakly" ordered media are frequently electrically dipolar. Asymmetric molecules, bound internally by covalent bonds, often show *permanent electric dipoles*. The dipoles occur when the covalent bond connects two different atoms. Figure 1.4 shows an example of an HCl molecule with a covalent bond formed by a spherically symmetric $1s$-orbital of hydrogen and an elongated $3p$-orbital of chlorine.

Although the molecule is electrically neutral, the mean positions of positive and negative charges do not coincide. The absolute value of the dipole moment $u = ql$ (a charge times a length) is measured in Debye units ($1D = 3.336.10^{-30}$ C.m). The dipole moment of two elementary charges separated by 1 Å is about 4.8 D. The dipole vector is directed from the negative charge to the positive one (in some chemical literature, an opposite direction is taken).

The existence of an electric dipole may also depend on the environment. Some phospholipids such as DMPC (dimyristoyl phosphatidylcholine) become ionized in the presence of water, whereas the charged entities do not separate from the molecule (Fig. 1.5). This is also the case with the 20 aminoacids (alanine, valine, leucine, etc., all chiral, except glycine), which are the basic building blocks of the primary (linear) structure of biological polymers (Fig. 1.6). We shall refer to the structure that is neutral overall but contains an equal number of locally charged centers of either sign as a *dipolar ion*, or a *zwitterion*.

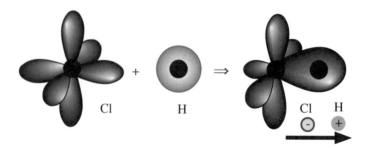

Figure 1.4. Dipole moment of the HCl molecule. The $3p$-orbital and the $1s$-orbital have one electron each. The resulting covalent bond with two electrons is polar: The electron cloud is denser about the Cl atom.

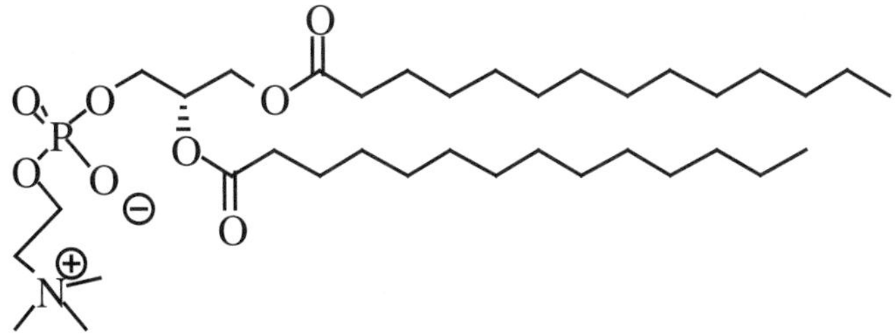

Dimyristoyl phosphatidylcholine (DMPC)

$$C_{36}H_{72}NO_8P$$
Mol. Wt.: 678

Figure 1.5. Zwitterion: surfactant molecule DMPC (dimyristoyl phosphatidylcholine) in water.

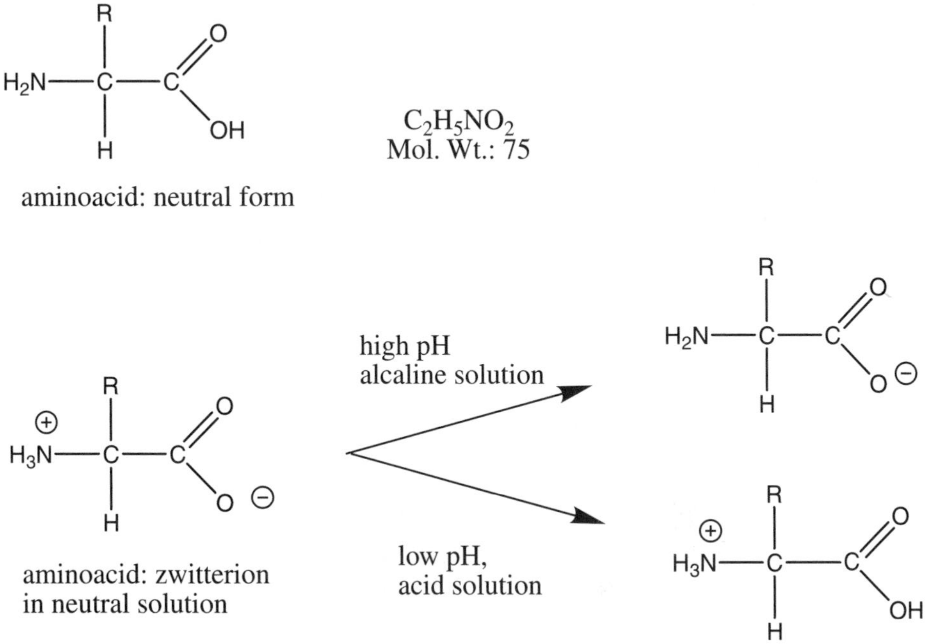

$$C_2H_5NO_2$$
Mol. Wt.: 75

aminoacid: neutral form

high pH
alkaline solution

aminoacid: zwitterion
in neutral solution

low pH,
acid solution

Figure 1.6. Aminoacids (that differ in side group R) exist mainly as zwitterions when in neutral water solutions. An increase in pH of the solution yields an excess of anionic forms, whereas a decrease in pH results in more cations.

Table 1.2. Dipole moments of some atomic groups and bonds.

Group/bond	Dipole, moment, D	Group/bond	Dipole moment, D
C_6H_6 (benzene)	0	H_2O	1.85
CO_2	0	NH_3	1.47
CCl_4	0	CH_3COOH	1.7
CH_3CN	3.9	CH	0.4
CN in cyanoethane		NH_2 in propylamine	
C_2H_5CN	4.02	$C_3H_5NH_2$	1.17
cis $C=C$	0.33	trans $C=C$	0
(in cis-2-butene)		(in trans-2-butene)	
OH in propanol		C–COOH	1.7
C_3H_7OH	1.68		

   Dipole moments of typical atomic groups are listed in Table 1.2. Note that the dipole moment depends on both the chemical structure of the molecule and on its conformation.

   The interaction ion-dipole plays an important role in the *solvation* of ions, i.e., in the phenomenon of clustering of solvent molecules-dipoles around the ion. If the solvent is water, then the ion is said to be *hydrated*. For example, a cation (e.g., Na^+) attracts and orients a neighboring dipole along the line joining the charge and the dipole, i.e., along the direction with angle $\theta = 0$ in Fig. 1.7. A hydrated ion might grow to large sizes: This is how the nucleation of rain droplets in clouds is explained. The gain of energy between

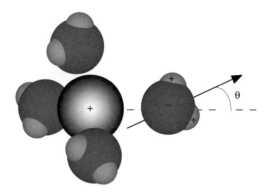

Figure 1.7. Ion-dipole interaction: hydrated cation.

the dipole oriented at random and the dipole $\theta = 0$ must be equal to or larger than $k_B T$ to stabilize the hydrated complex, which can be very stable if the dipolar ion is small and highly charged.

Attractive *dipole-dipole* and *free dipole-free dipole* (or *Keesom*) interactions are weaker than is the *ion-dipole* interaction, and lead to less stable constructions. A free dipole is a dipole capable of taking all directions in space. Therefore, the energy of interaction is calculated by taking an ensemble average over orientations, as in the Langevin model of paramagnetism:

$$w_{d-d}(r) = \frac{-u_1^2 u_2^2}{3(4\pi\varepsilon\varepsilon_0)^2 k_B T \, r^6},$$

(1.9)

when $k_B T > u_1 u_2/4\pi\varepsilon\varepsilon_0 r^3$. Note that $w_{d-d} \propto 1/T$: Higher temperatures enhance rotation of molecules and, thus, reduce orientational order of the dipoles. The Keesom interaction is one of the three attractive interactions between electrically neutral molecules. The other two already mentioned are *London interactions* between molecules with no permanent dipoles and *Debye interactions* between permanent and induced dipoles. All three scale as $1/r^6$ and have a generic name of *van der Waals forces*.

1.3.3. Induced Dipoles, Polarizability

An electric field E, either applied or caused by a neighboring molecule, induces an electric dipole $u_{\text{ind}} = \alpha_0 E$ on a neighboring molecule by separating the centers of the positive and negative charges; α_0 is the polarizability of the molecule. The model in Fig. 1.8 helps to

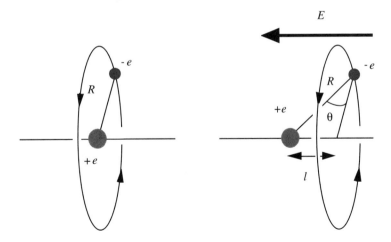

Figure 1.8. Electric field induces a dipole moment in a neutral system (an electron rotating around a proton).

estimate this induced moment u_{ind} in a simple case of an electron rotating around a proton. The electron is subjected to two forces: the Coulomb attractive force with the nuclear charge and the shifting force of the external field E. The change in the Coulomb force caused by the shift is roughly $F_c = \frac{e^2}{4\pi\varepsilon_0 R^2} \sin\theta \sim \frac{e^2 l}{4\pi\varepsilon_0 R^3} = \frac{eu_{ind}}{4\pi\varepsilon_0 R^3}$, and the field-induced force is $F_E = eE$. At equilibrium, $F_c = F_E$ and

$$u_{ind} = 4\pi\varepsilon_0 R^3 E = \alpha_0 E. \tag{1.10}$$

Here, we assume that the induced dipole is always oriented along the field; thus, $\alpha_0 = 4\pi\varepsilon_0 R^3$ is a scalar quantity. With $R = 1$ Å, $\varepsilon_0 = 8.85 \times 10^{-12}$ F/m, one gets an estimate $\alpha_0 = 10^{-40}$ C^{2}m^2/J for the polarizability of an atom or a small molecule in vacuum. However, larger molecules may have much larger polarizabilities, because $\alpha_0 \propto R^3$. Furthermore, if the freely rotating molecule has a permanent dipole moment u, an external field would restrict free rotation of such a molecule so that the time-average polarization would be different from zero. The resulting orientational polarizability equals $u^2/3k_B T$. For weakly polar liquid crystalline molecules such as N-(p-methoxybenzylidene)-p'-butylaniline (MBBA) and p-pentyl-p'-cyanobiphenyl (5CB), typical values of the total polarizabilities are in the range $(10 - 100) \times 10^{-40}$ C^{2}m^2/J. Note finally that MBBA, 5CB, and most of other molecules are anisometric. As a result, the induced dipole depends on the orientation of the molecule in the field, and the polarizability is a tensor.

The *ion-neutral molecule* and *dipole-neutral molecule* (or *Debye*) interactions are discussed in terms of field-induced dipoles. The Debye interaction scales as the Keesom interaction but is temperature independent. Finally, the *London* or *dispersion forces* are caused by fluctuative interactions between neutral atoms or molecules with no permanent dipoles. Actually, these forces are independent of the particular type of the molecule, because the fluctuations of the charge density are a universal quantum mechanic effect. Thus, the London interactions are present in any condensed matter system; they are treated in some detail below and in Section 1.4. The term "dispersion" originates in the dispersion of light in the visible, UV, and IR parts of the spectrum, and it should not be confused with the term "dispersion" describing colloidal systems: The frequencies at which an electromagnetic field causes a fluctuating dipole in a nonpolar molecule are absorption frequencies.

Fluctuations of charge densities induce mutual dipole moments in the neighboring molecules. The attractive character of interaction can be qualitatively derived from the tendency of the fluctuating polarizations to be in phase along the line joining the particles and in antiphase in the normal plane. London derived the dispersion energy of two neutral atoms treated as quantum oscillators, with frequency ν of the orbiting s-electrons, as[1]

$$w_L \approx -\frac{3}{2(4\pi\varepsilon_0)^2} \frac{\alpha_1\alpha_2}{r^6} \frac{h\nu_1\nu_2}{(\nu_1 + \nu_2)} = -\frac{C_{12}}{r^6}, \tag{1.11}$$

---

[1] F. London, Trans. Faraday Soc. **33**, 8 (1937).

Table 1.3. Pair interaction energies of neutral molecules separated by distance 5 Å (data compiled from J. A. Campbell, Chemical Systems: Energetics, Dynamics, Structure, W.H. Freeman and Company, San Francisco, 1970).

Molecule	Dipole moment, D	Energy (in J/mole) of interaction at $T = 298$ K and $a = 5$ Å		Boiling temp., K
		w_{d-d}	w_L	
He	0	0	3	4.2
CCl$_4$	0	0	7.5×10^3	350.9
CO	0.1	0.01	290	81.0
HCl	1	71	460	189.4

where α_1 and α_2 are the polarizabilities of the two molecules, $h = 6.63 \times 10^{-34}$ J s is the Planck constant. The positive constant C_{12} of the dimension J $\times$ m^6 is the "material" parameter of interacting particles independent on the separation distance r.

For atoms and very small molecules, the interactions are weak and unlikely to form stable, ordered phases at room temperatures. Noticing that $I_i = h\nu_i$ is some characteristic molecular energy that can be approximated by its first ionization potential, which is of the order of 10^{-18} J, and $\alpha_1 = \alpha_2 = 3 \times 10^{-40}$ C^{2}m^2/J (which would correspond to the molecular radii $R_i \approx 1.3$ Å in (1.10)), one estimates the constant $C = \frac{3\alpha^2 I}{4(4\pi \varepsilon_0)^2} \approx \frac{3R^6 I}{4}$ in (1.11) as $C \approx 0.6 \times 10^{-77}$ J m^6. At distance $r = 3$ Å, the corresponding attraction energy $|w_L| \approx 8 \times 10^{-21}$ J is larger than $k_B T$ at room temperature. Hence, fairly large nonpolar molecules can be kept in a condensed state exclusively due to the London forces. Even when the molecules are polar, the London forces still make a significant contribution, 20–99% of the total energy. The second strongest interactions of polar molecules are usually Keesom forces (see Table 1.3).

1.3.4. Repulsive Forces

Interactions between molecules include the attractive forces considered above and repulsive forces. Most of the time, the repulsive forces are caused by electrostatic repulsion between particles having charges of the same sign or by short-range steric repulsion. Steric repulsion originates largely from the Pauli exclusion principle and shows up when the electron clouds of two particles approach each other too closely; these interactions are hard to describe. In the simplest model of hard-core spherical particles of radius r_0, the energy of repulsion is assumed to be infinitely large when the separation between the centers is less than $2r_0$ and zero when the separation is larger than $2r_0$. In more realistic approaches, the repulsion potential at $r > 2r_0$ is often represented by an exponential decrease $w_{\text{rep}} \propto \exp(-r/\text{const})$, where const is positive, or, to simplify algebra in calculations involving both attractive (1.11), and repulsive forces, as

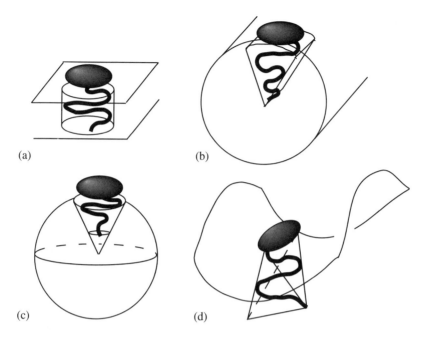

Figure 1.9. Molecular shape provokes different geometries of surfactant monolayer: (a) flat, (b) cylindrical, (c) spherical, and (d) saddle-like. The radii of curvature of the membranes can be different from the molecular length; see, e.g., (c).

$$w_{\text{rep}} = \frac{B}{r^{12}}, \tag{1.12}$$

where B is a positive constant.

Both electrostatic and steric repulsions appear in phases of interfaces, for example, in the lamellar phases of membranes. Because of the steric forces, the *shape* of the molecules has a pronounced effect on the nature of packing. Packing of amphiphilic molecules in monolayers or bilayers and cylindrical and spherical micelles depends in an extremely sensitive way on the molecular shape (Fig. 1.9). Note also that in the lamellar phases, spatial fluctuations of membranes are at the origin of a specific repulsive (Helfrich) potential that stabilizes the lamellar phases against attractive London forces, in the case of nonionic surfactants.

Table 1.4 summarizes schematic classification of molecular interactions that are assumed to take place in vacuum.

Table 1.4. Typical molecular interactions in vacuum.*

Type of interaction	Representative scheme	Potential of interaction	Attributed to	Comments
Ion 1 - Ion 2		$\dfrac{(ze)_1(ze)_2}{4\pi\varepsilon_0 r}$	Coulomb	Strongly directional; sign depends on the valences
Ion 1 - Permanent fixed dipole 2		$\dfrac{(ze)_1 u_2 \cos\theta}{4\pi\varepsilon_0 r^2}$	Coulomb	Strongly directional; sign depends on z and orientation of the fixed dipole
Ion 1 - Freely rotating permanent dipole 2		$-\dfrac{(ze)_1^2 u_2^2}{6(4\pi\varepsilon_0)^2 k_B T r^4}$		Always attractive
Permanent fixed dipole 1 - Permanent fixed dipole 2		$-\dfrac{u_1 u_2}{4\pi\varepsilon_0 r^3} \times (2\cos\theta_1\cos\theta_2 \\ -\sin\theta_1\sin\theta_2\cos\varphi)$	Coulomb	Sign depends on orientation; φ is the angle between the planes formed by each dipole and the line joining the dipoles

(continued)

15

Table 1.4. (*Continued*)

Type of interaction	Representative scheme	Potential of interaction	Attributed to	Comments
Permanent freely rotating dipole 1 Permanent freely rotating dipole 2		$-\dfrac{u_1^2 u_2^2}{3(4\pi\varepsilon_0)^2 k_B T r^6}$	Keesom	Always attractive
Ion 1 - Induced dipole 2		$-\dfrac{(ze)_1^2 \alpha_2}{2(4\pi\varepsilon_0)^2 r^4}$		Always attractive
Permanent fixed dipole 1-Induced dipole 2		$-\dfrac{u_1^2 \alpha_2}{2(4\pi\varepsilon_0)^2 r^6}(1 + 3\cos^2\theta)$		Always attractive
Permanent freely rotating dipole 1 - Induced dipole 2		$-\dfrac{u_1^2 \alpha_2}{(4\pi\varepsilon_0)^2 r^6}$	Debye	Always attractive

Induced dipole 1 - Induced dipole 2	$-\dfrac{3h\nu_1\nu_2\alpha_1\alpha_2}{2(\nu_1+\nu_2)(4\pi\varepsilon_0)^2 r^6}$	London	Always attractive; h is the Planck constant; ν is the characteristic vibration frequency of electron
Induced dipole 1 - Induced dipole 2 (retarded interaction)	$-\dfrac{23hc\alpha_1\alpha_2}{8\pi^2(4\pi\varepsilon_0)^2 r^7}$	Casimir and Polder	Always attractive; applies when $r > c/\nu$.
Hydrogen bond $\quad$ X – H⋯Y	Complicated, short-range, roughly proportional to $-\dfrac{1}{r^2}$, where r is the distance between the atoms X and Y.		
Repulsion	$\dfrac{B}{r^\beta}$		$B > 0$; β in range 9–15; usually, $\beta = 12$

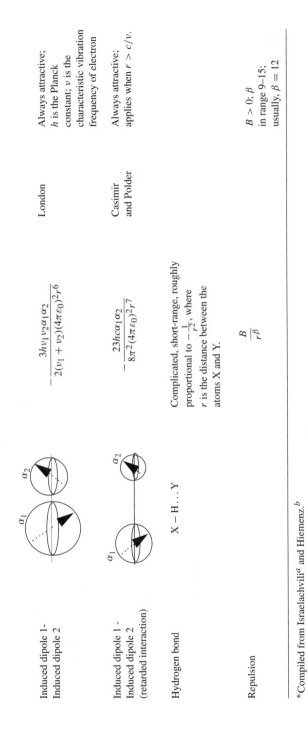

*Compiled from Israelachvili[a] and Hiemenz.[b]

[a] J. N. Israelachvili, Intermolecular & Surface Forces, 2nd edition, Academic Press, London, 1992, 450 pp.
[b] P. C. Hiemenz, Principles of Colloid and Surface Chemistry, 2nd edition, Marcel Dekker, Inc., New York, 1986, 790 pp.

17

1.3.5. Empirical Potentials of Interactions

Ad hoc potentials are frequently introduced in the calculations to simulate pair potentials. With the development of computers, it became possible to simulate the behavior of large ensembles of particles specifying their shape and repulsive and attraction forces. The most frequently used potentials are the following two:

- Hard spheres with weak attraction

$$w(r) = \begin{cases} -w_0 \left(\dfrac{r_0}{r}\right)^m, & \text{for} \quad r_0 \leq r < \infty, \\ \infty, & \text{for} \quad 0 \leq r < r_0. \end{cases} \tag{1.13}$$

- Lennard–Jones potential

$$w_{LJ}(r) = 4w_0 \left[\left(\frac{r_0}{r}\right)^{12} - \left(\frac{r_0}{r}\right)^{6} \right], \tag{1.14}$$

where w_0 is a positive constant that corresponds to the energy minimum located at separation distance $r = 2^{1/6} r_0$ (Fig. 1.10).

The potentials above refer to spherically symmetric interactions. Approaches to take into account orientational degrees of freedom of nonspherical particles have been recently

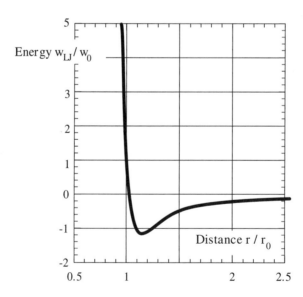

Figure 1.10. Lennard–Jones potential of interaction between two neutral particles.

developed, with the purpose of computer simulations of phases such as nematic liquid crystals. Among them are the Gay-Berne potential (a form of the Lennard–Jones potential with orientation-dependent parameters), and the Lebwohl-Lasher potential, in which the molecular positions are fixed, but deviations from parallel alignment of molecules cost energy.

1.3.6. Water, Hydrogen Bond, and Hydrophilic and Hydrophobic Effects

Water is the most abundant and the most important terrestrial liquid. It is an exceptional liquid by its chemical properties: very high melting and vapor temperatures, large latent heat of vaporization, lesser density in the solid (ice) than in the liquid state, and so on. All of these peculiarities are explained by the presence of the hydrogen bond between H_2O molecules. The hydrogen bond forms when a hydrogen atom covalently bonded to a strongly electronegative atom X acquires a partial positive charge δ^+ that is strong enough to form an essentially electrostatic bond with another electronegative atom Y (Fig. 1.11).

Hence, the symbol XH . . . Y. Typically, X, Y = O, N, F, Cl, S. The bond might connect either separate molecules (intermolecular hydrogen bond) or groups within the same molecule (intramolecular hydrogen bond). The hydrogen bond is of the order of 2 to 10 kcal/mole (0.1 to 0.5 eV/molecule), which is between the van der Waals bond (typically, 0.01 eV/molecule) and the covalent bond ($\approx$ 5 eV/molecule). The significance of the hydrogen bonding can be illustrated by a simple comparison of boiling temperatures of water (100°C) and methane CH_4 (−164°C) with no hydrogen bonds. The hydrogen bond plays a crucial role in establishing spatial geometry of molecular packing; e.g., in water, it leads to the local tetrahedral coordination of the oxygen atom (Fig. 1.12a).

The associative character of the hydrogen bond in water explains the *hydrophobic* effect, i.e., extremely weak solubility of nonpolar molecules (such as alkanes $CH_3(CH_2)_nCH_3$) in water. Nonpolar particles cannot participate in the formation of hydrogen bonds. When isolated individually in water, such particles are surrounded by a "cage" of water molecules (Fig. 1.12b). To preserve the tetrahedral network, the water molecules of the cage should choose only specific orientations that avoid interruption of hydrogen bonds by the particle. This additional *ordering* is entropically unfavorable and

$$\delta_X^-\ \ \delta^+ \qquad \delta_Y^-\ \ \delta_R^+$$
$$X - H \quad \cdots \quad Y - R$$

Figure 1.11. A hydrogen bond between two atomic groups XH and YR that contain strongly electronegative atoms (X and Y).

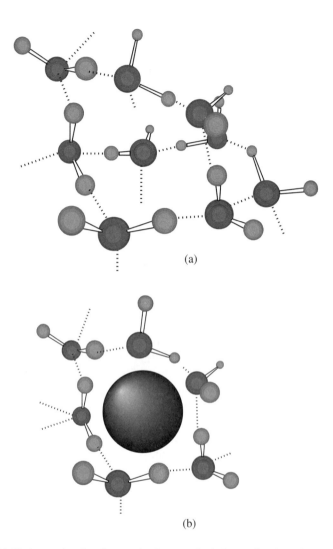

(a)

(b)

Figure 1.12. (a) Hydrogen bonds of water lead to tetrahedral coordination of oxygen atoms; the structure is labile because the molecules can reorient and reestablish new directions of hydrogen bonds. (b) Water molecules form a "cage" around the nonpolar particle. The hydrogen bonds preserve the tetrahedral coordination; however, the freedom of water molecules to reorient is restricted by the dissolved particle that cannot form hydrogen bonds.

explains why the nonpolar particles are hydrophobic: It costs a higher energy to immerse a nonpolar particle in water. It is the same hydrophobic effect that is responsible for the attraction between nonpolar molecules in water: The molecules aggregate to reduce the surface area of the cage, i.e., to reduce the loss of entropy.

Finally, it is usual to speak of a *hydrophilic effect* for the molecules that interact strongly with water. The balance of hydrophobic and hydrophilic interactions controls the behavior of amphiphilic molecules that contain both hydrophobic and hydrophilic groups.

From a more general point of view, one will notice that the forces of dispersion, nondirectional (or weakly directional, as in the case of liquid crystals), and attractive are responsible for the stability of molecular assemblies and their close-packed character, whereas the hydrogen bonds, which are directional, shape their configurations and their short-distance order.

1.4. van der Waals Forces Between Macroscopic Particles

1.4.1. Pairwise Summation of Molecular Forces; Hamaker Constant

Colloidal particles with size usually in the range between micrometers and nanometers often aggregate (coagulation and flocculation). The phenomenon suggests that the particles are attracted to each other by forces acting over a distance of the order of the particle size. It has long been recognized that these attractive long-range forces result from the van der Waals molecular interactions considered in the previous sections 1.3.2 and 1.3.3. The counterbalancing repulsive forces are of steric and electrostatic origin (see Chapter 14).

Although the molecular attractive potential (1.11), $w_L = -C_{12}/r^6$ decays quickly with molecular distances, scaling from microscopic to macroscopic distances leaves the van der Waals forces undiminished. The simplest way to calculate the van der Waals potential V_{vdW} for two macroscopic particles is to assume pairwise additivity of individual molecular interactions, as if they happen in the absence of any other molecules. Furthermore, one can replace the addition of individual molecular interactions (1.11) by integration; i.e., consider the matter as a continuum with an average number of molecules per unit volume, $\rho = $ const (dimension m^{-3}). The energy of interaction of two macroscopic bodies of densities ρ_1, ρ_2, occupying volumes V_1 and V_2 in free space, is then

$$V_{vdW} = -\int\limits_{V_1} dV_1 \int\limits_{V_2} dV_2 \frac{\rho_1 \rho_2 C_{12}}{|\mathbf{r}_1 - \mathbf{r}_2|^6};\tag{1.15}$$

here, $|\mathbf{r}_1 - \mathbf{r}_2|$ is the distance between volume elements dV_1 and dV_2 of the bodies 1 and 2. The constant C_{12} depends on molecular parameters such as polarizability, but not on the geometry of the bodies. Therefore, pairwise summation allows one to represent the van der Waals potential as a product of a geometrical factor embodied in the double volume integral

$$V_{v\,dW} = -\frac{A_{12}}{\pi^2} \iint\limits_{V_1,V_2} \frac{dV_1\,dV_2}{|\mathbf{r}_1 - \mathbf{r}_2|^6},$$ (1.16)

and a material factor $\rho_1\rho_2C_{12}$, often presented through the so-called Hamaker constant

$$A_{12} = \pi^2\rho_1\rho_2C_{12}$$ (1.17)

(dimension: Joules), a fundamental quantity in characterization of van der Waals forces. The consequence of the inversed sixth-power distance dependence of the potential (1.11) is that the energy of interaction (1.16) of two bodies remains constant when all distances (sizes and separations) are scaled by the same factor λ: $V \to \lambda^3 V$, $r \to \lambda r$ (see Problem 1.3). Two "molecules" of radii 0.2 nm at a distance of 1 nm would interact with the same energy as two spheres of radii 0.2 μm at a distance of 1 μm. The integral (1.16) has been calculated for a variety of simple geometries (see Table 1.5).

The typical values of the Hamaker constant for most condensed phases are in the range $(0.4 - 4) \times 10^{-19}$ J (see Chapter 11 in Israelashvili, 1992). The range is narrow, because according to (1.11) and (1.10), $C_{12} \propto \alpha_1\alpha_2 \propto R_1^3 R_2^3$, whereas $\rho_1\rho_2 \cong R_1^{-3} R_2^{-3}$ (R_1 and R_2 are molecular radii), so that $A_{12} \cong \rho_1\rho_2C_{12} \cong$ const. A very rough estimate (for the particles of the same sort) would be $A = \pi^2\rho^2 C \approx \pi^2 \left(\frac{4}{3}\pi R^3\right)^{-2} \frac{3}{4} R^6 I \approx 4 \times 10^{-19}$J.

Microscopic potential w_L (1.11) and macroscopic potential $V_{v\,dW}$ (1.16) were both derived under approximation that the information about the state of one fluctuating dipole is transferred to its counterpart (1) instantly and (2) without any distortions from the atoms nearby. Thus, two important comments are in order.

1.4.2. Retardation Effects

In reality, electromagnetic fields need a finite time to propagate. If the propagation time becomes comparable to the electron vibration period, then the fluctuating dipoles in the London picture become less correlated. With a typical period of electron vibrations $T = 10^{-16}$s and the speed of light $c = 3 \times 10^8$m/s, the characteristic length $r_r \cong cT$ at which the "retardation" becomes significant, is about 30 nm for free space; it can be significantly shorter in a medium where the propagation of electromagnetic waves is slower. The retarded London interactions have a steeper power law $1/r^7$ instead of $1/r^6$ when $r \geq r_r$. To see this qualitatively,[2] let us replace the frequency ν in the potential (1.11) by c/λ, where λ is the wavelength of fluctuation of the electron density: $w_{Lr} \cong -\frac{\alpha^2}{r^6}\frac{hc}{\lambda}$. For a given separation r, fluctuations with $\lambda \ll r$ are too fast to produce noticeable correlations, and there

---

[2]D. Tabor, Solids, Liquids and Gases, Cambridge University Press, 1979, p. 19–21.

Table 1.5. Nonretarded van der Waals attraction potentials for pairs of bodies of different geometry in free space.

Geometry	van der Waals potential (across vacuum)
Two spheres of radii a_1 and a_2; center-center distance r 	$V_{vdW} = -\dfrac{A_{12}}{6}\left[\dfrac{2a_1 a_2}{r^2-(a_1+a_2)^2} + \dfrac{2a_1 a_2}{r^2-(a_1-a_2)^2} + \ln\dfrac{r^2-(a_1+a_2)^2}{r^2-(a_1-a_2)^2}\right]$ H.C. Hamaker, Physics **4**, 1058 (1937)
Limiting case: $r \gg a_1, a_2$	$V_{vdW} \approx -\dfrac{16A_{12}}{9}\dfrac{a_1^3 a_2^3}{r^6}\left[1 + 3\dfrac{a_1^2+a_2^2}{r^2} + \dfrac{84 a_1^2 a_2^2}{5r^4} + \cdots\right]$
Limiting case: $h = r-(a_1+a_2) \ll a_1, a_2$ 	$V_{vdW} \approx -\dfrac{A_{12}}{12}\dfrac{\bar{a}}{h}\times\left\{1+\dfrac{h}{\bar{a}}\left[1-\dfrac{\bar{a}}{2(a_1+a_2)}\right]\left[1+\dfrac{\sqrt{3}h}{\bar{a}}-\dfrac{h}{2(a_1+a_2)}\right]+\dfrac{2h}{\bar{a}}\ln\dfrac{h}{\bar{a}}+\cdots\right\};$ $\bar{a} \equiv 2a_1 a_2/(a_1+a_2)$
Limiting case: sphere and semi-infinite block ($a_2 \to \infty$); surface-to-surface distance h 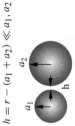	$V_{vdW} = -\dfrac{A_{12}}{6}\dfrac{a}{h}\left(1+\dfrac{h}{2a+h}+\dfrac{h}{a}\ln\dfrac{h}{2a+h}\right)$ Mahanty and Ninham

(continued)

23

Table 1.5. (*Continued*)

Geometry	van der Waals potential (across vacuum)
Square parallel plates of thicknesses δ_1, δ_2 and area L^2, separated by a gap of thickness d 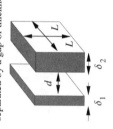	$V_{vdW} = -\dfrac{A_{12}L^2}{12\pi}\left[\dfrac{1}{d^2} + \dfrac{1}{(d+\delta_1+\delta_2)^2} - \dfrac{1}{(d+\delta_1)^2} - \dfrac{1}{(d+\delta_2)^2}\right]$ Mahanty and Ninham
Limiting case: two semi-infinite rods separated by distance d; δ_1, $\delta_2 \to \infty$ 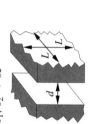	$V_{vdW} = -\dfrac{A_{12}L^2}{12\pi d^2}$ Mahanty and Ninham

Parallel cylinders of length L, radii a and axis-to-axis distance r; $L \gg r \gg a$

$$V_{vdW} \approx -\frac{3\pi A_{12}}{8}\frac{La^4}{r^5}\left(1 + \frac{25a^2}{4r^2} + 31.9\frac{a^4}{r^4} + 150.7\frac{a^6}{r^6} + \cdots\right)$$

Mahanty and Ninham

The same, but $r \gg L$

$$V_{vdW} \approx -A_{12}\frac{L^2 a^4}{r^6}\left(1 - \frac{L^2}{2r^2} + \cdots\right)$$

Mahanty and Ninham

Parallel cylinders, small surface-to-surface distance h; $a \gg h$

$$V_{vdW} \approx -\frac{A_{12}L}{24a}\left(\frac{a}{h}\right)^{3/2}\left(1 - \frac{h}{a} + \frac{1}{\sqrt{2\pi}}\ln\frac{h}{a} + \cdots\right)$$

Mahanty and Ninham

(continued)

Table 1.5. (*Continued*)

Geometry	van der Waals potential (across vacuum)

Two crossed cylinders;
$L \gg r \gg a$

$$V_{vdW} \approx -\frac{\pi A_{12}}{2} \left(\frac{a}{r}\right)^4 \left[1 + \frac{5a^2}{r^2} + 21.875 \frac{a^4}{r^4} + \cdots \right]$$

Mahanty and Ninham

Two crossed cylinders;
$L \gg a \gg h$; $h = r - 2a$

$$V_{vdW} \approx -\frac{A_{12}}{6} \frac{a}{h} \left(1 - \frac{3h}{2a} + \cdots \right)$$

Mahanty and Ninham

would be no contribution to the London interaction. However, if the fluctuations are slow enough, $\lambda \geq r$, the correlation between two electronic systems is good, and the "retarded" potential would scale as $w_L \cong -\frac{\alpha^2 hc}{r^7}$. This *retardation or Casimir-Polder*[3] *effect* is specific for the dispersion forces caused by electronic fluctuations. The range of the validity of the correction $1/r^6 \rightarrow 1/r^7$ is limited, not only from below (by r_r), but also from above. Really, at finite temperatures, the molecules experience other types of fluctuations, e.g., vibrational with characteristic times $T_v \approx 10^{-13}$s. The resulting attractive interactions are not retarded at scales of $cT_v \approx 30\,\mu$ m, which are much larger than r_r. In addition, other interactions, such as permanent dipole interactions, do not experience any retardation effects at all; they scale as $\propto 1/r^6$ and, thus, become predominant at $r \gg r_r$. A cumulative effect is that the r -dependence of the van der Waals interactions changes from $\propto 1/r^6$ to $\propto 1/r^7$ and then back to $\propto 1/r^6$ as the separation r between two particles increases. In practice, the retardation effect is of importance in the range 5 nm $< r < 100$ nm.

1.4.3. London Interactions in a Medium, Lifshitz Theory

Interactions between molecules or macroscopic particles immersed in a solvent are different from their interactions in vacuum. Effective charges, permanent dipole moments, and polarizabilities can be greatly modified by the separating medium. London's model and its extension (1.16)–(1.17) to macroscales through pairwise summation become prohibitively complicated when the "many-body" effects are taken into account. Both the many body problem and the problem of retardation effect are eliminated in the continuum Lifshitz theory[4] that describes the solution and interacting particles in terms of their bulk properties, the frequency-dependent dielectric permittivities, which replace the individual atomic polarizabilities. The theory is built on quantum electrodynamics, and it is not considered here. Qualitatively, the role of the intervening medium can be illustrated in the Lifshitz approach as follows.

A particle is represented as a dielectric sphere of radius R_i and dielectric constant ε_i. In free space, the polarizability of the particle would be related to ε_i as[5] $\alpha_i = 4\pi\varepsilon_0 R_i^3 \times \frac{\varepsilon_i-1}{\varepsilon_i+2}$. Note that $\alpha_i \rightarrow 0$ when $\varepsilon_i \rightarrow 1$. Similarly, a quantity that matters in interactions through a dielectric medium of permittivity ε_m, would be the *excess polarizibility* $\propto (\varepsilon_i - \varepsilon_m)$. When $\varepsilon_i = \varepsilon_m$, the particle is "lost" in the background. Therefore, the constant $C_{12} \propto \alpha_1\alpha_2$ and the Hamaker constant $A_{12} \propto C_{12} \propto \alpha_1\alpha_2$ should both scale as $(\varepsilon_1 - \varepsilon_m)(\varepsilon_2 - \varepsilon_m)$. The van der Waals forces that are always attractive in vacuum become repulsive in a solvent with dielectric constant ε_m intermediate between the dielectric permittivities of the two interacting particles, e.g., $\varepsilon_1 > \varepsilon_m > \varepsilon_2$. When the interacting particles have the same material properties, $\varepsilon_1 = \varepsilon_2$, the interaction remains attractive.

[3]H.G.B. Casimir and D. Polder, Phys. Rev. **73**, 360 (1948).

[4]E.M. Lifshitz, Zh. Eksp. Teor. Fiz. **29**, 94 (1955) [Sov. Phys. JETP **2**, 73 (1956)]; I.E. Dzyaloshinskii, E.M. Lifshitz, and L.P. Pitaevskii, Zh. Eskp. Teor. Fiz. **37**, 229 (1959) and Adv. Phys. **10**, 165 (1961).

[5]L.D. Landau and E.M. Lifshitz, Electrodynamics of Continuous Media, vol. 8, 2nd edition, Pergamon Press, Oxford, 1984.

In Lifshitz theory, the distance dependencies of the interparticles potentials are the same as in the theory based on pairwise summations; all results in Table 1.5 remain valid for interactions in vacuum (retardation effects are naturally incorporated into the Lifshitz theory). An important and advantageous feature is that the Lifshitz theory calculates the Hamaker constant from the frequency-dependent dielectric functions; the latter reflect the collective character of interactions and can be directly measured. Obviously, the Lifshitz theory should not be expected to work well on very small molecular scales, not only because of the atomic graininess of matter, but also because of the repulsive forces.

Note that we considered the dispersion forces for dielectrically isotropic media. When the dielectric properties are anisotropic, the van der Waals interaction would depend not only on the distance between the particles, but also on their mutual orientation; an illustration employing liquid crystalline droplets was given by de Gennes.[6]

London forces play an essential part in the phenomena of adhesion, surface tension, physical adsorption, flocculation, aggregation of particles in water, and in the conformation of condensed macromolecules such as proteins and polymers, because they are long-range, unscreened, and sum up to non-negligible quantities. Also, these attractive forces constitute one of the main contributions to the close-packed character of liquid crystal phases, in which anisometric molecules with anisotropic polarizability are orientationally ordered.

1.4.4. Casimir Interactions

In our above consideration of interactions in a medium, we did not consider the effects of boundaries and interfaces. The boundaries and interfaces impose certain restrictions on fluctuations in the system. These restrictions are not favored from the entropy point of view. Thus, the system should adjust to prevent the decrease of entropy. Suppose, for example, that the fluctuations in an initially infinite system are restricted by a pair of parallel plates separated by a distance d. In the semi-infinite regions outside of the plates, the spectrum of fluctuations is continuous, whereas between the plates, the modes of fluctuations become discrete. To make this restrictive space smaller, the system tends to decrease d. In other words, the geometrical restriction on fluctuations causes interaction between the plates. These geometrically imposed interactions of entropic origin are generally called Casimir interactions. In 1948, Casimir[7] considered the electromagnetic field between two parallel electrodes. Because the electric field must vanish at the conducting surfaces, the electromagnetic fluctuations in the cavity between the plates are restricted and the plates experience interaction (in this case, with an attraction potential $\propto 1/d^3$). Similar interactions, both attractive and repulsive, can occur in soft matter systems. In the latter case, one deals with thermal fluctuations of the field that describes some order in the system rather than with the quantum fluctuations of the electromagnetic field. Although both the Casimir and van der Waals (London) interactions are fluctuations-mediated, they might differ in

---

[6]P.G. de Gennes, C.R. Acad Sci. **271**, 469 (1970).
[7]H.B.G. Casimir, Proc. K. Ned. Akad. Wet. **51**, 793 (1948).

scaling properties (*d*-dependencies) and magnitudes. We will return to this question in Section 13.2.4; for a general review, see Kardar and Golestanian.[8]

1.5. Polymers and Biological Molecules

1.5.1. Synthetic Polymers

Polymers (or macromolecules) are molecules of high molecular mass composed of many small structural units connected by strong covalent bonds. Natural polymers are proteins, cellulose, and rubber. Polymers composed by identical structural units are called *homopolymers*. *Copolymers* contain more than one unit along the chain. The geometry of the macromolecule depends on the valency structure of the structural units. In the simplest case of *linear polymers*, each unit is connected to precisely two neighbors, such as in the linear polyethylene in Fig. 1.13a; the structural unit is $-CH_2-$. In branched macromolecules, a number of structural units show a valency greater than two and connect to three or more neighboring units (Fig. 1.13b). Finally, some polymers show three-dimensionally interconnected units and are called either cross-linked polymers or network structures (Fig. 1.13c). Vulcanization of rubber is an example of a linear polymer cross-linked into a network.

The polymers are synthesized by chemical reactions that connect low-molecular *monomers* into macromolecules. The structural unit of the macromolecule might have the same chemical formula as the monomer (so-called addition polymers) or lack certain atoms present in the monomer (condensation polymers). Polymerization then produces a byproduct, e.g., water, as in the case of aminoacids polymerized into peptides and proteins. Besides the covalent bond, other molecular interactions, especially hydrogen bonding, play an important role in organization of biological molecules and polymers, such as aminoacids and proteins, as considered in the next paragraph.

Obviously, the geometry of the macromolecule is of prime importance in macroscopic arrangements. If the macromolecule is sufficiently regular (linear homopolymers, for example), it often shows an orientational order, whereby macromolecules align parallel to each other, thanks to the dispersive interactions. This orientational order might be long-range; in which case, one deals with a polymer with liquid crystalline order (Chapters 2 and 3) at high temperatures. Kevlar is a well-known example of a nematic liquid crystalline polymer above 300°C, yielding materials of unusual strength when quenched at room temperature. The orientational order might also be only short-range: Because the stiffness of the polymer chain is finite, a macromolecule can fold or change its orientation in space to get around the other chains. As a result, the orientationally aligned domains are interrupted by nonoriented regions. On the other hand, often, the macroscopic order in linear polymers is even stronger than a simple parallel alignment of chains: Structural units that belong to parallel chains find themselves coordinated in a periodic lattice. The size of crystallites is usually substantially smaller than is the extended length of the macromolecule, and the

---

[8]M. Kardar and R. Golestanian, Rev. Mod. Phys. **71**, 1233 (1999).

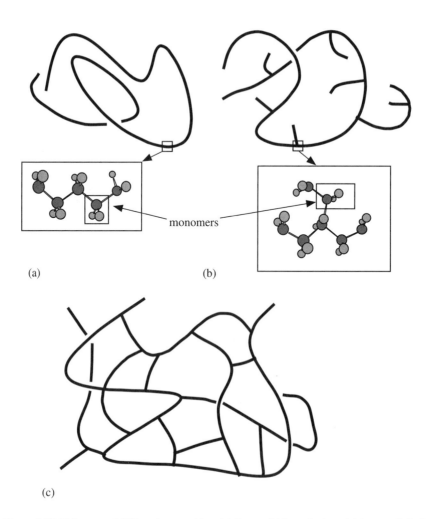

Figure 1.13. Polymers of different geometries: (a) linear, (b) branched, and (c) cross-linked.

Figure 1.14. Polymer often show short-range orientational and crystalline order.

whole structure is an alternating pattern of crystalline and amorphous regions (Fig. 1.14). At increased temperatures, this crystalline (or, more properly, semicrystalline) order melts. Upon decrease in the temperature, polymers with more or less extended amorphous regions transform into a glass state with immobilized chains.

Linear and branched polymers dissolve in suitable solvents. In contrast, their cross-linked counterparts cannot be readily dissolved; instead, they absorb the solvent and swell. Obviously, the properties of the polymers change as their molecular weight increases during polymerization. The most dramatic changes can be observed during cross-linking of linear polymers: With the increase of the molecular weight, at some well-defined *gel point*, the polymer transforms from a viscous fluid into an *elastic gel*. We will return to the physical properties of the polymers in the next chapters (mostly Chapter 15).

1.5.2. Aminoacids, Proteins

Aminoacids are nitrogen-containing monomers from which biopolymers such as peptides (polymers containing no more than 100 aminoacid subunits) and proteins (polymers with a larger number of subunits) are built. The generic form of all 20 aminoacids found in proteins is represented by a central carbon atom C (often denoted C_α) with four groups attached in a tetrahedral fashion. Three of these groups are common for all aminoacids: a hydrogen atom H, a carboxy group COOH, and an amino group NH_2 (Fig. 1.6). The only difference between the aminoacids is in the fourth group, the chain R. R-chain can be hydrophilic (e.g., $R = CH_2 - OH$, serine; $R = CH_2 - SH$, cysteine), hydrophobic ($R = CH_3$, alanine; $R = CH_2 - (CH_3)_2$, valine), negatively charged ($R = CH_2 - COO^-$, aspartic acid), or positively charged ($R = CH_2)_4 - NH_2^+$, lysine). In all examples, the four groups attached to the central carbon C_α are chemically different. Such tetrahedral

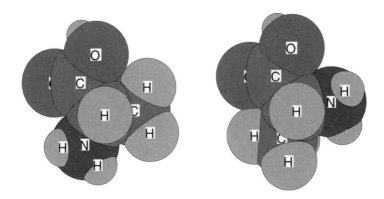

Figure 1.15. Chiral L (on the left side) and D (on the right side) enantiomers of alanine: No movement in space can result in overlapping of the two forms.

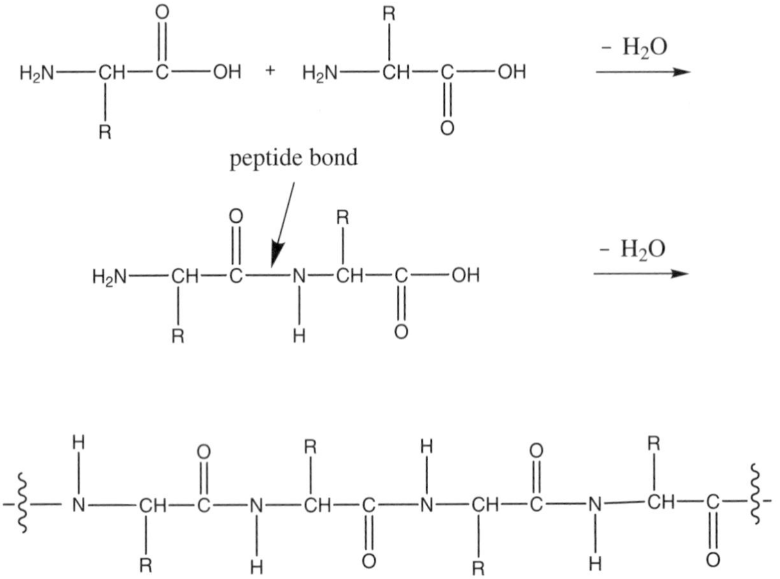

Figure 1.16. Polymerization of aminoacids results in peptides and proteins.

construction has no plane of symmetry; the C_α atom with four different groups is called an asymmetric carbon. There is only one aminoacid, glycine, in which R = H and, thus, two groups are identical. Therefore, all aminoacids, except glycine, can exist in two different forms, called L and D enantiomers, which are mirror images of each other (Fig. 1.15); they are called *chiral*. The property confers optical activity on chiral aminoacids (unless the material contains the L and D enantiomers in equal proportion). Proteins found in nature are composed exclusively of the L-aminoacids; the mechanism of the symmetry breaking during evolution remains unclear, but as Pasteur noticed long ago, the lack of mirror symmetry of biological objects, added to the fact that most biological objects are of L-type, indicates a possibly unique and rare source of life.

Aminoacids are joined together during protein synthesis by peptide bonds. One molecule of water per aminoacid is released during the reaction (Figs. 1.16 and 1.17).

Figure 1.17. A piece of a polymer chain of silk: Glycine R = H and alanine R = CH_3 groups are randomly distributed along the chain.

The four-atom peptide unit (CO–NH) is rigid and planar, because the bond between C and N has partial double character, which hinders free rotation of atoms around the main chain (Fig. 1.18). The rigidity of the peptide bond allows proteins to maintain a well-defined three-dimensional structure.

In contrast, the main-chain C–C_α bond and N–C_α bond are single bonds that allow rotations and, hence, folding of the protein chain in many different ways. This folding is characterized by a hierarchical structure. Most importantly, the seemingly irregular proteins have a remarkably regular feature called the *secondary structure* to contrast it with the *primary structure*, which is understood as the sequence of aminoacids. The secondary structure comes with two building elements, α-helices and β-sheets. The notations reflect the order in which these forms were discovered by Linus Pauling.[9]

The secondary structure occurs because of hydrogen bonding between the groups NH and C $=$ O that belong to different peptide units. In the α-helix, the C $=$ O group of the n-th aminoacid residue is bound to the NH group of the $(n+4)$-th residue (Fig. 1.19a). The main chain is coiled in a helicoidal fashion that puts the n-th unit close to the $(n+3)$-th and the $(n+4)$-th units in space. For example, α-keratin, which is a protein found in skin, nails, hair, and feathers, forms an α-helix. In β-sheets, the main chain remains almost fully stretched out; the hydrogen bonds occur between NH and C $=$ O groups that belong to different polypeptide strands (Fig. 1.19b). These strands can be chemically connected and belong to the same main-chain. The reverse in direction is provided by hairpin turns in which the n-th CO group is bound to the $(n+3)$-th NH group. β-sheets are formed, for example, in silk fibroin. Note that the β-sheets made of chiral aminoacids have their strands twisted. Usually, a single protein molecule contains both α-helix and β-sheets parts.

The secondary structure together with a complex interplay of hydrophobic, hydrophilic, hydrogen-bond, and van der Waals interactions result in higher levels of conformational hierarchy, the so-called *tertiary* and *quaternary structures*. As already indicated, the side R-groups of aminoacids come in different types, polar and nonpolar. A protein

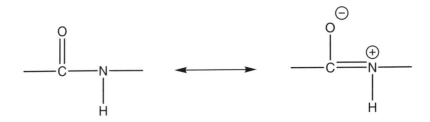

Figure 1.18. Rigid and planar geometry of a peptide unit is caused by a partial double character of the bond between atoms C and N.

[9]L. Pauling, The Nature of the Chemical Bond, 3rd edition, Cornell University Press, Ithaca, New York, 1960.

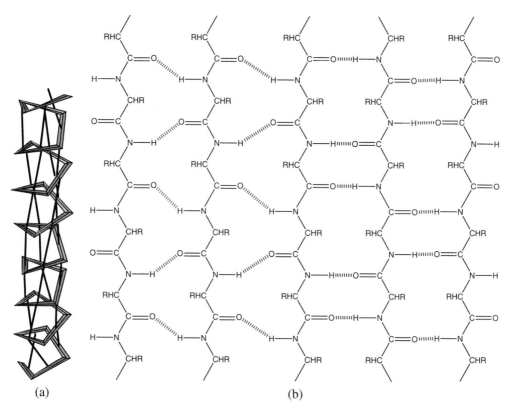

(a)   (b)

Figure 1.19. (a) Right-handed α-helix of a polypeptide main-chain stabilized by hydrogen bonds (shown by thin lines) between n-th and $(n+4)$ aminoacid residues. The diameter of the helix is about 5 Å. There are approximately 3.6 residues per turn of the helix. The side R-groups orient toward the outer region of the helix. (b) Schematic structure of β-sheet formed by hydrogen bonds between protein strands. The strands can be either parallel (three strands on the left side) or antiparallel (three strands on the right side).

globule soluble in water is organized to hide the hydrophobic R-groups inside and to expose the hydrophilic R-groups outside at the protein-water interface. On the contrary, proteins that transverse cell membranes have hydrophobic sides to fit properly the hydrophobic environment of the lipid bilayer (Fig. 1.20).

The complex three-dimensional hierarchy of proteins makes possible their biological functioning: Chemically and geometrically distinctive sites and pockets of the protein globule selectively react to the surrounding molecules.

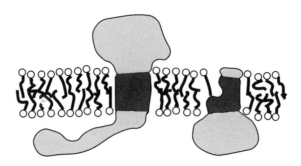

Figure 1.20. Membrane proteins have hydrophobic lateral sides to fit the hydrophobic part of the lipid bilayer; the parts with hydrophilic surfaces are outside the membrane.

1.5.3. DNA

Unlike the van der Waals interactions, the hydrogen-bond interaction is very specific. To form a hydrogen bond, two molecules must possess the needed groups, and they must have a proper shape and mutual orientation, as illustrated by the α-helix. Hydrogen bonding also shapes the most important biological molecules of DNA and RNA, which control genetic information in living systems. In these molecules, four bases, adenine, guanine, cytosine, and thymine (in DNA) or uracil (in RNA), are chemically attached to the backbone, either ribose (in RNA) or desoxyribose (in DNA). The bases of one chain form hydrogen bonding with the bases of a neighboring chain. Here, again, NH groups serve as a hydrogen-bond donor and the carbonyl $C = O$ groups serve as acceptors. Hydrogen bonding together with steric effects results in a highly specific pairing: Cytosine is paired with guanine, and adenine is paired with thymine (or uracil in RNA) (Fig. 1.21a,b). The best match between the complementary bases is provided when the two neighboring chains twist around each other (Fig. 1.21c). Note that the direction of the hydrogen bond is normal to the axis of the double helix. The sequence of bases along the strand keeps genetic information written in a four-letter alphabet.

1.5.4. Associations of Proteins: TMV, Microtubules

Folded proteins contain an array of chemically and geometrically distinctive sites that can selectively bind the surrounding molecules. Thus, proteins can serve as building blocks in supramolecular structures such as enzymes, ribosomes, and viruses. The aggregation is based on noncovalent bonds and, thus, is easy to control and modify. Interestingly, the tobacco mosaic virus (TMV) has a form of a hard rod; the length of TMV rods can be controlled in the laboratory. TMV solutions show liquid crystalline structures[10] and serve as an excellent model to study the nature of orientational order.

---

[10] A. Klug, Fed. Proceed. **31**, 30 (1972).

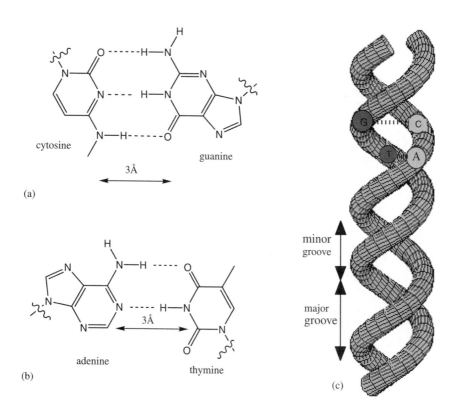

Figure 1.21. Hydrogen bonds between bases of two strands, (a) cytosine-guanine and (b) adenine-thymine, lead to the double helix structure of DNA (c).

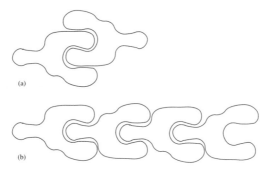

Figure 1.22. Assembling of protein globules into (a) dimers and (b) chains.

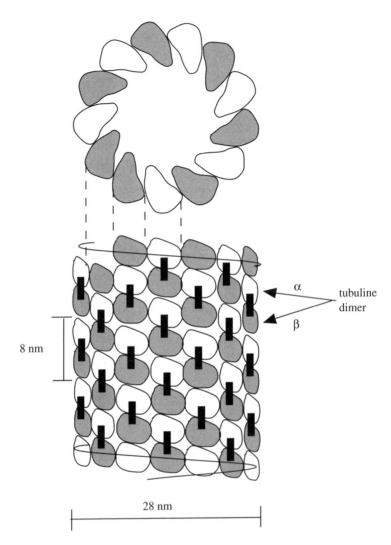

Figure 1.23. Microtubule: (a) normal cross section with 13 tubuline polypeptides; (b) side view shows tubuline dimers. After L. Stryer, Biochemistry, 4th edition, W.H. Freeman and Company, New York, 1064 pp., 1995, and B. Alberts, D. Bray, J. Lewis, M. Raff, K. Roberts, J.D. Watson, Molecular Biology of the Cell, Garland Publishing, Inc., New York, 1146 pp., 1983.

Proteins can form regular supramolecular units by themselves. Imagine, for example, that the protein globule has two complementary sites at the external surface. Bonding of such complementary sites might result in formation of protein dipoles, chains (Fig. 1.22), flat sheets, and even spherical and cylindrical shells. A hollow microtubule formed by dimers of α- and β-tubuline is shown in Fig. 1.23.

Problem 1.1. Calculate the second virial coefficient for

(a) a hard-core potential $w_{12}(r) = \begin{cases} 0, & r > 2r_0, \\ \infty, & r < 2r_0. \end{cases}$

(b) a square-well potential $w_{12}(r) = \begin{cases} \infty, & 0 < r < r_1, \\ -w_0, & r_1 < r < r_2, \\ 0, & r_2 < r. \end{cases}$

(c) a Lennard–Jones potential $w_{12}(r) = 4w_0 \left[\left(\frac{r_0}{r}\right)^{12} - \left(\frac{r_0}{r}\right)^6 \right]$.

Answers:

(a) $v = \frac{1}{2} \int\limits_{r=0}^{2r_0} [1 - \exp(-w_{12}/k_B T)] 4\pi r^2 dr = \frac{16\pi r_0^3}{3}$.

(b) $v = \frac{2\pi r_1^3}{3} \left[1 - \left(\frac{r_2^3}{r_1^3} - 1\right) \left(e^{w_0/k_B T} - 1\right) \right]$.

(c) $v = \frac{16\pi r_0^3}{3} \frac{4w_0}{k_B T} \int\limits_{x=0}^{\infty} \left[\frac{12}{x^{12}} - \frac{6}{x^6} \right] \exp\left\{ -\frac{4w_0}{k_B T} \left(\frac{1}{x^{12}} - \frac{1}{x^6}\right) \right\} x^2 dx$, where $x = r_0/r$; expanding

$\exp\left\{ -\frac{4w_0}{k_B T x^6} \right\}$, one gets $v = \frac{16\pi r_0^3}{3} \sum_{n=0}^{\infty} \alpha_n \left(\frac{4w_0}{k_B T}\right)^{\frac{2n+1}{4}}$, $\alpha_n = -\frac{2}{4n!} \Gamma\left(\frac{2n-1}{4}\right)$. (See Reichl.[11])

Problem 1.2. Real gases generally do not follow the Boyle-Mariotte law $pV = \text{const}$ derived for the ideal gas with no interactions. The van der Waals equation of state written for 1 mole of a "nonideal" gas as $(p + \frac{a}{V^2})(V - b) = RT$ takes into account both attractive and repulsive forces by introducing two phenomenological parameters a and b. Attractive forces are responsible for the gas-liquid phase transition that is predicted by the van der Waals equation. In some cases, the equation can be used to describe solutions, e.g., large (on molecular scale) colloidal particles dispersed in water; the two-dimensional version of the van der Waals equation is often applied to surfactant monolayers. (a) Write the van der Waals equation for α moles. (b) Find the second virial coefficient and the theta-temperature (known as the Boyle temperature T_B in gases). (c) Find the behavior of the product pV during isothermic compression when the temperature of the system is below and above the Boyle temperature T_B. (d) Write a van der Waals equation for a two-dimensional system of area S subjected to a surface pressure Π.

---

[11]L.E. Reichl, A Modern Course in Statistical Physics, Edward Arnold, Kent, U.K., 1980, p. 364.

Answers:

(a) $\left(p + \frac{\alpha^2 a}{V^2}\right)(V - \alpha b) = \alpha RT$; the pressure correction is proportional to α^2 because the probability of interaction of particles is proportional to the number of pair collisions, i.e., the square of their concentration.

(b) $v = \frac{k_B}{R}\left(b - \frac{a}{RT}\right)$; $T_B = \frac{a}{Rb}$.

(c) pV is monotonously increasing for $T > T_B$ and nonmonotonous (with a minimum) for $T < T_B$.

(d) $\left(\Pi + \frac{a}{S^2}\right)(S - b) = RT$.

Problem 1.3. How does the energy of interaction $-\frac{dV_1 dV_2}{|\mathbf{r}_1 - \mathbf{r}_2|^\alpha}$ [see (1.16)] between two volume elements dV_2 and dV_1 change when all distances (separation, particles sizes) are scaled by a factor λ?

Answers: The energy changes as $\lambda^{6-\alpha}$.

Problem 1.4. Plot the Lebwohl-Lasher potential $w_{LL} = -w_0 P_2(\cos\theta)$ as the function of the angle θ between the axes of two particles; $P_2(\cos\theta) = \frac{1}{2}(3\cos^2\theta - 1)$ is the second-order Legendre polynomial, and w_0 is a positive constant. Can this potential describe a ferroelectric state?

Answers: See Advances in the Computer Simulations of Liquid Crystals, Edited by P. Pasini and C. Zannoni, NATO Science Series, Ser. C: Math. and Phys. Sciences, **545**, 430 pp., Kluwer Acad. Publ. (2000), for a detailed discussion of Lebwohl-Lasher, Gay-Berne and other models.

Problem 1.5. Derive the van der Waals potential of interaction of two spheres and two parallel flat slabs of different thickness.

Answers: Table 1.5; for details of calculations, see J. Mahanty and B. W. Ninham (1976).

Problem 1.6. Consider the van der Waals potential (1.16) for two unlike particles 1 and 2 in a solvent (dielectric constants $\varepsilon_1 \neq \varepsilon_2$ and ε_m, respectively), and show that attraction between them is weaker than the arithmetic mean of attractions between the pair of like particles (1,1) and the pair of particles (2, 2); the interparticle distances are the same in all cases.

Answers: The dispersion energy (through the Hamaker constant) scales as $-(\varepsilon_1 - \varepsilon_m)(\varepsilon_2 - \varepsilon_m)$ for unlike particles, and $-(\varepsilon_1 - \varepsilon_m)^2$ and $-(\varepsilon_2 - \varepsilon_m)^2$ for the like particles; the stated result follows from the fact that $(\varepsilon_1 - \varepsilon_m)(\varepsilon_2 - \varepsilon_m) < \frac{(\varepsilon_1 - \varepsilon_m)^2 + (\varepsilon_2 - \varepsilon_m)^2}{2}$. In a multicomponent mixture, there is an effective van der Waals attraction between like particles that tends to aggregate them; see Israelachvili (1992) for further discussion. Note, however, that this result is not always correct (E.I. Kats, private communication). One can imagine, for example, that the $\varepsilon_1 \neq \varepsilon_2$ and $\varepsilon_1 = \varepsilon_2$ situations correspond to different regimes (retarded and nonretarded) of van der Waals interactions. In general, one has to remember that the interactions depend on ε defined by the whole range of the frequency spectrum and not by one particular frequency.

Further Reading

B. Alberts, D. Bray, J. Lewis, M. Raff, K. Roberts, and J.D. Watson, Molecular Biology of the Cell, Garland Publishing, Inc., New York, 1983.

C. Branden and J. Tooze, Introduction to Protein Structure, Garland Publishing, Inc., New York and London, 1991.

Paul J. Flory, Principles of Polymer Chemistry, Cornell University Press, Ithaca, 1953.

P.C. Hiemenz, Principles of Colloid and Surface Chemistry, 2nd edition, Marcel Dekker, Inc., New York, 1986.

J. N. Israelachvili, Intermolecular & Surface Forces, 2nd Edition, Academic Press, London, 1992.

J. Mahanty and B.W. Ninham, Dispersion Forces, Academic Press, London, 1976.

C. Tanford, The Hydrophobic Effect: Formation of Micelles and Biological Membranes, John Wiley & Sons, New York, 1980.

P. Schuster, G. Zundel, and C. Sandorfy, The Hydrogen Bond, vols. 1, 2, 3, North-Holland, Amsterdam, 1976.

L. Stryer, Biochemistry, 4th edition, W.H. Freeman and Company, New York, 1995.

E.J.W. Verwey and J.Th.G. Overbeek, Theory of the Stability of Lyotropic Colloids, Elsevier, New York, 1948.

CHAPTER 2

Atomic and Molecular Arrangements

This is not the place to discuss in detail the *crystalline* arrangements of atoms. Most text books on condensed matter physics describe the 14 Bravais lattices (groups of translations) and the 230 Schönflies–Fedorov groups (translations and point symmetries). These groups exhaust all possible symmetry groups with 3D discrete translations. In classic crystallography, the elements that build the symmetry of the group are atoms; they can be generally represented by points or, when the picture requires more complex characters, by the density of matter. Because of the large (compared with $k_B T$) binding energies, the atoms can be considered as fixed, and the symmetries can be easily visualized through rigid displacements (translations, rotations). These symmetries are not the main concerns of soft matter physics. In soft matter, the order is related to the specific peculiarities of molecular shapes, and fluidity, hence, entropy, are more relevant factors. We first describe how the concept of atomic order can be extended when taking into account the precise shapes of the atoms, no longer considered as points, and how these extensions enter naturally in the description of "complex" soft systems.

2.1. Atomic Order

2.1.1. Packing Densities

Our discussion of the chemical bond shows, in a first approximation, that condensed matter structures are packings whose density is limited by steric hindrances, as long as directional bonds (mostly covalent) are not predominant. Regular lattices of hard spheres with the densest possible packing are said to be *close packings*. As an example in two dimensions, the densest possible regular close packing of disks is provided by a triangular lattice with lattice parameter equal to the diameter of the disk; this is also the densest possible 2D packing, random and aperiodic packings included. This lattice is of hexagonal symmetry; each disk touches six others.

In three dimensions, the two types of regular atomic order that ensure close packing of highest known density are the face-centered cubic (FCC) and the hexagonal close packing (HCP) lattices. In both cases one has a stacking of hexagonally tiled dense two-dimensional

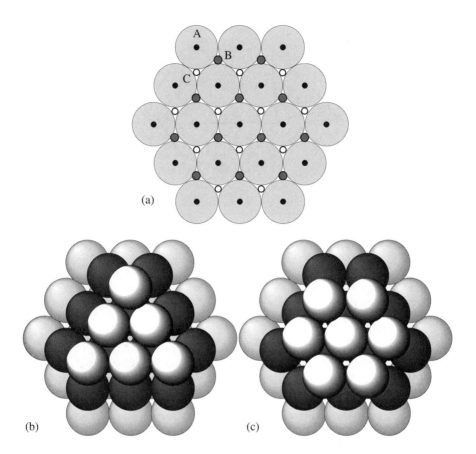

Figure 2.1. The close packing of hard disks on a plane is provided by a triangular lattice with hexagonal symmetry. In three dimensions, the close-packed structures can be built by stacking triangular layers of spheres. Each sphere in one layer is located directly above or below interstices among (a) three atoms in the adjacent layers. Two distinctive geometries are (b) FCC and (c) HCP lattices.

planes, arranged in the order either as ABCABC... (FCC lattices) or ...ABAB... (HCP lattices) (Fig. 2.1).

Each atom has $Z = 12$ neighbors in the first coordination shell. The dense planes are perpendicular to the ternary axis of the cube in the FCC case. The ratio between the volume occupied by spheres in contact over the total volume, the so-called *packing fraction*, is $p_{FCC} = p_{HCP} = 0.74048\dots$. Of course, for a system at finite temperature, this value is achieved only if the pressure is infinite. The centers of the 12 spheres in contact with a given sphere are at the vertices of a polyhedron of coordination, the cuboctahedron (Fig. 2.2), which has 14 faces (eight equilateral triangles and six squares); all edges are equal (the

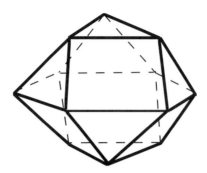

Figure 2.2. Cuboctahedron for . . . ABC . . . type of packing.

cuboctahedron is an example of a *semi-regular* polyhedron). Each atom has four nearest neighbors on the sphere of coordination.

Consider two successive planes A and B; the interstices between spheres that lie in the mid-surface are of two types:

1. The *tetrahedral sites* are the centers of four spheres packing (e.g., three in A, one in B), which are mutually in contact; the centers of the spheres are at the vertices of a regular tetrahedron (Fig. 2.3a).

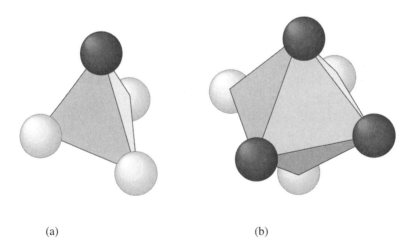

(a)   (b)

Figure 2.3. Interstitial sites of a (a) tetrahedral and (b) octahedral type in a close-packed lattice of spheres; the shown spheres are reduced in size.

2. The *octahedral sites* are the centers of six spheres packing (three in A, three in B), which are mutually in contact; the centers of the spheres are at the vertices of a regular octahedron (Fig. 2.3b).

It is important to recognize that tetrahedral sites yield the highest possible local density of any polyhedral site (defined as the packing fraction of the inner part of the polyhedron); therefore, the FCC and HCP lattices show spatial modulations of density at a microscopic characteristic length.

2.1.2. Liquids and Amorphous Media

In both liquids and amorphous media, the local order can be defined as the arrangement of the nearest (and the next nearest) neighbors. In a *liquid*, the density fluctuates only locally, but takes a well-defined spatially averaged value; this is a thermodynamically stable phase. On the contrary, the global arrangements in a *glass* or an *amorphous metal* depend critically on the conditions of fabrication (quench, irradiation, etc.), and these are not thermodynamically stable phases.

Diffraction techniques (X-ray, neutrons, etc.) have been used to obtain information about the local order by measuring the "radial distribution function," which determines how the matter density around an atom, averaged over all atoms, depends on the distance from the atom. The essential features of the experimental data are well understood within the framework of a model worked out by Bernal,[1] according to which liquids and amorphous media are dense *random*-packings (DRP) of hard spheres. The packing fraction of a well-relaxed DRP is about 0.637. Bernal has used the *polyhedral sites approach* to analyze his hand-made systems of equal spheres, showing that 86% of the polyhedra are tetrahedra, 6% are octahedra, and the remaining ones, in their great majority, have triangular faces (he calls them deltahedra). Among various types of deltahedra, there are large cages with 8, 9, and 10 vertices enclosing an "empty" space ("Bernal's holes").

A dual approach to the description of a packing is the averaged number $\langle Z \rangle$ of geometrical neighbors of each particle. It can be defined as the average number of faces of the so-called *Voronoi polyhedron* built around a particle (Fig. 2.4).

To construct the Voronoi polyhedron of a given particle, first connect this particle to all other particles of the set by line segments, then add the planes Π that are perpendicular bisectors to these segments, and finally, select those planes Π that set the bounds of the largest polyhedron not intersected by any other plane Π. In solid crystals, the Voronoi polyhedra are nothing other than the well-known Wigner–Seitz cells; in FCC and HCP lattices, as already stated, the number of faces is the number of atoms in contact with a given atom; i.e., $Z = 12$. More generally, the number of faces measures the number of nearest neighbors, whether they be in close contact or not in close contact in a hard sphere model.

---

[1] J.D. Bernal, Proc. Roy. Soc. A **280**, 299 (1964).

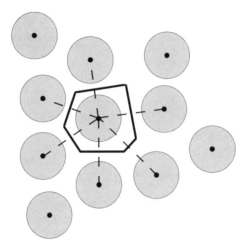

Figure 2.4. Voronoi polyhedron for a disordered system.

The Voronoi analysis emphasizes the average number of faces $\langle Z \rangle$ of the Voronoi cell and the frequency f_p of finding a face with p edges. Were the packing perfectly icosahedral (we shall soon see why the consideration of icosahedra is of interest), we would have $\langle Z \rangle = 12$ and $f_5 = 1$ (the Voronoi cells would then all be dodecahedral). Finney's relaxed DRP model gives $f_2 = 0.02$, $f_4 = 0.2$, $f_5 = 0.43$, $f_6 = 0.32$, and $f_7 = 0.03$. The average number $\langle p \rangle$ of edges per face is easily related to $\langle Z \rangle$ (see Coxeter[2]) by use of the Euler–Poincaré relation for a polyhedron (a Voronoi cell):

$$v - e + f = 2, \tag{2.1}$$

where v is the number of vertices, e is the total number of edges, and f is the number of faces. Let us apply (2.1) to a regular polyhedron with $q = 3$ edges at each vertex. One obtains easily a relation between the number of faces f and the number of edges p of each face, viz., $f = 12/(6 - p)$. We use the same relation as relating the *averaged values* $\langle Z \rangle$ and $\langle p \rangle$ in a set of Voronoi cells:

$$\langle Z \rangle = \frac{12}{6 - \langle p \rangle}. \tag{2.2}$$

A DRP model for a well-relaxed system of soft spheres yields $\langle p \rangle \approx 5.12$ and $\langle Z \rangle \approx 13.6$. A non-relaxed (less compact) DRP yields larger values than does the relaxed one. Note that for a body-centered cubic (BCC) crystal, the same calculation yields, with $Z =$

[2]H.S.M. Coxeter, Ill. J. Math. **2**, 746 (1958).

14, $\langle p \rangle \approx 5.15$, i.e., larger values than for the relaxed DRP. These data point to the fact that the packing fraction decreases on the average when the number of neighbors increases. Compacity of matter, which is at the origin of the (meta) stability of metallic glasses, requires high-packing fractions.

2.1.3. Geometrical Frustration

Tetrahedral arrangements, which are the most compact local close packings, and *defects* in these arrangements, which allow for diversity of coordination numbers, are the main ingredients to describing random packings of monoatomic fluids and metallic glasses. Because a centered, regular *icosahedron* is made of 20 equal *tetrahedra* having a common vertex, at the center of the icosahedron (Fig. 2.5), one expects local icosahedral symmetry.

Icosahedral order, in fact, slightly deviates from the close-packing in the sense that 12 equal spheres in contact with a sphere of the same radius are not in contact among themselves. Such an assembly of 13 spheres is stabilized by a balance between the (disfavorable) elastic distortion energy of the tetrahedron, and the (favorable) vibrational entropy. In effect, in a perfect icosahedral cluster of 13 atoms, the distance between the central atom and the atoms at the vertices is about 5% shorter than the distance between the neighboring atoms at the vertices, hence, a compressive stress at the central atom and a resulting elastic energy. But, contrarily, the larger distances between surface atoms produce some beneficial surface entropy.

Despite this divergence with close-packing, the *local density* of an icosahedral arrangement is larger than in a close-packed FCC or HCP structure. Therefore, there are reasons to believe that the icosahedral arrangement has a lower energy than the FCC or HCP ones. Lennard–Jones pair potential calculations for small aggregates confirm these conclusions.

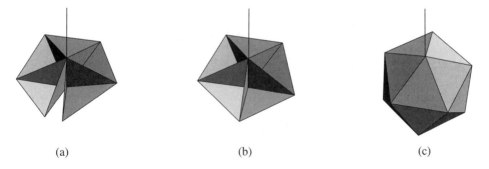

(a)   (b)   (c)

Figure 2.5. Tetrahedral and icosahedral order: (a) five regular tetrahedra around the common edge cannot fill space without frustration; the top faces of five tetrahedra are not shown in the picture; (b) five distorted tetrahedra with the axis of five-fold symmetry; (c) an icosahedron with one of the five-fold symmetry axes.

The problem is that one cannot tile space in a regular manner with icosahedra, because of their five-fold symmetry, an element of symmetry that is forbidden in a translation symmetric tiling of space. Trying to propagate icosahedral order coherently from a local icosahedral cluster results in irregularities (double coverings or vacancies, loss of positional correlations). This is an example of a frequent phenomenon in soft matter and complex systems, called *geometrical frustration*. In the present case, it is at the origin of a modulation of the coordination number around an average value Z, which is closer to 14 than to 12 (see above). Now, the reason why it is a positive extra-coordination, rather than a negative one, is obviously related to the sign of deviations to close packing alluded to above.

More elaborate quantitative treatments of this question have been proposed; they rely on the introduction of the concept of a curved crystal, see Kleman 1989. Assuming that

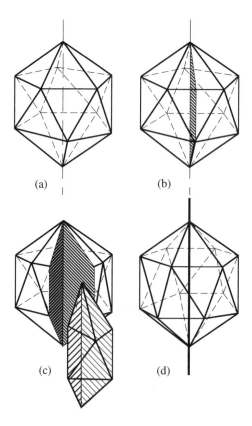

Figure 2.6. Introducing a disclination line into (a) an icosahedron with $Z = 12$. The body is (b) cut, and (c) a wedge of material bounded by symmetry-equivalent surfaces is added between the two lips of the cut; (d) the relaxed deltahedron has $Z = 14$ and a disclination line along the axis of the cut.

the structure is entirely tetrahedral, the theory of curved crystals shows that the tetrahedra pack mainly in the shape of centered icosahedra ($Z = 12$), except along certain atomic directions, which can be understood as defect lines of an ideal icosahedral structure. Figure 2.6 visualizes how these lines can be obtained from a perfect icosahedron, in a so-called "Volterra process" (to be explained in greater detail in Chapter 8). In the case shown here, the Volterra process transforms $Z = 12$ into $Z = 14$. Observe that the process amounts to an addition of a $2\pi/5$ angular section of an icosahedron between the lips of an opened cut surface Σ, afterward allowing the medium to relax elastically. This is our first encounter with the (badly defined) notion of a defect, here, a disclination line.

As a matter of fact, lines of defects of the above type were first advocated by Frank and Kasper[3] to describe *periodic crystals* of complex metallic alloys whose crystalline arrangements can be analyzed in terms of packing of non-regular tetrahedra: Given the different sizes of atoms that compose the phase, close-packing is the best description. The relation with liquids and amorphous metals is obvious. In these Frank and Kasper phases, the local coordination number takes the values $Z = 8, 9, 10, 12, 14, 15,$ or 16 for atoms located along the defect lines (Fig. 2.7).

2.1.4. Incommensurate Phases and Quasicrystals

In physics of crystals, phonons are small amplitude thermal vibrations of the atoms. As is well known, phonons in position $\mathbf{r}_i$ can be described in terms of eigenmodes, each mode $\mathbf{u_k}(\mathbf{r}_i, t) = \mathbf{U_k}(\mathbf{r}_i) \exp i(\mathbf{k} \cdot \mathbf{r}_i - \omega t)$ carrying a wavevector $\mathbf{k}$ and a frequency $\omega/(2\pi)$; $\mathbf{u_k}(\mathbf{r}_i, t)$ has the lattice periodicity (Bloch theorem), and $k_i = \frac{2\pi}{a_i} \frac{n}{N}$ ($n = 1, 2, \ldots, N$) is a reciprocal space vector, Na_i is the size of the sample in the i-direction, a_i is the lattice parameter (Born–von Karman boundary conditions). $\mathbf{k}$ is continuous with values in the first Brillouin zone when the sample is infinite. When the temperature varies, the frequency of a particular mode might tend to zero, whereas its amplitude increases, for a value of $\mathbf{k}$ that is incommensurate with the lattice periodicity. Such a "soft mode" is a precursor to a "displacive" phase transition in which the new periodicity $2\pi/|\mathbf{k}|$ coexists with the "old" one. The result is an "incommensurate phase" (for example: thiourea $SC(NH_2)_2$, the natural mineral calaverite, etc.). Such a phenomenon can also be understood as a *modulation* of the periodicity with an incommensurate period.

Figure 2.8 illustrates how a one-dimensional (1D) periodic structure with atoms regularly positioned at $x_n = na$ (a is the period, n is an integer) can be modulated in a commensurate and in an incommensurate manner. The atomic shifts in both modulated states are described by the same function $\sim \sin(2\pi\alpha x_n/a)$. When the number $\alpha < 1$ is rational, the modulated state is also periodic, with a larger unit cell. When α is irrational, the structure has no period and is said to be *incommensurate*. The incommensurate states might emerge when the forces responsible for cohesion play in opposite directions; e.g., the first neighbors show repulsive interactions, and the second neighbors show attractive

---

[3]F.C. Frank and J.S. Kasper, Acta Cryst. **11**, 84 (1958); Acta Cryst. **12**, 483 (1959).

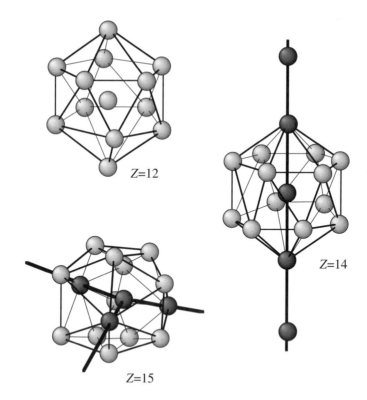

Figure 2.7. Frank–Kasper deltahedra as coordination shells with different Z surrounding an atom. Atoms depicted in dark grey are along the disclination lines.

Figure 2.8. (a) One-dimensional periodic structure of atoms located at $x_n = na$ and its two modulated versions with atomic shifts $\sim \sin(2\pi\alpha x_n/a)$: (b) commensurate modulation, $\alpha = 0.6$, periodic structure with a unit cell five times larger than that of the initial chain; (c) incommensurate modulation, $\alpha = 0.62\ldots$ close to the irrational number $\tau^{-1} = (\sqrt{5} - 1)/2$; no periodicity.

ones. In fact, incommensurate modulations are fairly frequent, either acting on the atoms directly, as above, or indirectly, through a coupling to the electronic density waves (Peierls instability) or to the spin density waves.

Modulated crystals in d dimensions (which are not crystals in the strict sense; there is some kind of "disorder," because the neighborhood of a given species of atom continuously changes in space) have been given a crystallographic description by using an embedding in a space of dimension $d + n$, where n is the number of modulations. This is illustrated by Fig. 2.9 for the case $d = 1, n = 1$. The "physical space" Σ (actually, Σ is a line in our example) is obtained as a cut of a 2D square lattice, and the 1D modulated structure along Σ is a *projection* of the atoms of the 2D crystal. The cut-and-project method is due to the Dutch school of crystallography.[4] If the slope of Σ with respect to the 2D lattice is rational, the projected structure is periodic along Σ. If the slope is irrational, the projected set is not periodic. Obviously, if all atoms of the 2D lattice are projected onto Σ that has an irrational slope, the resulting set on Σ would be dense: The Delaunay condition, which

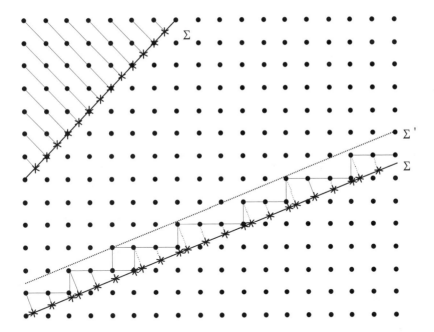

Figure 2.9. Illustration in $d + n = 2$ dimensions of the crystallography of periodically modulated and quasicrystalline materials. A 2D square lattice projects onto a line Σ either as a 1D periodic structure when the slope of Σ is rational (upper left corner) or as an incommensurate aperiodic structure when the slope is irrational (central part).

[4]P.M. deWolff, Acta Crystallogr. **A30**, 777 (1974); N.G. De Brujn, Proc. Konig. Ned. Akad. Weten **A84**, 39 (1981); T. Janssen, Acta Crystallogr. **A42**, 261 (1985).

states that there should be a nonvanishing minimum distance between atoms, would not be obeyed. To restrict this density, only those atoms are projected that are within a finite distance from Σ. In this case, there is a well-defined mean distance between the projected points on Σ, but the actual distances are modulated so that when the cut is irrational, the projected points repeat with two distances incommensurate with this mean distance, along an aperiodic sequence.

Prior to incommensurate atomic structures, it was recognized that some rare earths metals show a 1D helimagnetic structure,[5] incommensurate with the lattice parameter, due to a subtle competition between antiferromagnetic nearest neighbors and ferromagnetic next nearest neighbors interactions.

A particular type of incommensurability occurs in *quasicrystals*, a large family of complex metallic alloys, such as AlMn, AlFeCu, and AlLiCu, which have a local icosahedral order and no crystalline repeat distance, but keep long-range icosahedral correlations. Noncrystallographic symmetry enforces incommensurability. Other quasicrystalline symmetries met in nature are pentagonal, octagonal, decagonal, and dodecagonal. Quasicrystals with nearly free electrons, like AlLiCu, are stabilized seemingly by a special value of the number of valence electrons per atom, which yields to a depletion of the density of states near the Fermi level, according to a mechanism for stability discussed long ago by Jones and Hume-Rothery[6] for much simpler alloys. The prime experimental data for incommensurate phases or quasicrystalline alloys is their diffraction pattern. The pattern is indexed on an incommensurate basis, for example, as above, where $n = 1$ with two *incommensurate periods a* and *b*, by difraction spots in positions $q_{lm} = 2\pi(l/a + m/b)$. The integers l and m can evidently be chosen in such a way that q_{lm} takes a value as close as one wishes to any value given in advance: The diffraction pattern is dense. However, it does not appear to the experimentalist as continuous, because the Bragg spots are of unequal intensities, as can be inferred from a calculation of the diffraction pattern. Physical examples are (1) the Penrose pattern (Fig. 2.10), which mimics observed 2D decagonal crystal; the (dense) diffraction pattern can be obtained as a linear combination of base vectors along the edges of a regular pentagon; (2) the icosahedral pattern, whose diffraction pattern is a linear combination of base vectors along the edges of a regular icosahedron. The cut-and-project method introduced above for the incommensurate case can be extended to those quasicrystalline cases, with $d = 2, n = 2$ in the pentagonal case (hence, the pentagonal case is a 2D crystallographic cut of a four-dimensional crystal, whose atomic surfaces are 2D objects), and with $d = 3, n = 3$ in the icosahedral case.

Some of the Frank and Kasper phases can be thought of as *rational approximants* of quasicrystals, the cut being rational but close to the irrational fundamental one. Hence, local icosahedral symmetry is still present, but the medium is periodic.

---

[5]A. Yoshimori, J. Phys. Soc. Japan **14**, 807 (1951); J. Villain, Chem. Phys. Solids **11**, 303 (1959); T.A. Kaplan, Phys. Rev. **116**, 888 (1959).

[6]J. Friedel, Helv. Phys. Acta **61**, 538 (1988).

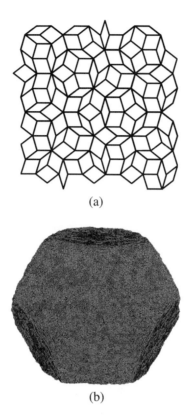

(a)

(b)

Figure 2.10. (a) Penrose tiling. (b) Simulation of a growing icosahedral quasicristal (courtesy V. Dmitrienko); notice the dodecahedral facetting.

2.2. Molecular Order

Everything said above about positional order referred to spherical particles. Because the molecules are not spherical, they establish not only *translational* but also *orientational* order. For particular ranges of temperature or concentration, the systems of nonspherical molecules produce *mesophases* (phases "intermediate" between regular crystals and isotropic fluids), in which either the positional order is kept but the orientational order is lost (*plastic crystals*) or the orientational order is preserved but the positional order is partially or completely lost (*liquid crystals*).

2.2.1. Plastic Crystals

Plastic crystals are made of almost spherical molecules. The nonsphericity allows one to define a "molecular axis" at each site, but the anisotropy of interactions is too small to pre-

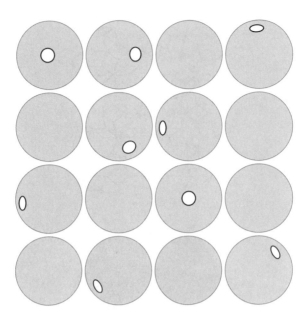

Figure 2.11. Plastic crystal: Positional order of particles with no orientational order. Spherical symmetry is restored on the average.

vent molecules from free rotation at increased temperatures: Although the molecules are correlated in position, the orientations of axes are not correlated (Fig. 2.11). Typical examples are Cl_2 and the recently discovered crystals, whose building units are Buckminster Fuller's "buckyballs" (*fullerenes*) of chemical formula C_{60}. The carbon atoms are located at the vertices of a polyhedron and have three bonds with neighboring carbon atoms on the same polyhedron; each bond can be thought of as an edge of the polyhedron, which is made of 20 regular hexagons and 12 regular pentagons, assuming that all edges have equal length.

2.2.2. The Building Blocks of Liquid Crystals

Liquid crystals are made of strongly anisometric molecules, either elongated (calamitic molecules) or disk-like (discotic molecules). As a rule, the inner part of mesogenic molecules is rigid (phenyl groups) and the outer part flexible (aliphatic chains). This double character explains altogether the existence of steric interactions (between rod-like or disk-like cores of the molecules) yielding orientational order and the fluidity of the mesomorphic phases. Typical examples follow.

2.2.2.1. p-Pentyl-p'-Cyanobiphenyl (5CB) and p-Octyl-p'-Cyanobiphenyl (8CB)

The most studied single-component nematic liquid crystals, cyanobiphenyls are now the prime materials used in liquid crystal display devices, due to their chemical stability, high dielectric, and optical anisotropy. They were first synthesized by Gray.[7] Figure 2.12 shows 5CB.

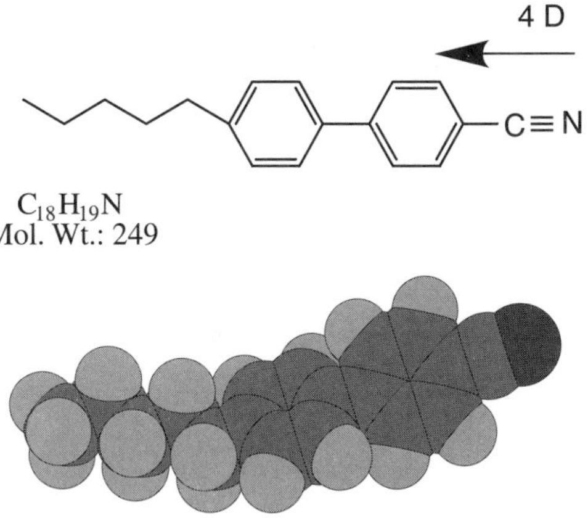

$C_{18}H_{19}N$
Mol. Wt.: 249

$$\text{solid} \xleftrightarrow{\;22.5^{O}C\;} \text{nematic} \xleftrightarrow{\;35^{O}C\;} \text{isotropic}$$

Figure 2.12. Chemical formula, molecular structure, and phase diagram of 5CB. Note that the benzene rings are located in different planes.

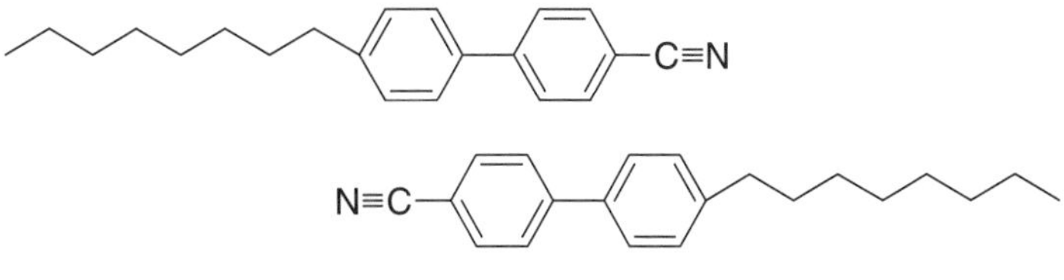

$C_{21}H_{25}N$
Mol. Wt.: 291

$$\text{solid} \xleftrightarrow{\;24^{O}C\;} \text{smectic A} \xleftrightarrow{\;34^{O}C\;} \text{nematic} \xleftrightarrow{\;42.6^{O}C\;} \text{isotropic}$$

Figure 2.13. Antiparallel dipole arrangements of 8CB molecules and phase diagram of 8CB.

[7]G.W. Gray, J. Phys. **36**, C1, 337 (1975).

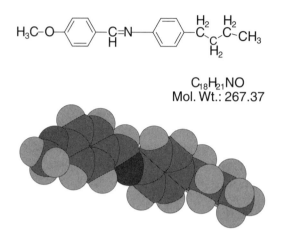

$C_{18}H_{21}NO$
Mol. Wt.: 267.37

Figure 2.14. Chemical formula and molecular structure of MBBA.

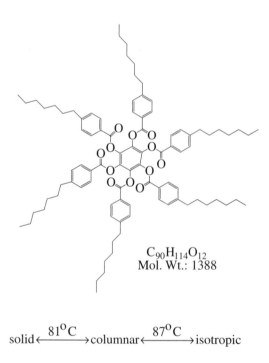

$C_{90}H_{114}O_{12}$
Mol. Wt.: 1388

$$\text{solid} \xleftrightarrow{\ 81^{\circ}C\ } \text{columnar} \xleftrightarrow{\ 87^{\circ}C\ } \text{isotropic}$$

Figure 2.15. Discotic molecules of hexa-heptyloxybenzoate of benzene form columnar phases. See S. Chandrasekhar, B.K. Sadashiva, K.A. Suresh, N.V. Madhusudana, S. Kumar, R. Shashidhar, and G. Venkatesh, J. Phys. (Paris) Colloq. **40**, C3–120 (1979).

The terminal cyanogroup results in a large longitudinal dipole moment ($\sim$ 4D). Because of strong dipole–dipole interactions, the molecules form antiparallel configurations. The effect is especially pronounced in the smectic phase of 8CB where an individual smectic layer is formed by a pair of molecules. The thickness of such a layer is approximately 1.4 of the length of a single extended 8CB molecule (Fig. 2.13).

2.2.2.2. N-(p-Methoxybenzylidene)-p'-Butylaniline (MBBA)

MBBA (Fig. 2.14) is another example of a well-studied substance forming a uniaxial nematic phase at room temperature. The molecule is less polar than is a 5CB molecule.

2.2.2.3. Discotic Molecules

Flat molecular cores tend to be parallel to each other; the flexible hydrocarbon chains are much more disordered (Figs. 2.15 and 2.16).

2.2.2.4. Amphiphilic Molecules, Polymers

The substances above produce *thermotropic mesophases*, i.e., phases with a single component, whose phase transitions can be induced by a change in temperature. *Lyotropic mesophases* occur when anisometric amphiphilic molecules (soaps, phospholipids, and various types of surfactant molecules, including those used in the cosmetic industry) are

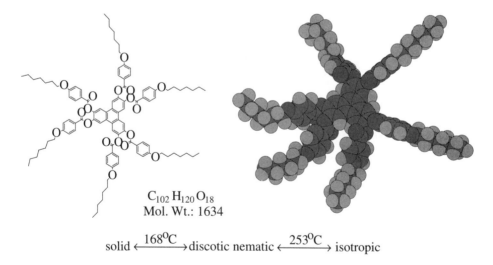

$$C_{102}\,H_{120}\,O_{18}$$
Mol. Wt.: 1634

$$\text{solid} \xleftrightarrow{\;168^{\circ}C\;} \text{discotic nematic} \xleftrightarrow{\;253^{\circ}C\;} \text{isotropic}$$

Figure 2.16. Hexa-n-hexyloxybenzoate-triphenylene: chemical structure, phase diagram, and one of the possible molecular configurations; note that the chains are not necessarily parallel to the central disk group. See N.H. Tinh, H. Gasparoux, and C. Destrade, Mol. Cryst. Liq. Cryst. **68**, 101 (1981).

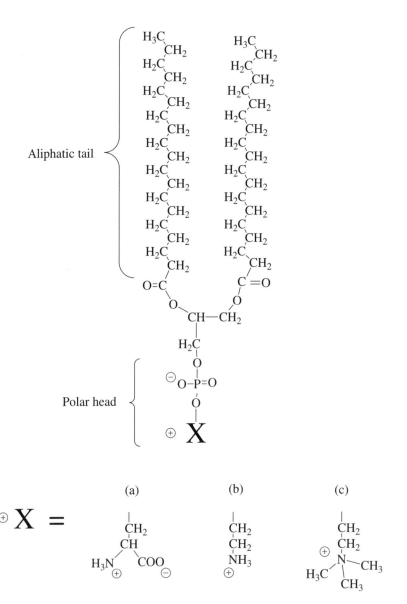

Figure 2.17. Molecular structure of phospholipids: (a) phosphatidylserine, (b) phosphatidylethanol-amine, and (c) phosphatidylcholine (lecithin). The number of carbon atoms in the aliphatic chains varies, usually between 16 and 20. The two chains might be of different length.

added to a solvent, such as water, brine, or oil. Amphiphiles have two distinct parts: a (polar) hydrophilic head and a (nonpolar) hydrophobic, oleophilic tail (generally aliphatic). Figure 2.17 depicts a general molecular structure of phospholipids that are present in biological cell membranes. Two *aliphatic chains* keep the cross section of the molecule from varying much from the polar head to the aliphatic part. The building units of lyotropic phases are aggregates of many amphiphilic molecules (micelles) rather than single molecules. The architecture of the molecule has consequences in the relative stability of different aggregates.

As an example, consider water solutions of a one-tailed amphiphile (such as sodium dodecyl sulphate, SDS). For concentrations above the critical micellar concentration (CMC), these molecules form aggregates of different shapes (Fig. 2.18), the simplest ones being spherical micelles, whose size scales with the size of the molecule, i.e., between 20 Å and 50 Å. The geometry prevents the hydrophobic tails from the contact with water. Mixing different amphiphilic substances might result in the formation of anisometric micelles.

At a higher concentration of the surfactant, one might get cylindrical micelles, infinite cylinders, bilayers, inverse cylinders, and inverse micelles. These complex elements are in turn building blocks for various phases with long-range order, considered later in this chapter.

Synthetic macromolecules, made of mesogenic monomers, attached either chain-like (Fig. 2.19a) or comb-like to a backbone (Fig. 2.19b), may also be building blocks for liquid crystalline phases. Biological polymers [DNA, PB(L or D)G, xanthane, etc.] form liquid crystal phases in solutions in vitro, due to the rigidity of their backbones. Some viruses with highly anisometric shape, such as tobacco mosaic virus (TMV), also form lyotropic liquid crystalline phases in solutions.

An interesting example of lyotropic mesomorphism is presented by the so-called chromonic liquid crystals.[8] The family embraces a range of dyes, drugs, nucleic acids, antibiotics, carcinogens, and anticancer agents. The molecules are plank-like or disk-like (rather than rod-like), with polar solubilizing groups at the periphery and an aromatic central core. Aggregation of molecules, caused primarily by face-to-face adhesion of aromatic cores, results in cylindrical stacks or other geometries, different from the micelles formed by rod-like surfactant molecules.

2.2.3. Classification of the Mesomorphic Phases

Mesomorphic phases (also called liquid crystals) are intermediary between liquids and solids. They show manifold possible structures; many can belong to the same compound (polymorphism). There are four basic types of liquid crystalline phases, classified according to the dimensionality of the translational correlations of building units: nematic (no translational correlations), smectic (1D correlations), columnar (2D correlations), and various 3D-correlated structures, such as cubic phases.

---

[8]J. Lydon, Curr. Opin. Colloid Interface Sci. **3**, 458 (1998).

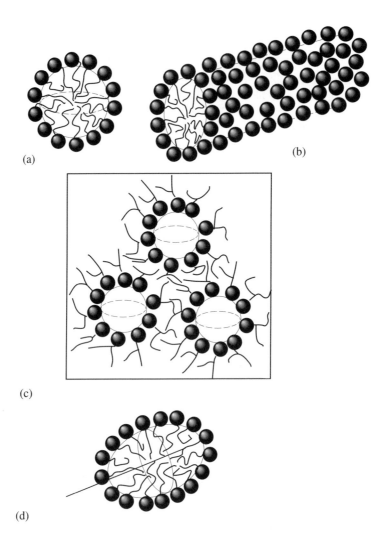

(a)

(b)

(c)

(d)

Figure 2.18. (a) Spherical micelle, (b) cylindrical micelles, (c) inverted spherical micelles, and (d) anisometric micelle.

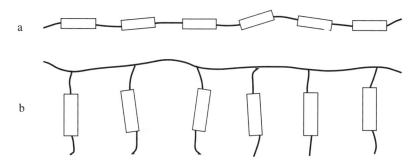

Figure 2.19. Polymers with mesogenic groups in (a) main-chain and (b) side chains are capable of forming thermotropic and lyotropic liquid crystalline phases.

2.2.3.1. Nematics

Uniaxial nematics, noted N, are optically uniaxial phases (Fig. 2.20). The unit vector **n** along the optic axis is called the *director*. Even when the building units are polar (such as 5CB molecules), molecular flip-flops and head-to-head overlapping establish centrosymmetric (average) arrangement in the nematic bulk. Thus, **n** and −**n** are equivalent notations, **n** ≡ −**n**. The director is an axis of continuous rotational symmetry: the symmetry point group of the N phase is the same as that of a homogeneous circular cylinder, viz. $D_{\infty h}$. The molecules, which are anisometric in shape, align *in average* parallel to **n**, this averaging being made over all "directions" of the individual molecules. The difficulty is to define unequivocally a relevant direction in each individual molecule. Nevertheless, the process makes sense, for the reason that **n** is unequivocally experimentally defined (for example, as an optical axis; see Chapter 3). A suitable choice for an individual direction can be along some chemically defined axis of the molecule, in the case of calamitics made of rod-like building units (Fig. 2.20a), or along some chemical "normal" to the disk in the case of discotics (Fig. 2.20b).

Another already mentioned property of N phases is their fluidity: the centers of gravity of the molecules are not correlated. Thus the continuous group of Euclidean translations R^3 belongs to the complete group of symmetry.

In *biaxial nematics* N_B, the symmetry point group is one of a prism (Fig. 2.21). Known N_B phases are rare: They are documented for anisotropic micelles and some mesogenic polymers, and it is most plausible that the symmetry is that of a prism with a rectangular cross section. An N_B phase is characterized by three directors, **n**, **t**, and **m** = **n** × **t**, such that **n** ≡ −**n**, **t** ≡ −**t**, and **m** ≡ −**m**.

When the building block (molecule or aggregate) is chiral, i.e., not equal to its mirror image, the nematic phase might show *twist* (Fig. 2.22). It is then called a *cholesteric* phase N*.

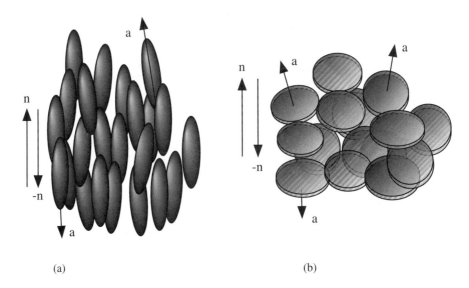

(a)   (b)

Figure 2.20. Uniaxial (a) calamitic and (b) discotic nematics can be viewed as a system of elongated rods or disks with axes **a** oriented preferentially along a common director **n**. Directions **n** and −**n** are equivalent even if the molecular axis **a** is a true vector. The units in the picture represent either individual molecules in the case of thermotropic nematics or micelles in the case of lyotropic nematics.

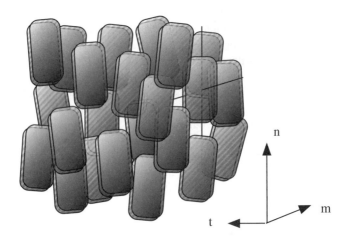

Figure 2.21. Biaxial nematic phase.

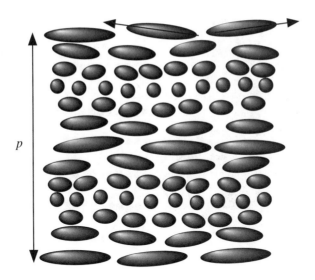

Figure 2.22. Cholesteric phase: a twisted nematic.

A rotation by an angle α about the cholesteric axis **c** is equivalent to a translation $p\alpha/2\pi$; p is here the pitch of the cholesteric twist, and it is twice the periodicity along the **c** axis. An N* phase can be characterized by three directors: **n**, "along" the local molecular axes, **c** along the axis of helicity (which is also the optic axis if the pitch is much smaller than the light wavelength), and $\mathbf{m} = \mathbf{n} \times \mathbf{c}$, "perpendicular" to both **n** and **c**. These three directors form a trihedron of directions ($\mathbf{n} = -\mathbf{n}$, $\mathbf{c} = -\mathbf{c}$, $\mathbf{m} = -\mathbf{m}$) that rotates with the cholesteric pitch. Both N_B and N* phases are liquid phases (no correlations in molecular positions).

2.2.3.2. Smectic Phases

Smectics are layered phases with quasi–long-range 1D translational order of centers of molecules in a direction normal to the layers. This positional order is not exactly the long-range order as in normal 3D crystals: As shown by Landau,[9] and Peierls,[10] the fluctuative displacements of layers in a 1D lattice diverge logarithmically with the linear size of the sample. However, the effect is noticeable only on scales 1 km and more; typical samples are thinner, (10–100) μm or even less. Within the layers, the molecules show fluid-like arrangement, bond-orientation order (discussed below), or solid-like arrangement. Although the structures of the latter type have non-zero shear elastic constants, these constants are much smaller than in regular 3D crystals. It is not clear if these structures, called smectics

[9]L.D. Landau, Phys. Z. Sowjet Union **2**, 26 (1937).

[10]R.E. Peierls, Annales de l'Institut Henri Poincaré **5**, 177 (1935).

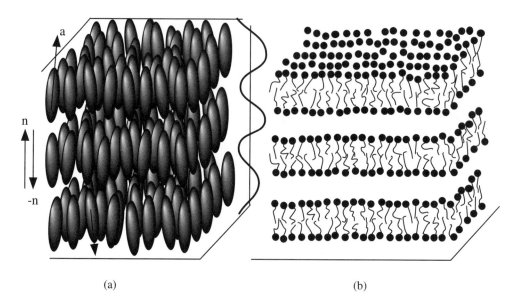

(a)   (b)

Figure 2.23. (a) Thermotropic smectic A phase with periodic modulation of density; (b) lyotropic L_α phase with surfactant bilayers separated by water; the layer of water might be much thicker than the surfactant bilayer.

of B, E, G, H, J, and K-types, maintain a true long-range order of the molecular positions in the direction normal to the layers. We will not consider these phases further.

Smectics with *liquid* layers are of three types, as described below.

- Smectic A (SmA) is a uniaxial medium with the optic axis perpendicular to the layers; the director **n** is along the normal to the layers (Fig. 2.23). There is no long-range positional order within the layers; each layer is a 2D fluid.

- Smectic C (SmC) is also composed of a 1D stack of fluid layers; however, it is a biaxial phase because the long axes **a** of the molecules are tilted with respect to the layers' normal **t** (Fig. 2.24). The axes **a** average to the "nematic" director **n**, if no attention is paid to the layers. The so-called tilt plane formed by **n** and **t** that contains the optical axes is a plane of mirror symmetry. Another operation of symmetry is a π-rotation around the axis C_2 that is perpendicular to the tilt plane. Because of the layered structure, the twofold symmetry axis C_2 lies either in the midplane of the smectic layers or in the plane between two layers. The combination of mirror reflection and twofold axis of rotation yields inversion symmetry.

The three operations of symmetry (mirror reflection, C_2 axis, and inversion) lead to interesting properties of the unit vector **v** specified by the projections of the molecules onto the smectic planes. This vector is not a director nor a vector. Viewed in a labo-

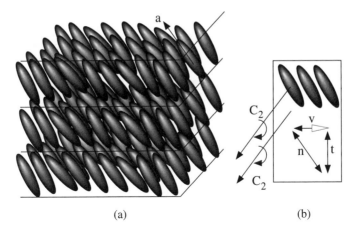

Figure 2.24. Smectic C phase (achiral): (a) general structure; (b) elements of symmetry; see text.

ratory coordinate frame, a π-rotation around the C_2 axis transforms **v** into $-$**v** (and **t** into $-$**t**). The vector $-$**v** is shifted along **t** after such a rotation, either by the half or the whole SmC layer's spacing, depending on the position of the C_2 axis. Because the SmC slab after such a rotation is not distinguishable from the original state, it implies a "director" character of **v**, **v** $\equiv$ $-$**v**. Consider now a π-rotation around the layers' normal **t**, which is not a symmetry axis. Such a rotation transforms **v** into $-$**v** in the laboratory coordinate frame, but the two states are clearly not identical, because the molecules are tilted in the opposite direction. Thus, this rotation does not identify the states **v** and $-$**v**. This dual character of **v** reflects in the nature of topological defects (see Section 12.1.7).

Another subtle point concerns the orientational order along C_2. Because of the layered structure of SmC, a π-rotation around the "nematic" director **n** is *not* an allowed operation of symmetry, in contrast to the N phase. Because of the mirror symmetry, however, there is no electric polarization along C_2. The situation changes when the molecules are chiral, as discussed in the next paragraph.

- Smectic C* (SmC*) composed of chiral molecules is a chiral version of SmC. The molecular tilt precesses around the normal to the layers (Fig. 2.25). The chirality suppresses the mirror and inversion symmetries, because mirror reflections and inversion would change the handedness (left vs right) of the helix. Therefore, as pointed out by R.B. Meyer,[11] the electric polarization **p** along the C_2 axis normal to the tilt plane is not cancelled by the symmetry. In the unperturbed helical SmC*, the local tilt plane rotates from layer to layer and so does the local polarization vector **p** (Fig. 2.25). The net polarization **P** averaged over distances much larger than the helicoidal pitch is thus zero. However, if one confines a very thin SmC* slab between two rigid plates (parallel

---

[11]R.B. Meyer, L. Liebert, L. Strzelecki, and P. Keller, J. Phys. (Paris) Lett. **36**, L69 (1975).

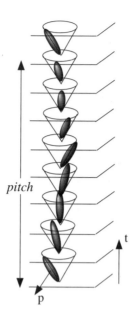

Figure 2.25. Chiral smectic C (SmC*) with periodic precession of the tilted molecules around the normal to the layers.

to the plane of drawing in Fig. 2.25) that favor an in-plane orientation of the molecules, the SmC* helix can be unwound. Such a "surface-stabilized" ferroelectric film has a nonzero net polarization **P** directed either toward or outward from the reader. Applying an electric field of certain polarity to the bounding surfaces, one reverses the direction of **P** and thus reverses the direction of the molecular tilt $\theta \leftrightarrow -\theta$. Such reorientation produces drastic optical changes when the film is wieved between crossed polarizers (see Section 3.3.3). This effect, discovered by Clark and Lagerwall,[12] is used to construct fast informational displays. Recent advances in chemical design (e.g., synthesis of banana-like molecules) have produced smectic phases with a rich variety of ferroelectric properties; some show antiferroelectricity and ferrielectricity; see the book *Chirality in Liquid Crystals* (2000).

Some smectic phases with liquid layers display hexagonal bond-orientational order in the layers. They are called hexatic smectics. For example, in hexatic smectic B, the molecules are normal to the layers and have no positional order within the layers, as in SmA. However, they show long-range hexagonal ordering of the directions that link the molecules (bond ordering) (see Chapter 4). Tilted versions of the hexatic B phase are hexatic smectics F and I. Note that the layers in A, C, and C* and the hexatic smectics are not correlated; these phases have been used as experimental models to verify theories of 2D media.

---

[12]N.A. Clark and S.T. Lagerwall, Appl. Phys. Lett. **36**, 899 (1980).

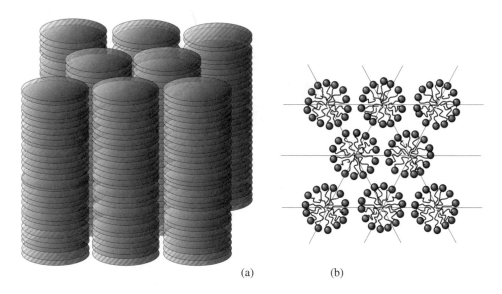

(a)   (b)

Figure 2.26. (a) Thermotropic hexagonal columnar phase formed by disk-like molecules. The intermolecular distances are random along the columns but fixed along a 2D crystalline lattice in the normal plane. (b) Lyotropic hexagonal columnar phase formed by cylindrical micelles; the picture shows a 2D cut normal to the cylinders.

2.2.3.3. Columnar Phases

Columnar phases show 2D long-range positional order with translational symmetries. *Hexagonal* columnar phases are often formed by columns of discotic molecules (Fig. 2.26) or by lyotropic cylinders. Other 2D symmetries have not been investigated thoroughly, because they are less frequent.

2.2.3.4. Tridimensional Phases

- *Lyotropic cubic phases* are formed of bilayers that extend along the three directions of space. The case of lyotropics has been studied in great detail: The mean surface of the bilayers is close to a periodic *minimal* surface, i.e., a surface whose principal curvatures σ_1 and σ_2 are everywhere equal and opposite: $\sigma_1 + \sigma_2 = 0$. This condition minimizes the bending energy of the bilayers (notions of surface geometry and elasticity theory will be studied in later chapters). The overall geometry of these cubic phases is complex to discuss in great detail; note only that the periodic arrangement refers to the layers rather than to the individual molecules, which are free to move within the layers. There are at least three types of lyotropic cubic phases known to date; one is shown in Fig. 2.27.

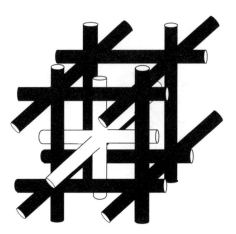

Figure 2.27. Lyotropic cubic phase.

- *Blue phases* are made of chiral molecules that organize in an inhomogeneous way on the following basis. Let $\mathbf{n}_0$ be some director, e.g., along the axis Z in Fig. 2.28. In the local state of the smallest energy, the chiral molecules in the vicinity of $\mathbf{n}_0$ have the tendency to rotate helically along all directions perpendicular to $\mathbf{n}_0$, not only along one direction $\mathbf{c}$, as in the N* phases. This geometry, which is called a *double-twist*, is energetically preferable to the 1D twist, at least for some chiral materials. However, as the distance from the director $\mathbf{n}_0 \| Z$ increases, the cholesteric cylindrical shells become flatter and the double twist smoothly disappears. The director far-field configuration comes closer to the 1D twist of the N* phase; the energy gain is reduced. Thus, the double twist cannot extend over the whole 3D space. A typical radius of the energy-gaining cylindrical region about the $\mathbf{n}_0$ axis is the half-pitch $p/2$. Now, these cylinders of finite radius cannot tile space continuously. The situation is reminiscent of the phenomenon of *frustration*, already met in Frank and Kasper phases. According to the most current models of BP's, this frustration is relieved by defect lines (of the disclination type), either regularly distributed or in disorder, also as in the Frank and Kasper phases or metallic glasses. Figure 2.28 illustrates how three cylinders of double twist generate a singularity in the region where they merge.

     The blue phases of types BPI and BPII are modeled as regular networks of disclination lines with periodicity of the order of p. Indeed, the 3D periodic structure of these phases is revealed in their nonzero shear moduli, ability to grow well-faceted monocrystals, and ability of Bragg reflection in the visible part of the spectrum (which is natural because p is of the order of a few tenths of a micron). The latter explains the name: When viewed under a polarizing microscope, the blue phases often appear blue. Usually, they exist in a very narrow temperature range (~ 1 K) between the isotropic and the N* phases. The third identified phase, BPIII, which normally occurs between

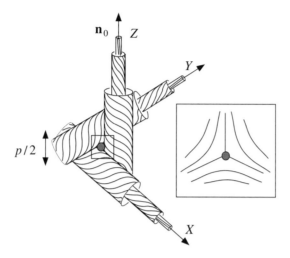

Figure 2.28. Blue phases are composed of regions with double twist; three such regions with a singularity that relieves frustration between them are shown in the picture. Two cylinders with double twist match at the contact point if the director tilt at their surfaces is $\pi/4$; however, the region where all three cylinders meet is singular. In current models, such singularities form a network of disclination lines. The circle marks the "core" of the disclination; the insert shows the director lines around the core.

the isotropic melt and BPII, is less understood. It might be a melted array of disclinations. Note that although most blue phases were observed in thermotropic systems, double-twist geometries can occur in solutions of biological polymers, such as DNA.

2.2.4. Isotropic Phases

We shall classify several systems under the heading of isotropic phases, such as follows.

- BPIII. This phase has the same molecular double-twisted arrangement as BPI and BPII at small scales, but it is believed that the cylinders form long, random, flexible, intertwined "worms."

- Phases of *associated colloids*. (i) Micelles, which are either roughly spherical or worm-like closed assemblies of surfactants, whose size, or size distribution, is in thermodynamic equilibrium with the solvent; (ii) The isotropic L_3 sponge phase, which is described according to the most current views as a bilayer of surfactant that extends through all space in a random fashion and divides the solvent into two connected continuous domains; this thermodynamic phase is sometimes referred to as "the plumber's

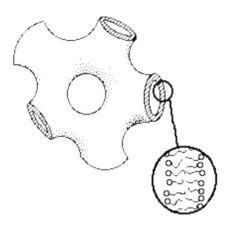

Figure 2.29. A model of lyotropic sponge phase: A bilayer that divides the solvent into two connected continuous domains.

nightmare" (see Fig. 2.29); (iii) Microemulsions, made of a monolayer of surfactant that extends through all space in a random fashion, separating two solvents of different natures, such as oil and water; this is again a thermodynamic phase.

- *colloidal solutions* (of macromolecules, proteins, or of biological polymers) and *molten polymers* (see Chapter 15).

The elements of the above classification are phases at thermodynamical equilibrium: The solutions are supposed to be in a "good" solvent. We shall not use the term of solution for metastable *dispersions* (in a "poor" solvent).

In the examples above, the solvent is an isotropic fluid. The last decade saw an explosive growth of interest in mesomorphic solvents. Mesomorphic solvents are very special because they impose their orientational (nematics) and positional (smectics) order onto the solute particles. For example, by orienting dye molecules in a nematic matrix, one can obtain a system with electrically-switchable dicroism. Adding a solute particle to a smectic host might lead to a nanoscale segregation: Depending on the chemical affinity to the host, the solute particles might segregate either inside the smectic layers or in the space between them.[13] The variety of phenomena becomes even richer when one dissolves a monomer into a liquid crystalline matrix and then polymerizes it in this matrix. Although polymer-liquid crystal composites are still lacking complete description, their widely known representatives, such as polymer-dispersed liquid crystals and polymer-stabilized liquid crystals[14] already found practical applications, mainly in the display industry.

---

[13] See, e.g., M.A. Glaser, In Advances in the Computer Simulations of Liquid Crystals, Edited by P. Pasini and C. Zannoni, NATO Science Series, Ser. C: Math. and Phys. Sciences, v.**545**, p. 263, sec. 4.2 (2000).

[14] Liquid Crystals in Complex Geometries Formed by Polymer and Porous Networks, Edited by G.P. Crawford and S. Zumer, Taylor & Francis, 1996, 506 pp.

2.3. Perturbations of the Crystalline Order

We shall make a distinction between the perturbations of small energy, which can relax in a finite time toward equilibrium, and those of large energy, which build singularities of the order parameter (a term to be defined later).

2.3.1. Weak Perturbations

Thermal fluctuations of small amplitude, large wavelength $l = 2\pi/|\mathbf{k}|$ compared with atomic or molecular distances, and small frequency compared with atomic or molecular frequencies are weak perturbations. Among these are phonons in crystalline solids, spin waves in ferromagnets, and fluctuations of the optical axis in nematics (Fig. 2.30). Fluctuations are analyzed as sums of *eigenmodes*

$$u_0(\mathbf{r}) = \sum_{\mathbf{k}} u_0(\mathbf{k}, \mathbf{r}) \exp i (\mathbf{k} \cdot \mathbf{r} - \omega t), \tag{2.3}$$

whose frequency $\omega/2\pi$ and wavevectors $\mathbf{k}$ depend on the nature of the order, in fact on the group of symmetry; the eigenmodes form a representation of this group.

Any weak perturbation imposed on the system can be analyzed as a sum of independent eigenmodes. Weak perturbations belong to the kingdom of *linear physics* and do not modify the order because $k = |\mathbf{k}|$ is much smaller than $2\pi/a$: the atoms, molecules, or spins do react cooperatively.

Soft modes occur when for some special value of k, say, k_c, and near some temperature T_c, the mode frequency $\omega \to 0$ and the amplitude remains finite; they are the precursors of second-order phase transitions with symmetry change. They were mentioned earlier.

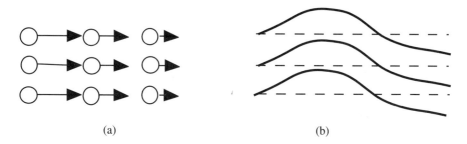

(a)   (b)

Figure 2.30. Fluctuations: (a) longitudinal phonon wave of atomic displacements; the arrows show the direction and the amplitude of displacement; (b) fluctuation of the director field (optic axis) in an uniaxial nematic.

2.3.2. Strong Perturbations

By definition, strong perturbations are perturbations that "break" the order parameter. They can be either point-like (point defects in solids, singular points in nematics, etc.), line-like (dislocations in solids, vortex lines in the superfluid phase of ^{3}He, disclinations in liquid crystals, focal conics in smectics, etc.), or surface-like (grain boundaries in solids or smectics, Bloch wall in ferromagnets, etc.). In the spirit of the approach to *defects* favored by the metallurgists, the analysis is still linear at some distance from the object. That is, it can be done in the framework of weak perturbations, which implies conservation of the order parameter and cooperative linear response, but needs a different description in the vicinity of the object, where the order parameter is truly *singular*. This is the *core* of the defect.

Strong perturbations can also be classified as thermodynamical (spontaneous) or imposed. Strong thermodynamical perturbations are, for example, the point defects of solids (vacancies or interstitials, Fig. 2.31). They are spread at random in the lattice, hence, their stabilization by the entropy of disorder, which competes with their (positive) internal energy (to create them). Another example is provided by some models of the solid - liquid transition, which assume that the liquid state is announced by a spontaneous multiplication of dislocation lines in the solid, again stabilized by their entropy of disorder. Other examples occuring in soft matter will appear in the course of this textbook. Finally, some periodic phases of the frustrated type, such as the Frank and Kasper phases and the blue phases, are often described in terms of defects, whose internal energy should be negative, because they do not carry entropy of disorder.

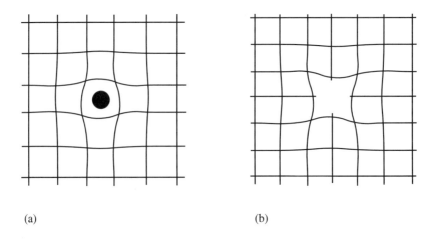

(a)   (b)

Figure 2.31. Point defects in a crystal caused by an (a) interstitial atom and (b) vacancy.

Imposed strong perturbations are, for example, nonequilibrium point defects nucleated as a result of radiation damage; dislocations in solids resulting from growth phenomena or from work-hardening; disclinations that nucleate, change shape, or disappear in liquid crystals under shear flow; Bloch walls in a ferromagnet that are involved in structural modifications under magnetic field action.

These strong perturbations affect a number of physical phenomena, in particular, nonreversible phenomena (stress-strain curves of solids, hysteresis of magnetic phenomena, dissipative phenomena in liquid crystals under flow), both at mesoscopic scales (at which these singularities show up) and at macroscopic scales (at which these singularities gather into more or less defined *textures*).

One of the purposes of this textbook is precisely to give the conceptual means to classify the various types of singularities of an ordered medium (this classification depends on the symmetry of the order parameter) and to understand their behavior under various actions.

Problem 2.1. (a) Show that the exact value of the packing fraction for the FCC lattice is $\pi\sqrt{2}/6$; (b) copper has an FCC structure. Calculate its density taking the atom radius equal to 1.28 Å.

Problem 2.2. Show that there are twice as many tetrahedral sites as octahedral sites in an ordered close-packed structure.

Answers: Each atom is surrounded by 12 atoms at the vertices of a cubooctahedron with six square facets and eight triangular facets. This is true whether the order is FCC or HCP. Hence, attached to each atom, there are six octahedra and eight tetrahedra. Each octahedron shares six vertices, and each tetrahedron shares four vertices. Therefore, the mean number of tetrahedra per atom is two and the mean number of octahedra per atom is one.

Problem 2.3.

(a) Show that for any polyhedron, $v - e + f = 2$ (the number of vertices v, of edges e, of faces f). Hint: Construct any polyhedron from any given one by removing or adding, according to the case, faces, edges, and vertices. This formula is known as the *Euler-Poincaré theorem*.

(b) Using the result of question (a), prove (2.2).

(c) Show that in a regular polyhedron, $v = \frac{4p}{6-p}$, $e = \frac{6p}{6-p}$, if the number of edges per vertex is $q = 3$.

Answers: (b) Let $\langle Z \rangle = \langle f \rangle$ be the mean number of faces on any Voronoi cell. We have $\langle Z \rangle \langle p \rangle = 2\langle e \rangle$ and $q\langle v \rangle = 2\langle e \rangle$. Hence, using the result of question (a), $\langle Z \rangle = 2 + \frac{\langle p \rangle \langle Z \rangle}{2}(1 - (2/q))$; i.e., $\langle Z \rangle = \frac{4q}{2q - q\langle p \rangle + 2\langle p \rangle}$; $q = 3$ yields (2.2).

Problem 2.4. Retrieve the geometry of the buckyball given in the text by employing the above Euler-Poincaré theorem.

Problem 2.5. Show that the diffracted spots intensity of the 1D quasicrystal of Fig. 2.9 reads as

$$I_q \cong A_q A_q^* = \left(\frac{2}{q_\perp d} \sin \frac{q_\perp d}{2} \right)^2,$$

where $q_\perp$ is the component in perpendicular space of $q = q_\parallel + q_\perp$ and d is the width of the strip.

Answers: We calculate the diffracted amplitude $A_q = \sum_N \exp i q_\parallel R_{N\parallel}$, where $R_{N\parallel}$ are the 1D coordinates in physical space of the vertices R_N belonging to the 2D hyper lattice. Now, the Bragg spots of the quasilattice are the projections of the Bragg spots of the 2D reciprocal lattice. Hence, $q R_N = q_\parallel R_{N\parallel} + q_\perp R_{N\perp} = 0 \pmod{2\pi}$; thereby, the amplitude can be written as $A_q = \sum_N \exp -i q_\perp R_{N\perp}$, where the $R_{N\parallel}$ are the components of the projections of the vertices R_N belonging to the stripe (Fig. 2.9). The vertices R_N fill the perpendicular space densely and homogeneously, by reason of the irrationality of the projection. Therefore, the sum can be replaced by an integral through the width d of the stripe:

$$A_q = \sum_N \exp -i q_\perp R_{N\perp} \sim \int_0^d \exp -i q_\perp y \, dy,$$

from which the expression of I_q is easily deduced. Although the spots are dense in the physical reciprocal space only, a few of them are intense enough to be visible, because I_q oscillates between null values and maxima $q_\perp \approx n/d$, $q_\parallel \approx m/d$, n and m integers, and decreases very fast with $q_\perp$. Of course, the same types of results apply, *mutatis mutandis*, for realistic quasicrystals [$d_\parallel = 2(D = 4)$; $d_\parallel = 3(D = 6)$].

Further Reading

N. Ashcroft and D. Mermin, Solid State Physics, Holt, Rinehart, and Winston, New York 1976.

Chirality in Liquid Crystals, Edited by H.-S. Kitzerow and C. Bahr, Springer, New York, 2000.

H. S. M. Coxeter, Regular Polytopes, Dover Publications, New York, 1973.

F.C. Frank, Supercooling of Liquids, Proc. Roy. Soc. **A215**, 43 (1952).

J. Friedel, Helv. Phys. Acta **61**, 538 (1988).

P.G. de Gennes and J. Prost, The Physics of Liquid Crystals, 2nd Edition, Clarendon Press, Oxford, 1993.

G.W. Gray and J.W. Goodby, Smectic Liquid Crystals. Textures and Structures, Leonard Hill, Glasgow, 1984.

J.-P. Hansen, Theory of Simple Liquids, Academic Press, New York, 1986.

C. Janot, Quasicrystals: A Primer, Clarendon Press, Oxford, 1992.

M. Kleman, Curved Crystals, Defects and Disorder, Adv. in Physics **38**, 605 (1989).

A.G. Petrov, The Lyotropic State of Matter. Molecular Physics and Living Matter Physics, Gordon and Breach Science Publishers, Amsterdam, 1999.

Physics and Chemistry of Finite Systems: From Clusters to Crystals, vols. I and II, edited by P. Jena, S.N. Khanna, and B.K. Rao, NATO ASI Series, Series C: Mathematical and Physical Sciences, Vol. 374, Kluwer Academic Publishers, 1992.

Physics of Complex and Supermolecular Fluids, S.A. Safran and N.A. Clark, eds. Wiley, New York, 1987.

S.A. Safran, Statistical Thermodynamics of Surfaces, Interfaces and Membranes, Addison-Wesley, Reading, MA, 1994.

R. Zallen, The Physics of Amorphous Bodies, John Wiley & Sons, 1983.

CHAPTER 3

The Order Parameter: Amplitude and Phase

The concept of an order parameter has appeared with the attempt to describe the order-disorder transition of alloys, specifically, to define a degree of disorder (see Section 3.1.6). First elaborated by Gorsky and Bragg and Williams to describe order-disorder transitions in alloys,[1] it has been developed in its modern form by Landau for the purpose of a phenomenological description of phase transitions. More specifically, it addresses the question of the description of the *long-range order* of the structural (crystallographic) or thermodynamic (magnetic, dielectric, etc) properties, which repeat uniformly in a given system. Related quantities are intensive thermodynamic variables. The opposite notion of *short-range order* refers to spatial thermodynamic fluctuations; these are particularly important near a second-order phase transition A ↔ B (see Chapter 4). Fluctuations that develop in, say, the high temperature phase A, announce the order properties of B. An important related quantity is the spatial extent ξ of these fluctuations, the coherence length. The same term of coherence length (or correlation length) is also used to conote the range of distortions induced in the order parameter of phase B by a local perturbation (Chapter 5, section 5.6), e.g., the extent of the central region of a defect, where the order parameter is "broken" (singular), the *core* of the defect.

It is possible to get an intuitive view of the order parameter through a series of very simple examples: Hence, the nematic phase is more ordered than is the isotropic phase into which it turns above some "clearing" temperature T_c. The choice of the measurable physical parameter that describes best the change of order and whose amplitude relates to the "degree of order" is in any case an important and delicate issue.

In addition to its *amplitude*, the order parameter possesses another characteristic, a *phase* (also called a *degeneracy parameter*), which is less often discussed. For reasons to appear later, we are as much interested in the degeneracy parameter as in the amplitude. The following examples will make the distinction clear.

---

[1] W. Gorsky, Z. Phys. **50**, 64 (1928); W.L. Bragg and E.J. Williams, Proc. Roy. Soc. London, **A 145**, 699 (1934).

3.1. The Order Parameter Space

3.1.1. Superfluid Helium

The order parameter is the wave function

$$\psi = |\psi_o|\exp i\phi. \tag{3.1}$$

The amplitude $|\psi_o(T, P)|$ measures the concentration of the constituent ^{4}He, which has "condensed"—these atoms are bosons—to a superfluid state. $|\psi_o(T, P)|$ depends on temperature and pressure and vanishes at the temperature of superfluid-normal fluid transition and above, in the normal state. The graph of Fig. 3.1 represents the Landau "condensation energy" $F_{\rm cond}(|\psi_o|)$, whose minimum takes the value of a thermodynamic potential at the equilibrium value of $|\psi_o(T, P)|$. It is well known from standard quantum mechanics that any spatial variation of the phase ϕ creates a current of matter $\mathbf{j} \propto i(\psi\nabla\psi^* - \psi^*\nabla\psi)$. Therefore, at equilibrium, the phase ϕ must be constant throughout the whole sample, but the thermodynamic potentials do not depend on ϕ: Two samples of superfluid ^{4}He with different ϕ's, taken at the same temperature and pressure, both assumed perfect, would have the same thermodynamic potentials. In other words, each value of the phase corresponds to another realization of the equilibrium. Thus, ϕ is a *degeneracy parameter*.

Because $\phi(\mathbf{r})$ is defined modulo 2π, it is an "angle" and its domain of variation can be represented by a circle S^1 of radius $|\psi_o(T, P)|$ (Fig. 3.1). Any point belonging to S^1

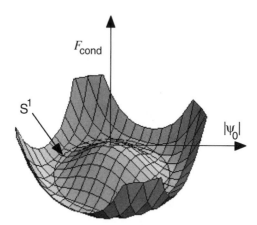

Figure 3.1. The circle S^1 is the order parameter space of the superfluid ^{4}He. It corresponds to the circular bottom of the condensation free energy plotted as the function of the order parameter $\psi = |\psi_o|\exp i\phi$. The states along the circle correspond to the same equilibrium value of the amplitude $|\psi_o(T, P)|$ and have the same energy, but differ in the phase ϕ.

represents a perfect state defined by the phase ϕ. S^1 is called the "*order parameter space*" (OP space) of the superfluid ^{4}He. The OP space is also called the "*degeneracy space*."

To conform to the terminology that will often be met in this textbook, a superfluid with spatial variations $\nabla\phi$ of the degeneracy parameter $\phi(\mathbf{r})$ will be said to be in a "deformed" state. As long as these deformations are "weak," i.e., vary significantly only on spatial scales much larger than the coherence length ξ, it is possible to consider the amplitude $|\psi_o(T, P)|$ as spatially constant, because ξ is precisely the minimum scale on which thermodynamic intensive quantities are to be defined. The variations of the amplitude, $\sim \nabla|\psi_o|$ that relax on small scales of the order of ξ cannot be included in any "macroscopic" description of a weakly deformed medium. Consequently, the function $\phi(\mathbf{r})$ maps any continuous path in a weakly deformed medium into some continuous path in S^1.

The case of superfluid ^{4}He provides us with the simplest example of a nontrivial OP space, but offers some difficulty in grasping the notion of "deformation" as it is included in the concept of phase. The examples that follow should shed some light on this question.

3.1.2. Heisenberg Ferromagnets

The order parameter is an intensive variable $\mathbf{M}(T, \mathbf{r})$, the magnetization per unit volume. In the Heisenberg model,[2] the magnetization is caused by pair interactions of spins arranged on a regular periodic lattice. The interactions align the spins parallel to each other. When the temperature is raised, the correlation between the spins becomes weaker and the amplitude $M(T)$ of $\mathbf{M}(T, \mathbf{r})$ gradually decreases. At and above the so-called Curie temperature, the entropy effects overcome the ordering interactions and the ferromagnetic phase transforms into the paramagnetic one, where the spins are not correlated and $M = 0$ (Fig. 3.2).

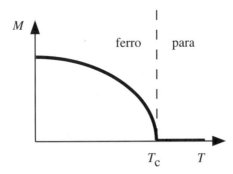

Figure 3.2. The magnetization amplitude M as a function of temperature T; T_c is the temperature of the second-order phase transition between the ferromagnetic and the paramagnetic states.

---

[2]W. Heisenberg, Z. Physik **49**, 619 (1928).

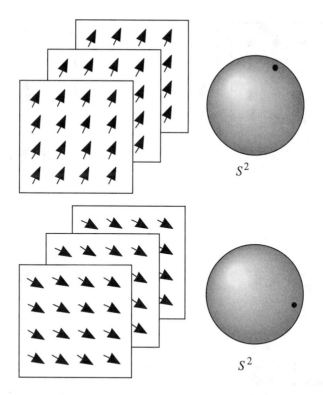

Figure 3.3. The Heisenberg isotropic ferromagnet: Different directions of the magnetization vector correspond to different points on the order parameter space S^2.

If there is no external field and no orientational coupling between the spins and the lattice, the direction of **M** in a ferromagnet is not fixed: Two uniform ferromagnets with the same M but different directions of **M** have the same thermodynamical potentials. Thus, the direction of **M** is the degeneracy parameter of a ferromagnet. Various possible orientations of **M** can be represented one-to-one by the points on a sphere of radius $M(T)$, which we denote as S^2. Thus, two states with a different orientation of **M** would map on two different points in S^2 (Fig. 3.3). When the magnetic state is "deformed," so that **M** changes continuously from point to point along any path γ in the sample, while M remains constant, the mapping of γ on S^2 is no longer a unique point but a smooth path Γ. The sphere S^2 is the OP space of the Heisenberg ferromagnet.

3.1.3. X-Y Ferromagnets

Consider a 2D-lattice in which each site carries a spin. In the ground state, all spins are parallel and confined to the (x, y) plane. The resulting order parameter is the 2D magneti-

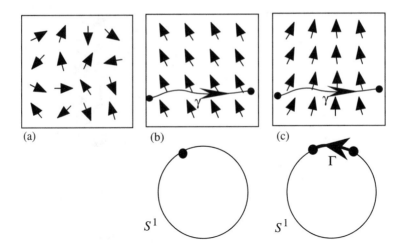

Figure 3.4. XY model: (a) paramagnetic, (b) uniform ferromagnetic, and (c) nonuniform ferromagnetic states. The path γ in the ferromagnet is mapped into a single point in the order parameter space S^1, when the state is uniform and into a path Γ when the state is nonuniform.

zation density, i.e., a 2D vector or, equivalently, a complex number. The *amplitude* of the order parameter depends on temperature; its *phase* $\phi(x, y)$ mod 2π is the degeneracy parameter and corresponds one-to-one to different orientations of the magnetization vector. The OP space is thus a circle S^1, as in the case of superfluid ^{4}He (Fig. 3.4).

3.1.4. Uniaxial Nematics

The amplitude of the order parameter will be defined at some length later in this chapter. At this stage, it is enough to accept that this amplitude is a temperature-dependent scalar $s(T)$ that vanishes in the isotropic phase, $T > T_c$ (Fig. 3.5) and is equal to unity when all molecules point rigidly in the same direction; the latter never happens at nonzero temperatures. In reality, $s(T_c) < s < 1$ because of the effect of thermal fluctuations on the order.

The degeneracy parameter is the director $\mathbf{n}(\mathbf{r})$, i.e., the direction of the molecules averaged in some macroscopic volume at least of the order of ξ^3 (coherence volume) about $\mathbf{r}$, small enough to consider that $\mathbf{n}(\mathbf{r})$ does not vary in this region (compare with our discussion of the ^{4}He superfluid phase). Hence, the order parameter could be written as $s\mathbf{n}(\mathbf{r})$, in a form analogous to that of a ferromagnet. However, the difference is considerable: Two perfect ferromagnets magnetized in opposite directions are two different realizations of the ground state, $\mathbf{M} \neq -\mathbf{M}$, whereas $s\mathbf{n}$ and $-s\mathbf{n}$ represent the same realization of a perfect nematic phase.

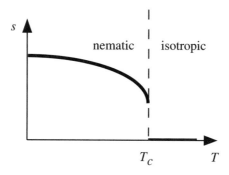

Figure 3.5. Behavior of the scalar order parameter in the nematic phase at the first-order N-I transition.

Start from the sphere S^2 of radius $s(T)$. Because $\mathbf{n} \equiv -\mathbf{n}$, two diametrically opposite points on this sphere represent the same perfect nematic, and we have to *identify* them to obtain the true OP space of the uniaxial nematic phase (Fig. 3.6a). All pairs of identified points represent the nematic phase one-to-one.

A way of visualizing this process is as follows (Fig. 3.6): Cut S^2 along any of its great circles, and conserve only one half-sphere. Except on its boundary, each point of the hemisphere is a one-to-one representative of a perfect nematic; on the boundary, we have still to identify opposite points, such as shown in Fig. 3.6b. This abstract process can be made

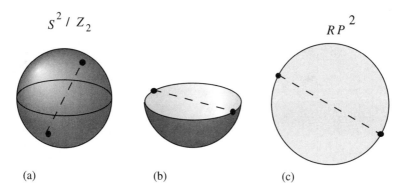

(a)   (b)   (c)

Figure 3.6. Order parameter space for the uniaxial nematic phase: (a) the S^2 sphere is twice too large, because each realization of the perfect N phase is represented by two opposite points; the sphere with pairs of antipodal points identified is noted S^2/Z_2, where Z_2 is the Abelian group with two elements; (b) half-sphere and opposite points identified; (c) half-sphere flattened to a full disk, without a change of topological properties.

more visual if one forgets the metric properties of the "manifold" that we are constructing and stresses only its "topological" properties, i.e., we take the liberty to deform the half-sphere by any process that does not introduce a cut or a tearing. The important result is that the mapping between the various realizations of the director field and the OP space follows these smooth deformations in a unique manner. We shall see indeed that the only interesting properties of the OP space are its topological properties.

We can transform in this way the half-sphere into a full disk, with the opposite points identified, and then bring those points one on the other, by pairs (Fig. 3.6c). The resulting *closed* manifold is called the Boyd surface; its embedding in 3D Euclidean space shows self-intersections,[3] but we can forget them when using its topological properties. In fact, had we done the foregoing operations in a 4D space, these self-intersections would not exist. The manifold we have obtained is called the projective plane, and it is noted RP^2. Notice, in Fig. 3.6c, that the dashed line is a closed line on RP^2. This remark will have far-reaching consequences later on.

3.1.5. Crystalline Solids

The order parameter space is the density $\rho(\mathbf{r})$ that is a triply periodic scalar quantity in a perfect crystal:

$$\rho(\mathbf{r}) = \sum_{\mathbf{k}} \rho_{\mathbf{k}} \exp i\mathbf{k} \cdot \mathbf{r}, \tag{3.2}$$

where the sum is done on the vectors $\mathbf{k}$ of the reciprocal space. The amplitude of the OP is given by the set of all $\rho_{\mathbf{k}}$'s. The degeneracy parameter can be understood as follows.

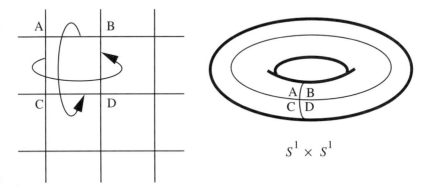

Figure 3.7. Torus $T^2 = S^1 \times S^1$ is the order parameter space of a 2D crystal.

---

[3] See D. Hilbert and S. Cohn-Vossen, Geometry and the Imagination, Chelsea Publishing Cy, New York, 1964, for a pictorial representation of Boyd's surface.

Displace a reference crystal by a constant quantity **u**. The new crystal thus obtained is a different realization of the same system at equilibrium, and it differs from the first by this translation only. The OP space is therefore the Bravais cell with identification of the boundaries points separated by a repeat vector $\mathbf{a}_i$. In the 2D case, which is illustrated in Fig. 3.7, the identification yields a manifold that is a 2D torus $T^2 = S^1 \times S^1$. This OP space describes various realizations of the crystal that differ by translations; we shall not discuss the symmetries of rotation, which make the complete OP space fairly complicated but bring no new physical insight in the question of OP space. A generalization to 3D is straightforward: The OP space limited to translations is the 3D torus T^3.

3.1.6. Order-Disorder Transitions in Alloys

At high temperatures, the $Cu_{50}Zn_{50}$ alloy (brass) has a BCC lattice with sites occupied by Cu and Zn with equal probabilities $\eta_{Cu} = \eta_{Zn} = 1/2$ (Fig. 3.8a). In the most ordered situation, at $0°$ K, the symmetry of the lattice is simple cubic: the Cu atoms occupy one sublattice and the Zn atoms occupy the other, Fig. 3.8b.

Let us define the OP as the quantity $\eta = |\eta_{Cu} - \eta_{Zn}|$. At a particular site, at 0 K, the probability of occupation by a Cu atom is either $\eta_{Cu} = 1$ or $\eta_{Cu} = 0$; hence, $\eta = 1$. Above 0 K and below the phase transition temperature T_c, η varies between 1 and 0. In the high temperature phase, $\eta \equiv 0$. Note that the phase change, although it occurs in a continuous fashion, brings a brutal change of symmetry at $T = T_c$. Furthermore, the symmetry group of the low temperature phase is a subgroup of the high temperature phase. These are characteristics of second-order phase transitions, the same as the normal-superfluid and the paramagnetic-ferromagnetic transitions we have already met.

Consider now a BCC lattice of *sites* occupied by Cu and Zn atoms in their most ordered state. There are two ways of filling the sites, yielding two kinds of domains, both with $\eta = 1$ but with different η_{Cu}: $\eta_{Cu} = 1$ or $\eta_{Cu} = 0$. The two types of domains within the same sample would be separated by wall defects (antiphase boundaries) (Fig. 3.8c). The OP space (restricted to the consideration of a fixed lattice of sites) is made of two points.

3.2. The Specific Order Parameter of Liquid Crystals: The Director

3.2.1. Microscopic Definition

Intuitively, the *director* $\mathbf{n}(\mathbf{r})$ of a *uniaxial nematic* can be imagined as the result of averaging over the axes **a** carried by individual molecules. Both directions $+\mathbf{a}$ and $-\mathbf{a}$ have to be taken with the same microscopic probability, because the nematic phase does not recognize different extremities of the molecules, on the average. This averaging is taken in a volume: (i) large compared with the molecular dimensions but (ii) small enough compared with the typical deformation lengths of the nematic phase, in order to assign a continuous value to

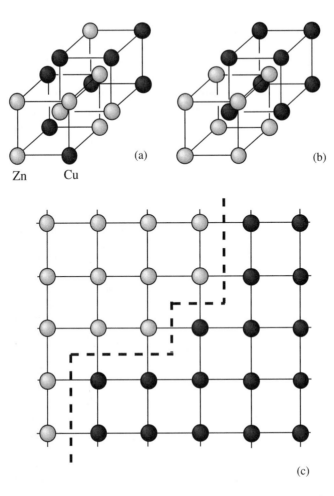

Zn   Cu

Figure 3.8. (a) Disordered and (b) ordered structures of $Cu_{50}Zn_{50}$ alloy; (c) two-dimensional cut of an antiphase boundary separating two domains in the ordered state.

the function $\mathbf{n}(\mathbf{r})$ when $\mathbf{r}$ varies. Conditions of the same type are always met in the description of macroscopic systems in quasi-equilibrium: The first condition makes precise what we mean by macroscopic, and the second one makes possible the consideration of a local thermodynamic equilibrium.

The intuitive picture above is useful, but not well-based, because the axis $\mathbf{a}$ does not have an unambiguous definition, except perhaps in the case in which the molecules are entirely rigid and have an axis of rotational symmetry, as in TMV. Besides, the central symmetry of the nematic phase requires the order parameter to be constructed from quantities that are even in **a**. The proper nematic order parameter is thus defined by the second-rank symmetric traceless tensor $\overline{\overline{Q}}$ (see Section 3.2.2). The director **n** specifies the direction of the principal axis of $\overline{\overline{Q}}$ (in the uniaxial nematic phase, $\overline{\overline{Q}}$ has a cylindrical symmetry). The director is a perfectly observable macroscopic variable, the optical axis of the uniaxial nematic.

The notion of director makes sense for other liquid crystalline phases. This is obvious in the cases of a smectic A and a cholesteric, which is nothing but a deformed, twisted nematic. The case of the SmC and SmC* phase (Figs. 2.24 and 2.25) is more delicate, as already discussed in Section 2.2.3.2.

This remark being made, let us come back to the uniaxial nematic phase. The director **n** bears no information about the degree of orientational order. We chose the axis Z along **n** and define spatial directions by a polar angle θ and an azimuthal angle φ. Let $f(\theta, \varphi)\, d\Omega$ be the probability of finding the molecular axis **a** within a solid angle $d\Omega = \sin\theta\, d\theta\, d\varphi$ about the direction (θ, φ). Because the directions $+\mathbf{a}$ and $-\mathbf{a}$ are equivalent in the nematic bulk, $f(\theta, \varphi) \equiv f(\pi - \theta, \varphi)$. Obviously, $f(\theta, \varphi)$ would be the same whatever the azimuth φ is; thus, $\partial f(\theta, \varphi)/\partial\varphi = 0$. Because the molecule is certainly oriented in the interval $0 \le \theta < \pi$, the function $f(\theta, \varphi) = f(\theta)$ is normalized:

$$\oiint f(\theta, \varphi)\, d\Omega = 2\pi \int_0^\pi f(\theta) \sin\theta\, d\theta = 1. \tag{3.3}$$

The distribution function $f(\theta)$ is flat above the nematic - isotropic transition, in the isotropic phase, and peaked for $\theta = 0$ below the transition. The degree of orientational order can be measured by the quantity s, first introduced by Tsvetkov:[4]

$$s = \frac{1}{2}\langle 3\cos^2\theta - 1 \rangle = 2\pi \int_0^\pi P_2(\cos\theta) f(\theta) \sin\theta\, d\theta, \tag{3.4}$$

where $P_2(\cos\theta) = \frac{1}{2}(3\cos^2\theta - 1)$ is the Legendre polynomial of the second order and $\langle \ldots \rangle$ means an average over all molecular orientations. In the isotropic phase, all orientations in space have equal probabilities; hence, $f(\theta, \varphi) = \frac{1}{4\pi}$ from (3.3), and (3.4) results in $s = 0$. For the most ordered state, the only molecular orientation is $\theta = 0$; hence, $f(\theta) = \frac{1}{4\pi}\delta(\theta)$, where $\delta(\ldots)$ is the Dirac function and $s = 1$. The scalar s can be taken as the modulus of the order parameter; it is often referred to as the scalar order parameter.

Note that if $f(\theta)$ is peaked at random in a direction perpendicular to **n**, i.e., $\theta = \pi/2$, one gets $s = -1/2$ from (3.4); therefore, s varies between $s = 1$ and $s = -1/2$. But the physical picture is different for $s < 0$; e.g., $s = -1/2$ corresponds to a layer of elongated molecules arranged randomly on a flat surface (e.g., a bounding plate) and remaining parallel to this surface (Fig. 3.9).

---

[4]V.N. Tsvetkov, Acta Physicochim. USSR **16**, 132 (1942).

Figure 3.9. Molecular distribution for the state with scalar order parameter $s = -1/2$.

Experimentally, for a typical thermotropic nematic material, s varies as indicated schematically in Fig. 3.5. s can be directly measured by a nuclear magnetic resonance (NMR) technique from analysis of spectra reflecting motion of nuclei (usually hydrogen or deuterium) that is averaged both on molecular (around the axis **a**) and macroscopic (around the director **n**) levels. Many other methods to measure s take advantage of the relations between the macroscopic properties of the nematic phase and the order parameter.

3.2.2. Macroscopic Properties

Macroscopic properties such as electric and magnetic susceptibilities of the nematic phase measured in different directions depend on both the director **n** and the scalar order parameter s. We consider a uniaxial nematic and use SI units,[5] keeping notations close to that of Jackson's (1998) textbook.

3.2.2.1. Dielectric Case

An electric field **E** applied to a dielectric medium creates local dipole moments and, thus, induces an electric polarization **P**. The functional form of **P(E)** depends on material properties and is generally unknown. One can expand **P** in terms of **E**

$$\mathbf{P} = \mathbf{P}_0 + \left(\frac{\partial \mathbf{P}}{\partial \mathbf{E}} \right)\bigg|_{\mathbf{E}=0} \cdot \mathbf{E} + \cdots, \tag{3.5}$$

where $\mathbf{P}_0$ is the vector of spontaneous polarization. In an orientationally ordered medium, $\mathbf{P}_0$ might be nonzero due to different mechanisms, such as the flexoelectric effect [polarization induced by orientational deformations, $\mathbf{n} = \mathbf{n}(\mathbf{r})$], order-electricity [polarization induced by spatial changes of the order parameter $s = s(\mathbf{r})$], or surface polarization (caused

---

[5]The SI units are meter, kilogram, second, and Ampere for electric current; thus, it is the MKSA system. The system defines the magnetic permeability of free space (magnetic constant) as $\mu_0 = 4\pi \times 10^{-7}$ Henry/m and the permittivity of free space (electric constant) as $\varepsilon_0 = 8.8541878\ldots \times 10^{-12}$ Farada/m from the relation $1/(\varepsilon_0 \mu_0) = c^2$, where $c = 2.99792458 \ldots \times 10^8$ m/s is the speed of light in vacuum.

by broken polar symmetry at the bounding surface, $\mathbf{n} \neq -\mathbf{n}$). We assume $\mathbf{P}_0 = 0$. For weak fields, the expansion (3.5) can be cut at the linear term:

$$P_i = \varepsilon_0 \alpha_{ij} E_j, \tag{3.6}$$

where α_{ij} are the components of the *dielectric susceptibility* tensor $\overline{\overline{\alpha}}$ and ε_0 is the electric constant, the permittivity of free space (see Footnote 5). Note that the vectors $\mathbf{P}$ and $\mathbf{E}$ have generally different orientations. If the medium is isotropic, then all directions are equivalent, and $\mathbf{P}$ is parallel to $\mathbf{E}$. Recalling that the displacement is a vector sum of the electric field and the induced polarization, $\mathbf{D} = \varepsilon_0 \mathbf{E} + \mathbf{P}$, one obtains

$$D_i = \varepsilon_0 E_i + \varepsilon_0 \alpha_{ij} E_j = \varepsilon_0 (\delta_{ij} + \alpha_{ij}) E_j = \varepsilon_0 \varepsilon_{ij} E_j, \tag{3.7}$$

where ε_{ij} are the components of the *relative electric permittivity* tensor $\overline{\overline{\varepsilon}}$ and δ_{ij} is the Kronecker delta.

The tensor $\overline{\overline{\varepsilon}}$ is symmetric; it is diagonal in the Cartesian coordinates with the axis Z along $\mathbf{n}$. The two eigenvalues of the diagonalized tensor relative to the directions perpendicular to $\mathbf{n}$ are degenerate, because of the cylindrical symmetry of the uniaxial nematic. Thus, in this coordinates system:

$$\overline{\overline{\varepsilon}} = \begin{pmatrix} \varepsilon_\perp & 0 & 0 \\ 0 & \varepsilon_\perp & 0 \\ 0 & 0 & \varepsilon_{||} \end{pmatrix}. \tag{3.8}$$

In an arbitrary coordinate system, one can write $\varepsilon_{ij} = \varepsilon_\perp \delta_{ij} + \varepsilon_a n_i n_j$, where $\varepsilon_a = \varepsilon_{||} - \varepsilon_\perp$ is the *dielectric anisotropy* of the liquid crystal. The dielectric constants $\varepsilon_{||}$ and $\varepsilon_\perp$ are frequency dependent. At optical frequencies, $\varepsilon_{||}$ and $\varepsilon_\perp$ are directly related to the indices of refraction of the medium (Section 3.3.2). At low frequences, $\varepsilon_{||}$ and $\varepsilon_\perp$ are of the order (1-10) and the dielectric anisotropy ε_a can be either positive or negative, depending on the molecular structure. For example, $\varepsilon_a \approx 13$ for 5CB (Fig. 2.12), whereas $\varepsilon_a \approx -0.7$ for MBBA at 25°C; there are materials in which ε_a changes sign with the temperature; see the book by Blinov and Chigrinov.

3.2.2.2. Magnetic Case

An external magnetic field of *induction* $\mathbf{B}$ applied to a liquid crystal induces a magnetization $\mathbf{M} = \mu_0^{-1} \mathbf{B} - \mathbf{H}$, where μ_0 is the permeability of vacuum, $\mathbf{H}$ is the analog of the electric displacement $\mathbf{D}$, called the magnetic field or the magnetic field strength; $\mathbf{B}$ is often called the *magnetic flux density*. If the field is small, the magnetization can be assumed linearly dependent on the field, $M_i = \chi_{ij} H_j$. The magnetization can be written through the components $\mathcal{B}_i$ of $\mathbf{B}$:

$$M_i = \mu_0^{-1} \chi_{ij} \mathcal{B}_j, \tag{3.9}$$

where $i, j = x, y, z$, $\chi_{ij} = \chi_{ji}$ are the components of the symmetric magnetic susceptibility tensor $\overline{\overline{\chi}}$ (calculated per unit volume). The last expression neglects the local field
effects, which is justified by the smallness of χ_{ij} ($\approx 10^{-5}$ in SI units; in CGS units, the
values of χ_{ij} are 4π times smaller).

As in the electric case, two eigenvalues of $\overline{\overline{\chi}}$ are degenerate, due to the cylindrical symmetry. Hence, in the frame of its eigenvectors along $\mathbf{n}$ and along two arbitrary directions
perpendicular to $\mathbf{n}$, $\overline{\overline{\chi}}$ writes:

$$\overline{\overline{\chi}} = \begin{pmatrix} \chi_\perp & 0 & 0 \\ 0 & \chi_\perp & 0 \\ 0 & 0 & \chi_{||} \end{pmatrix}. \tag{3.10}$$

In an arbitrary coordinate system, $\chi_{ij} = \chi_\perp \delta_{ij} + \chi_a n_i n_j$, where $\chi_a = \chi_{||} - \chi_\perp$ is the
anisotropy of the magnetic susceptibility. In the usual nematics such as 5CB or MBBA,
$\chi_{||} < 0$ and $\chi_\perp < 0$: The materials are diamagnetic; i.e., the field-induced magnetization
is opposite to the applied field. Paramagnetic liquid crystals may be prepared by synthe-
sizing molecules with metal atoms. No single-component ferromagnetic mesophases have
been reported so far; however, a ferromagnetic mesophase can be obtained when the liquid
crystal is doped with small ferromagnetic particles. In what follows, we restrict the consideration to the diamagnetic materials. As a rule, the diamagnetic anisotropy is positive,
$\chi_a > 0$, and $\mathbf{n}$ orients along the field. The reason is that the circular electric currents in the
aromatic rings of molecules, such as 5CB (Fig. 2.12) and MBBA (Fig. 2.14), create a large
negative component of the diamagnetic susceptibility in the direction perpendicular to the
plane of the rings (and to the long axis of the molecule); i.e., $|\chi_\perp| > |\chi_{||}|$.

The anisotropies of macroscopic properties are caused by the orientational order of liquid crystals. Measuring these anisotropies allows one to establish the degree of the orientational order s. The magnetic measurements are especially convenient compared with their
electric counterparts, because in the magnetic case, the local field acting on the molecules
differs very little from the external field. It can be seen by recalling that the magnetic
susceptibilities ($\approx 10^{-5}$) are much smaller than 1 and by writing

$$\mathcal{B}_i = \mu_0(\delta_{ij} + \chi_{ij})H_j = \mu_0(1 + \chi_\perp)H_i + \mu_0 \chi_a n_i n_j H_j. \tag{3.11}$$

The relation between s and anisotropic magnetic properties can be established by in-
troducing an effective magnetic susceptibility tensor $\overline{\overline{\kappa'}}$ for each individual molecule. Because magnetic interactions between molecules are small, the macroscopic susceptibility
can be obtained from the sum of molecular susceptibilities with appropriate averaging over
the distribution function $f(\theta)$ (the procedure of summation is less obvious for the dielectric case because the molecular polarizabilities are strongly influenced by the surrounding
dipoles). One finds (Problem 3.2):

$$\chi_a = N(\kappa'_{||} - \kappa'_\perp)s, \tag{3.12}$$

where N is the number of molecules per unit volume; $\kappa'_{\parallel}$ and $\kappa'_{\perp}$ are the molecular susceptibilities calculated along the molecular axis and the (degenerate at zero frequency) directions normal to it, respectively.

Thus, the amount of order in the nematic system can be defined in terms of a macroscopic property. In the isotropic phase, $\chi_a = 0$ and $s = 0$. Let $\mathrm{Tr}\,\overline{\overline{\chi}} = \chi_{xx} + \chi_{yy} + \chi_{zz}$; one can construct a traceless symmetric tensor

$$Q_{ij} = Q\left(\chi_{ij} - \tfrac{1}{3}\delta_{ij}\,Tr\overline{\overline{\chi}}\right). \tag{3.13}$$

One easily checks that $\mathrm{Tr}\,\overline{\overline{Q}} = Q_{xx} + Q_{yy} + Q_{zz} = 0$. Each component vanishes in the isotropic phase, and it is proportional to s in the uniaxial nematic phase:

$$\overline{\overline{Q}} = Q\begin{pmatrix} -\chi_a/3 & 0 & 0 \\ 0 & -\chi_a/3 & 0 \\ 0 & 0 & 2\chi_a/3 \end{pmatrix}. \tag{3.14}$$

The value of the constant Q is not important, and we can choose $Q = N^{-1}(\kappa'_{\parallel}-\kappa'_{\perp})^{-1}$ for convenience. With $\chi_{ij} = \chi_{\perp}\delta_{ij} + \chi_a n_i n_j$, one obtains

$$Q_{ij} = s(T)\left(n_i n_j - \tfrac{1}{3}\delta_{ij}\right). \tag{3.15}$$

The tensor order parameter allows us to describe the biaxial nematic phase as well:

$$Q_{ij} = s\left(n_i n_j - \tfrac{1}{3}\delta_{ij}\right) + p(l_i l_j - m_i m_j), \tag{3.16}$$

where $(\mathbf{n}, \mathbf{l}, \mathbf{m} = [\mathbf{nl}])$ are three orthogonal unit directors and p is the "biaxiality parameter": $p = 0$ in the uniaxial phase.

Anisotropy of liquid crystals and the possibility to orient the director by an applied electric or magnetic field leads to numerous practical applications. Any actual liquid crystal cell is confined; say, by a pair of parallel glass plates. Orienting action of the substrates might prevent director reorientation if the external field is weak. However, if the field is higher than some threshold value, it eventually overcomes both the "anchoring" at the surfaces and the elasticity of the nematic bulk and reorients the director. This is the *Frederiks effect*, first discovered for the magnetic case;[6] see also Section 5.4.3 and 13.2.3. When the field is removed, the surface anchoring restores the original director structure. Thus, one can use the external field and surface anchoring to switch the liquid crystal orientation back and forth. The Frederiks effect, mainly its dielectric version, is used in many electrooptic

---

[6]V. Frederiks and V. Tsvetkov, Sov. Phys. **6**, 490 (1934); V. Freedericksz and V. Zolina, Trans. Faraday Soc. **29**, 919 (1933).

devices (displays, optical shutters, etc.). The liquid crystal is usually sandwiched between two transparent electroconductive plates (for example, glass covered with indium tin oxide) coated with a suitable alignment layer (for example, a buffed polymer film). The voltage across the cell controls the director configuration and, thus, the optical properties of the cell. Note that a number of factors, such as nonlocal character of the electric field distribution in the distorted liquid crystal, finite electroconductivity, flexoelectricity (Section 5.3), surface polarization (Section 13.2.3), and so on, make the electric effects in nematics much more complicated than their magnetic counterparts.

3.3. Light Propagation in Anisotropic Media; Application to Director Fields

Probing anisotropic media such as crystals and liquid crystals with light historically greatly contributed to our understanding of the structure and properties of condensed matter. O. Lehmann's popular studies of microscopic textures[7] and determination of mesomorphic phases and structures by G. Friedel[8] were based on such observations. Thin (microns) liquid crystal samples are easy to prepare between glass plates; large birefringence of these materials results in good contrast of regions with different orientation of the optical axis (which coincides with the director in uniaxial nematic and smectic A phases). Most liquid crystal structures can be recognized readily by the observation of textures and defects between crossed polarizers. Furthermore, by applying an external field to a liquid crystal cell, one changes the director configuration and, consequently, the related optical properties, as just indicated above. It is therefore of great importance to gain some background on the basic principles of light propagation in anisotropic media.

3.3.1. Fresnel Equation

The propagation of light in a homogeneous medium is described by Maxwell's equations, which in the absence of currents and charges write in SI units as

$$\nabla \times \mathbf{E} = -\partial \mathbf{B}/\partial t, \tag{3.17}$$

$$\nabla \times \mathbf{H} = \partial \mathbf{D}/\partial t, \tag{3.18}$$

$$\nabla \cdot \mathbf{B} = 0, \tag{3.19}$$

$$\nabla \cdot \mathbf{D} = 0. \tag{3.20}$$

These equations should be supplemented by the relations that describe how the material responds to the electromagnetic field of the propagating wave. Because we are in-

---

[7]O. Lehmann, Flüssige Kristalle; W. Engelmann, Leipzig, 1904; Flüssige Kristalle und die Theorie des Lebens; J. Ambr. Barth, 1908.

[8]G. Friedel, Les états mésomorphes de la matière, Ann. de Phys. **18**, 273 (1922).

terested in the linear effects only, for which the polarization produced by the field is proportional to that field, then (3.7) and (3.11) can serve as these *constitutive equations*. Note that the problem of light propagation corresponds to electromagnetic fields of high frequences $\sim 10^{15}$ Hz. The dielectric permittivities are frequency dependent, and their values at optical frequencies (of the order of unity) are often substantially smaller than the static or low frequency ($\varpi \sim 10^3$ Hz) values, especially when the material is composed of polar molecules. The orientation of the permanent dipole moments of polar molecules contributes significantly to the permittivity at low frequencies; however, it is small at optical frequencies (e. g., in water, the static permittivity is $\varepsilon \approx 78$, but at optical frequencies, $\varepsilon = 1.78$). Besides, in nonabsorbing media, all components ε_{ij} are real. Furthermore, the magnetization effects can be neglected, so that $\mathbf{M} = 0$ and (3.11) simplifies to $\mathbf{B} = \mu_0\mathbf{H}$.

Consider a plane monochromatic wave $\mathbf{E}(\mathbf{r}, t) = \mathbf{E}_0 \exp(i\mathbf{k}\cdot\mathbf{r} - i\varpi t)$, $\mathbf{H}(\mathbf{r}, t) = \cdots$, and so on. Here, $\mathbf{r}$ is the radius vector directed from the source of the wave to the point of observation, ϖ is the frequency, and $\mathbf{k}$ is the wavevector, which is real, $k = 2\pi/\lambda = \varpi/v$, because the dielectric media are nonconductive, i.e., transparent; finally, λ is the wavelength and v is the phase velocity of the wave. The Maxwell's equations simplify to

$$\mathbf{k} \times \mathbf{E} = \varpi\mathbf{B}, \tag{3.21}$$

$$-\mathbf{k} \times \mathbf{H} = \varpi\mathbf{D}, \tag{3.22}$$

$$\mathbf{k} \cdot \mathbf{H} = 0, \tag{3.23}$$

$$\mathbf{k} \cdot \mathbf{D} = 0. \tag{3.24}$$

The last set of equations implies that the vectors $\mathbf{k}$, $\mathbf{D}$, and $\mathbf{H}$ are mutually perpendicular. Besides, $\mathbf{H} \perp \mathbf{E}$. Because $\mathbf{H}$ is normal to three vectors $\mathbf{k}$, $\mathbf{D}$, and $\mathbf{E}$, these three form a plane. The energy transfer vector is defined by the Poynting vector $\mathbf{S} = [\mathbf{E} \times \mathbf{H}]$. Therefore, $\mathbf{S}$ and $\mathbf{k}$ generally have different directions in an anisotropic medium (Fig. 3.10). The phase of the waves moves in the direction $\mathbf{k}$; the energy is transferred in the direction $\mathbf{S}$. These two directions coincide in isotropic media.

Eliminating $\mathbf{H}$ from (3.21) and (3.22), one obtains

$$\varpi^2\mu_0\mathbf{D} = k^2\mathbf{E} - \mathbf{k}(\mathbf{Ek}). \tag{3.25}$$

This expression must be compatible with the constitutive equation (3.7), $\mathbf{D} = \bar{\bar{\varepsilon}}\mathbf{E}$; hence,

$$(N^2\delta_{ij} - N_iN_j - \varepsilon_{ij})E_j = 0, \tag{3.26}$$

where we introduce a "refractive index" vector $\mathbf{N} = \mathbf{k}\dfrac{1}{\varpi\sqrt{\varepsilon_0\mu_0}}$ oriented along $\mathbf{k}$. The absolute value N is called the refractive index of a given wave in a given medium. N is inversely proportional to the phase velocity v of the propagating wave in the medium. Because $k = \varpi/v$ and $1/\sqrt{\varepsilon_0\mu_0}$ is the speed c of light in vacuum, then $N = c/v$.

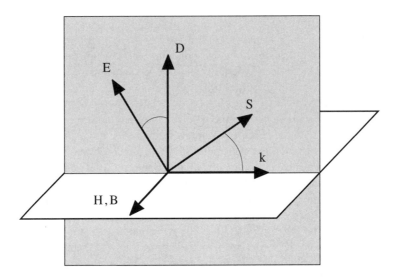

Figure 3.10. Orientation of vectors that characterize a plane monochromatic wave in an anisotropic dielectric where the electric field **E** and the dielectric displacement **D** have different directions. The dashed plane contains the vectors **E**, **D**, **S**, and **k**; the vectors **H**||**B** are normal to this plane. The direction **k** along which the wave phase moves is different from the direction **S** = [**E** × **H**] of the energy transfer; **S**⊥**E** and **k**⊥**D**.

The three homogeneous equations (3.26) have a nontrivial solution in **N** only if their discriminant vanishes. This requirement results in the famous *Fresnel equation*, which reads as

$$\text{Det}\,(\mathbf{k}, \omega) = N^2(\varepsilon_x N_x^2 + \varepsilon_y N_y^2 + \varepsilon_z N_z^2) - [\varepsilon_x N_x^2(\varepsilon_y + \varepsilon_z) + \varepsilon_y N_y^2(\varepsilon_x + \varepsilon_z)$$

$$+ \varepsilon_z N_z^2(\varepsilon_x + \varepsilon_y)] + \varepsilon_x \varepsilon_y \varepsilon_z = 0 \qquad (3.27)$$

in the coordinate system with diagonalized $\bar{\bar{\varepsilon}}$. In a uniaxial medium, where $\bar{\bar{\varepsilon}}$ contains only two independent entries, $\varepsilon_{||}$ and $\varepsilon_\perp$,

$$\text{Det}\,(\mathbf{k}, \omega) \equiv (N^2 - \varepsilon_\perp)(\varepsilon_{||} N_z^2 + \varepsilon_\perp(N_x^2 + N_y^2) - \varepsilon_{||}\varepsilon_\perp) = 0. \qquad (3.28)$$

3.3.2. Ordinary and Extraordinary Waves

Equation (3.28) shows that in a uniaxial medium, the surface of refractive indices $\sum$ (also called the surface of wavevectors), of degree four in the space of refractive indices, splits into a sphere of radius $N^2 = \varepsilon_\perp$ and an ellipsoid of revolution $\frac{N_z^2}{\varepsilon_\perp} + \frac{N_x^2 + N_y^2}{\varepsilon_{||}} = 1$. There are two points where the two surfaces touch each other; the axis that connects these two points is the optic axis of the uniaxial nematic. There are two waves propagating in an uniaxial medium. For one of them, the so-called *ordinary wave* (o-wave), the medium behaves as an isotropic medium with the index of refraction $n_o = \sqrt{\varepsilon_\perp}$. The second wave is *extraordinary* (e-wave) with an index of refraction N that depends on the direction of propagation. Introducing an angle θ between the optic axis and the wavevector $\mathbf{k}$, $N_z^2 = N^2 \cos^2 \theta$, $N_x^2 + N_y^2 = N^2 \sin^2 \theta$, one gets the effective refractive index for the extraordinary wave as the function of the ray direction θ:

$$\frac{1}{N^2(\theta)} = \frac{\cos^2 \theta}{n_o^2} + \frac{\sin^2 \theta}{n_e^2}, \text{ or } N(\theta) = \frac{n_o n_e}{\sqrt{n_e^2 \cos^2 \theta + n_o^2 \sin^2 \theta}}, \tag{3.29}$$

where $n_e = \sqrt{\varepsilon_{||}}$ is the extraordinary refractive index.

If $n_e < n_o$, the sphere is outside of the ellipsoid (an optically negative nematic, composed usually of disk-like molecules); if $n_e > n_o$, the sphere is inside the ellipsoid (an optically positive nematic composed usually of elongated molecules) (Fig. 3.11). If propagation is along the optic axis, then both ordinary and extraordinary waves have the same

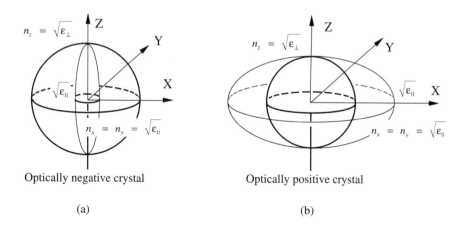

Optically negative crystal   Optically positive crystal

(a)   (b)

Figure 3.11. Surfaces of refractive indices for (a) an optically negative and (b) an optically positive uniaxial crystal.

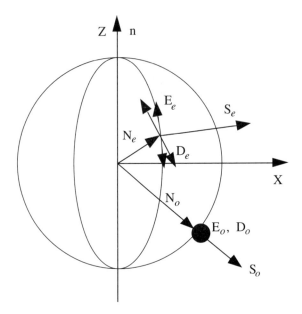

Figure 3.12. Surfaces of refractive indices for an optically negative uniaxial crystal with indicated directions of wavevectors **k**, ray vectors **S**, electric fields **E**, and displacement vectors **D** for ordinary (circle) and extraordinary (ellipse) components.

velocity and are indistinguishable. The refractive indices n_o and n_e are inversely proportional to the phase velocity v of the propagating wave in the medium. Because $k = \varpi/v$, then $n_o = c/v_o$ and $n_e = c/v_e$. A typical birefringence $(n_e - n_o)$ of a thermotropic nematic material is $\approx (0.1 \div 0.2)$; it can be somewhat smaller in lyotropic systems.

Using Eqs.(3.26), one can find the polarizations of the ordinary and extraordinary waves (Fig. 3.12). Without loss of generality, we assume that the XZ coordinate plane contains the parallel vectors **k** and **N**; i.e., $N_y = 0$. Then, the two solutions of (3.26) have the following polarizations:

1. The o-wave is linearly polarized along the axis Y, which is normal to both the optic axis **n** and the wavevector **k**:

$$E_x^{(o)} = 0, \; E_y^{(o)} \neq 0, \; E_z^{(o)} = 0. \tag{3.30}$$

Because $D_x = \varepsilon_\perp E_x^{(o)} = 0$, $D_y = \varepsilon_\perp E_y^{(o)} \neq 0$, and $D_z = \varepsilon_{||} E_z^{(o)} = 0$, then $\mathbf{D}||\mathbf{E}$ and $\mathbf{S}||\mathbf{k}$ in the o-wave. Both ray and phase velocities are equal to $c/\sqrt{\varepsilon_\perp} = c/n_o$ and do not depend on the direction of propagation.

2. The e-wave has the components

$$E_x^{(e)} \neq 0, \ E_y^{(e)} = 0, \ E_z^{(e)} \neq 0, \tag{3.31}$$

where

$$\frac{E_z^{(e)}}{E_x^{(e)}} = \frac{N_z^2 - \varepsilon_\perp}{N_x N_z} = -\frac{\varepsilon_\perp}{\varepsilon_{||}} \frac{N_x}{N_z}. \tag{3.32}$$

Equation (3.32) shows that the **E**-vector in the e-wave is tangential to the refractive index ellipsoid $\frac{N_z^2}{\varepsilon_\perp} + \frac{N_x^2}{\varepsilon_{||}} = 1$. We recall that the equation of the tangent line to a curve specified by an implicit equation $\varphi(N_x, N_z) = 0$ is $\tan \alpha = -\frac{\partial \varphi / \partial N_x}{\partial \varphi / \partial N_z} = -\frac{\varepsilon_\perp}{\varepsilon_{||}} \frac{N_x}{N_z}$, where α is the angle between the tangent line and the axis X. The direction of **S** is easily found as

$$\frac{S_x^{(e)}}{S_z^{(e)}} = \frac{\varepsilon_\perp}{\varepsilon_{||}} \frac{N_x}{N_z}, \tag{3.33}$$

because **S** is normal to **E**.

3.3.3. Observations in Polarized Light. Microscopy

Note that the plane of polarization of the e-wave always contains the director **n**, whereas the o-wave is always polarized normally to **n**. This rule is of use in observations with polarized light.

3.3.3.1. Transmitted Intensity; the Schlieren Texture

Consider a nematic slab sandwiched between two glass plates and placed between two crossed polarizers. The director **n** is in plane of the slab (Fig. 3.13) and depends on the in-plane coordinates (x, y). We assume that it does not depend on the vertical coordinate z. The light beam impinges normally on the cell, along the axis z. A polarizer placed between the source of light and the sample makes the impinging light linearly polarized. In the nematic, the linearly polarized wave of amplitude A and intensity $I_0 = A^2$ splits into the ordinary and extraordinary waves with mutually perpendicular polarizations and amplitudes $A \sin \beta$ and $A \cos \beta$, respectively; $\beta(x, y)$ is the angle between the local $\mathbf{n}(x, y)$ and the polarization of incident light. The vibration of the electric vectors at the point of entry are in phase. However, the two waves take different times, $n_o d/c$ and $n_e d/c$, respectively, to pass through the slab. At the exit point, the electric vibrations $A \sin \beta \cos \left(\omega t - \frac{2\pi}{\lambda_0} n_o d \right)$ and $A \cos \beta \cos \left(\omega t - \frac{2\pi}{\lambda_0} n_e d \right)$ gain a phase shift $\Delta \varphi = \frac{2\pi d}{\lambda_0} (n_e - n_o)$, where λ_0 is the wave-

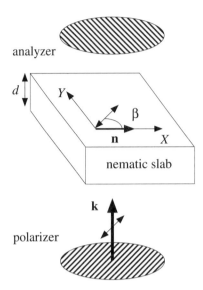

Figure 3.13. Propagation of light through a polarizer, uniaxial slab, and analyzer.

length in vacuum. The projections of these two vibrations onto the polarization direction of the analyzer behind the sample are

$$a = A \sin \beta \cos \beta \cos \left(\omega t - \frac{2\pi}{\lambda_0} n_o d \right),$$

$$b = -A \sin \beta \cos \beta \cos \left(\omega t - \frac{2\pi}{\lambda_0} n_e d \right). \tag{3.34}$$

When two harmonic vibrations $A_1 \cos(\omega t + \varphi_1)$ and $A_2 \cos(\omega t + \varphi_2)$ of the same frequency occur along the same directions, then the resulting vibration $\overline{A} \cos(\omega t + \overline{\varphi})$ has an amplitude defined from $\overline{A}^2 = A_1^2 + A_2^2 + 2A_1 A_2 \cos(\varphi_1 - \varphi_2)$. The analyzer, thus, transforms the pattern of (x, y)-dependent phase difference into the pattern of transmitted light intensity $I(x, y) = \overline{A}^2$. The intensity of light passed through the crossed polarizers and the nematic slab between them follows from (3.34) as

$$I = I_0 \sin^2 2\beta \sin^2 \left[\frac{\pi d}{\lambda_0} (n_e - n_o) \right]. \tag{3.35a}$$

The last formula refers to the case when **n** is perpendicular to the axis z. If **n** makes a constant angle θ with the axis z, then (3.35a) becomes [see (3.29)]

$$I = I_0 \sin^2 2\beta \sin^2 \left[\frac{\pi d}{\lambda_0} \left(\frac{n_o n_e}{\sqrt{n_e^2 \cos^2 \theta + n_o^2 \sin^2 \theta}} - n_o \right) \right]. \qquad (3.35b)$$

Of course, the treatment can be further extended to describe the optical properties of complex director configurations in nontrivial geometries (e.g., inside liquid crystalline droplets) or in field-driven cells; we refer interested readers to numerous reviews; see, e.g., Blinov and Chigrinov (1994).

Equations (3.35) are fundamental for understanding liquid crystal textures. First, note that the phase shift and, thus, I depend on λ_0. As a result, when the sample is illuminated with a white light, it would show a colorful texture (Fig. 3.14). The interference colors are especially pronounced when $(n_e - n_o) d \approx (1 \div 3)\lambda_0$. With typical $(n_e - n_o) \approx 0.2$, $\lambda_0 \approx 500$ nm, the "colorful" range of thicknesses is $d \approx (1 \div 10)\mu$m. Second, the director tilt θ greatly changes the phase shift. When $\mathbf{n} \| z$ (the so-called homeotropic orientation, $\theta = 0$), the sample looks dark: Only the ordinary wave propagates and according to (3.35b), $I = 0$. Third, if $\theta > 0$ but $\beta = 0, \pm\pi/2, \ldots$, one might still observe dark textures, $I = 0$, even in nonmonochromatic light. In a sample with in-plane director distortions $\mathbf{n}(x, y)$, wherever $\mathbf{n}$ (or its horizontal projection) is parallel or perpendicular to the polarizer, the propagating

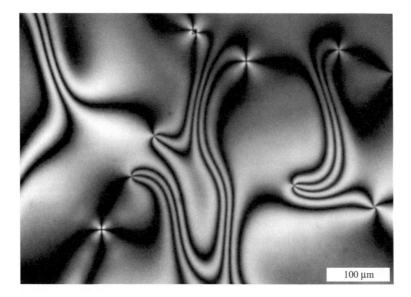

Figure 3.14. Schlieren texture of a thin ($\approx 1\mu$m) film of the nematic 5CB on a glycerin substrate. Note the interference colors; dark brushes mark the regions in which the director is parallel to either the polarizer or the analyzer. Nodes in which four brushes meet are cores of topological point defects.

mode is either pure extraordinary or pure ordinary and the corresponding region of the texture appears dark. Figure 3.14 is an example of such a texture with dark "brushes of extinction." Points at which the brushes converge are centers of topological defects. The texture with defects and dark brushes is called the Schlieren texture.

In principle, rotating the crossed polarizers, one can reconstruct the whole in-plane director pattern of the Schlieren texture. However, there is an ambiguity: because $I \sim \sin^2 2\beta$, the reconstruction produces two director patterns, $\beta(x, y)$, and $\beta(x, y) + \pi/2$. To map the in-plane director configuration unambiguously, one can use optical compensators, or quartz wedges.

Compensators are plates with a known phase shift. For example, Red Plate- I gives a phase shift equal to 575 nm. The plate has two well-defined directions, with a minimum refractive index n_{min} (called "the fast direction") and n_{max} (called "the slow direction"). The compensator is inserted into the slot between the sample and the analyzer, at 45° to the polarizer and analyzer directions. The black brushes of the texture become red, because the sample's phase retardation in these regions is zero and the total phase retardation is defined entirely by the compensator. The region where the director is parallel to the fast direction and the region where the director is parallel to the slow direction would acquire different interference colors and, thus, can be distinguished. Without the compensator, these regions would be equally bright because for both of them, $I \propto \sin^2 2\beta = \sin^2 \frac{\pi}{2} = 1$ in (3.35).

3.3.3.2. Fermat Principle and the Path of the Extraordinary Light

The extraordinary light, when observed separately, fluctuates in intensity due to the thermal vibrations of the director field, which cause variation of $N(\mathbf{k})$ in time (flickering); no flickering is associated with the ordinary light.

Fermat's principle tells us that the integral

$$\mathcal{F} = \int_A^B \mathbf{N}(\mathbf{k}, \mathbf{r}) \cdot d\mathbf{r} \tag{3.36}$$

is minimized along the path actually followed by the energy (i.e., the envelope of the Poynting vector $\mathbf{S}$) with respect to any neighboring path; here, $d\mathbf{r}$ is a vector tangent to the path. We now develop some considerations on its use to calculating the path of the extraordinary ray, restricting to a uniaxial medium.

According to (3.36), the *ordinary ray*, for which $N(\mathbf{k}) = n_o$, propagates along a straight line in a slightly deformed nematic or smectic, *slightly* meaning that the gradient of the director is small compared with the inverse wavelength of light, with the result that the conditions of geometrical optics are satisfied. The wavevector $\mathbf{k}$, the Poynting vector $\mathbf{S}$, and the tangent to the path $d\mathbf{r}$ are colinear. On the other hand, the *extraordinary ray* propagates in a more complex way, because $\mathbf{S}$ and $d\mathbf{r}$ are no longer aligned with $\mathbf{k}$, and the use of (3.36) is more involved. In fact, the situation simplifies immediately if one notices that Fermat's principle has a straightforward interpretation in terms of velocities: Fermat's integral $\mathcal{F}$ is indeed proportional to the duration of the path between A and B (for an ordinary ray, $n_0 = c/v_0$; see below), and Fermat's principle tells, indeed, that this duration is minimized. Therefore, Fermat's principle can also be expressed as the minimization of

$$\mathcal{F} = \int_A^B \frac{1}{s} \, d\ell, \tag{3.37}$$

where $s = s(\mathbf{t}, \mathbf{r})$ is the ratio of the (scalar) group velocity to the velocity of light (i.e., is proportional to the Poynting vector) and $\mathbf{t} = d\mathbf{r}/d\ell$ is the unit vector tangent to the path. Let us introduce the ray vector $\mathbf{s}$ with the direction parallel to the Poynting vector $\mathbf{S}$ and the absolute value defined from the condition $\mathbf{s} \cdot \mathbf{N} = 1$. We can easily show that $|\mathbf{s}| = s$. This relation $\mathbf{s} \cdot \mathbf{N} = 1$ stresses the conjugated character of the two quantities. $\mathbf{S}$ is perpendicular to the surface of indices (Fig. 3.12):

$$\frac{N_z^2}{\varepsilon_\perp} + \frac{N_x^2 + N_y^2}{\varepsilon_{||}} = 1. \tag{3.38}$$

Therefore, $s_x = N_x/\varepsilon_{||}$, $s_y = N_y/\varepsilon_{||}$, $s_z = N_z/\varepsilon_\perp$, which satisfies the relation $\mathbf{s} \cdot \mathbf{N} = 1$. By substitution into (3.38), one gets

$$\varepsilon_\perp s_z^2 + \varepsilon_{||}(s_x^2 + s_y^2) = 1, \tag{3.39}$$

from which the conjugate of (3.29) can be written as

$$\frac{1}{s^2} = n_o^2 \cos^2 \phi + n_e^2 \sin^2 \phi. \tag{3.40}$$

Here, ϕ is the angle of the Poynting vector with the optic axis. Note that $\mathbf{s}$, $\mathbf{k}$, and the optic axis are in the same plane.

The expression of s given in (3.40) is written in the principal axes of the surface of indices; in a situation in which the liquid crystal is deformed, it has to be written in the laboratory frame, and expressed in the form $s = s(\mathbf{t}, \mathbf{r})$. The equation of the path is then obtained by integrating the Euler-Lagrange equations, which minimize Fermat's integral:

$$\frac{\partial(s^{-1})}{\partial x_i} - \frac{d}{d\ell} \frac{\partial(s^{-1})}{\partial t_i} = 0, \, i = 1, 2, 3, \tag{3.41}$$

with the appropriate boundary conditions (see Problem 3.3).

3.3.3.3. Miscellaneous

The observations in polarized light discussed above are greatly affected by twist deformations. We consider briefly the so-called cholesteric planar texture: The director is uniform in the plane (x, y), but twists along the normal axis z (Fig. 2.22). A simple "wave-guide" regime, called the *Mauguin regime*,[9] occurs when the pitch of this helicoidal structure is much larger than the wavelength of light, $p \gg \lambda_0/(n_e - n_o)$. When the incident light is polarized along the director (at the entry) or perpendicular to it, the transmitted light is still linearly polarized, but the direction of polarization is rotated by an angle γ; γ is the angular difference in the director orientations at the opposite plates. If the cholesteric pitch is of the order of λ_0, $p = \lambda_0/\bar{n}$ ($\bar{n}$ is some average value of the refractive index), then one observes Bragg reflection; the cholesteric sample appears colored in the reflected light. Finally, for shorter pitches, the cholesteric behaves as an optically active medium with a huge rotatory power. All of these effects are intensively used in numerous optical applications of twisted nematics and cholesterics; for example, the electrically controlled Mauguin regime is at the heart of the so-called "twisted" nematic displays.

Interaction of light and soft matter structures is not limited by a passive modification of the properties of light by the medium, as briefly illustrated above. There are optical effects of nonlinear nature, in which the propagating light causes changes in the medium. The well-known examples are self-focusing and self-diffraction of light beams propagating in the nonlinear medium, harmonic generation and other wave mixing effects, and soon. These effects can be observed both in isotropic and anisotropic media. There are also nonlinear optical effects intimately related to the orientational order. We list only few, referring the readers to reviews by Khoo (1995) and Simoni (1997).

1. Optical Frederiks effect, i.e., director reorientation in the electric field of the propagating light beam.[10]

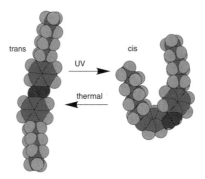

Figure 3.15. *Trans* isomer and *cis* isomer of the mesogenic molecule diheptylazobenzene. *Trans-to-cis* isomerization can be triggered by short-wavelength (ultraviolet) irradiation. Thermal relaxation or long-wavelength irradiation restores the *trans* configuration.

---

[9] C. Mauguin, Bull. Soc. fr. Minér. Cristallogr. **34**, 3 (1911).

[10] A. Saupe, Deformation of a nematic liquid crystal by polarized light, subm. to Phys. Rev. Lett. on July 17, 1969; published in "Dynamics and Defects in Liquid Crystals," edited by P.E. Cladis and P. Palffy-Muhoray, Gordon and Breach Science Publishers, Amsterdam, p. 441 (1998); B. Ya. Zel'dovich, N.F. Pilipetskii, A.V. Sukhov, N.V. Tabiryan, JETP Lett. **31**, 263 (1980); A.S. Zolot'ko, V.F. Kitaeva, N. Kroo, N.I. Sobolev, L. Csillag, *ibid.* **32**, 158 (1980); I.C. Khoo and S.L. Zhuang, Appl. Phys. Lett. **37**, 3 (1980).

2. Thermal effects of light propagation. Their importance is enhanced by a relatively narrow temperature range of stability of the liquid crystalline phases; small temperature variations cause drastic changes in the refractive indices or even trigger phase transitions.

3. Light-induced conformational changes of the molecules, such as *trans-cis* isomerization (Fig. 3.15), which might lead to drastic changes in the phase diagram of the material. For example, rod-like *trans* isomers of the diheptylazobenzene shown in Fig. 3.15 are capable of forming the liquid crystal phase, whereas their cis-counterparts are not.

4. Light-induced polymerization and photoinduced orientational order in polymers irradiated with polarized light.

5. "Guest-host" effect: anisometric foreign molecules or particles can be aligned by the liquid crystal. When the guest is dye, the effect can be used in electro-optical devices. A fluorescent dye allows one to image 3D director structures (so-called fluorescence confocal polarizing microscopy[11]), Fig. 11.17.

Problem 3.1.

(a) Find the order parameter space for a SmC.

(b) Compare the order parameter space of SmC and 2D nematic with molecules parallel to the boundary.

(c) How does the order parameter space of a ferromagnet change in the strong magnetic field?

Problem 3.2.   Prove the relationship $\chi_a = \chi_{\|} - \chi_{\perp} = N(\kappa'_{\|} - \kappa'_{\perp})s$, Equation (3.12).

Answers: Suppose the molecules are rigid cylinders with axis **a** oriented along the axis Z' of some coordinate system (X', Y', Z'). In this system, the components κ'_{ij} of the molecular tensor read simply as $\kappa'_{x'x'} = \kappa'_{y'y'} = \kappa'_{\perp}$, $\kappa'_{z'z'} = \kappa'_{\|}$. In the laboratory system (X, Y, Z) with the axis Z oriented along **n**, the components of the molecular susceptibility tensor are found from the transformation law $\kappa_{\alpha\beta} = \cos(\alpha, \alpha')\cos(\beta, \beta')\kappa'_{\alpha'\beta'}$, where $\cos(\alpha, \alpha')$ are direction cosines of angles between the coordinate axes of the two systems. Macroscopic magnetic susceptibilities $\chi_{\alpha\beta} = N \int_0^{2\pi} d\varphi \int_0^{\pi} \kappa_{\alpha\beta}(\theta, \varphi) f(\theta, \varphi) \sin\theta \, d\theta$ of N molecules will be defined by the molecular susceptibilities averaged over the distribution function $f(\theta, \varphi)$. Using (3.4), one finds $\chi_a = \chi_{\|} - \chi_{\perp} = N(\kappa'_{\|} - \kappa'_{\perp})s$.

Problem 3.3.   Calculate the paths of the extraodinary rays about a radial disclination line of strength $k = 1$. The plane of incidence and direction of light polarization are perpendicular to the line (Fig. 3.16) (adapted from Grandjean[12]).

Answers: The solution is translation invariant along the disclination. Let $r = f(\omega)$ be the equation of the path in a plane perpendicular to the line, in polar coordinates, the origin of the coordinates

---

[11]I.I. Smalyukh, S.V. Shiyanovskii, and O.D. Lavrentovich, Chem. Phys. Lett. **336**, 88 (2001).

[12]F. Grandjean, Bull. Soc. Franç. Minéralogie, **42**, 42 (1919).

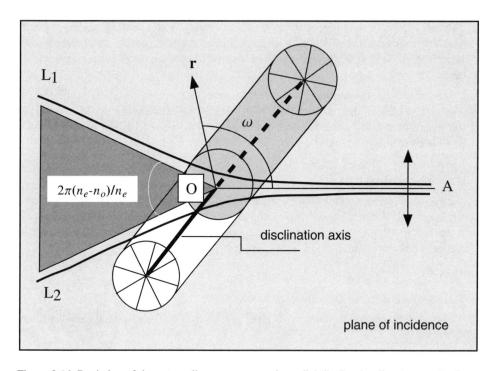

Figure 3.16. Deviation of the extraordinary rays around a radial disclination line (perpendicular to the plane of the drawing). We assume that the director in $(\mathbf{r}, \boldsymbol{\omega})$ is along the $\mathbf{r}$-direction (disclination $k = 1$, see Section 11.1).

being taken on the line. We have $d\ell^2 = dr^2 + r^2 d\omega^2$ and $\cos\phi\, d\ell = dr$; ϕ is the angle of the Poynting vector with the optical axis. Hence, putting $r' = \frac{dr}{d\omega}$, one gets $\mathcal{F} = \int (n_o^2 r'^2 + n_e^2 r^2)^{1/2}\, d\omega$. The associated Euler-Lagrange equation is $n_e^2 r^2 + n_o^2 (2r'^2 - r\, r'') = 0$, whose solutions can be written, with two arbitrary constants of integration a and b, as

$$r = \left(a \cos \frac{n_e}{n_o}\omega + b \sin \frac{n_e}{n_o}\omega \right)^{-1}.$$

Consider all solutions that correspond to rays coming from a point $\mathbf{A}$ at infinity, angle $\omega = 0$ (Fig. 3.16). This condition yields $a \equiv 0$, and the set of paths can then be written as $r = 1/b \sin \frac{n_e}{n_o}\omega$. Therefore, all paths have another asymptotic direction parallel to the directions OL_1 or OL_2, according to the position of the incoming ray with respect to the dividing line AO. We have angles $AOL_1 = AOL_2 = \pi \frac{n_o}{n_e}$. Figure 3.16 assumes that the nematic is uniaxial positive; the region between OL_1 or OL_2 does not receive any ray.

Problem 3.4. Optical properties of biaxial media, in which all three principal dielectric constants are different, e.g., $\varepsilon_x < \varepsilon_y < \varepsilon_z$ can be deduced from (3.27). Observe that for the wave propagating

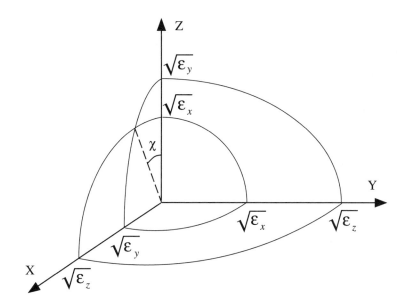

Figure 3.17. Surface of refractive indices for a biaxial medium. Dashed line shows the direction of the optical axis.

in the XY plane, $N_z = 0$, (3.27) becomes $(N^2 - \varepsilon_z)(\varepsilon_x N_x^2 + \varepsilon_y N_y^2 - \varepsilon_x \varepsilon_y) = 0$ with two solutions, a circle $N^2 = \varepsilon_z$, and an ellipse $\frac{N_x^2}{\varepsilon_y} + \frac{N_y^2}{\varepsilon_x} = 1$. Because $\varepsilon_x < \varepsilon_y < \varepsilon_z$, the ellipse is located inside the circle. In a similar way, one finds solutions for the XZ and YZ planes. In the YZ plane, the ellipse is outside of the circle, and in the XZ plane, the ellipse and the circle intersect. The surface of wave vectors has four intersections in the plane XZ (one in each quadrant). In these points, there is only one value of the refractive index; the corresponding directions are called the optical axes of biaxial crystal (Fig. 3.17). Find the angle χ between the two optical axes and the axis Z.

Answers: $\tan \chi = \pm \sqrt{\frac{\varepsilon_z (\varepsilon_y - \varepsilon_x)}{\varepsilon_x (\varepsilon_z - \varepsilon_y)}}$.

Problem 3.5. Find the intensity of light transmitted through a pair of parallel polarizers and a nematic slab with **n** in the plane of the slab.

Answers: $I = I_o \{1 - \sin^2 2\beta \sin^2[\frac{\pi d}{\lambda_0}(n_e - n_o)]\}$; the notations are the same as in (3.35).

Problem 3.6. Calculate the components ε_{ij} of the dielectric permittivity tensor for the chiral nematic liquid crystal with the director field $\mathbf{n} = \{\cos qz, \sin qz, 0\}$, where $q = 2\pi/P$ and P is the pitch of the director helicoid with an axis along the z-axis.

Answers: With $\varepsilon_{ij} = \varepsilon_\perp \delta_{ij} + \varepsilon_a n_i n_j$, one finds $\varepsilon_{xx} = \varepsilon_\perp + \varepsilon_a \cos^2 qz = \varepsilon(1 + \delta \cos 2qz)$; $\varepsilon_{yy} = \varepsilon_\perp + \varepsilon_a \sin^2 qz = \varepsilon(1 - \delta \cos 2qz)$; $\varepsilon_{zz} = \varepsilon_\perp$; $\varepsilon_{xy} = \varepsilon_{yx} = \varepsilon_a \sin qz \cos qz = \varepsilon\delta \sin 2qz$, and $\varepsilon_{xz} = \varepsilon_{zx} = \varepsilon_{yz} = \varepsilon_{zy} = 0$; here, $\varepsilon = (\varepsilon_{||} + \varepsilon_\perp)/2$ and $\delta = (\varepsilon_{||} - \varepsilon_\perp)/(\varepsilon_{||} + \varepsilon_\perp)$; $\varepsilon_{||}$ and $\varepsilon_\perp$ are the permittivities parallel and perpendicular to the director, respectively. For chiral smectic C^*, similar calculations have been performed by Berreman.[13]

Further Reading

L.M. Blinov and V.G. Chigrinov, Electrooptic Effects in Liquid Crystal Materials, Springer series on partially ordered systems, New York, 1994.

P.M. Chaikin and T.C. Lubensky, Principles of Condensed Matter Physics, Cambridge University Press, 1995.

D. Demus and L. Richter, Textures of Liquid Crystals, Verlag Chemie, Weinheim, New York, 1978.

P.G. de Gennes and J. Prost, The Physics of Liquid Crystals, Oxford Science Publication, Clarendon Press, Oxford, 1993.

J.D. Jackson, Classical Electrodynamics, 3rd Edition, John Wiley & Sons, Inc., New York, 1998.

N.H. Hartshorne and A. Stuart, Crystals and the Polarising Microscope, 2nd edition, Edward Arnold & Co., London, 1950, 476 pp.

I.-C. Khoo, Liquid Crystals: Physical Properties and Nonlinear Phenomena, John Wiley & Sons, Inc., New York, 1995, 298 pp.

L.D. Landau and E.M. Lifshitz, Electrodynamics of Continuous Media, Pergamon Press, New York, 1960.

L.D. Landau and E.M. Lifshitz, Statistical Physics, Addison-Wesley, Reading, MA, 1969.

F. Simoni, Nonlinear Optical Properties of Liquid Crystals and Polymer Dispersed Liquid Crystals, World Scientific, Singapore, 1997, 260 pp.

R.E. Stoiber and S.A. Morse, Microscopic Identification of Crystals, Ronald, 1972.

The Optics of Thermotropic Liquid Crystals, Edited by S. Elston and R. Sambles, Taylor & Francis Ltd., London, 1998.

---

[13]D.W. Berreman, Mol. Cryst. Liq. Cryst. **22**, 175 (1973).

CHAPTER 4

Phase Transitions

Physical matter might exist in different forms, or phases, that differ by the type of order, mass density, and so on. Transitions between the phases are described by the behavior of thermodynamic potentials, such as the Gibbs G or the Helmholz F free energies, and by the derivatives of these potentials. The phase transition is called first order if the first derivatives of the potentials (such as entropy, order parameter) exhibit a discontinuity, a jump at the transition point. These transitions involve nonzero latent heat. If the first derivatives are continuous but the second derivatives (such as heat capacities) diverge at the transition, it is a second-order phase transition. It is customary to lump into the second category of continuous phase transitions all transitions with no latent heat. Although the thermodynamic potentials in the second-order transitions change gradually, the symmetry of the system changes discontinuously.

There are three types of theories to describe the appearance of orientational and partial translational order in soft matter systems during phase transitions. Phenomenological theories employ the idea of Landau that the free energy in the vicinity of the transition can be expanded in power expansion as a function of a small amplitude order parameter. These theories rely on symmetry considerations; it is often hard to assign a physical meaning to the coefficients of the expansion, i.e., to connect them to parameters of molecular interactions. The theories of the second category are molecular-statistical in nature and start with an appropriate model of molecular interactions. For example, Onsager's model of hard rods emphasizes repulsive interactions and the associated effects of the orientational-dependent excluded volume. The model of Maier and Saupe, in contrast, is based on anisotropic van der Waals forces of attraction. Finally, the techniques of the renormalization group allows a conceptual extension of the theories of phase transitions with a full treatment of precritical fluctuations (for some indications on this type of approach, see Chapter 7). As indicated in Chapter 3, such fluctuations also relate to the question of short range order.

105

4.1. Landau–de Gennes Model of the Uniaxial Nematic-Isotropic Phase Transition

The phenomenological Landau theory has three essential steps: (1) finding a proper order parameter; (2) expanding the free energy in the vicinity of the transition with respect to the (small) order parameter; and (3) finding the minima of the free energy at each temperature, pressure, and so on, as functions of the order parameter.

The tensor $\overline{\overline{Q}}$ has all of the symmetries of the order parameter that are required to describe the nematic-to-isotropic phase transition N ⇔ I. It has been employed by de Gennes to build the Landau expansion of the (Gibbs) free energy density in the vicinity of the transition:

$$g(T) = g_0 + \tfrac{1}{2} A(T) Q_{\alpha\beta} Q_{\beta\alpha} - \tfrac{1}{3} B(T) Q_{\alpha\beta} Q_{\beta\gamma} Q_{\gamma\alpha}$$
$$+ \tfrac{1}{4} C(T) Q_{\alpha\beta} Q_{\alpha\beta} Q_{\gamma\delta} Q_{\gamma\delta} + \cdots, \tag{4.1}$$

where g_0 is the free energy density of the isotropic phase and summation over repeated indices is implied. Note that we use lowercase, e.g., $g(T)$, to conote an energy density, and uppercase, e.g., $G = \int g \, dV$, to conote an energy. An important note about the thermodynamic meaning of G that would be relevant to many other situations discussed in this book is in order here. Very often, dealing with N ⇔ I transition or other phenomena in soft matter, one connects the density g to either the Gibbs G or the Helmholtz F free energies, or just to an abstract "free energy," without distinguishing the two. We recall that if the system is maintained at a constant temperature T and pressure p, the quantity that evolves to its minimum is G. If T and volume V are kept constant, then the quantity to minimize is $F = G - pV$. Real experiments are usually done under the constraints $p = $ const and $T = $ const, whereas computer modeling is easier for $V = $ const and $T = $ const. If the changes in the density (volume) of the experimental system are small, the difference in the theoretical results based on G or F minimization would be small as well. The N ⇔ I transition falls into this category because the transition-induced jump of the density at atmospheric pressure is small, about 0.3%; thus, there is no density-dependent terms in the expansion (4.1) and g might be treated as the density of either G or F.

Each term of the expansion (4.1) is invariant not only under the operations of symmetry of the nematic phase, but also under any operation that changes the "phase" of the order parameter (in the sense of Section 3.1), such as a global rotation in space. A remarkable feature of this expansion is the presence of the cubic term, which is odd in the scalar order parameter s and, thus, not invariant under the transformation $Q_{\alpha\beta} \rightarrow -Q_{\alpha\beta}$; i.e., $s \rightarrow -s$. As discussed above, the nematic states described by s and by $-s$ are not degenerate, and therefore, the transformation $s \rightarrow -s$ is not allowed. Consequently, according to Landau's theorem, the N ⇔ I transition is of the first order, as confirmed by numerous experiments. This is in contrast with the Heisenberg ferromagnet, where the states $|\mathbf{M}|$ and $-|\mathbf{M}|$ are degenerate; hence, the cubic term is forbidden in the Landau expansion of the ferromagnet.

Landau's theorem states that a *necessary* (but not sufficient) condition for the transition to be second order is that the terms of the third order in the order parameter free energy expansion vanish identically; i.e., $B \equiv 0$. In fact, as it is well known, the Curie point is a second-order phase transition.

The coefficients A, B, and C in the Landau–de Gennes expansion (4.1) are temperature dependent. Apparently, A should be positive in the high-temperature region in the isotropic phase and negative in the low-temperature ordered nematic phase: $A > 0$ would allow us to obtain the energy minimum at $s = 0$, and $A < 0$ would allow us to obtain the minimum at $s \neq 0$. The simplest possible form of $A(T)$ is a linear one: $A(T) = a(T - T^*)$. If the transition were of the second order, its temperature would be precisely T^*. In the present case of a first-order transition, the meaning of T^* is slightly different: It marks the limit of metastability of the isotropic phase upon cooling, as we shall see below. Simplifying the model further, we assume that a, B, and C are temperature-independent positive constants.

Substitution of (3.15) into expansion (4.1) yields

$$g = g_0 + \tfrac{1}{3}a(T - T^*)s^2 - \tfrac{2}{27}Bs^3 + \tfrac{1}{9}Cs^4, \qquad (4.2)$$

which should be minimized with respect to s. Equation $\partial g / \partial s = 0$ may be written as

$$a(T - T^*)s - \tfrac{1}{3}Bs^2 + \tfrac{2}{3}Cs^3 = 0, \qquad (4.3)$$

and it has three solutions, two of which are

$$s_{\text{iso}} = 0 \ (\text{the isotropic phase}), \qquad (4.4a)$$

$$s_{\text{nem}} = \frac{B}{4C}\left[1 + \sqrt{1 - \frac{24a(T - T^*)C}{B^2}}\right] > 0 \ (\text{the nematic phase}). \qquad (4.4b)$$

The third solution

$$s_3 = \frac{B}{4C}\left[1 - \sqrt{1 - \frac{24a(T - T^*)C}{B^2}}\right]$$

should be disregarded. It corresponds either to an energy maximum or, at temperatures $T < T^*$ ($s_3 < 0$), to a relative minimum (Fig. 4.1). Because at $T < T^*$ the solution $s_{\text{nem}} > 0$ provides the absolute free energy minimum, the state $s_3 < 0$ is not achieved in equilibrium.

The transition temperature T_c and the corresponding value $s_c > 0$ of the order parameter are defined from the condition that the free energy densities of the two phases are equal,

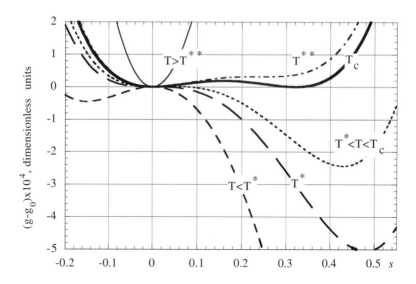

Figure 4.1. Free energy density vs scalar order parameter: (4.2) in the dimensionless form $(g - g_0) \times 10^4 = 1.32(\frac{T-T^*}{T_c})s^2 - 0.0181 \times s^3 + 0.0282 \times s^4$ with $T_c = 319.0\,\text{K}$, $T^* = 318.3\,\text{K}$, and $T^{**} = 319.1\,\text{K}$ (see Problem 4.1).

$g(s \neq 0) = g_0$, and the condition of stability $\partial g/\partial s = 0$:

$$T_c = T^* + \frac{B^2}{27aC}, \tag{4.5}$$

$$s_c = B/3C. \tag{4.6}$$

There is another temperature T^{**} of importance that marks the absolute limit of overheating of the nematic phase. It is achieved when the solution $s_{nem} > 0$ ceases to provide a minimum of the free energy density, i.e., when the expression under the square root in (4.4b) vanishes:

$$T^{**} = T_c + \frac{B^2}{216aC}. \tag{4.7}$$

Likewise, the temperature T^* is the limit of metastability of the isotropic phase upon cooling; note that $T_c = T^* = T^{**}$ when $B = 0$ (a second-order phase transition).

At $T > T^{**}$, the isotropic state is absolutely stable. At $T_c < T < T^{**}$, a relative minimum at $s_{nem} > 0$ appears. If the system is initially in the isotropic phase, it will remain isotropic. If the system is initially in the nematic state, it can be superheated above T_c.

At $T^* < T < T_c$, the situation is reversed: It is the isotropic state that is metastable, and the nematic state that is stable. If the system is cooled down from the isotropic phase, in the region $T^* < T < T_c$, one would observe nucleation and growth of nematic droplets.

A careful reader might raise an objection about the very applicability of the expansion (4.1) in terms of the order parameter to the first-order transition. Landau expansions are well suited for the second-order phase transitions, in which the order parameter changes continuously from 0 to a small value in the ordered phase. Partial justification of the expansion (4.1) comes from the fact that the N $\Leftrightarrow$ I transition is of the *weak first order*, close to the second-order phase transition. Experimentally, the weakness of the transition manifests in a small latent heat (less than $k_B T$ per molecule) and in strong pretransitional effects (e.g., in light scattering) over a relatively wide temperature range. Theoretically, the weakness of the transition is related to the smallness of the coefficient B, a feature that makes the expansion (4.1) meaningful. The relative smallness of B is illustrated using dimensionless energy units. For the volume densities of energy $[g] = [\text{J/m}^3]$, the normalizing factor is $[RT\rho] = [(\text{J} \cdot \text{mol}^{-1} \cdot K^{-1}) \cdot K \cdot (\text{mol/m}^3)] \equiv [\text{J/m}^3]$, where R is the molar gas constant and ρ is the number of moles per unit volume. Usually, experimental systems such as magnets fit the theory with dimensionless Landau coefficients of the order of unity. In liquid crystals, experiments yield $B, C \approx 0.1$ or less (see Problem 4.1). As a result, the short-range effects are important above T_c, because the transition is nearly second order (B is small), and the coherence length is large. This yields macroscopically visible effects, like magnetic– and electric field–induced birefringence (Kerr effect).[1] The physical origin of the smallness of the coefficients B and C remains a puzzle.[2]

4.2. Nematic Order and Statistical Theory of Rigid Rod-Like Particles

The approximate theory is due to Onsager[3] and marks a keystone in the study of solutions. It deals with rigid, elongated particles and applies to rigid polymer solutions (polybenzyl-glutamate in dioxane) or anisotropic viruses (tobacco mosaic virus). The phase transition is described as a result of steric repulsion (effect of the excluded volume).

4.2.1. Free Energy of a Solution of Spherical Particles

This simpler case will familiarize us with the statistical treatment of excluded volume effects. Let us consider a solution, total volume V. There are n noninteracting hard spheres floating in a solvent that is composed of N particles (atoms or molecules), $N \gg n$. The pure solvent has a total Helmholtz free energy $F_{\text{solv}}(T)$. The solute adds to this free energy the following contributions:

---

[1]P.G. de Gennes, Mol. Cryst. Liq. Cryst., **12**, 193 (1971).

[2]See review by P.H. Mukherjee, J. Phys.: Condens. Matter **10**, 9191 (1998).

[3]L. Onsager, Ann. N.Y. Acad. Sci. **51**, 627 (1949).

- A term of entropy of disorder $k_B T \ln n! \approx n k_B T \ln \frac{n}{e}$.

- A term of chemical potential $n \mu(T)$ due to the added spheres; this term is proportional to n.

- A perturbation of the entropy due to the fact that a given particle of the solute may occupy only that part of the whole volume V that is left free by the other particles. This perturbation implies steric interactions between two particles, three particles, and so on, and should take the form of an expansion with terms $\propto n^2$, $\propto n^3$, and so on. The largest term of this expansion can be written as $n^2 k_B T v / 2$, where v has the dimension of a volume and is related to the excluded volume, as we shall see below.

4.2.1.1. The Excluded Volume

Let v_s be the volume of one particle of solvent, and $v_p = \frac{4\pi}{3} r^3$ be the volume of one particle of solute. The total volume is $V = N v_s + n v_p$. We shall introduce the particles of solute one after the other, assuming $n v_p \ll N v_s$, so that the total volume in which they are introduced is equal to V at each step.

The introduction of the first particle of solute can be done in η_1 different ways with

$$\eta_1 = AV. \tag{4.8}$$

The introduction of the second particle is made in a partially occupied volume, $V - u$, where u is the volume excluded by the first particle; viz., $u = \frac{4\pi}{3}(2r)^3$ (Fig. 4.2). Hence,

$$\eta_2 = A(V - u), \tag{4.9}$$

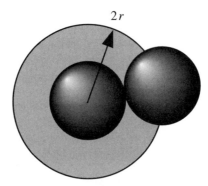

$2r$

Figure 4.2. Excluded volume for spherical particles of radius r: The separation between the centers of the two hard spheres cannot be smaller than $2r$.

and subsequently,

$$\eta_3 = A(V - 2u), \ldots, \eta_i = A(V - (i - 1)u). \tag{4.10}$$

The total number of ways to introduce n particles is, therefore,

$$W = \frac{1}{n!} \prod_{k=1}^{n} \eta_k = \frac{A^n V^n}{n!} \prod_{i=1}^{n-1} \left(1 - \frac{iu}{V}\right), \tag{4.11}$$

where $n!$ is the number of arrangements of n identical and indistinguishable particles.
The corresponding entropy of mixing is given by

$$S_{\text{mix}}/k_B = \ln W = n \ln A + n \ln V - \frac{u}{V} \sum_{i=1}^{n-1} i - \ln n!, \tag{4.12}$$

where we have replaced $\ln \left(1 - \frac{iu}{V}\right)$ by $\left(-\frac{iu}{V}\right)$. For $n \gg 1$, (4.12) simplifies to

$$S_{\text{mix}}(n, V) = k_B \left[n \ln AV - \frac{u}{2V} n^2 - n \ln \frac{n}{e}\right]. \tag{4.13}$$

We want to consider only the variation $\Delta S_{\text{mix}}(n, V) = S_{\text{mix}}(n, V) - S_{\text{mix}}(0, V) - S_{\text{mix}}(n, nv_p)$ of the entropy between the final state (n, V) and the sum of the initial states $(0, V)$ and (n, nv_p). The state (n, nv_p) mostly has the effect of removing the $n \ln \frac{n}{e}$ term. One obtains, after an easy calculation,

$$\Delta S_{\text{mix}}(n, V) = k_B \left[n \ln \frac{V}{nv_p} + \frac{u}{2v_p} n - \frac{u}{2V} n^2\right]. \tag{4.14}$$

For the case of an ideal solution, there is no heat of mixing. We assume that the more dilute the solution is, the smaller the energy of interaction is between the individual particles of the solute and the solvent. The free energy per particle of solute is

$$F_1 = F_{01}(T) - \frac{T \Delta S_{\text{mix}}}{n} = F_{01}(T) + k_B T \left(\ln v_p c + \frac{1}{2} uc + \text{const}\right), \tag{4.15}$$

where $c = n/V$ is the particle concentration. The last equation allows us to illustrate further the idea of osmotic pressure introduced in Chapter 1.

4.2.1.2. The Osmotic Pressure

In a typical osmotic pressure experiment, a solution with concentration c is separated from the pure solvent by a *hemipermeable membrane* that is permeable to the solvent but not

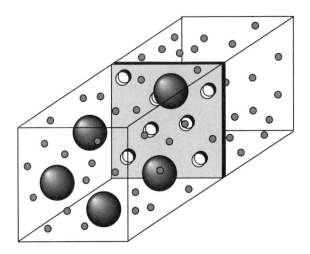

Figure 4.3. Hemipermeable membrane transparent for the solvent molecules but not for the solute molecules.

to the particles of the solute (Fig. 4.3). Assume that the membrane is shifted so that the volume of the solution varies by a quantity dV. The variation of the free energy $F_{tot} = nF_1 = cVF_1$ can be written as

$$dF_{tot} = -p_{os}\, dV, \tag{4.16}$$

where p_{os} is of evidence given by the expression

$$p_{os} = -\left(\frac{\partial F_{tot}}{\partial V}\right)_{T,n}. \tag{4.17}$$

The obvious interpretation of (4.17) is that p_{os} is a pressure acting on the semipermeable membrane; p_{os} is the osmotic pressure. Note that it is not counterbalanced by any pressure in the pure solution behind the membrane (Fig. 4.3), because $F_{behind} \equiv 0$ according to the origin of energies we have chosen in (4.17). We obtain an expected expression

$$p_{os} = k_B T \left(c + \tfrac{1}{2} u c^2\right), \tag{4.18}$$

with $v = \tfrac{1}{2} u = 4v_p$ being the excluded volume calculated per one particle, as introduced by (1.5) and (1.6).

Remark: Physicochemists are used to expressing the thermodynamical mixing functions in terms of *partial quantities* $(\frac{\partial G}{\partial n_i})_{n_{j,T}}$, where G is any thermodynamic function. The chemical potentials are such partial quantities: They are given by $\mu_{\text{solvent}} = \mu_{0,\text{solvent}}(T) - T\frac{\partial \Delta S_{\text{mix}}}{\partial N}$; $\mu_{\text{solute}} = \mu_{0,\text{solute}}(T) - T\frac{\partial \Delta S_{\text{mix}}}{\partial n}$. It is easy to verify that the free energy nF_1 is the sum of these partial quantities

$$n[F_1 - F_{01}(T)] = -T\left(\frac{\partial \Delta S_{\text{mix}}}{\partial N}N + \frac{\partial \Delta S_{\text{mix}}}{\partial n}n\right). \tag{4.19}$$

4.2.2. Free Energy of a Solution of Rigid Rods

Equation (4.13) for the entropy of a system of *hard spheres* shows that the excluded volume $\sim u$ is entropically unfavorable. However, when the particles are elongated, the excluded volume effects become *orientation dependent* (Fig. 4.4). Obviously, the excluded volume of parallel rods is smaller than that of perpendicular rods. As a result, the entropy effect of the excluded volume would drive the system to the orientationally ordered state. This is the basic physical idea behind the Onsager's theory of nematic ordering.

Consider cylindrical rigid rods of length L and diameter D, such that $L \gg D$. Such a solution, when subjected solely to repulsion forces, would present a N $\Leftrightarrow$ I transition when $\phi L/D \geq 4$, where $\phi = c\pi LD^2/4$ is the volume fraction of the solute. As above, $c = n/V$.

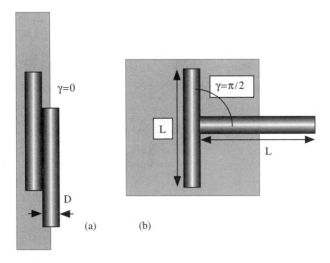

Figure 4.4. The excluded volume in a system of two hard rods depends on the angle γ between their axes: The volume is minimum for parallel alignment, $\gamma = 0$, and maximum for perpendicular alignment, $\gamma = \pi/2$.

We specify the *angular distribution* of the rods by the density

$$f(\theta, \varphi) \, d\Omega = f_{\mathbf{a}} \sin\theta \, d\varphi \, d\theta,$$

which expresses the probability for a given rod to point along the direction $\mathbf{a}$ with polar coordinates (θ, φ) within a small solid angle $d\Omega = \sin\theta \, d\varphi \, d\theta$. By definition,

$$\int_0^{2\pi} \int_0^{\pi} f(\theta, \varphi) \sin\theta \, d\varphi \, d\theta = 1. \tag{4.20}$$

A natural generalization of (4.15) is

$$F_1 = F_{01}(T) + k_B T \left[\int f_{\mathbf{a}} \ln(4\pi \, v_p c f_{\mathbf{a}}) \, d\Omega + \frac{1}{2} c \int \int f_{\mathbf{a}} f_{\mathbf{a}'} \beta(\mathbf{a}, \mathbf{a}') \, d\Omega \, d\Omega' \right]. \tag{4.21}$$

The first integral stands for the generalization of $\ln c$; $c f_{\mathbf{a}}$ is indeed the concentration of rods in the direction $\mathbf{a}$, and the factor 4π gives the correct limiting value $\ln c$ of the integral when $f_{\mathbf{a}}$ is equally distributed along all directions (in the isotropic phase). The second integral generalizes the excluded volume effects; $\beta(\mathbf{a}, \mathbf{a}')$ is the volume excluded by a rod of direction $\mathbf{a}$ to a rod of direction $\mathbf{a}'$. In the limit $L \gg D$, one finds

$$\beta(\mathbf{a}, \mathbf{a}') = 2L^2 \, D | \sin\gamma |, \tag{4.22}$$

where $\gamma = \arccos(\mathbf{a}, \mathbf{a}')$. Note that $\beta(\mathbf{a}, \mathbf{a}') = \beta(\mathbf{a}', \mathbf{a})$, as expected.

The normalization condition (4.20) for $f_{\mathbf{a}}$ can be directly introduced in (4.21) under the form of a Lagrange condition, by adding the null term

$$\lambda \left(\int f_{\mathbf{a}} \, d\Omega - 1 \right), \tag{4.23}$$

where λ is a Lagrange multiplier. Let us now minimize the sum of (4.21) and (4.23) with respect to $f_{\mathbf{a}}$, which is an unknown function; one finds

$$\ln(4\pi \, v c_p f_{\mathbf{a}}) = -\lambda - 1 - \frac{c}{2} \int \beta(\mathbf{a}, \mathbf{a}') f_{\mathbf{a}'} \, d\Omega', \tag{4.24}$$

where the physical quantities L, c, D appear in a dimensionless form in the product

$$c\beta(\mathbf{a}, \mathbf{a}') = 2cL^2 \, D | \sin\gamma | = \frac{8}{\pi} \phi \frac{L}{D} | \sin\gamma |. \tag{4.25}$$

Equation (4.24) has an obvious solution, viz. $f_{\mathbf{a}} = \frac{1}{4\pi}$, which is satisfied for λ taking some constant value. This solution describes the isotropic phase. Other solutions are more difficult to determine, and no exact solution is known. Onsager has employed a variational method that starts from the trial function $f_{\mathbf{a}} = \frac{\alpha}{4\pi \sinh \alpha} \cosh(\alpha \cos \theta)$ that satisfies the normalization condition (4.20). Here, θ is the angle between $\mathbf{a}$ and the nematic axis, and α is the variational parameter; α vanishes in the isotropic phase and is large in the nematic phase; $\alpha_{\mathrm{nem}} \cong 18.84$ at coexistence. One will notice that larger α makes the trial function more peaked at $\theta = 0$ and π. The nematic and isotropic phases have, respectively, the volume fractions $\phi_{\mathrm{nem}} = 4.5D/L$ and $\phi_i = 3.3D/L$ at coexistence; these values demonstrate that the transition occurs for rather large aspect ratios L/D. Furthermore, estimates show that with $L/D \leq 10$, one should take into account higher order terms such as $\propto c^2$ in the virial expansion. Therefore, the Onsager model cannot claim an accurate description of a low-molecular weight thermotropic nematic phases, where normally $L/D < 10$. The Onsager system is athermal: ϕ_{nem}, ϕ_i do not depend on temperature.

Onsager's theory has been refined by Flory for the description of semirigid molecules: They are placed on a lattice, whose parameter is equal to the persistence length, and they are allowed to take zigzag shapes. Flory's theory is more successful than is Onsager's for concentrated solutions, and it yields coexistence values $\phi_{\mathrm{nem}}\frac{L}{D} \approx 12.5$, $\phi_i \frac{L}{D} \approx 8$ that are larger than those of the Onsager's theory.

Experiments conducted on poly-benzyl-glutamate (PBG) in dioxane confirm that the phase transition I $\rightarrow$ N happens for a critical volume fraction ϕ_c, which depends only on the aspect ratio L/D; i.e., $\phi_c L/D = \mathrm{const}$. However, the numerical coefficients do not compare well with the theory. The discrepancies come mostly from the fact that the theory neglects attractive interactions (e.g., van der Waals forces) and effects of polydispersity.

4.3. Maier–Saupe Mean Field Theory of the Isotropic-Nematic Transition

The Maier–Saupe theory for nematics is the analog of the Weiss molecular field theory for ferromagnets. It is assumed that molecular interactions are of van der Waals type. The repulsive forces and excluded volume effects are not taken into consideration. The pairwise potential for two molecules located at $\mathbf{r}$ and $\mathbf{r}'$, is given by a product $U_{rr'} = -\frac{B}{|\mathbf{r}-\mathbf{r}'|^6} P_2(\cos \gamma)$, where γ is the angle between the two molecular axes $\mathbf{a}$ and $\mathbf{a}'$. The potential favors parallel alignment. Instead of calculating all pairwise interactions of a given molecule, the Maier–Saupe theory supposes that each molecule is submitted to some *mean potential* that is averaged over the positions and orientations of all other molecules. Averaging over the positions produces a constant $b = \sum_i \langle B/|\mathbf{r} - \mathbf{r}_i|^6 \rangle|_{r_i}$, whereas orientational averaging over the distribution function $f(\theta, \varphi)$ leads to $\langle P_2(\cos \gamma) \rangle|_{\Omega_i} = s P_2(\cos \theta)$, where (θ, φ) are the polar angles of $\mathbf{a}$ in the coordinate system with the axis Z along the director, and s is the scalar order parameter as usual. The effective potential, thus, writes as

$$U = -bs\, P_2(\cos\theta). \tag{4.26}$$

With the known field acting on the molecule, the probability distribution function is

$$f(\theta, \varphi) = C \exp[-U(\theta, s)/k_B T], \tag{4.27}$$

where

$$C^{-1} = 2\pi \int_0^\pi \exp[-U(\theta, s)/k_B T] \sin\theta\, d\theta.$$

Using (4.27), we are now in the position to find all values of s that yield a minimum of the free energy $F_1(s \neq 0) - F_{01}(s = 0) = E - TS$ per molecule. E is the internal energy calculated by averaging the effective potential (4.26) over $f(\theta, \varphi)$:

$$E = -\tfrac{1}{2}bs^2, \tag{4.28}$$

where the coefficient $1/2$ compensates for counting each interaction twice. The orientational entropy per molecule is

$$S = -k_B \int f(\theta, \varphi) \ln[4\pi f(\theta, \varphi)]\, d\Omega = -bs^2/T - k_B \ln 4\pi C, \tag{4.29}$$

so that

$$\Delta F = F_1 - F_{01} = \tfrac{1}{2}bs^2 + k_B T \ln 4\pi C. \tag{4.30}$$

The condition of an extremum of ΔF, $\partial\, \Delta F/\partial s = 0$, yields

$$s = 2\pi C \int_0^\pi P_2(\cos\theta) \exp[-U(\theta, s)/k_B T] \sin\theta\, d\theta, \tag{4.31}$$

and it coincides with the definition (3.4) of s, which demonstrates the self-consistency of the theory: The distribution function, expressed through s, should produce the same value of the order parameter when inserted in (3.4). The stability condition for the minimum, $\partial^2\, \Delta F/\partial s^2 > 0$, reduces to

$$\frac{\partial}{\partial s}\left\{ 2\pi C \int_0^\pi P_2(\cos\theta) \exp[-U/k_B T] \sin\theta\, d\theta \right\} < 1. \tag{4.32}$$

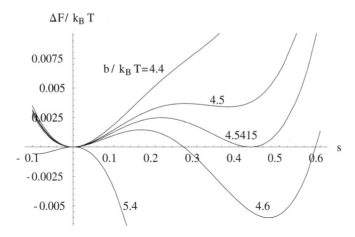

Figure 4.5. Free energy $\Delta F/k_B T$ per molecule as a function of the scalar order parameter for different ratios $b/k_B T = 4.4, 4.5, 4.5415, 4.6, 5.4$ (from the top curve to the bottom one).

The equilibrium values of s can be found from (4.31) and (4.32). One can draw plots of the function $Y_1(s) \equiv 2\pi C \int_0^\pi P_2(\cos\theta) \exp[-U/k_B T] \sin\theta \, d\theta$ versus s for different $b/k_B T$ and find the intersections of these curves $Y_1(s)$ with the straight line $Y_2(s) \equiv s$, which makes an angle $45°$ with the s-axis [see (4.31)]. The intersection points, in which $Y_1(s)$ is tilted less than $45°$ from the s-axis, correspond to the equilibrium state, in accordance with (4.32). Another way to clarify the predictions of the theory is to draw directly the dependencies $\Delta F(s)$ for different temperatures, as in Fig. 4.5.

The temperature, at which the two energy minima at $s = 0$ and $s \neq 0$ are of equal depth, is $T_{NI} \approx b/(4.5415 k_B)$; the corresponding scalar order parameter $s_{NI} \approx 0.429$ is similar to many experimental values reported in the literature.

4.4. The Smectic A–Nematic Transition

4.4.1. Order Parameter

As already stated, the order parameter of the SmA phase has two different components:

- A nematic component, with amplitude s and phase $\mathbf{n}$.
- A component relating to the modulation of matter density along the z-axis

$$\rho = \sum \rho_k \exp(-i\mathbf{k} \cdot \mathbf{z}), \qquad (4.33)$$

where $\mathbf{k} = m\frac{2\pi}{d_0}\boldsymbol{\nu}$, m is an integer, $\boldsymbol{\nu}$ is a unit vector along the normal to the layers, and d_0 is the smectic layer spacing. We shall define $q_0 = \frac{2\pi}{d_0}$ as the corresponding elementary wave vector.

Consider a slightly deformed smectic. The deformation is described by a displacement field $\mathbf{u} = u\mathbf{z}$ such that the material that was at $\mathbf{r}'$ before deformation is now at $\mathbf{r}$, with $\mathbf{r}' = \mathbf{r} - \mathbf{u}$. Hence, the density at $\mathbf{r}$ can be written as

$$\rho(\mathbf{r}) = \sum_k \rho_k \exp[-i\mathbf{k}(\mathbf{r} - \mathbf{u})] = \rho_0 + \rho_1 \exp[-i\mathbf{q}_0(\mathbf{r} - \mathbf{u})] + \cdots$$
$$= \rho_0 + \rho_1 \exp(i\varphi)\exp(-i\mathbf{q}_0 \cdot \mathbf{r}) + \cdots,$$

(4.34)

where we have introduced the phase $\varphi = \mathbf{q}_0 \cdot \mathbf{u}$. This phase is a degeneracy parameter; when varying in a range $[0, 2\pi]$, it describes all possible positions of the smectic phase with respect to a reference smectic $\varphi = 0$. Therefore, the smectic order parameter (restricted to the layers) is a complex number:

$$\psi = \psi_0 \exp i\varphi.$$

(4.35)

4.4.2. Ginzburg–Landau Expansion

The formation of smectic clusters in the nematic phase, whose size diverges when $T \to T_c$, leads to drastic changes of certain material parameters, such as divergence of the elastic coefficients for twist and bend deformations of the director field, or divergence of the cholesteric pitch if the nematic is chiral. We now build a phenomenological Ginzburg–Landau picture of the transition.

Let us first consider the smectic-order parameter alone; it enters the Ginzburg–Landau expansion of the free energy density

$$f_{\mathrm{SmA}} = \alpha |\psi|^2 + \frac{\beta}{2}|\psi|^4 + \frac{1}{2M_{||}}\left|\frac{\partial\psi}{\partial z}\right|^2 + \frac{1}{2M_\perp}|\nabla_\perp\psi|^2 + \cdots,$$

(4.36)

where $\alpha = a(T - T_c)$; the positive coefficients a, β, $M_{||}$, and $M_\perp$ are temperature independent; $\nabla_\perp = (\frac{\partial}{\partial x}, \frac{\partial}{\partial y}, 0)$, $|\nabla_\perp\psi|^2 = |\frac{\partial\psi}{\partial x}|^2 + |\frac{\partial\psi}{\partial y}|^2$. There are no odd terms in (4.36), so that the transition can be second order. The most notable feature of the expansion (4.36) is the presence of the gradient terms, which reflects a possibility of spatial variations of ψ. The coefficients $1/M_{||}$ and $1/M_\perp$ describe the (anisotropic) rigidity of the smectic phase for deformations along the normal and in the layers. The fluctuative deformations we consider have a small amplitude and a long wavelength, so that the amplitude ψ_0 of the order parameter is assumed not to vary. Hence,

$$\frac{1}{2M_{||}}\left|\frac{\partial\psi}{\partial z}\right|^2 \equiv \frac{1}{2M_{||}}\frac{\partial\psi}{\partial z}\frac{\partial\psi^*}{\partial z} = \frac{1}{2M_{||}}q_0^2\psi_0^2\left(\frac{\partial u}{\partial z}\right)^2,$$

(4.37)

which is the compressibility term, written usually as $\frac{1}{2}B(\frac{\partial u}{\partial z})^2$ in the classic free energy density of the smectic phase (see Section 5.2.3), with the Young modulus

$$B = \frac{1}{M_\parallel}\psi_0^2 q_0^2. \tag{4.38}$$

We shall now see that the $\frac{1}{2M_\perp}$ term is modified when the nematic order parameter is introduced in (4.36).

In the nematic phase, s reaches its maximum value precisely at the transition to the smectic phase, because the smectic phase usually appears at lower temperatures. We can safely assume that the nematic phase is well ordered just above T_c and that s is of the order of 1. Therefore, s is practically temperature independent in the smectic phase immediately below T_c. In contrast, the director $\mathbf{n}_0 = (0, 0, 1)$ may suffer fluctuations $\delta \mathbf{n} = (\delta n_x, \delta n_y, 0)$, $\mathbf{n} = \mathbf{n}_0 + \delta \mathbf{n}$, which analogously to fluctuations of ψ have to be taken into consideration, as follows.

The Ginzburg–Landau free energy of the smectic phase must be invariant with respect to simultaneous rotations of the director $\mathbf{n}$ and the normal $\boldsymbol{\nu}$ to the layers:

$$\mathbf{n} = (\delta n_x, \delta n_y, 1), \quad \boldsymbol{\nu} = \left(-\frac{\partial u}{\partial x}, -\frac{\partial u}{\partial y}, 1\right); \tag{4.39}$$

these variables are taken here as independent. Under the rotation by a small angle $\theta \ll 1$, the director tilt δn_x is equivalent to the displacement of layers $u = \theta x = -x \, \delta n_x$ along the z-axis. In a fixed coordinate frame, this displacement is equivalent to a phase change $\varphi \rightarrow \varphi - q_0 x \, \delta n_x$, so that $\psi \rightarrow \psi_0 \exp(i\varphi) \times \exp(-i q_0 x \, \delta n_x)$ and

$$\frac{\partial}{\partial x}\psi \rightarrow \left(\frac{\partial}{\partial x} - i q_0 \, \delta n_x\right)\psi. \tag{4.40}$$

The Ginzburg–Landau free energy with the correct gradient term is, thus,

$$f_{\text{SmA}} = \alpha \, |\psi|^2 + \frac{\beta}{2} \, |\psi|^4 + \frac{1}{2M_\parallel}\left|\frac{\partial \psi}{\partial z}\right|^2 + \frac{1}{2M_\perp} \, |(\nabla_\perp - i q_0 \, \delta \mathbf{n})\psi|^2. \tag{4.41}$$

Note that the angle between $\mathbf{n}$ and $\boldsymbol{\nu}$, which is small, can be measured by the components of the vector product

$$\boldsymbol{\nu} \times \mathbf{n} = \left(-\frac{\partial u}{\partial y} - \delta n_y, \frac{\partial u}{\partial x} + \delta n_x, 0\right).$$

Let us apply the operator $\boldsymbol{\nu} \times \mathbf{n}$ to $iq_0\psi$:

$$\boldsymbol{\nu} \times \mathbf{n}\,iq_0\psi = \left(-\frac{\partial}{\partial y} - iq_0\,\delta n_y, \frac{\partial}{\partial x} + iq_0\,\delta n_x, 0 \right)\psi.$$

The quantity $\mid (\nabla_\perp - iq_0\,\delta \mathbf{n})\psi \mid^2$ is, thus, nothing else than the value of the angle between $\mathbf{n}$ and $\boldsymbol{\nu}$, squared, times q_0^2.

The total free energy density is obtained by adding to f_{SmA} the free energy of deformation of the nematic *director*, i.e., the Frank–Oseen energy:

$$f_{FO} = \tfrac{1}{2}K_1(\text{div}\,\mathbf{n})^2 + \tfrac{1}{2}K_2(\mathbf{n} \cdot \text{curl}\,\mathbf{n})^2 + \tfrac{1}{2}K_3(\mathbf{n} \times \text{curl}\,\mathbf{n})^2, \tag{4.42}$$

where K_1, K_2, and K_3 are the elastic constants of splay, twist, and bend deformations, respectively (see Section 5.1.1). For small director distortions in the vicinity of the transition, the total energy density $f_{tot} = f_{SmA} + f_{FO}$ is

$$f_{tot} = \alpha \mid \psi \mid^2 + \frac{\beta}{2}\mid \psi \mid^4 + \frac{1}{2M_\parallel}\left| \frac{\partial \psi}{\partial z} \right|^2 + \frac{1}{2M_\perp}\mid (\nabla_\perp - iq_0\,\delta\mathbf{n})\psi \mid^2$$

$$+ \frac{1}{2}K_1(\text{div}\,\delta\mathbf{n})^2 + \frac{1}{2}K_2(\mathbf{n} \cdot \text{curl}\,\delta\mathbf{n})^2 + \frac{1}{2}K_3\left(\frac{\partial}{\partial z}\,\delta\mathbf{n} \right)^2. \tag{4.43}$$

4.4.3. Analogy with Superconductors

As pointed out by de Gennes,[4] the sum $f_{tot} = f_{SmA} + f_{FO}$ (4.43) is most remarkably analogous to the Ginzburg–Landau functional describing a superconductor-normal metal phase transition:

$$f_{super} = f_{normal} + \alpha \mid \psi \mid^2 + \frac{\beta}{2}\mid \psi \mid^4 + \frac{\hbar^2}{4m}\left| \left(-i\nabla - 2e\frac{\mathbf{A}}{\hbar c} \right)\psi \right|^2$$

$$+ \frac{(\text{curl}\,\mathbf{A})^2}{8\pi} - \frac{\mathbf{H}_0 \cdot \text{curl}\,\mathbf{A}}{4\pi}. \tag{4.44}$$

We use the cgs (centimeter, gram, second) units and present the functional f_{super} as the Gibbs free energy density, which facilitates comparison with f_{tot}. In (4.44), the superconductor's order parameter ψ is the wavefunction of the coherent ensemble of Cooper pairs, $\mathbf{A}$ is the magnetic vector potential, $\mathbf{H}_0$ is the external magnetic field, different from the local value of the magnetic field (magnetic induction) $\mathbf{B} = \text{curl}\,\mathbf{A}$ at a given point of the superconductor, m and e are the electron mass and the electron charge, and c is the speed

---

[4]P.G. de Gennes, Solid State Comm. **10**, 753 (1972).

of light. Under the assumption $M = M_{||} = M_{\perp}$, the identification is

$$\hbar^2/(2m) \leftrightarrow 1/M, \quad 2e/\hbar c \leftrightarrow q_0, \quad \mathbf{A} \leftrightarrow \delta\mathbf{n}; \tag{4.45}$$

the term $(\operatorname{curl}\mathbf{A})^2/(8\pi)$ corresponds to the twist and bend terms in the Frank–Oseen elastic energy, $1/(4\pi) \leftrightarrow K_2, K_3$. Of course, the analogy is incomplete. For example, f_{super} has no terms reflecting the elastic anisotropy $K_2 \neq K_3$ and no term corresponding to the divergence term $(\operatorname{div}\delta\mathbf{n})^2$. On the other hand, f_{tot}, as expressed in (4.43), has no term analogous to the external field term $\mathbf{H}_0 \cdot \operatorname{curl}\mathbf{A}$; therefore, the smectic-nematic transition is similar to the superconductor–normal metal transition in a zero magnetic field $\mathbf{H}_0 = 0$.

An immediate analogy that comes to mind while comparing (4.43) and (4.44) is that there is a smectic analog of the Meissner–Ochsenfeld effect, according to which the magnetic induction is zero in the superconductor; i.e., $\mathbf{B} = \operatorname{curl}\mathbf{A} = 0$. In a similar way, the smectic phase does not allow twist and bend deformations that are both associated with curl$\mathbf{n}$. The reason is simple: Twist and bend violate equidistance of the smectic layers (and splay does not). A thick stack of paper sheets is a good working model of the effect.

To determine the smectic order parameter and the director fluctuations, the free energy (4.43) should be minimized with respect to two functions: $\psi(\mathbf{r})$ and $\delta\mathbf{n}(\mathbf{r})$. The two corresponding equations (called the Ginzburg–Landau equations in the case of superconductors) for bulk equilibrium allow one to specify two types of characteristic lengths: the coherence lengths universally denoted ξ and the penetration lengths universally denoted λ. As in superconductors, the predictions of the Ginzburg–Landau model for smectics depend strongly on the relative values of these characteristic lengths.

4.4.4. Characteristic Lengths

4.4.4.1. Coherence Lengths

The gradient terms of the type $|\nabla\psi|^2$ in (4.43) and (4.44) prevent the order parameter amplitude from changing too quickly in space. Let us minimize $\int (f_{\text{SmA}} + f_{\text{FO}})\, dV$ with respect to $\psi^*(\mathbf{r})$, the complex conjugate of $\psi(\mathbf{r})$. The variational problem yields the following equation for the bulk equilibrium (Problem 4.6):

$$\alpha\psi + \beta|\psi|^2\psi - \frac{1}{2M_{||}}\left(\frac{\partial\psi}{\partial z}\right)^2 - \frac{1}{2M_{\perp}}(\nabla_{\perp} - iq_0\,\delta\mathbf{n})^2\psi = 0. \tag{4.46}$$

The coherence lengths suggested by the last equation are

$$\xi_{||} = \frac{1}{\sqrt{2|\alpha|M_{||}}} \quad\text{and}\quad \xi_{\perp} = \frac{1}{\sqrt{2|\alpha|M_{\perp}}}. \tag{4.47}$$

For $T > T_c$, these lengths are the sizes of the smectic clusters in the nematic bulk, $\xi_{\parallel}$ is measured along the normal to the layers, and $\xi_{\perp}$ is the transverse length. At $T < T_c$, these lengths are those along which a strong perturbation of the amplitude of the order parameter relaxes; for example, the coherence lengths characterize the size of the core of smectic dislocations, which are considered later in this section. As $T \to T_c$, the coherence lengths diverge.

4.4.4.2. Penetration Lengths

We assume now that at constant ψ_0, the director field suffers a perturbation, either a splay, twist, or bend deformation. Director splay is compatible with the smectic layering. Twist and bend involve curl $\mathbf{n} \neq 0$ and are not compatible with smectic layering. The length on which the twist or bend penetrate the smectic phase is called the penetration length (Fig. 4.6).

The equation that minimizes $\int (f_{SmA} + f_{FO}) \, dV$ with respect to $\delta \mathbf{n}$ can be significantly simplified if one assumes that $M_{\parallel} = M_{\perp}$ and writes the Frank–Oseen free energy density in the reduced form $f_{FO} = \frac{1}{2} K (\text{curl } \mathbf{n})^2$, which reflects our interest in twist and bend only. The variation of $\int_V [\frac{1}{2M} (\nabla_{\perp} - i q_0 \, \delta \mathbf{n}) \psi + \frac{1}{2} K (\text{curl } \delta \mathbf{n})^2] \, dV$ with respect to $\delta \mathbf{n}$ produces the second equation of bulk equilibrium for the director fluctuations $\delta \mathbf{n}$ (Problem 4.6):

$$\mathbf{j} = -\frac{i q_0}{2M} (\psi^* \nabla \psi - \psi \nabla \psi^*) - \frac{q_0^2}{M} \delta \mathbf{n} |\psi|^2. \tag{4.48}$$

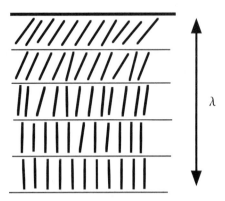

Figure 4.6. Bend penetration depth λ in the smectic A phase. The layers are parallel to the bounding plate. A weak bend deformation imposed at the surface penetrates into the smectic bulk over the distance λ.

Here, the vector

$$\mathbf{j} = K \, \text{curl}(\text{curl}\, \delta\mathbf{n}) \qquad (4.49)$$

is an analog of the current density in the superconductor: $\mathbf{j}$ is parallel to the director in the absence of coupling with the smectic order parameter.

Equation (4.48) suggests the characteristic length

$$\frac{1}{q_0}\sqrt{\frac{MK}{|\psi|^2}} = \frac{1}{q_0}\sqrt{\frac{MK\beta}{|\alpha|}},$$

where the amplitude $|\psi|$ of the order parameter is replaced by its value $\sqrt{|\alpha|/\beta}$ in the undeformed smectic. In fact, there are four penetration lengths, accounting for different combinations of $M_\parallel$, $M_\perp$, K_2, and K_3,

$$\lambda_{2,3}^{\parallel,\perp} = \frac{1}{q_0}\sqrt{\frac{M_{\parallel,\perp} K_{2,3}\beta}{|\alpha|}}. \qquad (4.50)$$

The fifth important penetration length is the one that measures the relative importance of the splay term versus the compressibility term; i.e.,

$$\lambda = \frac{1}{q_0}\sqrt{\frac{M_\parallel K_1}{|\psi|^2}} = \sqrt{\frac{K_1}{B}}. \qquad (4.51)$$

4.4.5. Anomalies of K_2 and K_3 Coefficients

As already stated, director splay within the smectic clusters appearing in the nematic bulk at $T > T_c$ does not break the equidistance and parallelism of the layers. Consider now twist (K_2) and bend (K_3) deformations of some amplitude $|\delta\theta|$, where $\delta\theta$ is the angle between the normals to the layers at the two ends of the cluster (Fig. 4.7). A nonzero $|\delta\theta|$ causes streching/compressing of the layers. Therefore, the energy cost of the distortions within the cluster can be estimated either as the compressibility term, $B(\delta\theta)^2\xi^3 \sim M_\parallel^{-1}\psi_0^2 q_0^2(\delta\theta)^2\xi^3$, or, equivalently, as the curvature elasticity term, $\delta K[\text{curl}(\delta\theta)]^2\xi^3 \sim \delta K(\delta\theta)^2\xi^3$, where δK is the increase of the Frank elastic constant K caused by the presence of smectic layers. Comparing the two estimates, one concludes that $\delta K \sim B\xi^2 \sim \frac{q_0^2}{M_\parallel}\psi_0^2\xi^2$. To find ψ_0^2, we notice that the fluctuative appearance of a smectic cluster of volume ξ^3 requires an energy cost $|\alpha||\psi|^2\xi^3 \sim k_B T$, i.e. $|\psi|^2 \sim \xi^{-3}k_B T/|\alpha| \propto \xi^{-1}$ [see (4.47)]. Therefore, $\delta K \sim \frac{q_0^2}{M_\parallel}\psi_0^2\xi^2 \propto \xi$. More careful calculations that take into account the difference between $\xi_\parallel$

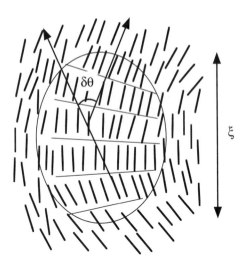

Figure 4.7. A smectic cluster in a nematic matrix that imposes bend deformation of the director.

and $\xi_\perp$ show the following renormalization of K_2 and K_3:

$$K_{2,3} \rightarrow K_{2,3}^{\text{bare}} + \delta K_{2,3}; \quad \delta K_2 \propto \xi_\perp^2 / \xi_{||}, \quad \delta K_3 \propto \xi_{||}. \tag{4.52}$$

The model, thus, predicts an indefinite growth of the twist and bend constants when the nematic phase approaches the smectic phase; in contrast, the splay elastic constant K_1 is not renormalized by the appearance of the smectic clusters.

4.4.6. Abrikosov Phases with Dislocations

Let us return to the Meissner–Ochsenfeld effect. The magnetic induction remains zero in the superconductor bulk, $\mathbf{B} = \text{curl}\,\mathbf{A} = 0$, even in the presence of a weak external field $\mathbf{H}_0$. However, if the field is high enough, it penetrates the superconductor. It may happen in two different ways. In type-I materials, the strong field penetrates the whole bulk, destroying superconductivity above the so-called thermodynamic critical field. In type II superconductors, before the material becomes normally conducting, the field penetrates partially, through line-like regions, called the *vortex lines* (Fig. 4.8). The lattice of vortices is thermodynamically stable and forms a special phase, called either the mixed state or the Abrikosov phase. The mixed state shows second-order transitions to the low-field superconducting phase and to the high-field normal phase.

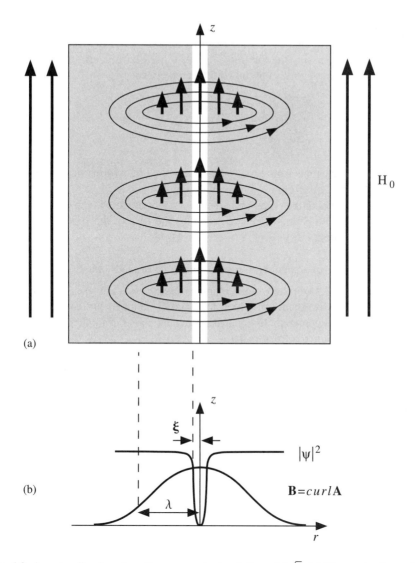

Figure 4.8. A vortex line in a type II superconductor, $\lambda/\xi > 1/\sqrt{2}$. (a) The vortex is parallel to the external magnetic field $\mathbf{H}_0$; the field penetrates the bulk of the superconductor in the core of the defect, along the axis z; circular lines show the particle currents circulating around the cylindrical core. (b) Wave function amplitude $|\psi|^2$ and magnetic field $\mathbf{B} = \mathrm{curl}\,\mathbf{A}$ changes in the core region of the defect, over two characteristic distances $\xi(T)$ and $\lambda(T)$, respectively.

The "core" of the vortex is in a normal conducting state. Away from the core, the magnetic field quickly decreases and the material regains its superconductivity. The total magnetic flux that is crossing the interior Σ of any closed loop γ surrounding the singular region (and containing it entirely) of the vortex is quantized:

$$\int_\Sigma \mathbf{B} \cdot d\Sigma \equiv \oint_\gamma \mathbf{A} \cdot d\mathbf{l} = N \frac{2\pi\hbar c}{2e}, \tag{4.53}$$

where the first identity stems from Stokes theorem, which transforms surface integrals into line integrals. The flux quantum is $\frac{2\pi\hbar c}{2e}$, and M is an integer.

The phase diagrams of the type I and type II superconductors in the (H_0, T) plane are, thus, very different. The parameter that determines the type of the superconductor is the temperature-independent ratio $\kappa = \lambda/\xi$:

- $\kappa < 1/\sqrt{2}$; type I superconductors; no intermediate (mixed) phase.
- $\kappa > 1/\sqrt{2}$; type II superconductors; magnetic field above the so-called lower critical field causes the mixed state in which the normal metal in the core of the vortices coexists with the superconducting matrix; above the upper critical field, the Abrikosov phase transforms into a normal metal.

The analogs of vortex lines in smectics are screw and edge dislocations. The dislocations introduce bend and twist into the system of smectic layers. For example, the elementary edge dislocation shown in Fig. 4.9 is associated with a bent director field. In the presence of dislocations, the phase φ ceases to be a single-valued function of coordinates.

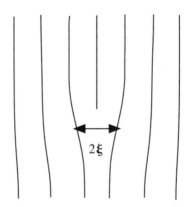

Figure 4.9. An edge dislocation in a SmA phase; a coherence length ξ is the distance over which a local perturbation affects the amplitude of ψ.

In going once around an elementary dislocation, φ changes by 2π, as in the case of the vortex. The equivalent of the flux quantization in a superconductor is the condition

$$\oint_{\gamma} \delta\mathbf{n}\, d\mathbf{l} = d_0, \qquad (4.54a)$$

d_0 being the smectic layer spacing. More generally,

$$\oint_{\gamma} \delta\mathbf{n}\, d\mathbf{l} = N\, d_0, \qquad (4.54b)$$

where N is the number (with sign) of elementary dislocations crossing the area Σ. Similarly, the cores of the dislocations are regions where curl $\mathbf{n} \neq 0$; the "normal" core of the dislocation is in the state with $|\psi| = 0$.

We come now to an intriguing question: With all superconductor-smectic similarities between the order parameters ψ's and defects (vortices versus dislocations), is there a liquid crystal analog of the Abrikosov phase, in which bend or twist can coexist with smectic layering? To answer the question, one needs to first find a liquid crystal analog of the external magnetic field $\mathbf{H}_0$ in superconductors (there is no Abrikosov phase at $\mathbf{H}_0 = 0$). As already discussed, the free energy density (4.42) of a nematic liquid crystal has no term corresponding to $\mathbf{H}_0 \cdot \text{curl}\, \mathbf{A}$ in (4.44). The missing link should be proportional to curl $\mathbf{n}$ [rather than to $(\text{curl}\, \mathbf{n})^2$]. An elegant and effective way to create an intrinsic source of deformation curl $\mathbf{n} \neq 0$ is to transform the nematic into a cholesteric liquid crystal, by simply adding chiral molecules.

If the nematic liquid crystal is chiral (chiral mesogenic molecules or chiral dopant), the equilibrium director structure is usually a helicoid with a pitch P and an axis, say, along the z-axis:

$$\mathbf{n} = (\sin 2\pi z/P, \cos 2\pi z/P, 0). \qquad (4.55)$$

Chirality leads to an additional term in the Frank–Oseen free energy density of the cholesteric:

$$f_{\text{Ch}} = f_{FO} - K_2 k_0 (\mathbf{n} \cdot \text{curl}\, \mathbf{n}), \qquad (4.56)$$

where f_{FO} is given by (4.42), and $k_0 = 2\pi/P$ at equilibrium (see Section 5.1). Thus, the cholesteric phase is an analog of the normal metal in an external magnetic field, with the correspondence $h \equiv K_2 k_0 \leftrightarrow \mathbf{H}_0$ and $K_2 k_0 (\mathbf{n} \cdot \text{curl}\, \mathbf{n}) \leftrightarrow \frac{\mathbf{H}_0 \cdot \text{curl}\, \mathbf{A}}{4\pi}$. The chirality h is a "field" conjugate to twist $(\mathbf{n} \cdot \text{curl}\, \mathbf{n})$. The Ginzburg–Landau parameter of interest is the ratio of the twist penetration length to the smectic coherence length: $\kappa = \lambda_{\text{twist}}/\xi = \frac{M}{q_0}\sqrt{2 K_2 \beta}$. When $\kappa < 1/\sqrt{2}$ (type I smectics), the mean-field theory based on the free

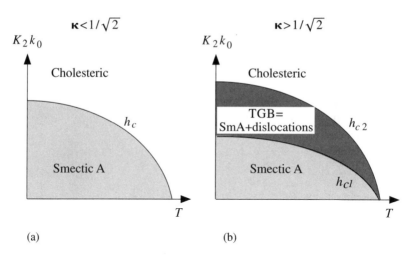

Figure 4.10. Phase diagram of (a) the SmA phase of type I, $\kappa < 1/\sqrt{2}$, and (b) the SmA phase of second type II, $\kappa > 1/\sqrt{2}$. Redrawn from Chaikin and Lubensky (1995).

energy density $f_{SmA} + f_{Ch} = f_{SmA} + f_{FO} - K_2 k_0 (\mathbf{n} \cdot \text{curl } \mathbf{n})$ [see (4.41), (4.42), and (4.56)] predicts a first-order smectic–cholesteric transition, at the thermodynamic critical field $h_c = \sqrt{K_2 \alpha^2 / \beta}$ (Problem 4.7). The twist is either completely expelled from the system or the system is phase separated into smectic and cholesteric regions. If $\kappa > 1/\sqrt{2}$, the new phase intervenes between the smectic and the cholesteric states on the (h, T) plane (Fig. 4.10). The new phase has been called the twist-grain boundary (TGB) phase by Renn and Lubensky who predicted it theoretically.[5]

The coexistence of twist and smectic order in the TBG phase is provided by the lattice of screw dislocations. As seen in Fig. 4.11, a row of parallel screw dislocations with spacing l_d form a TGB. Two smectic blocks on the opposite sides of the boundary are slightly tilted with respect to each other by an angle $2\pi\gamma = 2\arcsin(\frac{d_0}{2l_d}) \approx d_0/l_d$. The smectic blocks of width l_b (which is the distance between two consecutive grain boundaries) are free of dislocations. The twist is just concentrated mainly in the region of grain boundaries. The spatially average twist along the normal to grain boundaries is $\bar{k}_0 = 2\pi\gamma/l_b \approx d_0/(l_b l_d)$; the pitch $P = 2\pi/\bar{k}_0$ of the structure is $P = l_b/\gamma$. Interestingly, one can classify commensurate and incommensurate TGB phases, depending on whether $\gamma = l_b/P$ is rational or irrational. Among the commensurate versions of TGB, one might find quasicristalline symmetries, when the TGB phase is invariant under a rotation (around the twist axis) by $2\pi/Q$, where $Q = 5$ or $Q > 6$.

---

[5]S.R. Renn and T.C. Lubensky, Phys. Rev. **A 38**, 2132 (1988).

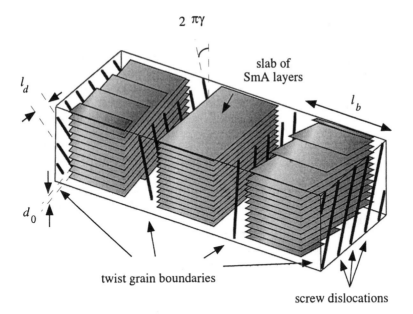

Figure 4.11. TGB phase: Smectic slabs separated by grain boundaries that accommodate the twist between two adjacent slabs. The twist grain boundaries can be composed of screw dislocations. Redrawn from Chaikin and Lubensky (1995).

4.5. Kosterlitz–Thouless Model of Phase Transitions

A system with positional or orientational order loses this order during a phase transition into the isotropic phase. However, the order can be partially spoiled even before the transitions by defects such as dislocations or disclinations. It is tempting to consider the process of melting as the nucleation and proliferation of defects. Of course, there are limitations to such an idea. For example, when the number of defects is really large, the very definition of the "defect" is questionable because the "ordered" background ceases to exist. Nevertheless, if one is interested primarily in the beginning of "melting," the idea is physically appealing. Kosterlitz and Thouless[6] have developed a 2D model of dislocation melting, a simplified version of which we discuss below.

Without going into details of dislocation properties (those are considered in Chapters 8 and 9), we observe that in a 2D lattice, a point dislocation causes a strain field $\nabla u \propto 1/r$ and bears an elastic energy

$$E = K \int_{r_{\text{core}}}^{R} \frac{dr}{r} = K \ln \frac{R}{r_c} + E_{\text{core}}, \tag{4.57}$$

[6]J.M. Kosterlitz and D.J. Thouless, J. Phys. C **5**, L124 (1972); C **6**, 1181 (1973).

where K is some elastic constant; the limits of integration are the radius r_{core} of the "core" of the defect and some macroscopic distance R, which is either the size of the system when the dislocation is isolated or the separation between dislocations; and E_{core} is the energy of the core. If N is the number of defects per unit area of the plane, then $R = 1/\sqrt{N}$.

Now, let us consider an ensemble of defects as a "gas" with the density N per unit area. There are $N_{sites} = 1/r_{core}^2$ sites available for these defects. The entropy of the gas of defects is defined by the number of configurations N defects can create at N_{sites} sites:

$$S = k_B \ln \frac{N_{sites}!}{N!(N_{sites} - N)!} \approx k_B N \ln \frac{N_{sites}}{N} \tag{4.58}$$

(the approximation is valid for $N_{sites} \gg N$, in accordance with Stirling's formula).

The free energy of the gas of defects is then

$$F = E - TS = N \ln \sqrt{\frac{N_{sites}}{N}} (K - 2k_B T) + N E_{core}. \tag{4.59}$$

It is easy to see that starting with some critical temperature, $T > T_c = K/2k_B$, the free energy becomes negative and the crystal melts. Above the melting temperature T_c, there is a finite *equilibrium* number of dislocations:

$$N = \frac{1}{r_c^2} \exp\left[\frac{-E_{core}}{2k_B(T - T_c)} \right], \tag{4.60}$$

while below it there are no equilibrium dislocations at all.

Problem 4.1. Find the Landau coefficients in (4.2) in dimensionless units (see text), T^{**} and T^* for the nematic material MBBA using $T_c = 319.0$ K, mass density 1.09 g $\cdot$ cm^{-3} at T_c, and the experimental data:[7] $aT_c = 43.1$ J cm^{-3}, $B = 2.66$ J cm^{-3}, and $C = 2.76$ J cm^{-3}.

Answers: The normalized factor for volume energy density is $RT_c\rho_c$, where ρ_c is the ratio of the mass density to the relative molecular weight of MBBA (see Fig. 2.14): $\rho_c \approx \frac{1.09 \text{ g·cm}^{-3}}{2} 67.37$ g $\cdot$ mol$^{-1} \approx 4.1 \times 10^3$ mol/m^3. Thus, in the dimensionless units $aT_c/3 = 1.32$, $2B/27 = 0.0181$, $C/9 = 0.0282$, $T^* = 318.3$ K, and $T^{**} = 319.1$ K (see Fig. 4.1).

Problem 4.2. The Landau coefficients presented in Problem 4.1 are determined by measuring the transition temperature, order parameter, latent heat, and so on. On the basis of Landau–de Gennes theory, find (a) s_{nem} versus $T^{**} - T$; (b) specific heat c_P versus $T^{**} - T$ in the nematic phase and its jump at the transition; (c) entropy discontinuity; and (d) the latent heat of the transition.

---

[7]Y. Poggi, J.C. Filippini, and R. Aleonard, Phys. Lett. **57A**, 53 (1976).

Answers:

(a) $s_{\text{nem}} = \frac{B}{4C} + \sqrt{\frac{3a(T^{**}-T)}{2C}}$.

(b) $c_P = -T\left(\frac{\partial^2 g}{\partial T^2}\right)_P = \frac{a^2 T}{2C}\left[1 + \frac{B}{2\sqrt{6aC}}(T^{**}-T)^{-1/2}\right]$ and $2a^2 T_c/C$.

(c) $S_{\text{iso}} - S_{\text{nem}} = -\left.\frac{\partial(g_0-g)}{\partial T}\right|_{T=T_c} = \frac{aB^2}{27C^2}$.

(d) $H_{\text{iso}} - H_{\text{nem}} = T_c(S_{\text{iso}} - S_{\text{nem}}) = \frac{aB^2 T_c}{27C^2}$; for more results, see Anisimov (1991).

Problem 4.3. Find the scalar order parameter in Onsager's model as the function of the parameter α.

Answers: $s = 1 - 3\frac{\coth\alpha}{\alpha} + \frac{1}{\alpha^2}$; $s = 0.85$ at $\alpha = 18.84$.

Problem 4.4. (a) Find the latent heat of phase transition in the Maier–Saupe theory. (b) Assuming that the order parameter is small, find the Landau–de Gennes coefficients A, B, and C in terms of the Maier–Saupe parameter b (see Stephen and Straley[8]).

Answers:

(a) $bs_c^2/2 \approx 0.42 k_B T_c$ per molecule.

(b) $A = \frac{3b}{2T}\left(T - \frac{b}{5k_B}\right)$, $B = \frac{9b^3}{70(k_B T)^2}$, and $C = \frac{9b^4}{700(k_B T)^3}$.

Problem 4.5. Usually, a racemic mixture with equal number of L and D enantiomers shows no macroscopic chirality. However, if the interaction between molecules of the same chirality substantially differs from that between molecules of opposite chirality, one might expect a phase separation and formation of large chirality-pure domains. Andelman and de Gennes[9] considered a model of pair clustering in chiral monolayers.

Consider a chiral molecule with four different groups attached to a central carbon atom in a tetrahedral fashion. One of the groups is an aliphatic chain that sticks out of the water; three other groups, A, B, and C (for example, A, B, C = NH_2, CH_3, Cl, F, CN, etc.) are in contact with water (Fig. 4.12a).

Two molecules interact through two pairs of groups that face each other as shown by dashed lines in Fig. 4.12b,c. If the pair interaction energies w_{AB}, w_{AA}, w_{AC}, and so on, between the groups are known, then the tendency of the system to form a heterochiral (L,D) or homochiral (L,L) dimers can be found by comparing the corresponding pair partition functions Z_{LD} and Z_{LL}.

(a) Using notation $f_{ij} = \exp(-w_{ij}/k_B T)$, find the two partition functions for (L,L) and (L,D) pairs.

[8]M.J. Stephen and J.P. Straley, Rev. of Mod. Phys. **46**, 617 (1974).

[9]D. Andelman and P.G. de Gennes, C.R. Acad. Sci. **307**, Sér. II, 23 (1988).

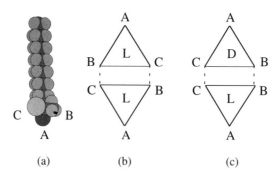

Figure 4.12. Chiral surfactant molecules [L-enantiomer is shown in (a)] at the surface of water might form either (b) homochiral or (c) heterochiral pairs, depending on the type of interactions among three different water-contacting atomic groups A, B, and C.

(b) Suppose that the groups B and C have acquired opposite electric charges, so that $f_{BB} = f_{CC} = 0$, $f_{BC} \gg 1$, and that the particle A is indifferent to both charged groups, $f_{AB} = f_{AC} = v$. Find the preferable type of dimers.

(c)–(g) Find the type of dimers when (c) identical groups prefer not to interact with each other, $f_{ii} = 0$; (d) group A is indifferent; i.e., $f_{Aj} = f = \text{const}$; (e) groups A, B, and C bear nonzero charges a, b, and c, respectively; $a+b+c = 0$; (f) all interactions are of London type with potential $w_{ij} = -U\alpha_i\alpha_j$, where U is a positive constant such that $U\alpha_i\alpha_j/k_BT \ll 1$; (g) temperature is very high, $k_BT \gg w_{ij}$.

Answers:

(a) $Z_{LL} = f_{BC}^2 + f_{AC}^2 + f_{AB}^2 + 2f_{AB}f_{CC} + 2f_{AA}f_{BC} + 2f_{AC}f_{BB}$, $Z_{LD} = f_{BB}f_{CC} + f_{AA}f_{BB} + f_{AA}f_{CC} + 2f_{AC}f_{BC} + 2f_{AB}f_{BC} + 2f_{AB}f_{AC}$.

(b) Denoting $f_{AA} = u$, one finds $\Delta Z = Z_{LL} - Z_{LD} = f_{BC}^2 + 2f_{BC}(u - 2v)$, and the dimers will be homochiral ($\Delta Z > 0$) for any $f_{BC} > 2(2v - u)$. If, on the other hand, attraction forces occur only between alike groups, so that $f_{ij} = 0$ for $i \neq j$, then $\Delta Z = -f_{BB}f_{CC} - f_{AA}f_{BB} - f_{AA}f_{CC} < 0$ and heterochiral dimers are formed.

(c) The heterochiral region is inside a cone of revolution that is tangent to the planes $f_{AB} = 0$, $f_{BC} = 0$, $f_{AC} = 0$.

(d) homochiral if $f_{BC} \gg f_{BB}, f_{CC}$.

(e) homochiral.

(f) $\Delta Z = -\left(\frac{|U|}{k_BT}\right)^3 \frac{(\alpha_A - \alpha_B)^2(\alpha_B - \alpha_C)^2(\alpha_C - \alpha_A)^2}{2}\left[1 + \frac{|U|}{3k_BT}(\alpha_A + \alpha_B + \alpha_C)^2 + \cdots\right] < 0$, heterochiral dimers.

(g) $\Delta Z \to 0$: Fast, thermally activated rotations of the molecules make them effectively nonchiral.

Problem 4.6. Derive (4.46) and (4.48).

Answers: Variation of $\int (f_{\mathrm{SmA}} + f_{\mathrm{FO}})\, dV$ with respect to $\psi^*(\mathbf{r})$ first produces an expression

$$\int_V dV \left[\alpha\psi\, \delta\psi^* + \beta|\psi|^2\psi\, \delta\psi^* + \frac{1}{2M_{||}} \left(\frac{\partial\psi}{\partial z}\right)\left(\frac{\partial\, \delta\psi^*}{\partial z}\right) \right.$$

$$\left. + \frac{1}{2M_\perp}(\nabla_\perp - iq_0\, \delta\mathbf{n})\psi(\nabla_\perp + iq_0\, \delta\mathbf{n})\, \delta\psi^* \right] = 0.$$

To modify terms of the type $\nabla\, \delta\psi^*$, we introduce $\eta = (\nabla_\perp - iq_0\, \delta\mathbf{n})\psi$ and use the identity $\nabla(\eta\, \delta\psi^*) = \eta\nabla\, \delta\psi^* + \delta\psi^*\nabla\eta$. Then,

$$\int_v \eta\nabla\, \delta\psi^*\, dV = \int_v \nabla(\eta\, \delta\psi^*)\, dV - \int_v \delta\psi^*\nabla\eta\, dV$$

$$= \oint_S \eta\, \delta\psi^*\, dS - \int_v \delta\psi^*\nabla\eta\, dV,$$

and the variation is

$$\int_V \left[\alpha\psi + \beta|\psi|^2\psi + \frac{1}{2M_{||}} \left(\frac{\partial\psi}{\partial z}\right)^2 + \frac{1}{2M_\perp}(\nabla_\perp - iq_0\, \delta\mathbf{n})^2\psi \right] \delta\psi^*\, dV$$

$$+ \oint_S \left[\frac{1}{2M_{||}} \left(\frac{\partial\psi}{\partial z}\right) + \frac{1}{2M_\perp}(\nabla_\perp - iq_0\, \delta\mathbf{n})\psi \right] \delta\psi^*\, dS = 0.$$

The requirement of zero variation results in (4.46) for the bulk and in the boundary condition $[\frac{1}{2M_{||}}(\frac{\partial\psi}{\partial z}) + \frac{1}{2M_\perp}(\nabla_\perp - iq_0\, \delta\mathbf{n})\psi] \cdot \mathbf{t} = 0$, where $\mathbf{t}$ is the unit vector normal to the smectic surface.

A similar procedure applied to a simplified version of the free energy

$$\int_V \left[\frac{1}{2M}(\nabla - iq_0\, \delta\mathbf{n})\psi + \frac{1}{2}K\,(\mathrm{curl}\, \delta\mathbf{n})^2 \right] dV$$

results in (4.48); to simplify an intermediate expression $K\,\mathrm{curl}\,\mathrm{curl}\,\delta(\delta\mathbf{n})$, one might use the identity $\mathbf{b}\,\mathrm{curl}\,\mathbf{a} - \mathbf{a}\,\mathrm{curl}\,\mathbf{b} = \mathrm{div}\,[\mathbf{a} \times \mathbf{b}]$, where $\mathbf{a} = \delta(\delta\mathbf{n})$ and $\mathbf{b} = \mathrm{curl}\, \delta\mathbf{n}$.

Problem 4.7. Find the thermodynamic critical field for the first-order smectic A - nematic transition in type I materials with $\kappa < 1/\sqrt{2}$.

Answers: $h_c = \sqrt{K_2\alpha^2/\beta}$. For calculations of the lower and upper critical fields in type II smectics, see Chaikin and Lubensky (1995).

Further Reading

M.A. Anisimov, Critical Phenomena in Liquids and Liquid Crystals, Gordon and Breach, Philadelphia, 1991.

P.M. Chaikin and T.C. Lubensky, Principles of Condensed Matter Physics, Cambridge University Press, 1995.

P.G. de Gennes and J. Prost, The Physics of Liquid Crystals, Oxford Science Publication, Clarendon Press, Oxford, 1993.

L.D. Landau and E.M. Lifshitz, Statistical Physics, Addison-Wesley, Reading, MA, 1969.

Phase Transitions in Liquid Crystals, Edited by S. Martellucci and A.N. Chester, NATO ASI series. Series B, Physics, vol. 290.

V.V. Schmidt, The Physics of Superconductors, Springer-Verlag, Berlin, 1997.

H.E. Stanley, Introduction to Phase Transitions and Critical Phenomena, Oxford University Press, New York, 1987.

G. Vertogen and W.H. de Jeu, Thermotropic Liquid Crystals, Fundamentals, Springer Series in Chem. Phys., vol. 45, Springer-Verlag, Berlin, 1988.

CHAPTER 5

Elasticity of Mesomorphic Phases

5.1. Uniaxial Nematics and Cholesterics

5.1.1. The Free Energy Density

We describe the state of deformation of a nematic or a cholesteric phase, at a fixed temperature T, by the director field $\mathbf{n}(\mathbf{r})$. The free energy associated with the deformation depends necessarily on the gradient of the director, $\nabla \mathbf{n}$, whose components $\frac{\partial n_i}{\partial x_j}$ will be noted $n_{i,j}$. We assume that the distortions are small,

$$|n_{i,j}| \ll \frac{1}{a}, \tag{5.1}$$

where a is a typical molecular length. This assumption has some advantages, as follows:

- There is a well-defined "tangent" *perfect* (liquid) crystal at each point $\mathbf{r}$, with orientation $\mathbf{n}(\mathbf{r})$, whose spatial extention is large enough to make a *continuous* description possible.

- Therefore, the order parameter is a locally well-defined constant $s[T(\mathbf{r})]$, which depends on temperature uniquely.

- At any point $\mathbf{r}$, the symmetry properties of the tangent liquid crystal should reflect in its free energy density $f(\mathbf{r})$, which however depends not only on $\mathbf{n}(\mathbf{r})$, but also on its derivatives $\nabla \mathbf{n}$. It is stated that f must be invariant under any change of the orientation of $\mathbf{n}$ and of the values of its derivatives $\nabla \mathbf{n}$, which are allowed by the symmetries. This invariance is by no means trivial, and it should be considered as a principle, to which one could attach the name of Noll (principle of material invariance); it goes much farther than does the invariance of the Landau expansion, which does not depend on the derivative.

Therefore, the free energy density of a nematic or a cholesteric specimen must be invariant by any operation that preserves the local orientation of $\mathbf{n}$:

1. Invariance under the operation $\mathbf{n} \to -\mathbf{n}$, an operation that is common to both nematic and cholesteric phases. Note that this operation of symmetry is *not* a space transformation that transports matter, nor a central inversion; it is an operation in the order parameter space that does not affect the physical molecules.

   In addition, in the case of nematics, there is:

2. Central inversion about any point.

3. Invariance under any rotation about $\mathbf{n}$.

   And in the case of cholesterics, there is

2* *No* central inversion. Such an operation would transform a cholesteric from its right-handed form, say, into a left-handed one.

3* Invariance under rotations by π about the director $\mathbf{n}$, the axis of helicity χ, and the transverse axis $\boldsymbol{\tau} = \chi \times \mathbf{n}$.

Note that the only symmetry among the above that subsists in a double-twisted blue phase is the operation $\mathbf{n} \to -\mathbf{n}$.

In both nematic and cholesteric cases, it will be enough to restrict the free energy density expansion to terms quadratic in $\nabla\mathbf{n}$, because $n_{i,j}$ is so small.

The only *scalar* invariants linear in $n_{i,j}$ and invariant under any rotation are $\mathrm{div}\,\mathbf{n} = n_{i,i}$ and $\mathbf{n}\cdot\mathrm{curl}\,\mathbf{n} = \varepsilon_{ijk}n_{k,j}n_i$. Here, ε_{ijk} form a completely antisymmetric unit tensor (the Levi–Civita tensor) with $\varepsilon_{123} = \varepsilon_{231} = \varepsilon_{312} = 1$, $\varepsilon_{132} = \varepsilon_{213} = \varepsilon_{321} = -1$; ε_{ijk} is zero when any two indices are alike. In index notations, the components of a vector product are $[\mathbf{a}\times\mathbf{b}]_i = \varepsilon_{ijk}a_jb_k$; hence, $\mathrm{curl}_i\,\mathbf{n} = [\nabla\times\mathbf{n}]_i = \varepsilon_{ijk}n_{k,j}$. Because $\mathrm{div}\,\mathbf{n}$ is odd in $\mathbf{n}$, it must be excluded from the free energy density expansion. On the other hand, $(\mathrm{div}\,\mathbf{n})^2$ is allowed for both nematics and cholesterics. Furthermore, $\mathbf{n}\cdot\mathrm{curl}\,\mathbf{n}$ changes its sign under inversion $(x \to -x, y \to -y, z \to -z)$; so it can appear only in the cholesteric free energy density, but not in the nematic one. Finally, the quadratic scalar invariant $(\mathrm{curl}\,\mathbf{n})^2$ appears in both phases.

Because in the identity $(\mathrm{curl}\,\mathbf{n})^2 = (\mathbf{n}\cdot\mathrm{curl}\,\mathbf{n})^2 + (\mathbf{n}\times\mathrm{curl}\,\mathbf{n})^2$, both terms on the right-hand side are symmetry invariant, the free energy density limited to first-order derivatives writes in the nematic case as

$$f_{\mathrm{FO}} = \frac{1}{2}K_1(\mathrm{div}\,\mathbf{n})^2 + \frac{1}{2}K_2(\mathbf{n}\cdot\mathrm{curl}\,\mathbf{n})^2 + \frac{1}{2}K_3(\mathbf{n}\times\mathrm{curl}\,\mathbf{n})^2. \tag{5.2}$$

The free-energy density (5.2) is referred to as the Frank–Oseen energy density with Frank elastic constants K_1, K_2, and K_3 (all three are necessarily positive). We already used it in the analysis of the N $\leftrightarrow$ SmA transition (Section 4.4.2). In the cholesteric phase, the presence of the term $\mathbf{n}\cdot\mathrm{curl}\,\mathbf{n}$ leads to

$$f_{\mathrm{FO}} = \frac{1}{2}K_1(\mathrm{div}\,\mathbf{n})^2 + \frac{1}{2}K_2(\mathbf{n}\cdot\mathrm{curl}\,\mathbf{n} + q_0)^2 + \frac{1}{2}K_3(\mathbf{n}\times\mathrm{curl}\,\mathbf{n})^2 \tag{5.3}$$

where $q_0 = \frac{2\pi}{p}$, and p is the cholesteric pitch, as we shall see below; in our notations, q_0 is positive for a right-handed cholesteric, and negative for a left-handed cholesteric, provided the frame of coordinates is right handed.

The expressions (5.2) and (5.3) are complete at the level of the expansion we assumed, but only if one neglects the so-called divergence terms:[1]

$$f_{13} + f_{24} = K_{13} \operatorname{div}(\mathbf{n} \operatorname{div} \mathbf{n}) - K_{24} \operatorname{div}(\mathbf{n} \operatorname{div} \mathbf{n} + \mathbf{n} \times \operatorname{curl} \mathbf{n}). \tag{5.4}$$

The K_{24} term can be reexpressed as a quadratic form of the first derivatives (see (5.7) below), whereas the K_{13} term is proportional to the *second* derivatives $n_{i,jk}$ and thus might in principle be comparable to $f_{\mathrm{FO}} \propto n_{i,j} n_{k,l}$. The divergence nature of these terms allows one to represent the volume integral $\int (f_{13} + f_{24})\, dV$ as a *surface* integral by virtue of the Gauss theorem:

$$\int K_{13} \operatorname{div} \mathbf{g}_{13}\, dV - \int K_{24} \operatorname{div} \mathbf{g}_{24}\, dV = K_{13} \oiint_A \boldsymbol{\nu} \cdot \mathbf{g}_{13}\, dA$$

$$- K_{24} \oiint_A \boldsymbol{\nu} \cdot \mathbf{g}_{24}\, dA, \tag{5.5}$$

where $\mathbf{g}_{13} = \mathbf{n} \operatorname{div} \mathbf{n}$, $\mathbf{g}_{24} = \mathbf{n} \operatorname{div} \mathbf{n} + \mathbf{n} \times \operatorname{curl} \mathbf{n}$, and $\boldsymbol{\nu}$ is the unit vector of the outer normal to the surface A. However, the divergence terms (5.4) must not be neglected on the grounds of transformation (5.5). The energy integrals (5.5) scale in general linearly with the size of the deformed system, as do the integrals $\int f_{\mathrm{FO}}\, dV$. Note also that the transformation (5.5) is valid only when K_{13} and K_{24} are constants. The basic difference between f_{FO} and $(f_{13} + f_{24})$ shows up when one seeks for equilibrium director configurations by minimizing the total free energy functional $\int (f_{\mathrm{FO}} + f_{13} + f_{24})\, dV$: the K_{13} and K_{24} terms do not enter the Euler–Lagrange variational derivative for the bulk. However, they can contribute to the energy and influence the equilibrium director through boundary conditions at the surface A.

5.1.2. Geometrical Interpretations of Director Deformations

5.1.2.1. Bulk Terms

K_1 is called the *splay* elastic modulus. The splay deformation $\operatorname{div} \mathbf{n}$ is nonvanishing in the two geometries depicted in Fig. 5.1a and b: $\operatorname{div} \mathbf{n} = \frac{1}{r}$ for 2D splay ($n_x = \cos\varphi, n_y = \sin\varphi, n_z = 0$), we employ cylindrical coordinates (r, φ, z) (Fig. 5.1a) and $\operatorname{div} \mathbf{n} = \frac{2}{r}$ for 3D splay ($n_x = \frac{x}{r}, n_y = \frac{y}{r}, n_z = \frac{z}{r}$), $r = \sqrt{x^2 + y^2 + z^2}$, (Fig. 5.1b). The K_2 and K_3 terms both vanish because $\operatorname{curl} \mathbf{n} \equiv 0$.

---

[1] J. Nehring and A. Saupe, J. Chem. Phys. **54**, 337 (1971); **56**, 5527 (1972).

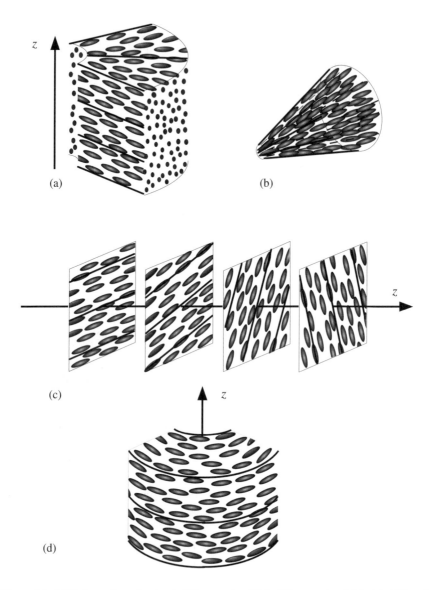

Figure 5.1. (a) Deformation of splay in 2D geometry, (b) in 3D geometry, (c) twist, (d) bend.

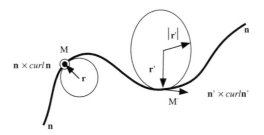

Figure 5.2. Geometrical meaning of $\mathbf{n} \times$ curl $\mathbf{n}$.

K_2 is called the *twist* elastic modulus: $\mathbf{n} \cdot$ curl $\mathbf{n}$ is the only nonvanishing deformation in the geometry of Fig. 5.1c that illustrates a right-handed rotation of the director ($n_x = \cos qz, n_y = \sin qz, n_z = 0$). Here, $q = \frac{\alpha}{d} = \frac{2\pi}{p} > 0$, α is the angle of twist through the slab, d is the thickness of the slab, and p is the pitch. Note that $\mathbf{n} \cdot$ curl $\mathbf{n} + q = 0$; $q = -\mathbf{n} \cdot$ curl $\mathbf{n}$ is the wavenumber attached to the pitch; hence, the interpretation of q_0 in the cholesteric free energy density (5.3), which vanishes for $\mathbf{n} \cdot$ curl $\mathbf{n} \equiv -q_0$ in a perfect cholesteric phase.

This is the place to make an important comment about the geometrical meaning of the equation $\mathbf{n} \cdot$ curl $\mathbf{n} \equiv 0$. This equation is a necessary and sufficient condition for the envelopes of the director to be *perpendicular* to a family of surfaces. The "sufficient" part is easy: Assume that $\mathbf{n}$ is such that one can write

$$\mathbf{n} = \pm \nabla\phi/|\nabla\phi|, \tag{5.6}$$

i.e., $\mathbf{n}$ is perpendicular to the set of surfaces $\phi(x, y, z) = $ const, letting the constant vary; $1/|\nabla\phi|$ is a renormalizing factor, introduced to satisfy the condition $\mathbf{n}^2 = 1$. Equation (5.6) immediately yields $\mathbf{n} \cdot$ curl $\mathbf{n} \equiv 0$: Twist vanishes. Consequently, twist is absent in lamellar phases, even when they are distorted.

K_3 is called the *bend* elastic modulus: $\mathbf{n} \times$ curl $\mathbf{n}$ is the only nonvanishing deformation in the geometry of Fig. 5.1d ($n_x = \sin\varphi, n_y = \cos\varphi, n_z = 0$) and $|\mathbf{n} \times$ curl $\mathbf{n}| = \frac{1}{r}$. The bend term has a precise geometrical meaning in any distorted configuration: $\mathbf{n} \times$ curl $\mathbf{n}$ is a vector along the principal normal to the line that envelops the directors at the point M. The length of this vector is the curvature $\frac{1}{r}$ at M (Fig. 5.2). If $\mathbf{n} \times$ curl $\mathbf{n} \equiv 0$, the envelopes are straight lines.

5.1.2.2. Divergence Terms

The divergence terms are of much more complex geometrical (and physical) meaning.

K_{24} is called the *saddle-splay* elastic modulus for the reason that will become clear when a similar modulus is discussed for lamellar phases. If $\mathbf{n}$ depends only on one Cartesian coordinate, the K_{24} term vanishes identically. Saddle-splay is absent in 2D splay, (Fig. 5.1a), but not in the 3D splay: div $(\mathbf{n}\,\text{div}\,\mathbf{n} + \mathbf{n} \times \text{curl}\,\mathbf{n}) = \frac{2}{r^2}$ in Fig. 5.1.b. It is important to realize that another saddle-splay is hidden in the density f_{FO} (5.2), as clear from the identity

$$(\text{div}\,\mathbf{n})^2 + (\text{curl}\,\mathbf{n})^2 = n_{i,j}n_{i,j} + \text{div}\,(\mathbf{n}\,\text{div}\,\mathbf{n} + \mathbf{n} \times \text{curl}\,\mathbf{n}). \tag{5.7}$$

K_{13} is often called a *mixed splay-bend* elastic modulus. Its geometrical meaning can be further clarified by considering the surface densities of the divergence terms:

$$s_{13} + s_{24} = K_{13}(\boldsymbol{\nu} \cdot \mathbf{n})\text{div}\,\mathbf{n} - K_{24}\boldsymbol{\nu} \cdot (\mathbf{n}\,\text{div}\,\mathbf{n} + \mathbf{n} \times \text{curl}\,\mathbf{n}). \tag{5.8}$$

The surface density of the K_{24} term contains only derivatives along directions tangent to the surface A, whereas the K_{13} term, besides the tangential derivatives, necessarily contains a derivative along the normal to A.[2]

The scalar div $\mathbf{n}$ has no simple meaning for a generic distortion of a nematic or cholesteric phase, but it has a geometrical interpretation in lamellar phases.

5.1.3. Material Elastic Constants

5.1.3.1. Small Molecules Liquid Crystals (SMLC)

The coefficients K_1, K_2, K_3, K_{13}, and K_{24} have the dimension of a force; henceforth, they can be expressed as ratios of an energy U to a length a. The only typical energies in an ordered medium are the interaction energies between atoms or molecules, the most representative being the interactions between nearest neighbors. Obviously, such energies cannot be much larger than $k_B T_c$, where T_c is the "clearing" temperature at which the nematic phase melts. In the case of nematics of small molecules, T_c is in the room temperature range or slightly above, say, 400 K, and a reasonable guess is $U \approx 5 \times 10^{-21}$ J $(5 \times 10^{-14}$ erg). The only typical lengths are molecular or atomic lengths. Taking a ~ 1 nm and U as above, one gets $K_i \approx \frac{U}{a} \approx 0.5 \times 10^{-11}$ N $(0.5 \times 10^{-6}$ dyn). This magnitude is most often measured (e.g., by the Frederiks transition method, Section 5.4.3) for K_1, K_2, K_3 in thermotropic nematics. For example, for 5CB,[3]

$$K_1 = 0.64 \times 10^{-11}\,\text{N}; \quad K_2 = 0.3 \times 10^{-11}\,\text{N}; \quad K_3 = 1 \times 10^{-11}\,\text{N}.$$

In most cases, K_2 is the smallest coefficient. Consequently, twist shows up often in deformed nematic samples, such as droplets that demonstrates twisted structures (see Section 11.1.6).

---

[2] V.M. Pergamenshchik, Phys. Rev. **E48**, 1254 (1993); **E49**, 934 (E) (1994).

[3] For a review of the most recent values, see L.M. Blinov and V.G. Chigrinov, Electrooptic Effects in Liquid Crystal Materials, Springer series on partially ordered systems, New York, 1994.

Very little is known about K_{24}, see Crawford,[4] and practically nothing is known for sure about K_{13}. As shown by Ericksen,[5] the requirement $\mathbf{n}$ = const in a nematic with $q = 0$ and $K_{13} = 0$ leads to the restriction $0 < K_{24} < K_1$ or $0 < K_{24} < K_2$ (whichever is smaller).

5.1.3.2. Liquid Crystal Polymers (LCP)

One does not find in this case the simplicity of SMLCs, but on the other hand, one expects that the coefficients would relate in an interesting way to *molecular conformations*; this field of research is still open to investigation. Usually, K_2 keeps smaller than K_1 and K_3. This result is intuitive, because the molecular length L does not play *a priori* a large role in a pure twist deformation. On the other hand, K_1 and K_3 are strongly modified. There are, however, few exact theoretical and not enough experimental results on the issue of K_1, K_2, K_3 in different polymers. Two cases should be distinguished (for details, see the review papers[6]).

Rigid polymers in solution.   As already stated, this is the case of numerous polymers of biological origin in solution, or viruses like TMV in water; polyamids in sulphuric acid have a nematic phase whose elements are used as fibers for fabrics. In these cases, the transition to the nematic phase fits the Onsager model. In the Onsager model of the rigid rods, K_1 and K_2 are affected little when the solution is dilute, because the rods do not interfere much with each other under splay and twist. The bend deformation, on the contrary, is expected (see Lee and Meyer footnote) to show strong interference effects. Measurements of K_i/χ_a by the Frederiks technique (χ_a is a diamagnetic or dielectric anisotropy) and K_i/η by inelastic Rayleigh scattering (η is a viscosity) show that K_1 and K_3 increase with the molecular weight $M_w \propto \frac{L}{d}$ and the volume fraction $\Phi = \frac{\pi}{4}d^2 Ln$ (L and d are the length and the diameter of the molecule; n is the number of molecules per unit volume). An analytical solution,[7] obtained in the limit of a high scalar order parameter $s \rightarrow 1$, $K_1 = \frac{7}{8\pi}\frac{k_B T}{d}\Phi\frac{L}{d} \approx 3K_2$, $K_3 = \frac{4}{3\pi^2}\frac{k_B T}{d}\Phi^3(\frac{L}{d})^3$, shows that K_3 increases much faster than does K_1 with the volume fraction.
   Numerical results obtained within the framework of a Flory–Onsager model, modified in order to take into account the variation of excluded volume with deformation, are shown in Table 5.1.

Semiflexible polymers.   Two features, characteristic of both lyotropic and thermotropic LCPs, appear when the molecular weight increases, as follows

---

[4]G.P. Crawford, in Physical Properties of Liquid Crystals: Nematics, Edited by D.A. Dunmur, A. Fukuda, and G.R. Luckhurst, INSPEC, The Institution of Electrical Engineers, London, U.K. (2001), p. 230.

[5]J.L. Ericksen, Inequalities in liquid crystal theory, Phys. Fluids **9**, 1205 (1966).

[6]S.D. Lee and R.B. Meyer, in "Liquid Crystallinity in Polymers," Edited by A. Ciferri, p. 343, VCH Pub. 1991; Liq. Cryst. **7**, 15 (1990).

[7]T. Odjik, Liq. Cryst. **1**, 553 (1986).

Table 5.1. Numerical values of the Frank coefficients in the Flory–Onsager model* (in units of 10^{-11}N or 10^{-6} dynes).

	$s = 0.7$	$s = 0.8$	$s = 0.9$
K_1	1.8	2.1	3.3
K_2	0.6	0.69	1.1
K_3	8.0	14.0	43.0
K_1/K_2	3	3	3
K_3/K_2	13	21	39

*G. Strajer, S. Fraden, and R.B. Meyer, Phys. Rev. A **39**, 4828 (1989); R.B. Meyer et al., Faraday Discuss. Chem. Soc. **79**, 125 (1985).

- The density of *chain ends* decreases with L. This has a direct effect on the splay coefficient K_1. The chain ends contribute to the total energy by the elastic (solid type) deformation they carry and by their entropy, which can be estimated assuming that the chain ends form a perfect gas. The elastic contribution, calculated by de Gennes,[8] yields a contribution to K_1 proportional to L^2. On the other hand, one can speculate on the possibility of a splay deformation at vanishing mass density variation (no elastic deformation), as in Fig. 5.3a: The V-shaped void between two rods is healed by inserting the end of the third rod. The main effect is then due to the variation of entropy, which gives rise to a contribution to K_1[9] of the order of $\frac{k_B T}{d} \frac{L}{d} \Phi$ (Φ is the volume fraction in the case of a lyotropic LCP and is a constant in the case of a thermotropic LCP). Another way of relaxing splay deformation is by the appearance of "hairpins," as in

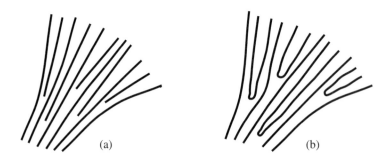

Figure 5.3. (a) Effect of chain ends on the splay modulus; (b) hairpins in polymeric chains.

[8]P.G. de Gennes, Mol. Cryst. Liq. Cryst. Lett. **34**, 177 (1977).

[9]R.B. Meyer, in *Polymer Liquid Crystals*, Edited by A. Ciferri, W.R. Krigbaum, and R.B. Meyer, Academic Press, New York, 1982, p. 133.

Fig. 5.3b. In the limit when the chains become infinite in length, any splay deformation at constant polymer density is forbidden, and K_1 becomes increasingly large.

- The bend deformations, on the contrary, do not require density changes. The corresponding bend modulus relates to the persistence length ℓ_p, $\kappa = k_B T \ell_p$ (see Section 15.1). Scaling arguments allow one to find expressions for the Frank elastic constants. The main prediction[7] is that $K_3 \approx \frac{k_B T}{d} \Phi \frac{\ell_p}{d}$ becomes independent of the

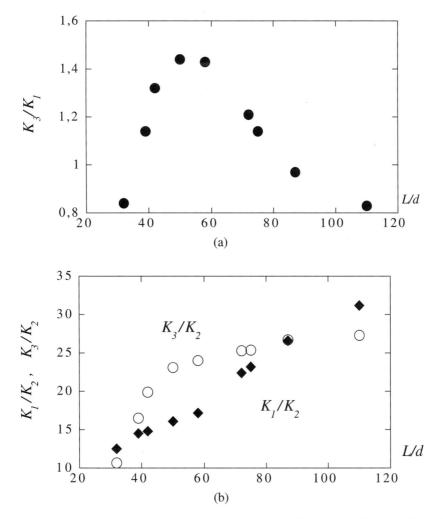

Figure 5.4. (a) K_3/K_1, (b) K_1/K_2, and (c) K_3/K_2 as functions of the chemical length L/d of the polymer chain; PBG in a solvent mixture composed of 18% of dioxane and 82% dichloromethane (adapted from Lee and Meyer).

chain length, assuming $L \gg \ell_p$. Figure 5.4 shows experimental data for poly-γ-benzylglutamate (PBG). The liquid crystalline phases are formed by PBG molecules in their extended α-helical conformation in different organic solvents, such as m-cresol, dioxane, benzene, chloroform, and so on. Typically, in these PBG solutions, $\ell_p \approx 10^3$ Å, whereas $d \approx (15 - 25)$ Å and $\Phi = 0.1 - 0.2$. Note that a maximum in the dependence K_3/K_1 in Fig. 5.4a happens at $L/d \approx 50$, or $L \approx 10^3$ Å, a value that is in excellent agreement with the direct measurements of ℓ_p. However, the measured ratio K_2/K_1 does not satisfy the theoretical ratio $1/3$.

Very few measurements have been carried on thermotropic LCPs. Let us cite Zheng-Min and Kleman,[10] in which it has been shown that in a mainchain polyester, K_1 is very large (by a factor of ≈ 10) compared with SMLCs nematics, whereas K_2 and K_3 have more conventional values. The cores of defects (disclinations) in this compound display molecular arrangements that corroborate the large anisotropy of the Frank coefficients.[11]

5.2. Lamellar Phases

5.2.1. Free Energy Density

The order parameter of an SmA phase includes a nematic contribution (coming from the normal $\mathbf{n}(\mathbf{r})$ to the layers) and a 1D solid contribution. The free energy density follows immediately. The nematic "bulk" part (5.2) yields the same free energy as above, viz.

$$f_1 = \tfrac{1}{2} K_1 (\operatorname{div} \mathbf{n})^2 + \tfrac{1}{2} K_3 (\mathbf{n} \times \operatorname{curl} \mathbf{n})^2, \tag{5.9}$$

with no twist term, as already stated. The solid part yields

$$f_2 = \frac{1}{2} B \left(\frac{d - d_0}{d_0} \right)^2 \tag{5.10}$$

$$= \frac{1}{2} B \gamma^2, \tag{5.11}$$

where d_0 is the equilibrium repeat distance, d is the actual layer thickness measured along $\mathbf{n}$ (in a continuous model, d is a continuous function of $\mathbf{r}$), and B is the Young modulus for the 1D solid.

Let us comment in more detail on the K_2 and K_3 terms. Consider a closed loop drawn in a slightly deformed smectic. The total number of traversed layers is exactly zero and can

---

[10] Sun Zheng-Min and M. Kleman, Mol. Cryst. Liq. Cryst, **111**, 321 (1984).
[11] G. Mazelet and M. Kleman, Polymer **27**, 714 (1986).

Errata

Soft Matter Physics: An Introduction

Maurice Kleman   Oleg D. Lavrentovich

Equations from Eq. (5.12) on page 145 to Eq. (5.106) on page 174 are incorrectly labeled; the numeration within this range should be adjusted by subtracting one unit (thus, "(5.12)" should be labeled "(5.11)", etc.). Label (5.11) on page 144 should be deleted. All references to equations in the text are correct.

Page 497, line 12 from bottom of page: $(E_{DC}, 0, 0)$ should read $(0, 0, E_{DC})$.

be expressed as $\frac{1}{d_0} \oint \mathbf{n} \cdot d\ell$, if the layers do suffer bend deformation, but no compression or dilatation. Hence, by virtue of Stokes's theorem, $\oint \mathbf{n} \cdot d\ell = \int \text{curl}\,\mathbf{n} \cdot d\mathbf{S} = 0$, which yields curl $\mathbf{n} = 0$; i.e.,

$$\mathbf{n} \cdot \text{curl}\,\mathbf{n} \equiv 0 \qquad (5.12a)$$

(no twist of the director field), and

$$\mathbf{n} \times \text{curl}\,\mathbf{n} \equiv 0 \qquad (5.12b)$$

(no bend of the director field). Identity (5.11a) has already been studied, and it subsists even if the layers are compressed or dilated. Identity (5.11b) tells us that the envelopes of the director field (the field of normals to the layers) are straight lines $\frac{1}{r} \equiv 0$, and that the layers are parallel (see Problem 5.1). If they are not, there is some contribution to f_2 that varies from point to point. Let us compare the f_2 and K_3 contributions. We have

$$\frac{\{K_3\}}{\{f_2\}} = \frac{K_3}{B} \frac{(\mathbf{n} \times \text{curl}\,\mathbf{n})^2}{\gamma^2} \cong \frac{K_3}{B} \frac{1}{r^2}.$$

Note $\frac{K_3}{B} = \lambda_3^2$, where λ_3 is a material length that must be comparable to the layer separation. Hence $\frac{\{K_3\}}{\{f_2\}} \cong (\frac{\lambda_3}{r})^2$, a very small quantity indeed, because r is macroscopic. The K_3 contribution, which is a deformation associated with the layers compression, is negligible compared with the B contribution, which is of the same nature.

The free energy density, (5.9) and (5.10), reduces eventually to

$$f = \frac{1}{2} K_1 (\text{div}\,\mathbf{n})^2 + \frac{1}{2} B\gamma^2. \qquad (5.13)$$

The ratio of K_1 to B defines an important length scale

$$\lambda = \sqrt{K_1/B} \qquad (5.14)$$

called "the penetration length"; λ is of the order of the layer separation but diverges when the system approaches the SmA-nematic transition. One expects that a SMLC SmA would have K_1 of the same order as in a nematic phase stable at higher temperatures. With $\lambda \sim d_0 \approx (1 \div 3)$ nm, and $K_1 \approx 10^{-11}$ N, one finds $B \approx 10^6 \div 10^7$ N/m^2, a value of the compressibility modulus that is 10^3 to 10^4 times smaller than in a solid.

5.2.2. Splay and Saddle-Splay Deformations

We now discuss another expression of the free energy density,

$$f = \frac{1}{2} K (\sigma_1 + \sigma_2)^2 + \overline{K}\sigma_1\sigma_2 + \frac{1}{2} B\gamma^2, \qquad (5.15)$$

which with respect to the former one (5.12) is supplemented by the divergence saddle-splay term $\overline{K}$; the splay constant is noted K instead of K_1. The parameters $\sigma_1 = \frac{1}{R_1}$ and $\sigma_2 = \frac{1}{R_2}$ are the principal curvatures of the smectic layer in $\mathbf{r}$ and allow one to interpret the layers deformation in terms of the mean curvature $H = \frac{1}{2}(\sigma_1 + \sigma_2)$ and the Gaussian curvature $G = \sigma_1\sigma_2$. Expression (5.14) is valid for any large bending of the layers and small "solid" deformations. To ensure that the free energy density (5.14) is positive definite for the lamellar phase, $\overline{K}$ must be within the range $-2K < \overline{K} \leq 0$; K is always positive.

We first show that

$$\operatorname{div}\mathbf{n} = \pm(\sigma_1 + \sigma_2). \tag{5.16}$$

A few concepts in the theory of surfaces are in order here. The layers are symbolized by surfaces, and we consider such a surface Σ in the neighborhood of a point M (Fig. 5.5). The orthogonal axes MX_1, MX_2 are taken in the plane tangent to Σ at M, and the axis MZ is along the normal (arbitrarily oriented). The equation of the surface in the vicinity of M is

$$z = ax_1^2 + 2bx_1 x_2 + cx_2^2 + 0(3), \tag{5.17}$$

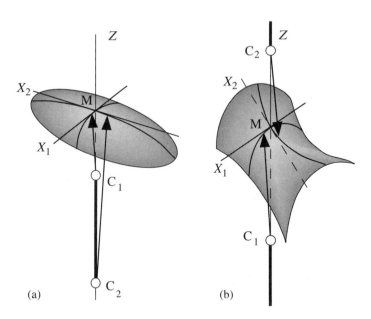

(a)   (b)

Figure 5.5. Element of area in the vicinity of its tangent plane at M: (a) elliptic point; (b) hyperbolic or saddle point.

but we can choose the directions X_1 and X_2 such that the cross-term vanishes:

$$z = \frac{1}{2}\left(\frac{x_1^2}{R_1} + \frac{x_2^2}{R_2}\right) + 0(3). \tag{5.18}$$

This choice defines the *principal directions* at M. Any plane containing MZ intersects Σ along a parabola (up to third order) whose apex is in M. The principal planes $x_2 = 0$ and $x_1 = 0$ yield parabolae whose curvatures in M are $1/R_1$ and $1/R_2$, respectively. It is easy to show that R_1 and R_2 are extrema of the radii of curvature of all parabolae in M, which are obtained by letting $x_2 = \mu x_1$ in (5.17), where μ is a variable parameter. We distinguish two cases, as follows.

- If R_1 and R_2 have the same sign, the element of surface lies on one side of the tangent plane, and all centers of curvature lie between C_1 and C_2, the two principal centers of curvature in M. The two radii of curvature $MC_1 = R_1$ and $MC_2 = R_2$ are signed lengths. The Gaussian curvature $G = \sigma_1\sigma_2$ is positive. M is called an *elliptic point*. In the neighborhood of M, the surface looks like an elliptic paraboloid (Fig. 5.5a).
- If R_1 and R_2 have opposite signs, the element of surface intersects the tangent plane, and all centers of curvature lie outside of the segment C_1C_2. M is called a hyperbolic point, and the surface looks locally like a saddle (Fig. 5.5b). The Gaussian curvature $G = \sigma_1\sigma_2$ is negative.

The director is normal to the surface,

$$\mathbf{n} = \pm\nabla\phi/|\nabla\phi|, \quad \text{where} \quad \phi = -z + \frac{1}{2}\left(\frac{x_1^2}{R_1} + \frac{x_2^2}{R_2}\right),$$

and it has the components

$$\mathbf{n} = \pm\left\{\frac{x_1}{R_1}, \quad \frac{x_2}{R_2}, \quad -1\right\}\left(1 - \frac{x_1^2}{2R_1^2} - \frac{x_2^2}{2R_2^2}\right)^{-\frac{1}{2}}, \tag{5.19}$$

and at the point M, we have the result we were seeking for:

$$\mathrm{div}\,\mathbf{n} = \pm\left(\frac{1}{R_1} + \frac{1}{R_2}\right). \tag{5.20}$$

Note that div $\mathbf{n}$ does not depend on the distribution of $\mathbf{n}$ outside of Σ.

A classic result in surface theory is that the Gaussian curvature reads as

$$G = \sigma_1\sigma_2 = \frac{1}{2}\mathrm{div}\,(\mathbf{n}\cdot\mathrm{div}\,\mathbf{n} + \mathbf{n}\times\mathrm{curl}\,\mathbf{n}). \tag{5.21}$$

This expression relates the Gaussian curvature term in smectics to the nematic divergence terms (5.4). It explains why saddle-splay disappears in the geometry of 2D splay: The surfaces normal to **n** have zero Gaussian curvature.

It is usual, in the physics of lamellar media, to note $K_1 = K$ and $-2K_{24} = \overline{K}$, as in (5.14).

Let us now consider a single lamella, or membrane, of thickness d_0, and introduce the bend moduli of this membrane:

$$\kappa = K\, d_0, \quad \bar{\kappa} = \bar{K}\, d_0. \tag{5.22}$$

The free energy per unit area can be written as

$$f_c = \frac{1}{2}\kappa(\sigma_1 + \sigma_2)^2 + \bar{\kappa}\sigma_1\sigma_2, \tag{5.23}$$

assuming that the membrane is decoupled from other lamellae, so that there are no B-terms. This expression of the free energy will prove of importance in the study of surfactant systems in solvents. Equation (5.22) has long been known as the expression of the energy per unit area in the mechanical theory of shells,[12] with the correspondence:

$$\kappa \Leftrightarrow \frac{Eh^3}{12(1-v^2)}, \quad \bar{\kappa} \Leftrightarrow \frac{-Eh^3}{12(1+v)}, \tag{5.24}$$

where E is the Young modulus and h is the thickness of the shell. Note that in this theory κ and $\bar{\kappa}$ have the same order of magnitude. This remark holds as well for surfactant membranes, whose material constants can be discussed in terms of mechanical interactions at the molecular level[13] (see Chapter 14). Finally, because $\sigma_1\sigma_2 \sim H^2$, the Gaussian curvature term brings to the free energy density f a contribution comparable to that one of the mean curvature.

5.2.3. Free Energy Density for Small Deformations

Consider now the case in which the layers are but slightly bent, so that their deformation can be described by a single scalar variable, the component of the displacement $v(x, y, z_0)$ of the layers along the normal of the unperturbed layers, taken as the z-axis. A layer formerly at position z_0 is in $z = z_0 + v(x, y, z_0)$ after deformation. Hence, $z_0 = z - v(x, y, z_0)$, which we rewrite using a new variable $u(x, y, z)$ for displacement:

$$z_0 = z - u(x, y, z). \tag{5.25}$$

---

[12] A.E.H. Love, The Mathematical Theory of Elasticity, Dover, 1944, article 298.
[13] A.G. Petrov and A. Derzhanski, J. Physique Fr. **3**, C3-15 (1976).

The equation of the deformed layer is $\phi(x, y, z) = z_0$, where $\phi(x, y, z) = z - u(x, y, z)$, and the director $\mathbf{n} = \pm \nabla \phi / |\nabla \phi|$ writes

$$\mathbf{n} = \pm \left\{ -\frac{\partial u}{\partial x}\left(1 + \frac{\partial u}{\partial z}\right), \quad -\frac{\partial u}{\partial y}\left(1 + \frac{\partial u}{\partial z}\right), \quad 1 - \frac{1}{2}\left[\left(\frac{\partial u}{\partial x}\right)^2 + \left(\frac{\partial u}{\partial y}\right)^2\right] \right\}$$

(5.26)

to the second order. To the same order (Problem 5.4),

$$\gamma = \frac{\partial u}{\partial z} - \frac{1}{2}(\nabla_\perp u)^2,$$

(5.27)

where the term $(\nabla_\perp u)^2 \equiv (\frac{\partial u}{\partial x})^2 + (\frac{\partial u}{\partial y})^2$ accounts for the effective layer compressions caused by layers' tilting by small angles $\partial u/\partial x$, $\partial u/\partial y$. The free energy density, restricted to quadratic terms, is

$$f = \frac{1}{2}K_1(\Delta_\perp u)^2 + \frac{1}{2}B\left(\frac{\partial u}{\partial z}\right)^2,$$

(5.28)

where $\Delta_\perp = \frac{\partial^2}{\partial x^2} + \frac{\partial^2}{\partial y^2}$. However, there are many important phenomena, such as field- or strain-induced layers undulations (see Section 5.5.1) that cannot be properly described by the harmonic form (5.27) and one has to retain the nonlinear terms originating in (5.26):

$$f = \frac{1}{2}K_1(\Delta_\perp u)^2 + \frac{1}{2}B\left[\frac{\partial u}{\partial z} - \frac{1}{2}(\nabla_\perp u)^2\right]^2.$$

(5.29)

The saddle-splay term vanishes in most cases of practical interest (including the layer undulations), either because one of the principal curvatures is zero or because the regions with alternating signs of the Gaussian curvature compensate each other.

5.3. Free Energy of a Nematic Liquid Crystal in an External Field

Electric field effects are usually studied in the geometry when the liquid crystal is confined between two glass plates. The inner surfaces of the plates are covered with a transparent electroconductive material such as indium tin oxide (ITO). Furthermore, the electrodes are often coated with a special material (a polymer such as polyimide, surfactant layer, etc.) to align the director along some particular direction called "direction of anchoring" or "easy axis." There are four basic mechanisms of interactions between the applied electric field and the liquid crystal: (a) dielectric, (b) flexoelectric, (c) through surface polarization originating from electric double layers of ions or from the symmetry breaking $\mathbf{n} \neq -\mathbf{n}$ at the boundary, and (d) through the motion of electric carriers—ions. To simplify things, we neglect the flexoelectric effect (to be considered later) and assume that the liquid crystal is a perfect insulator, and that there is no surface polarization of type (c) at the bounding plates.

Suppose the cell is connected to an external source that keeps the voltage across the cell constant. The dielectric coupling implies that the liquid crystal molecules tend to orient in such a way that the component of the tensor of molecular polarizability along the field is maximum. To find the equilibrium orientation of the director field in an external electric field, one has to minimize the elastic energy $F_{\mathrm{FO}} = \int f_{\mathrm{FO}}\, dV$ supplemented by:

- The energy of the electric field:[14]

$$F_E = \int f_E\, dV = \frac{1}{2} \int \mathbf{E} \cdot \mathbf{D}\, dV, \qquad (5.30a)$$

whose variation in a small reorientation of the molecules can be written as:

$$\delta F_E = \frac{1}{2} \int \mathbf{E} \cdot \delta \mathbf{D}\, dV. \qquad (5.30b)$$

- The free energy associated with the change of the charge at the electrodes when $\mathbf{n}$ reorients. To keep the voltage across the cell constant, there should be indeed a supply of energy δF_G from the electric source. δF_G is equal (with the opposite sign) to the work needed to maintain the fixed voltage when the electric displacement $\delta \mathbf{D}$ changes and modifies the surface-charge density at the plates by the quantity $(-\delta D_z)$ (the subscript "z" denotes the component of $\mathbf{D}$ along the normal to the plates):

$$\delta F_G = \iint_A \psi\, \delta D_z\, dA = \int \operatorname{div}\left(\psi\, \delta \mathbf{D}\right) \mathbf{dV}, \qquad (5.31a)$$

where we have employed the Gauss theorem. Here, ψ is the electric potential, and dA is the surface element of the electrode. Because $\operatorname{div}\left(\psi\, \delta \mathbf{D}\right) = \psi \operatorname{div} \delta \mathbf{D} + \delta \mathbf{D} \cdot \nabla \psi$, $\mathbf{E} = -\nabla \psi$, and $\operatorname{div} \delta \mathbf{D} = 0$ (there are no free electric charges), one gets

$$\delta F_G = -\int \mathbf{E} \cdot \delta \mathbf{D}\, dV = -2\delta F_E. \qquad (5.31b)$$

---

[14] J.D. Jackson, Classical Electrodynamics, John Wiley & Sons, Inc., New York, 3rd edition, 1999, Chapter 4; V.G. Sugakov and E.M. Verlan, Hydrodynamics of Liquid Crystals, Kiev State University, 1978.

The minimum of the total bulk free energy is achieved when

$$\delta F_{FO} + \delta F_E + \delta F_G = \delta(F_{FO} - F_E) = 0. \tag{5.32}$$

F_E can be written in terms of the field strength and the components of the dielectric permittivity tensor $\bar{\bar{\varepsilon}}$ (Chapter 3). Representing the electric displacement as a sum of its two components, normal to $\mathbf{n}$ and along $\mathbf{n}$, i.e., $\mathbf{D} = \varepsilon_0 \varepsilon_\perp \mathbf{E}_\perp + \varepsilon_0 \varepsilon_\| \mathbf{E}_\| = \varepsilon_0 \varepsilon_\perp \mathbf{E} + \varepsilon_0 \varepsilon_a (\mathbf{E} \cdot \mathbf{n}) \mathbf{n}$, where $\varepsilon_a = \varepsilon_\| - \varepsilon_\perp$, one finds

$$\mathbf{E} \cdot \mathbf{D} = \varepsilon_0 \varepsilon_\perp E^2 + \varepsilon_0 \varepsilon_a (\mathbf{n} \cdot \mathbf{E})^2. \tag{5.33}$$

Dropping the orientation-insensitive first term on the right-hand side, one finally arrives at the expression for the nematic free energy density supplemented by the dielectric term:

$$f = f_{FO} - f_E = f_{FO} - \frac{1}{2} \varepsilon_0 \varepsilon_a (\mathbf{n} \cdot \mathbf{E})^2. \tag{5.34}$$

In a uniaxial nematic with $\varepsilon_a > 0$, $\mathbf{n}$ orients along $\mathbf{E}$; when $\varepsilon_a < 0$, $\mathbf{n}$ is degenerate in the plane perpendicular to $\mathbf{E}$.

Similar considerations hold for the magnetic field. Let

$$\mathbf{M} = \mu_0^{-1} \chi_\perp \mathbf{B} + \mu_0^{-1} \chi_a (\mathbf{B} \cdot \mathbf{n}) \mathbf{n} \tag{5.35}$$

be the magnetization generated by the magnetic field of induction $\mathbf{B}$. The corresponding Zeeman energy density is the integral $-\int_0^B \mathbf{M} \cdot d\mathbf{B}$. The free energy density in the presence of the magnetic field becomes

$$f = f_{FO} - \frac{1}{2} \mu_0^{-1} \chi_a (\mathbf{n} \cdot \mathbf{B})^2. \tag{5.36}$$

Here again, the $\mathbf{n}$-independent term is omitted. The susceptibility anisotropy χ_a is positive in the calamitic nematic phases, but it is negative in discotic nematics, so that the director is degenerate in the plane perpendicular to $\mathbf{B}$.

As already indicated, the effects of the electric field are complicated by mechanisms different from the dielectric coupling: flexoelectricity, surface polarization, and free ions.

Flexoelectricity[15] in liquid crystals is an analog of piezoelectricity in solid crystals: Director curvature causes electric polarization of the medium. The effect can be qualitatively illustrated by deformations in a nematic medium composed of anisometric (pear-like or banana-like, Fig. 5.6) molecules that have permanent dipole moments. For example, splay of pear-like molecules with a permanent dipole along the long axis leads to macroscopic

---

[15]R.B. Meyer, Phys. Rev. Lett. **22**, 918 (1969).

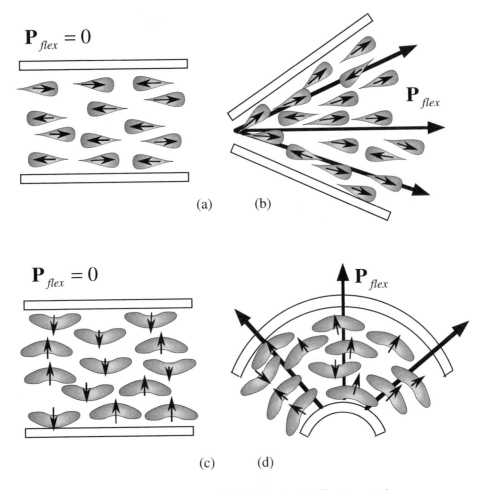

Figure 5.6. Molecular models for flexoelectric effects (see text).

electric polarization $P_1 = e_1 n(\text{div } n)$. Bend of banana-like molecules, whose permanent dipole is perpendicular to the director, causes polarization $P_3 = e_3(\text{curl } n \times n)$, where the vector $\text{curl } n \times n$ is along the radius of curvature of the envelopes of the director. The coefficients e_1 and e_3 have dimension μ/a^2, where μ is a dipole moment. One expects $e_i \approx 10^{-11} \text{C/m}$ ($\approx 10^{-4}$ CGS units). Flexoelectric polarization P_f couples to the external field; the contribution to the free energy density is linear in E (as opposed to the dielectric term quadratic in E):

$$f_{\text{flex}} = -P_f \cdot E = -[e_1 n \, \text{div } n - e_3(n \times \text{curl } n)] \cdot E. \tag{5.37}$$

A similar effect is that of order electricity,[16] in which the source of polarization is the gradient of the *scalar* part of the order parameter. Another linear electric effect occurs when the "heads" and "tails" of molecules have different affinity with the bounding plates. If these molecules have longitudinal dipole moments, the broken surface symmetry results in a surface polarization[17] P_s and a contribution $(-P_s \cdot E)$ to the free energy density. Surface polarization of a different type results from the electric double layers formed by selective ion adsorption at interfaces. Ions are always present in liquid crystals, either as the residuals of the chemical synthesis or because of the injection from the electrodes under applied voltage. A variety of electrohydrodynamic effects occurs when the current of ions drags the liquid crystal into motion.

5.4. Standard Applications of the Elasticity of Nematics

5.4.1. Minimization of the Free Energy in the Generic Case

Determination of the equilibrium director by minimizing the free energy functional belongs to the classic field of variational analysis.[18] We first give a general scheme and then illustrate it by examples.

5.4.1.1. General Scheme

Let $F = \int_V f \, dV$ be the total free energy of a nematic or a cholesteric phase in a volume V. In the sequel, f is given by (5.2), with a possible addition of the field terms (5.33), (5.35), or other terms. The minimization problem consists in looking for a distribution of the director $n(r)$, which minimizes F under physical constraints such as the surface orientation of the director, shape of the sample, and so on.

---

[16]G. Barbero, I. Dozov, I. Palierne, and G. Durand, Phys. Rev. Lett. **56**, 2056 (1986).

[17]P. Guyot-Sionnest, H. Hsiung, and Y.R. Shen, Phys. Rev. Lett. **57**, 2963 (1986).

[18]Calculus of variations: see, e.g., G. Arfken, "Mathematical Methods for Physicists," Academic Press, Inc., San Diego, 1985, Chapter 17.

Consider an *arbitrary* variation $\delta\mathbf{n}(\mathbf{r})$ that is such that $\mathbf{n}' = \mathbf{n}(\mathbf{r}) + \delta\mathbf{n}(\mathbf{r})$ yields a higher total free energy F'. The method consists, therefore, in calculating $\delta F = F' - F$:

$$\delta F = \int_V \left[\frac{\partial f}{\partial n_i} - \frac{\partial}{\partial x_j} \frac{\partial f}{\partial n_{i,j}} \right] \delta n_i \, dV + \int_\Sigma \frac{\partial f}{\partial n_{i,j}} \delta n_i \, dA_j, \qquad (5.38)$$

and writing that $\delta F = 0$ for *any* choice $\delta\mathbf{n}(\mathbf{r})$. Equation (5.37) obtains after integration by parts, which has the advantage of letting the δn_i field appear, and not its derivatives; but this is done at the expense of the appearance of a surface term.

Furthermore, in the present case, one has to introduce a supplementary term in the variation of F, viz:

$$\delta F_L = - \int \lambda n_i \, \delta n_i \, dV, \qquad (5.39)$$

which originates in the condition of normalization of $\mathbf{n}$

$$-\frac{1}{2}\lambda(\mathbf{n}^2 - 1) \equiv 0, \qquad (5.40)$$

where the so-called Lagrange multiplier λ is an unknown function of coordinates (that has nothing in common with the penetration length in lamellar phases).

The volume and the surface terms in (5.37) have to be treated on different footings.

In the bulk.   Because $\delta\mathbf{n}(\mathbf{r})$ can be chosen at will (but still obeying the normalization condition $\mathbf{n}'^2 = 1$, which reads to the first order $\mathbf{n} \cdot \delta\mathbf{n} = 0$), let us consider a field $\delta\mathbf{n}$ that vanishes everywhere on the boundary and in the bulk, except in an infinitesimally small volume dV located in a particular point $\mathbf{r}$, where $\delta\mathbf{n}$ is chosen constant. Therefore, one gets

$$h_i \equiv -\frac{\partial f}{\partial n_i} + \frac{\partial}{\partial x_j} \frac{\partial f}{\partial n_{i,j}} = \lambda n_i. \qquad (5.41)$$

These equations must be true at any point $\mathbf{r}$ of the sample, and they provide us with the so-called Euler–Lagrange minimization differential equations. The vector $\mathbf{h}$ is called the molecular field. Equations (5.40) have to be supplemented by the normalization condition (5.39) and the boundary conditions, which we discuss now.

On the boundaries.   Conditions on the boundaries can be defined either by data related to $\mathbf{n}$ on the surface or by data related to the external torques and forces that are applied on the surface. The most frequent experimental case is the first one, with $\mathbf{n}$ fixed on the boundary by physicochemical conditions. One necessarily has then $\delta\mathbf{n}_\Sigma \equiv 0$, and the surface terms of (5.37) vanish identically; they play no role in the mathematical expression of the boundary conditions. Consider now the case of external torques $\mathbf{C}$ applied on the surface director, per unit area, so that a torque $\mathbf{C} dA$ acts on a surface element. The work done by this torque when the director rotates by an angle $\delta\boldsymbol{\omega}$ is $\mathbf{C} \cdot \delta\boldsymbol{\omega} \, dA$. But $|\delta\boldsymbol{\omega}|$ is the angle between $\mathbf{n}$ and $\mathbf{n}'_\Sigma = \mathbf{n} + \delta\mathbf{n}_\Sigma$; i.e.,

$$\delta\mathbf{n}_\Sigma = \boldsymbol{\omega} \times \mathbf{n}_\Sigma. \tag{5.42}$$

Hence, the free energy is increased by the quantity

$$\delta F_{\text{ext}} = \int_\Sigma \omega_j C_j \, dA. \tag{5.43}$$

Writing $dA_j = \nu_j \, dA$, where $\boldsymbol{\nu}$ is a unit vector along the outward normal to the surface, and introducing (5.41) in the surface terms of (5.37), we get

$$\omega_p \left\{ C_p + \frac{\partial f}{\partial n_{i,j}} \varepsilon_{ipq} n_q \nu_j \right\} = 0, \tag{5.44}$$

which is true for any value of ω_p, which is a virtual rotation. Hence,

$$C_p = -\frac{\partial f}{\partial n_{i,j}} \nu_j \varepsilon_{ipq} n_q, \tag{5.45}$$

which expresses the fact that the total torque acting on the director at the surface vanishes at equilibrium. Such equations, which describe the director elastic relaxation on the surface, can take into account the K_{13} and K_{24} divergence terms if they are introduced in the free energy density f (the Euler–Lagrange equations for the bulk are not altered by these terms). Incorporation of the K_{13} term is not simple because this term contains derivatives along the normal to the surface: The procedure should take into account that the liquid crystal properties such as scalar order parameter and density change over some nonzero distance near the surface.[19] In the rest of this chapter, we assume $K_{13} = 0$.

5.4.1.2. Special Cases and Simplifying Assumptions

Equation (5.40) has been solved under some special assumptions, such as that of (1) *isotropic elasticity* with equal Frank constants $K_1 = K_2 = K_3 = K$, or (2) *planar geometry* of distortions (**n** staying parallel to a fixed plane). These two assumptions are often made simultaneously. Although they are far from reality, they make easier a qualitative under-

---

[19] V.M. Pergamenshchik, Phys. Rev. E**58**, R16 (1998); Phys. Lett. A**243**, 167 (1998); V.M. Pergamenshchik and S. Zumer, Phys. Rev. E**59**, R531 (1999).

standing of the specific aspects of the present curvature elasticity, as well as some properties of defects. Note that the assumption (1) is better obeyed in SMLCs than in the LCPs: Very often, $K_1 \approx K_3$ indeed, except in the vicinity of the N-SmA phase transition, where K_3 diverges. Also, K_2 is generally smaller than are K_1 and K_3. Hence, twist deformations are favored;[20] because these deformations define locally an axis of helicity, the assumption (2) is also somewhat obeyed on macroscopic lengths. The situation (2) is considered in Problem 5.2. Below, we briefly discuss the isotropic elasticity approach.

The free energy density in the presence of the magnetic field writes for $K_1 = K_2 = K_3 = K$ as

$$f = \frac{1}{2}K[(\text{div }\mathbf{n})^2 + (\text{curl }\mathbf{n})^2] - \frac{1}{2}\mu_0^{-1}\chi_a(\mathbf{B}\cdot\mathbf{n})^2, \tag{5.46}$$

or still,

$$f = \frac{1}{2}K\left\{n_{q,p}n_{q,p} + (n_p n_{q,q})_{,p} - (n_p n_{q,p})_{,q}\right\} - \frac{1}{2}\mu_0^{-1}\chi_a \mathcal{B}_i n_i \mathcal{B}_j n_j, \tag{5.47}$$

where we use the notation $\mathcal{B} = |\mathbf{B}|$. There are two divergence terms in (5.46):

$$(n_p n_{q,q})_{,p} = \text{div}\,(\mathbf{n}\cdot\text{div }\mathbf{n}), \qquad -(n_p n_{q,p})_{,q} = \text{div}\,(\mathbf{n}\times\text{curl }\mathbf{n})$$

[see also (5.4)], which play no role in the minimization of $F = \int f\,dV$ in the bulk, because, as repeatedly stated, they can be transformed into surface terms by virtue of the Gauss theorem. Hence, as long as we are not interested in surface terms, we can write

$$f = \frac{1}{2}K n_{q,p}n_{q,p} - \frac{1}{2}\mu_0^{-1}\chi_a(\mathbf{B}\cdot\mathbf{n})^2 = \frac{1}{2}K(\nabla\mathbf{n}\cdot\nabla\mathbf{n}^T) - \frac{1}{2}\mu_0^{-1}\chi_a(\mathbf{B}\cdot\mathbf{n})^2, \tag{5.48}$$

and the Euler–Lagrange equation takes a simple form

$$K\,\Delta\mathbf{n} + \mu_0^{-1}\chi_a(\mathbf{B}\cdot\mathbf{n})\cdot\mathbf{B} = \lambda\mathbf{n}. \tag{5.49}$$

5.4.1.3. Director Parameterization

A practical way to take into account the constraint $\mathbf{n}^2 = 1$ in the minimization problem is to parameterize the director through the polar θ and azimuthal φ angles

$$\mathbf{n} = (\sin\theta\cos\varphi, \sin\theta\sin\varphi, \cos\theta), \tag{5.50}$$

---

[20] C. Mauguin, Bull. Soc. Franç. Minér. Crist. **34**, 71 (1911).

instead of employing the Lagrange multiplier. Generally, both θ and φ depend on all three Cartesian coordinates, but the condition $\mathbf{n}^2 = 1$ is always satisfied. In the next section, we use the angular parametrization to find the equilibrium $\mathbf{n}(\mathbf{r})$ for a so-called hybrid-aligned nematic film.

5.4.2. Hybrid-Aligned Nematic Film

5.4.2.1. Fixed Boundary Conditions

A nematic slab (Fig. 5.7a) is confined between two flat plates $z = 0$ and $z = d$, at which the director orientation is strongly fixed by physicochemical conditions:

$$\theta(z = 0) = \bar{\theta}_0, \theta(z = d) = \bar{\theta}_d. \tag{5.51}$$

The director is assumed to lie in the vertical XZ plane, $\varphi = 0$:

$$\mathbf{n} = [\sin\theta(z), 0, \cos\theta(z)]. \tag{5.52}$$

Because (5.51) implies no twist, the assumption of isotropic elasticity, in this case reduced to $K_1 = K_3 = K$, is not far from reality, and we employ it here.

With θ depending only on z, the Frank–Oseen elastic energy per unit area is

$$F_{\text{FO}} = \int_{z=0}^{z=d} f_{\text{FO}}[\theta, \theta', z]\, dz, \tag{5.53}$$

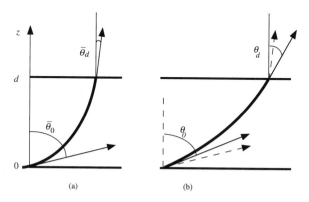

(a)   (b)

Figure 5.7. Hybrid aligned film: (a) fixed boundary conditions; (b) soft boundary conditions with director distortions relaxed at the expense of director deviations from the anchoring directions (5.50) at the boundaries.

where $\theta' = \frac{\partial \theta}{\partial z}$. With $K_1 = K_3 = K$, it simplifies to

$$F_{\text{FO}} = \frac{1}{2} K \int_{z=0}^{z=d} (\theta')^2 \, dz. \tag{5.54}$$

Because the integrand in (5.53) does not depend on θ explicitly, the Euler–Lagrange equation for bulk equilibrium (see Appendix A),

$$\frac{\partial f_{\text{FO}}}{\partial \theta} - \frac{d}{dz} \frac{\partial f_{\text{FO}}}{\partial \theta'} = 0, \tag{5.55}$$

reduces to $\theta'' = 0$, with the solution $\theta(z) = c_1 z + c_2$. The two constants c_1 and c_2 are defined from the boundary conditions (5.50) so that the equilibrium director configuration is found as

$$\theta(z) = (\bar{\theta}_d - \bar{\theta}_0)\frac{z}{d} + \bar{\theta}_0. \tag{5.56}$$

The elastic energy per unit area of a nematic film with director configuration (5.55),

$$F_{\text{FO}} = \frac{1}{2} K \frac{(\bar{\theta}_0 - \bar{\theta}_d)^2}{d}, \tag{5.57}$$

depends on the difference $|\bar{\theta}_0 - \bar{\theta}_d|$ in director orientation at the boundaries.

5.4.2.2. Soft Boundary Conditions

Deriving (5.55), we assumed that the surface interactions were infinitely strong and kept the director surface orientation fixed and independent of the elastic distortions in the bulk [see (5.50)]. In reality, molecular interactions at the plates are of finite strength. To reduce the total energy, strong elastic distortions in the bulk can be relaxed by deviating the director from the surface axes $\bar{\theta}_0$ and $\bar{\theta}_d$. The energy carried by these surface deviations is described by an "anchoring potential" f_s that is a function of the magnitude of deviations; e.g., $f_{s0}(\theta_0 - \bar{\theta}_0)$. The free energy per unit area of the hybrid cell becomes

$$F = \int_0^d f_{\text{FO}}[\theta, \theta'] \, dz + f_{s0}(\theta_0 - \bar{\theta}_0) + f_{sd}(\theta_d - \bar{\theta}_d). \tag{5.58}$$

The find the equilibrium $\theta(z)$ that minimizes the functional (5.57), one still starts with the Euler–Lagrange equation (5.54), which is again of the form $\theta'' = 0$. Its solution can be conveniently written as [compare to (5.55)]

$$\theta(z) = (\theta_d - \theta_0)\frac{z}{d} + \theta_0. \tag{5.59}$$

The two constants of integration θ_0 and θ_d are the actual polar angles of the director at the two substrates, different from $\bar{\theta}_0$ and $\bar{\theta}_d$, as these angles are defined by the balance of elastic and anchoring forces through the new boundary conditions (see Appendix A):

$$\left[-\frac{\partial f_{FO}}{\partial \theta'} + \frac{df_{s0}}{d\theta}\right]_{z=0} = 0 \quad \text{and} \quad \left[\frac{\partial f_{FO}}{\partial \theta'} + \frac{df_{sd}}{d\theta}\right]_{z=d} = 0. \qquad (5.60)$$

Suppose that deviations from the anchoring directions $\bar{\theta}_0$ and $\bar{\theta}_d$ are small so that only the first term can be preserved in the series expansion of the anchoring potentials:

$$f_{si} = \frac{1}{2}W_i(\theta_i - \bar{\theta}_i)^2, \quad i = 0, d. \qquad (5.61)$$

Here, W is the so-called anchoring coefficient characteristic of a given liquid crystal–substrate pair (Section 13.2.2). The approximation (5.60) works well when the cell is thick, $d \gg K/W_i$. The boundary conditions (5.59) then reduce to

$$K(\theta_0 - \theta_d) + W_0 d(\theta_0 - \bar{\theta}_0) = 0 \quad \text{and} \quad K(\theta_d - \theta_0) + W_d d(\theta_d - \bar{\theta}_d) = 0 \qquad (5.62)$$

and yield

$$\theta_0 = \bar{\theta}_0 - \frac{L_0}{d + L_0 + L_d}(\bar{\theta}_0 - \bar{\theta}_d) \quad \text{and} \quad \theta_d = \bar{\theta}_d + \frac{L_d}{d + L_0 + L_d}(\bar{\theta}_0 - \bar{\theta}_d).$$

$$(5.63)$$

Here, $L_0 = K/W_0$ and $L_d = K/W_d$ have the dimension of a length and are known as the *anchoring extrapolation lengths*. As easy to see, the finite anchoring allows the director distortions to relax so that the total energy

$$F = \frac{1}{2}K\frac{(\bar{\theta}_0 - \bar{\theta}_d)^2}{d + L_0 + L_d} \qquad (5.64)$$

is reduced compared with that of (5.56).

Equation (5.63) interprets the effect of anchoring as an effective increase of the film thickness, $d \rightarrow (d + L_0 + L_d)$. Hence, the weakening of distortions,

$$|\theta'| = |\theta_d - \theta_0|/d < |\bar{\theta}_d - \bar{\theta}_0|/d$$

(Fig. 5.7b), and the decrease of the total energy. Geometrical interpretation of extrapolation lengths L_0 and L_d and their effect on director configuration is shown in Fig. 5.8.

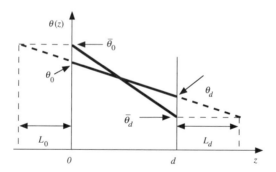

Figure 5.8. Geometrical interpretation of the anchoring extrapolation lengths.

As an order of magnitude, take $W \sim \frac{U_{SN}}{a^2}$, where U_{SN} is the orientational part of the energy of interaction of a molecule with the substrate and a is a molecular length. Then,

$$L = \frac{K}{W} \sim a \frac{U}{U_{SN}}. \tag{5.65}$$

The situation when L is of the order of a few molecular lengths will be referred to as *strong anchoring*: $W \sim (10^{-3} - 10^{-2})\,\mathrm{J/m^2} = (1 - 10)\,\mathrm{erg/cm^2}$. *Weak anchoring* refers to L in the "optical range": 0.1–1 μm. Experimental situations are in the range $W \sim (10^{-6} - 10^{-3})\,\mathrm{J/m^2}$.

5.4.2.3. Balance of Torques

Coming back to the surface equations (5.59), we wish to interpret them in terms of balance of torques. For the sake of simplicity, we again use $K_1 = K_3 = K$, and start, as above, from the expression of $\delta F = \delta F_{\mathrm{vol}} + \delta F_{\mathrm{surf}}$:

$$\delta F_{\mathrm{vol}} = -K \int \frac{\partial^2 \theta}{\partial z^2} \delta\theta \, dz; \tag{5.66}$$

$$\delta F_{\mathrm{surf}} = -K \left(\frac{\partial \theta}{\partial z}\right)_0 \delta\theta_0 + K \left(\frac{\partial \theta}{\partial z}\right)_d \delta\theta_d$$
$$+ W_0(\theta_0 - \bar{\theta}_0)\,\delta\theta_0 + W_d(\theta_d - \bar{\theta}_d)\,\delta\theta_d, \tag{5.67}$$

where $\delta\theta = \alpha\eta(z)$ is the virtual director rotation, as before. We recognize in δF_{surf} the products of surface torques by angles of rotation; for example, $W_0(\theta_0 - \bar{\theta}_0)$ is the external torque exerted by the bottom substrate on the nematic. Because $W_0(\theta_0 - \bar{\theta}_0) - K(\frac{\partial\theta}{\partial z})_0 = 0$, then $K(\frac{\partial\theta}{\partial z})_0$ must be interpreted as the torque exerted by the internal stresses on the director at $z = 0$. This quantity is exactly $C_y = \frac{\partial f}{\partial n_{x,z}} n_z - \frac{\partial f}{\partial n_{z,z}} n_x$ of (5.44) with $\boldsymbol{\nu} = (0, 0, -1)$. Similarly, $-K(\frac{\partial \theta}{\partial z})_d$ is the torque exerted by the internal stresses on the director at $z = d$, where $\boldsymbol{\nu} = (0, 0, 1)$. Note that these two elastic torques are exactly opposite. Consider a slice of material between z and $z + \delta z$. Clearly, the total variation of the energy of this slice $\delta F \mid_z^{z+dz}$ is

$$\delta F \bigg|_z^{z+dz} = -K \left(\frac{\partial \theta}{\partial z} \right)_z \delta \theta_z + K \left(\frac{\partial \theta}{\partial z} \right)_{z+\delta z} \delta(\theta_{z+\delta z}). \tag{5.68}$$

Because $\frac{\partial^2 \theta}{\partial z^2} = 0$ by minimizing δF_{vol}, $\frac{\partial \theta}{\partial z}$ is a constant. Hence, the torque exerted by the internal stresses generated in the region below z on the surface director $\boldsymbol{\nu} = (0, 0, -1)$ is $K(\frac{\partial \theta}{\partial z})$. Similarly, the torque exerted by the internal stresses generated in the region below $z + \delta z$ on the surface director above $z + \delta z$, $\boldsymbol{\nu} = (0, 0, -1)$, is still $K(\frac{\partial \theta}{\partial z})$: *Nematics transmit torques*. We shall come back to this concept of "stress," which has been alluded to here.

5.4.3. External Field Effects: Characteristic Lengths and Frederiks Transitions

In the geometry of Fig. 5.9, depicting a semi-infinite nematic volume, we assume that the anchoring is infinitely strong at the bounding plane $z = 0$ so that $\mathbf{n} = (1, 0, 0)$ is the boundary condition for $z = 0$. The magnetic field $\mathbf{B} = (0, \mathcal{B}, 0)$ tends to twist the director. The Euler–Lagrange equation is

$$K_2 \frac{\partial^2 \theta}{\partial z^2} + \mu_0^{-1} \chi_a \mathcal{B}^2 \sin \theta \cos \theta = 0 \tag{5.69}$$

(see Problem 5.2). The same equation follows from the balance of torques exerted on a slice of material $z, z + dz$. The total torque exerted by the internal stresses on the material is

$$\left(0, 0, K_2 \frac{\partial^2 \theta}{\partial z^2} \right). \tag{5.70}$$

Similarly, the total torque exerted by the magnetic field is

$$(\mathbf{M} \times \mathbf{H}) = \left(0, 0, \mu_0^{-1} \chi_a \mathcal{B}^2 \sin \theta \cos \theta \right), \tag{5.71}$$

directed along the z-axis, hence (5.68).

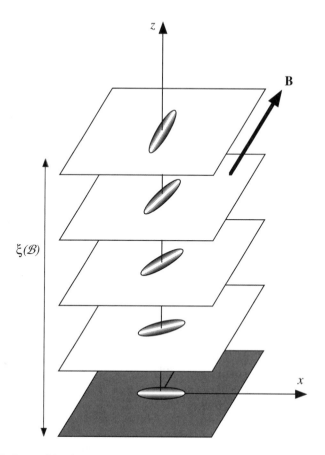

Figure 5.9. Competition between strong planar anchoring and an applied magnetic field.

A characteristic length ξ, called the magnetic coherence length, appears in (5.68):

$$\xi = \frac{1}{B}\sqrt{\frac{K_2}{\mu_0^{-1}\chi_a}}. \tag{5.72}$$

Integrating (5.68) once, one gets

$$\xi \frac{\partial \theta}{\partial z} = \pm \cos \theta, \tag{5.73}$$

where the constant of integration has been chosen to obey $\theta = \pm\frac{\pi}{2}$, $\frac{\partial \theta}{\partial z} = 0$ for $z \to \infty$. Let us now look for the solution of (5.72), which satisfies $\theta = 0$ for $z = 0$. One gets two solutions:

$$z = +\infty \qquad \theta = \frac{\pi}{2} \qquad \tan\left(\frac{\theta}{2} + \frac{\pi}{4}\right) = \exp\frac{z}{\xi} \tag{5.74a}$$

$$z = -\infty \qquad \theta = -\frac{\pi}{2} \qquad \tan\left(\frac{\theta}{2} + \frac{\pi}{4}\right) = \exp\frac{-z}{\xi}, \tag{5.74b}$$

which differ only by the sign of the twist. The director rotates helically from $\theta = 0$ to $\theta = \pm\frac{\pi}{2}$ (these two boundary conditions are physically the same) either along a right-handed helix or a left handed one. Note the typical values: For $K_2 = 10^{-11}$ N, $\chi_a = 10^{-6}$ (SI units), and $B = 1$T, one obtains $\xi \approx 3\mu$m, i.e., a macroscopic length. The effect of

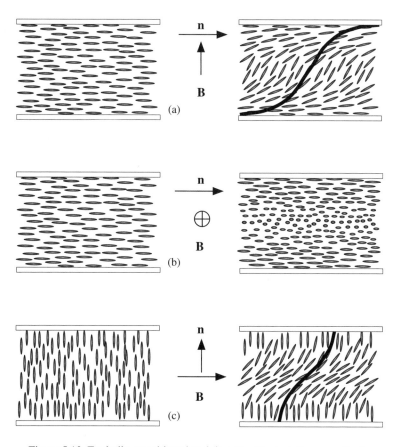

Figure 5.10. Frederiks transitions involving K_1, K_2, and K_3, respectively.

the strong anchoring on $z = 0$ adverse to the magnetic field is cured on a macroscopic, observable length.

One expects, more generally, that any perturbation of the director by the magnetic field has a typical range $\xi \cong \frac{1}{B}\sqrt{\frac{K_i}{\mu_0^{-1}\chi_a}}$, where K_i is the relevant Frank coefficient or a combination of Frank coefficients.

An important application is the Frederiks transition (see Fig. 5.10 and Problem 5.8a). A nematic slab is located between two parallel slides at a distance d and is strongly anchored ($W \to \infty$), so that **n** is oriented unidirectionally in the plane of the cell. The magnetic field is applied normally to the plates to disturb this uniform orientation ($\chi_a > 0$). When the magnetic field reaches a value $\mathcal{B}_c$ such that $\xi(\mathcal{B}_c) = \frac{d}{\pi}$, the nematic shows a transition toward a nonuniform state. This transition is reminiscent of a second-order Landau phase transition, with $\mathcal{B}$ playing the role of temperature; the order parameter is the angular deviation of the director from the uniform unperturbed orientation, measured at some z different from 0 and d, in the middle of the sample (say, at $z = d/2$).

The three sample geometries of Fig. 5.10 allow for the measurement of K_1, K_2, and K_3 respectively, according to the formula:

$$\mathcal{B}_{ci} = \frac{\pi}{d}\sqrt{\frac{K_i}{\mu_0^{-1}\chi_a}}, \tag{5.75}$$

where $\mathcal{B}_{ci}$ is the critical field above which the order parameter differs from zero. The formula is derived in the approximation of infinitely strong surface anchoring; finite anchoring decreases $\mathcal{B}_{ci}$ (Section 13.2.3).

5.5. Standard Applications of the Elasticity of Smectics

5.5.1. Smectic Phase with Small Deformations

We consider in this section a smectic phase that is only but slightly deformed with respect to the planar ground state. The distortions are described by a displacement function $u(x, y, z)$, and the free energy density is given by (5.27). The relating Euler–Lagrange differential equation for the function $u(x, y, z)$ can be written as

$$K_1 \Delta_\perp^2 u - B\frac{\partial^2 u}{\partial z^2} = 0, \tag{5.76}$$

where $\Delta_\perp^2 = (\frac{\partial^2}{\partial x^2} + \frac{\partial^2}{\partial y^2})^2$. We notice that the two terms of (5.75) differ in the following: The curvature elasticity term contains partial derivatives of the fourth order; on the other hand, the position elasticity term contains only second order derivatives, as in classic elasticity. The presence of this curvature term yields a remarkable elasticity.

5.5.1.1. Long-Range Effect of Layers Fluctuations

Consider a semi-infinite smectic, limited by a plane at $z = 0$, occupying the $z < 0$ half-space, whose boundary displays a static deformation $u(x, y, 0)$ that we Fourier analyze as

$$u(x, y, 0) = \int u_q \exp i\mathbf{q} \cdot \mathbf{r} d^2\mathbf{q}, \quad \mathbf{q} = (q_x, q_y, 0). \tag{5.77}$$

Because (5.75) is linear in $u(x, y, z)$, the Fourier components are independent. Consider, therefore, only one q_x component that we note $u = u_q \cos qx$, and look for a solution of (5.75) of the type

$$u = u_q \cos qx \exp z/L, \tag{5.78}$$

which obeys the boundary conditions $u(x = 0) = u_q \cos qx$ and $u(z \to -\infty) = 0$. We easily find

$$L = \sqrt{\frac{B}{K_1}} q^{-2} = \frac{1}{\lambda q^2}. \tag{5.79}$$

Let $\Lambda \cong q^{-1}$ be a typical length of the deformation of the boundary. Equation (5.78) tells one that the long-range effect of the perturbation is much larger than is Λ ($L \cong \Lambda \frac{\Lambda}{\lambda}$): λ is a microscopic length; thus, $L \gg \Lambda$. This is at variance with what happens in an usual solid, where a surface perturbation of typical size Λ is felt over a distance $L \cong \Lambda$ inside the material, i.e., over a volume $\approx \Lambda^3$; it is at once clear that this difference takes its origin in the order of the Euler differential equation. The result $L \cong \Lambda$ for solids was long known as a "principle" of elasticity theory, the principle of Saint-Venant. Layered media with curvature elasticity do not obey the principle of Saint-Venant.

5.5.1.2. Undulation Instability of Dilatation

Consider a macroscopic smectic specimen inserted between two strictly parallel rigid plates. The layers are parallel to the plates, and the anchoring is taken infinite for convenience. The system is assumed to be devoid of stresses when the gap between the plates is D. Increase now the gap by a small quantity Δ such that $D \to D + \Delta$, without increasing the number of layers. In a pure elastic process, the defomation is a uniform dilatation of layers described by a displacement function $u(z) = \alpha z$, $\alpha = \frac{\Delta}{D}$. The layers store (positional) elastic energy $\frac{1}{2} B \alpha^2 D = \frac{1}{2} B \frac{\Delta^2}{D}$ per unit area of plate. However, above some critical value Δ_c of Δ, a nonuniform curvature deformation would have less energy than would uniform dilatation. We want to calculate Δ_c and the wavelength of the curvature deformation, assuming, again for convenience, that it is a 1D deformation (Fig. 5.11a,b).

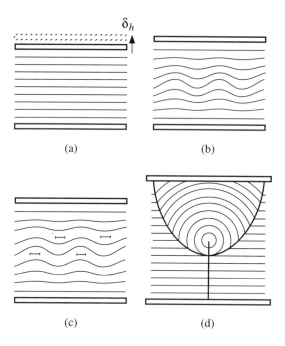

δ_h

(a)   (b)

(c)   (d)

Figure 5.11. Relaxation of a (a) dilated smectic sample by (b) undulation of layers; further dilation can cause (c) nucleation of dislocations and, eventually, (d) focal conic domains.

The total deformation corresponding to periodic undulation of layers can be written as:

$$u = \alpha z + u_0 \cos qx \sin kz, \qquad (5.80)$$

where u_0 is small. In order to calculate the amplitude of undulations, it appears that we have to introduce the nonlinear terms in the free energy density [see Problem 5.4 and (5.28)] as

$$f = \frac{1}{2} K_1 \left(\frac{\partial^2 u}{\partial x^2} \right)^2 + \frac{1}{2} B \left[\left(\frac{\partial u}{\partial z} \right) - \frac{1}{2} \left(\frac{\partial u}{\partial x} \right)^2 \right]^2. \qquad (5.81)$$

Calculating the total energy per unit length in the x-direction, i.e., $F = \frac{q}{2\pi} \int_0^{2\pi/q} f \, dx$ $\int_0^D dz$, with $D = \frac{\pi}{k}$, one gets an expression in u_0 and α that has to be minimized with respect to u_0. Note that the minimization of F with respect to α is meaningless because the expression of F does not show the external action that displaces the plates, one with respect to the other. Considering only the leading term $F \propto u_0^2$, one gets

$$\frac{4}{D} \frac{\partial F}{\partial u_0} = (K_1 q^4 + Bk^2 - B\alpha q^2) u_0, \qquad (5.82)$$

which yields two solutions when $\frac{\partial F}{\partial u_0} = 0$. The solution $u_0 = 0$ (pure dilatation) is stable as long as $\frac{\partial^2 F}{\partial u_0^2}$ is positive; i.e., $\alpha < \frac{k^2}{q^2} + \lambda^2 q^2$, where $\lambda = \sqrt{\frac{K_1}{B}}$. The solution $\alpha = \frac{k^2}{q^2} + \lambda^2 q^2$ yields $\frac{\partial^2 F}{\partial u_0^2} = 0$, but any relative displacement Δ of the plates such that $\alpha > \alpha_c$ yields $\frac{\partial^2 F}{\partial u_0^2} < 0$. In other words, the layers planar geometry is unstable against curvature distortions when $\alpha > \alpha_c$. One gets α_c by minimizing the expression $\alpha = \frac{k^2}{q^2} + \lambda^2 q^2$ with respect to q ($k = \frac{\pi}{D}$ is fixed):

$$\alpha_c = 2k\lambda = 2\pi \frac{\lambda}{D}. \tag{5.83}$$

The critical displacement Δ_c at $\alpha = \alpha_c$ is

$$\Delta_c = D\alpha_c = 2\pi\lambda, \tag{5.84}$$

which is independent of the sample thickness and λ is of the order of a molecular length. For $D = 100\mu m$, a typical value is $\alpha_c = 6 \times 10^{-5}$, which yields a stress $\sigma_c = B\alpha_c \approx 6 \times 10^2 \, \text{N/m}^2$. Thus, the mechanical instability appears for weak applied stresses.

When $\Delta < \Delta_c$, we have $u_0 = 0$. The calculation of u_0 above the threshold requires a higher order term $\propto u_0^4$ in the expansion of $F(u_0)$ [see (5.80)] and yields

$$u_0 = \frac{8}{3}\lambda \left(\frac{\alpha - \alpha_c}{\alpha_c} \right)^{1/2}, \tag{5.85}$$

a slow increase of u_0 with $\Delta > \Delta_c$. But as soon as u_0 reaches a value of the order of the layers thickness, a second mechanism of relaxation might appear, which is the nucleation of edge dislocations in the areas where the dilatation is maximum (Fig. 5.11c) (see Helfrich[21]). At still higher stresses, the dislocations transform into parabolic focal conic domains, which are described in Chapter 11.

The dilatation instability has been studied experimentally by elastic light scattering; it allows for a direct measurement of $\lambda = k/q_c^2$. For details, see Delaye et al.[22]

5.5.2. Smectic Phase with Large Deformations and Topological Deformations

The two examples studied above put on equal footings curvature elasticity $f_c = \frac{1}{2}K_1(\text{div } \mathbf{n})^2$ and "position" elasticity $f_p = \frac{1}{2}B\gamma^2$: One will easily check that they con-

---

[21] W. Helfrich, Appl. Phys. Lett. **17**, 531 (1970); J.P. Hurault, J. Chem. Phys. **59**, 2086 (1973).

[22] M. Delaye, R. Ribotta, and G. Durand, Phys. Lett. **A44**, 139 (1973); N.A. Clark and R.B. Meyer, Appl. Phys. Lett. **30**, 3 (1973).

tribute by equal amounts to the total deformation in the first example, and in the second example, above the instability threshold.

Scaling arguments show that this parity is not generic. Let L be a typical length of the deformation, and let one assume that the same typical length shows up in the three directions of space. The corresponding energies are

$$F_c \cong K_1 L \quad \text{and} \quad F_p \cong BL^3; \tag{5.86}$$

i.e., $F_c/F_p \cong (\lambda/L)^2$. Hence, it appears that the energy of curvatures is the larger the smaller L is, the contrary being true for position elasticity. One can therefore infer, a contrario, that in real situations, curvature distortions are predominant at large scales ($L \gg \lambda$), position distortions are predominant at small scales ($L < \lambda$), and they both contribute at microscopic scales ($L \cong \lambda$). This also points to the importance of instability processes at all scales, because, in the two limits indicated:

- At large scales, one expects $\gamma \equiv 0$, i.e., the layers being all of equal thickness and therefore parallel one to the other. Generally, the boundary conditions can be satisfied only by the appearance of large scale defects, which in this case are *focal conic domains*.

- At small scales, one expects $\sigma_1 + \sigma_2 \equiv 0$, i.e., the layers are curved in the shape of minimal surfaces. This constraint also yields specific instabilities, such as *screw dislocations*, as we shall see later.

This analysis does not take into account the K_{13} and K_{24} terms. As already mentioned, in lamellar phases, $\mathbf{n} \times \text{curl } \mathbf{n}$ is quasi zero; thus, the K_{13} and K_{24} terms of the form div $(\mathbf{n} \text{div } \mathbf{n})$ cannot be separated from each other. One expects that the corresponding modulus $\overline{K}$, which governs the Gaussian curvature term in (5.14) and should vary in the range $-2K < \overline{K} \le 0$ in the lamellar phase, might be of a reasonable value $\sim (-K)$. The energy contribution of $\overline{K}$ is a topological property of the volume under consideration, it takes a different value each time a strong deformation or a phase transition yield a change in the topology of the layers. Examples are drawn in Fig. 5.12. For instance, the geometry of Fig. 5.12a shows layers that are slightly deformed around the strict planar geometry. Each layer has bumps and hollows that create regions of opposite Gaussian curvatures. According to the fact that $\sigma_1 \sigma_2$ is a saddle-splay-like term, the total Gaussian curvature should be vanishing:

$$F_{24} \equiv \bar{K} \int \sigma_1 \sigma_2 \, d\Sigma = 0. \tag{5.87}$$

In Fig. 5.12b, the fluctuations of the sphere likewise sum up to zero, and the same integral can now be calculated on a perfect sphere, yielding

$$F_{24} \text{ (sphere)} = 4\pi \bar{K}, \tag{5.88}$$

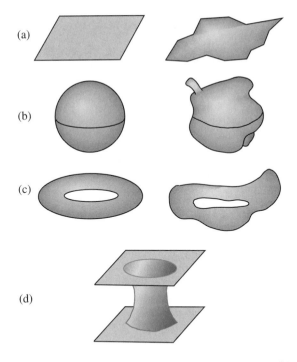

Figure 5.12. To illustrate the Gauss-Bonnet theorem; (a) a fluctuating layer with vanishing Gaussian curvature on the average, (b) a fluctuating sphere, (c) a fluctuating torus, (d) half-torus extended to infinity.

whereas, by the same argument, one gets on a full fluctuating torus (Fig. 5.12c):

$$F_{24} \text{ (torus)} = 0, \tag{5.89}$$

because regions of positive and negative Gaussian curvature compensate on the average. Note that the fluctuations are not restricted to be small: Any surface "homotopic" to a sphere, like, say, a very elongated ellipsoid, yields the same Gauss-Bonnet integral as the sphere. This topological property of the integral is true as well for the torus, or for any closed surface.

Consider now a half-torus of positive Gaussian curvature. It can be transformed into a sphere (by gluing two full circles along the lips of the half-torus) without changing the integral $\int \sigma_1 \sigma_2 \, d\Sigma = 4\pi$.

Therefore, the fluctuating half-torus of negative Gaussian curvature, extended to infinity, Fig. 5.12d, is such that

$$F_{24} \text{ (neg. half-torus)} = -4\pi \, \bar{K}. \tag{5.90}$$

Figure 5.13. A torus with two holes, $g = 2$, $\chi = -2$.

More generally, the Gauss-Bonnet theorem relates to the Gaussian curvature integral of a closed manifold to the *Euler characteristic* χ of the manifold:

$$\oint \sigma_1 \sigma_2 \, d\Sigma = 2\pi \chi. \tag{5.91}$$

This quantity is defined in Problem 5.5. Instead of the Euler characteristic, one also introduces the *genus*, which is the number of handles of a closed manifold: $g = 1$ for the torus, which is also a sphere with one handle, $g = 2$ for the two-holes torus (a pretzel, or a sphere with two handles, Fig. 5.13), and so on. We have the relation

$$\chi = 2(1 - g). \tag{5.92}$$

Coming back to physical considerations, one will notice that the sign of $\bar{K}$ plays an excessively important role in layered systems and, in particular, in lyotropics, where the favored presence of inner micelles or half-toruses of negative Gaussian curvature can lead to spectacular phase transitions. Important examples are the cubic phase and the sponge phase L_3, which are phases in which the bilayers have everywhere negative Gaussian curvature. They will be discussed later.

5.6. Thermodynamic Fluctuations

At thermodynamic equilibrium and at $T \neq 0$ K, any system suffers fluctuations of energy $\frac{1}{2}k_B T$ per degree of freedom. This is the principle of equipartition of energy. The amplitude of these fluctuations can be calculated easily, starting from an expansion of the free energy of the system in the Fourier components of the *phase* of the order parameter. It is indeed assumed that the amplitude of the order parameter is not affected by these fluctuations, as long as the system is far from any phase transition. Any uniform change of the phase component of the order parameter (i.e., at wavevector $\mathbf{q} = 0$) in a perfect system does not change the free energy. Consequently, one expects that in an elastically deformed system, long wavelength fluctuation modes of these "breaking-symmetry continuous variables" would be continuously excited by an energy as small as $\frac{1}{2}k_B T$ and would continuously disappear. This is equivalent to say that we study these fluctuations at a frequency of $\omega = 0$.

More will be said about their slow dynamics under the action of an external field (which would inject energies $\gg \frac{1}{2}k_B T$ per degree of freedom) in the next chapters.

5.6.1. Thermodynamic Fluctuations in Nematics

Nematics are turbid: They scatter light strongly, because of the spontaneous fluctuations of the director alignment.[23] The continuous theory of these fluctuations suffices to explain the observed effects.[24]

We assume that the sample is uniformly aligned along the z-axis. We are interested in the (small) fluctuations $\delta\mathbf{n} = (n_x(\mathbf{r}), n_y(\mathbf{r}), 0)$ of the director $\mathbf{n} \cong (n_x(\mathbf{r}), n_y(\mathbf{r}), 1)$. We expand $n_x(\mathbf{r}), n_y(\mathbf{r})$ in Fourier components (see Appendix B) and express the free energy density (5.2) as a function of these components. One gets, for example:

$$\text{div } \mathbf{n} = i V^{-1} \sum_{\mathbf{q}} \exp(i\mathbf{q}.\mathbf{r})[q_x n_x(\mathbf{q}) + q_y n_y(\mathbf{q})] \tag{5.93}$$

and

$$(\text{div } \mathbf{n})^2 = (\text{div } \mathbf{n}) \cdot (\text{div } \mathbf{n})^*$$
$$= V^{-2} \sum_{\mathbf{q},\mathbf{q}'} \exp[i(\mathbf{q} - \mathbf{q}') \cdot \mathbf{r}][q_x n_x(\mathbf{q}) + q_y n_y(\mathbf{q})][q'_x n_x(\mathbf{q}') + q'_y n_y(\mathbf{q}')]^*,$$

i.e.,

$$\int_V (\text{div } \mathbf{n})^2 d\mathbf{r} = V^{-1} \sum_{\mathbf{q}} \left| q_x n_x(\mathbf{q}) + q_y n_y(\mathbf{q}) \right|^2. \tag{5.94}$$

One eventually gets, after calculating all terms in the same way,

$$\int_V f_{\text{FO}} d\mathbf{r} = \frac{1}{2} V^{-1} \sum_{\mathbf{q}} \left\{ K_1 \left| q_x n_x(\mathbf{q}) + q_y n_y(\mathbf{q}) \right|^2 + K_2 \left| q_x n_y(\mathbf{q}) - q_y n_x(\mathbf{q}) \right|^2 \right.$$
$$\left. + K_3 q_z^2 \left(\left| n_x(\mathbf{q}) \right|^2 + \left| n_y(\mathbf{q}) \right|^2 \right) \right\}.$$

Let $\delta\mathbf{n}_2$ be the component along $\mathbf{q}$ of $\delta\mathbf{n}$ ($\delta\mathbf{n} = \delta\mathbf{n}_1 + \delta\mathbf{n}_2, \delta\mathbf{n}_1.\delta\mathbf{n}_2 = 0$). Clearly, $\left| q_x n_x(\mathbf{q}) + q_y n_y(\mathbf{q}) \right|^2 = (q_x^2 + q_y^2)\delta\mathbf{n}_2^2 = q_\perp^2 \delta\mathbf{n}_2^2$. Similarly, $q_x n_y(\mathbf{q}) - q_y n_x(\mathbf{q})$ is the z-component of the cross product $\delta\mathbf{n} \times \mathbf{q}$ times $q_\perp$; i.e., $\left| q_x n_y(\mathbf{q}) - q_y n_x(\mathbf{q}) \right|^2 = q_\perp^2 \delta\mathbf{n}_1^2$.

[23] Orsay Liquid Crystals Group, Phys. Rev. Lett. **22**, 1361 (1969).

[24] P.-G. de Gennes, C.R. Acad. Sci. Paris **266**, 15 (1968).

Therefore,

$$\int_V f_{\mathrm{FO}} d\mathbf{r} = \frac{1}{2} V^{-1} \sum_{\mathbf{q}} \left\{ q_\perp^2 (K_1 \delta \mathbf{n}_1^2 + K_2 \delta \mathbf{n}_2^2) + q_{\parallel}^2 K_3 (\delta \mathbf{n}_1^2 + \delta \mathbf{n}_2^2) \right\}. \qquad (5.95)$$

This diagonalization of $\int_V f_{\mathrm{FO}} d\mathbf{r}$ (cross terms have disappeared) allows for the application of the theorem of equipartition of energy and yields

$$\left\langle \delta \mathbf{n}_1^2 \right\rangle = V k_B T / (q_\perp^2 K_1 + q_{\parallel}^2 K_3), \qquad (5.96a)$$

$$\left\langle \delta \mathbf{n}_2^2 \right\rangle = V k_B T / (q_\perp^2 K_2 + q_{\parallel}^2 K_3), \qquad (5.96b)$$

where the brackets $\langle \, \rangle$ denote a thermal average.

These expressions are useful to compute the light intensity scattered by the fluctuations of orientation. The propagation of light is sensitive to the electric field radiated by the microscopic dipoles, which are induced on the molecules by the electric field of the incoming light. We shall not repeat here this analysis, which is developed in de Gennes and Prost textbook.

The equations above are easily extended to take into account the presence of a magnetic field. They are therefore useful to discuss the range of the correlations $\langle \mathbf{n}_i(\mathbf{r}) \mathbf{n}_i(\mathbf{r}') \rangle$ of the orientation. With $\mathbf{B}$ along the z-direction, one gets

$$\left\langle \delta \mathbf{n}_1^2 \right\rangle = V k_B T / \left(q_\perp^2 K_1 + q_{\parallel}^2 K_3 + \mu_0^{-1} \chi_a B^2 \right), \qquad (5.97a)$$

$$\left\langle \delta \mathbf{n}_2^2 \right\rangle = V k_B T / \left(q_\perp^2 K_2 + q_{\parallel}^2 K_3 + \mu_0^{-1} \chi_a B^2 \right). \qquad (5.97b)$$

Now, the calculation of the correlation functions

$$\langle \mathbf{n}_i(\mathbf{r}) \mathbf{n}_i(\mathbf{r}') \rangle = V^{-2} \sum_{\mathbf{q},\mathbf{q}'} \langle \mathbf{n}_i(\mathbf{q}) \mathbf{n}_i(-\mathbf{q}') \rangle \exp \left[i(\mathbf{q} \cdot \mathbf{r} - \mathbf{q}' \cdot \mathbf{r}') \right] \qquad (5.98)$$

is made easy by noticing that the Fourier components of the director are uncorrelated for different values of the wavevector $\mathbf{q}$, corresponding to different degrees of freedom. Hence,

$$\langle \mathbf{n}_i(\mathbf{q}) \mathbf{n}_i(-\mathbf{q}') \rangle = \left\langle | \mathbf{n}_i(\mathbf{q}) |^2 \right\rangle \delta_{\mathbf{q}+\mathbf{q}',0},$$

and

$$\langle \mathbf{n}_i(\mathbf{r}) \mathbf{n}_i(\mathbf{r}') \rangle = V^{-2} \sum_{\mathbf{q}} \left\langle | \mathbf{n}_i(\mathbf{q}) |^2 \right\rangle \exp \left\{ i \mathbf{q} \cdot (\mathbf{r} - \mathbf{r}') \right\}. \qquad (5.99)$$

The calculation proceeds by substituting the values $\langle \delta \mathbf{n}_1^2 \rangle$ and $\langle \delta \mathbf{n}_2^2 \rangle$ calculated above in the quantities $\langle \mathbf{n}_i(\mathbf{r})\mathbf{n}_i(\mathbf{r}') \rangle$, and it yields eventually to a correlation of the form (assuming that the three Frank coefficients are equal):

$$\langle \mathbf{n}_i(0)\mathbf{n}_i(\mathbf{R}) \rangle \propto \frac{k_B T}{K R} \exp \left\{ -\frac{R}{\xi} \right\}, \tag{5.100}$$

where $\xi = \frac{1}{B} \sqrt{\frac{K}{\mu_0^{-1} \chi_a}}$ is the magnetic coherence length introduced in (5.71). The correlations decrease slowly with distance at zero field.

5.6.2. Thermodynamic Fluctuations in Smectics

The calculation of the thermal fluctuations of an undeformed smectic proceeds along the same lines. One gets, starting from (5.27),

$$\left\langle |u(\mathbf{q})|^2 \right\rangle = \frac{k_B T}{B q_z^2 + K q_\perp^2 \left(q_\perp^2 + \xi^{-2} \right)}, \tag{5.101}$$

where a contribution from a magnetic field along the z-direction has been introduced (the ξ dependent term). The mean square fluctuation can be easily calculated from this expression:

$$\left\langle u^2(0) \right\rangle = \frac{k_B T}{4\pi \sqrt{B K}} \ln \frac{\xi}{d_0}, \tag{5.102}$$

where d_0 is the lower cutoff, taken equal to the repeat distance. As the magnetic field decreases to zero, the logarithmic term diverges; in fact for a sample of size L, and in zero field, the last expression is replaced by

$$\left\langle u^2(0) \right\rangle = \frac{k_B T}{4\pi \sqrt{B K}} \ln \frac{L}{d_0}. \tag{5.103}$$

The mean-squared thermal fluctuation of the layers diverges. This is the so-called Peierls-Landau instability characteristic of systems with 1D positional order. Despite the fact that the divergence is substantial only for very large systems (with $K = 10^{-11}$ N, $k_B T = 4 \times 10^{-21}$ J $= 4 \times 10^{-14}$ erg, $\sqrt{K/B} \sim d_0 \sim 1$ nm, one obtains $\sqrt{\langle u^2(0) \rangle} \sim d_0$ only when $L \sim 10$ km), we conclude that smectics do not possess true long-range translational order.

The X-ray structure factor shows up a power-law decay, whereas it is a delta peak in 3D systems. The scattering intensity is the Fourier transform of the Debye–Waller factor[25]

---

[25] A. Caillé, C.R. Acad. Sci. Paris **274B**, 891 (1972).

$$G(\mathbf{R}) = \exp\left[-\frac{1}{2}q_z^2 \left\langle |u(\mathbf{R}) - u(0)|^2 \right\rangle\right],$$
(5.104)

where the displacement correlation function writes:

$$\left\langle |u(\mathbf{R}) - u(0)|^2 \right\rangle = \frac{2}{(2\pi)^3} \int d\mathbf{q} \left\langle |u(\mathbf{q})|^2 \right\rangle [1 - \exp(-i\mathbf{q}\cdot\mathbf{R})]$$

$$\cong \frac{k_B T}{4\pi\sqrt{KB}} \times \begin{bmatrix} \ln R_{||}, & R_\perp = 0, \\ \ln R_\perp^2, & R_{||} = 0. \end{bmatrix}$$
(5.105)

The calculation of the scattered intensity yields

$$I(\mathbf{q}) \propto \begin{bmatrix} |q_z - q_0|^{-2+\eta}, & q_\perp = 0, \\ |q_\perp|^{-4+2\eta}, & q_z = 0, \end{bmatrix}$$
(5.106)

with $\eta = \frac{k_B T q_0^2}{8\pi\sqrt{KB}}$, $q_0 = 2\pi/d_0$.

5.A. Appendix A: One-Dimensional Variational Problem

Minimization of the functional of the type (5.52),

$$F = \int_{z=0}^{z=d} f[\theta, \theta', z]\, dz,$$
(5A.1)

is the simplest standard problem of variational analysis (see Arfken [18]), because the integrand depends only on one unknown function $\theta(z)$, and this function depends only on one independent variable z. Below we consider both fixed boundary conditions

$$\theta(z = 0) = \bar{\theta}_0, \quad \theta(z = d) = \bar{\theta}_d$$
(5A.2)

and soft boundary conditions, without restriction on the particular physical content of the problem.

5.A.1. Fixed Boundary Conditions

All possible functions $\theta(z)$ can be represented by paths that connect two points $[z = 0, \theta = \bar{\theta}_0]$ and $[z = d, \theta = \bar{\theta}_d]$ (Fig. 5.14a). Let $\theta_{eq}(z)$ be the path that minimizes (5A.1). Any

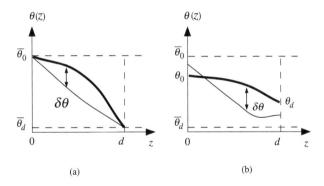

Figure 5.14. A hypothetical function (thick line) that minimizes the energy functional (5A.1) for (a) fixed boundary conditions and the functional (5A.11) for (b) soft boundary conditions. The thin line shows any other function close to the equilibrium one.

other close path $\theta(z)$ can be written as $\theta_{eq}(z) + \delta\theta(z)$, where $\delta\theta(z)$ is a virtual director rotation equivalent to the arbitrary variation δn_i in the general scheme (Section 5.4.1). The variation $\delta\theta(z)$ is conveniently represented as a product $\alpha\eta(z)$, where the new arbitrary function $\eta(z)$ describes the deformation of the path and the small scale parameter α describes the magnitude of variation:

$$\theta(z) = \theta_{eq}(z) + \alpha\eta(z). \tag{5A.3}$$

The function $\eta(z)$ must be differentiable and vanish at the fixed endpoints, $\eta(z = 0) = \eta(z = d) = 0$. The free energy functional for the path $\theta(z)$ is

$$F[\theta(z)] = \int_0^d f[\theta_{eq}(z) + \alpha\eta(z), \theta'_{eq}(z) + \alpha\eta'(z), z]\,dz. \tag{5A.4}$$

The extremum of the last integral, considered as a function of α, obtains when

$$\left[\frac{\partial F(\alpha)}{\partial\alpha}\right]_{\alpha=0} = 0. \tag{5A.5}$$

Differentiation of $F(\alpha)$ with respect to α gives

$$\frac{\partial F(\alpha)}{\partial\alpha} = \int_0^d \left[\frac{\partial f}{\partial\theta}\frac{\partial\theta}{\partial\alpha} + \frac{\partial f}{\partial\theta'}\frac{\partial\theta'}{\partial\alpha}\right] dz = \int_0^d \left[\frac{\partial f}{\partial\theta}\eta(z) + \frac{\partial f}{\partial\theta'}\frac{d\eta(z)}{dz}\right] dz.$$

Note that the two α-dependent functions θ and θ' are treated as independent variables. Integrating the second term by parts and recalling that $\eta(z = 0) = \eta(z = d) = 0$,

$$\int_0^d \frac{d\eta(z)}{dz} \frac{\partial f}{\partial \theta'} \, dz = \underbrace{\eta(z) \frac{\partial f}{\partial \theta'} \Big|_{z=0}^{z=d}}_{=0} - \int_0^d \eta(z) \frac{d}{dz} \frac{\partial f}{\partial \theta'} \, dz, \tag{5A.6}$$

we obtain

$$\int_0^d \left[\frac{\partial f}{\partial \theta} - \frac{d}{dz} \frac{\partial f}{\partial \theta'} \right] \eta(z) \, dz = 0. \tag{5A.7}$$

Note the correspondence between (5A.5) and the requirement $\delta F = 0$ in the general scheme of Section 5.4.1.1. Multiplying (5A.7) by α, one gets

$$\int_0^d \left[\frac{\partial f}{\partial \theta} - \frac{d}{dz} \frac{\partial f}{\partial \theta'} \right] \delta\theta \, dz = \alpha \left[\frac{\partial F(\alpha)}{\partial \alpha} \right]_{\alpha=0} = \delta F = 0. \tag{5A.8}$$

The extremum condition, either in the form (5A.7) or (5A.8), can be satisfied only if

$$\frac{\partial f}{\partial \theta} - \frac{d}{dz} \frac{\partial f}{\partial \theta'} = 0, \tag{5A.9}$$

which is the differential Euler–Lagrange equation for our 1D problem. Its solution is a family of functions $\theta = \theta(z, c_1, c_2)$ with two constants of integration determined from the two boundary conditions (5A.2).

When f does not depend on z explicitly, the first integral of the Euler–Lagrange equation (5A.9) can be found immediately as

$$\theta' \frac{\partial f}{\partial \theta'} - f = \text{const} \tag{5A.10}$$

(see Problem 5.3).

5.A.2. Soft Boundary Conditions

The energy to minimize contains surface terms:

$$F = \int_0^d f[\theta, \theta', z] \, dz + f_{s0}(\theta_0 - \bar\theta_0) + f_{sd}(\theta_d - \bar\theta_d). \tag{5A.11}$$

Again, the difference $\delta\theta = \theta(z) - \theta_{\text{eq}}(z) = \alpha\eta(z)$ can be characterized by a new function $\eta(z)$ that is not necessarily zero at the boundaries (Fig. 5.14b). Because

$$\frac{df_{s0}(\theta_0)}{d\alpha} = \frac{d}{d\alpha} f_{s0}[\theta_{0,\text{eq}} + \alpha\eta(z=0)] = \eta(0)\frac{df_{s0}}{d\theta_0},$$

then

$$\frac{\partial F(\alpha)}{\partial\alpha} = \int_0^d \left[\frac{\partial f}{\partial\theta}\frac{\partial\theta}{\partial\alpha} + \frac{\partial f}{\partial\theta'}\frac{\partial\theta'}{\partial\alpha}\right] dz + \frac{df_{s0}}{d\theta_0}\eta(0) + \frac{df_{sd}}{d\theta_d}\eta(d). \tag{5A.12}$$

Integration of the second term in the volume integral by parts yields the extremum condition in the form

$$\int_0^d \left[\frac{\partial f}{\partial\theta} - \frac{d}{dz}\frac{\partial f}{\partial\theta'}\right]\eta(z)\,dz + \left[-\frac{\partial f}{\partial\theta'} + \frac{df_{s0}}{d\theta}\right]_{z=0}\eta(0)$$

$$+ \left[\frac{\partial f}{\partial\theta'} + \frac{df_{sd}}{d\theta}\right]_{z=d}\eta(d) = 0. \tag{5A.13}$$

Therefore, the Euler–Lagrange equation (5.54) should still be used to find the family of solutions in the form $\theta = \theta(z, c_1, c_2)$, but now the two integration constants c_1 and c_2 are defined from the boundary conditions in the form

$$\left[-\frac{\partial f}{\partial\theta'} + \frac{df_{s0}}{d\theta}\right]_{z=0} = 0 \quad \text{and} \quad \left[\frac{\partial f}{\partial\theta'} + \frac{df_{sd}}{d\theta}\right]_{z=d} = 0. \tag{5A.14}$$

One can also insert the function $\theta = \theta(z, c_1, c_2)$ into F (5A.11), integrate it, and then minimize the integral with respect to c_1 and c_2.

5.B. Appendix B: Formulae for Fourier Transforms

Let $m(\mathbf{r})$ be a continuous observable defined in a volume V. Most generally, it is assumed that this volume is a parallelopiped of edges L_1, L_2, L_3, with periodic boundary conditions. Hence, $V = L_1L_2L_3$, with

$$m(x, y, z) = m(x + L_1, y, z) = \cdots. \tag{5B.1}$$

The following formulae are of constant use:

$$m(\mathbf{r}) = V^{-1}\sum_{\mathbf{q}}\exp(i\mathbf{q}\cdot\mathbf{r})m(\mathbf{q}), \tag{5B.2a}$$

$$m(\mathbf{q}) = \int_V d\mathbf{r}\exp(-i\mathbf{q}\cdot\mathbf{r})m(\mathbf{r}), \tag{5B.2b}$$

$$\sum_{\mathbf{q}} \exp[-i\mathbf{q} \cdot (\mathbf{r} - \mathbf{r}')] = V\delta(\mathbf{r} - \mathbf{r}'), \qquad (5B.2c)$$

$$\int_V d\mathbf{r} \exp[i\mathbf{r} \cdot (\mathbf{q} - \mathbf{q}')] = V\delta_{\mathbf{qq}'} = (2\pi)^3\delta(\mathbf{q} - \mathbf{q}'), \qquad (5B.2d)$$

where

$$\mathbf{q} = \left(\frac{2\pi}{L_1}n_1, \frac{2\pi}{L_2}n_2, \frac{2\pi}{L_3}n_3\right)$$

is the set of reciprocal wavevectors such that (5B.1) is satisfied (n_1, n_2, n_3 are integers); $\delta_{\mathbf{qq}'}$ is the Kronecker delta, equal to 1 when $\mathbf{q} = \mathbf{q}'$ and 0 otherwise; $\delta(\mathbf{q} - \mathbf{q}')$ is the 3D Dirac delta function.

In the $V \to \infty$ limit, the sum over $\mathbf{q}$ is transformed into an integral:

$$\sum_{\mathbf{q}} \to V \int \frac{d\mathbf{q}}{(2\pi)^3}, \qquad (5B.3)$$

and (5B.2a) and (5B.2c) become

$$m(\mathbf{r}) = \int \frac{d\mathbf{q}}{(2\pi)^3} \exp(i\mathbf{q} \cdot \mathbf{r})m(\mathbf{q}), \qquad (5B.4)$$

$$\int d\mathbf{q} \exp[-i\mathbf{q} \cdot (\mathbf{r} - \mathbf{r}')] = (2\pi)^3\delta(\mathbf{r} - \mathbf{r}'). \qquad (5B.5)$$

All of these expressions are written in dimension 3. The dimensionality appears in the above as the exponent of 2π. The extension to any dimension is straightforward.

Problem 5.1. Prove that $\mathbf{n} \times \text{curl}\,\mathbf{n}$ is along the principal normal of the envelope of the director and has modulus $1/r$, i.e., equal to the curvature.

Answers: In components notations, $\mathbf{n} \times \text{curl}\,\mathbf{n}$ can be written as $(\mathbf{n} \times \text{curl}\,\mathbf{n})_i = \varepsilon_{ijk}n_j\varepsilon_{kpq}n_{q,p}$. Because $\varepsilon_{ijk}\varepsilon_{kpq} = \delta_{ip}\delta_{jq} - \delta_{iq}\delta_{jp}$, one gets $(\mathbf{n} \times \text{curl}\,\mathbf{n})_i = -n_j n_{i,j}$. Hence, $\mathbf{n} \times \text{curl}\,\mathbf{n} = -\mathbf{n} \cdot \nabla\mathbf{n} = -\frac{d\mathbf{n}}{ds}$, where s is the curvilinear abscissa along the envelope of the directors oriented along the direction of $\mathbf{n}$. According to the Frenet formulae, one also has $\frac{d\mathbf{n}}{ds} = \frac{\nu}{r}$.

Problem 5.2. Planar Director Distortions. Assume that the nematic cell with the director configuration

$$\mathbf{n} = (n_x, n_y, 0) = (\cos\theta, \sin\theta, 0)$$

is subjected to a 2D magnetic field $\mathbf{B} = (\mathcal{B}_x, \mathcal{B}_y, 0)$. Here, $\theta = \theta(x, y, z)$ is the angle between the x-axis and the director. Write (a) the elastic free energy density and (b) the Euler–Lagrange equation for the problem, assuming $K_1 = K_3 = K$.

Answers:

(a) The elastic part of the free energy density is

$$f_{FO} = \frac{1}{2}(K_1 \sin^2 \theta + K_3 \cos^2 \theta) \left(\frac{\partial \theta}{\partial x} \right)^2 + \frac{1}{2}(K_1 \cos^2 \theta + K_3 \sin^2 \theta) \left(\frac{\partial \theta}{\partial y} \right)^2$$

$$+ \frac{1}{2}(K_3 - K_1) \sin 2\theta \frac{\partial \theta}{\partial x} \frac{\partial \theta}{\partial y} + \frac{1}{2}K_2 \left(q_0 - \frac{\partial \theta}{\partial z} \right)^2. \tag{5.106}$$

It takes a simpler form when $K_1 = K_3 = K$, so that the total free energy density that includes the contribution of the 2D magnetic field $\mathbf{B} = (\mathcal{B}_x, \mathcal{B}_y, 0)$ becomes

$$f = \frac{1}{2}K \left[\left(\frac{\partial \theta}{\partial x} \right)^2 + \left(\frac{\partial \theta}{\partial y} \right)^2 \right] + \frac{1}{2}K_2 \left(q_0 - \frac{\partial \theta}{\partial z} \right)^2$$

$$- \frac{1}{2}\mu_0^{-1} \chi_a (\mathcal{B}_x \cos \theta + \mathcal{B}_y \sin \theta)^2. \tag{5.107}$$

(b) The free energy density (5.107) integrates to a total energy of the form

$$\int f \left(\theta, \frac{\partial \theta}{\partial x}, \frac{\partial \theta}{\partial y}, \frac{\partial \theta}{\partial z} \right) dx \, dy \, dz.$$

Any variation of the orientation θ makes f, not the element of volume, vary. Therefore, the Euler–Lagrange equation can be written as

$$\frac{\partial f}{\partial \theta} - \left[\frac{\partial}{\partial x} \frac{\partial f}{\partial \theta_x} + \frac{\partial}{\partial y} \frac{\partial f}{\partial \theta_y} + \frac{\partial}{\partial z} \frac{\partial f}{\partial \theta_z} \right] = 0;$$

i.e.,

$$\mu_0^{-1} \chi_a \left[(\mathcal{B}_x^2 - \mathcal{B}_y^2) \sin 2\theta - 2\mathcal{B}_x \mathcal{B}_y \cos 2\theta \right]$$

$$- 2 \left[K \left(\frac{\partial^2}{\partial x^2} + \frac{\partial^2}{\partial y^2} \right) + K_2 \frac{\partial^2}{\partial z^2} \right] \theta = 0. \tag{5.108}$$

Problem 5.3. Prove that the first integral of the Euler–Lagrange equation for the functional (5.52) has the form (5A.10) when the integrand does not depend on the coordinate z explicitly.

Answers: Denote $G = \theta' \frac{\partial f}{\partial \theta'} - f$. Then,

$$\frac{dG}{dz} = \theta'' \frac{\partial f}{\partial \theta'} + \theta' \frac{d}{dz} \left(\frac{\partial f}{\partial \theta'} \right) - \frac{df}{dz}$$

$$= \theta'' \frac{\partial f}{\partial \theta_z} + \theta' \frac{d}{dz} \left(\frac{\partial f}{\partial \theta'} \right) - \frac{\partial f}{\partial \theta} \theta' - \frac{\partial f}{\partial \theta'} \theta''$$

$$= \theta' \left\{ \frac{d}{dz} \left(\frac{\partial f}{\partial \theta'} \right) - \frac{\partial f}{\partial \theta} \right\}$$

and $G = \text{const}$ if the Euler–Lagrange equation holds.

Problem 5.4. Show that in the smectic A phase, the dilatation of a layer can be written to the second order as

$$\gamma = \frac{d - d_0}{d_0} = \left(\frac{\partial u}{\partial z} \right) - \frac{1}{2} (\nabla_\perp u)^2 .$$

Answers: Let $z_0 = z - u(x, y, z)$ be the equation of a deformed layer of coordinate z_0 before deformation. This quantity varies by an amount d_0 when one goes from a deformed layer to the next, along a path of length d (the value of d_0 after deformation) along $\mathbf{n}$. Applying Taylor's expansion theorem to such a displacement, which brings a point on the z_0-layer to the $z_0 + d_0$-layer, one gets $z_0 + d_0 = z_0 + d\mathbf{n} \cdot \nabla z_0$. Hence,

$$\gamma = \frac{d - d_0}{d_0} = -1 + \frac{1}{|\nabla z_0|} \approx -1 + \frac{1}{n_z - \mathbf{n} \cdot \nabla u} .$$

The desired result follows from the expression of

$$\mathbf{n} = \left\{ -\partial u/\partial x (1 + \partial u/\partial z), \, -\partial u/\partial y (1 + \partial u/\partial z), \, 1 - \frac{1}{2} (\nabla_\perp u)^2 \right\}$$

written to the second order.

Problem 5.5. Starting from the Euler-Poincaré theorem demonstrated in Problem 2.3, Chapter 2, prove that for a polyhedron of genus g, one has the following generalization:

$$\chi_g = v - e + f = 2(1 - g).$$

Hint: cut the original polyhedron obeying the Euler-Poincaré theorem along two polygons having no edge in common, and introduce a polygonized handle joining the boundaries of these polygons. "Polygonize" it in such a manner that the polygons along the cut match along the edges and the vertices. It is then easy to show that the Euler characteristic $\chi_1 = 0$, independently of the way the polygonization of the handle has been performed.

The identity between this result and (5.91) requires one to delve deeply into the differential geometry of surfaces.

Problem 5.6. Starting from (5.22) $f_c = \frac{1}{2}\kappa(\sigma_1 + \sigma_2)^2 + \bar{\kappa}\sigma_1\sigma_2$, establish the conditions of stability, expressed as inequalities imposed on the curvature moduli κ and $\bar{\kappa}$, of the lamellar phase, the micellar phase, and the sponge phase.

Answers: Equation (5.22) can be written as the sum of two terms with positive expressions for the curvatures:

$$f_c = \frac{1}{2}\left[\left(\kappa + \frac{1}{2}\bar{\kappa}\right)(\sigma_1 + \sigma_2)^2 - \frac{1}{2}\bar{\kappa}(\sigma_1 - \sigma_2)^2\right].$$

(a) If both coefficients are positive, viz. $\kappa + \frac{1}{2}\bar{\kappa} > 0$, $\bar{\kappa} < 0$, the ground state is when $\sigma_1 + \sigma_2 = 0$, $\sigma_1 - \sigma_2 = 0$; i.e., it is the lamellar phase.

(b) If $\kappa + \frac{1}{2}\bar{\kappa} < 0$, $\bar{\kappa} < 0$, there is no ground state and the energy decreases without limit for $|\sigma_1 + \sigma_2|$ large, $|\sigma_1 - \sigma_2| = 0$. One, therefore, expects some stabilization for $G > 0$, e.g., the micellar phase.

(c) If $\kappa + \frac{1}{2}\bar{\kappa} > 0$, $\bar{\kappa} > 0$, there is no ground state and the energy decreases without limit for $\sigma_1 + \sigma_2 = 0$, $|\sigma_1 - \sigma_2|$ large, i.e., when the curvature $\sigma = |\sigma_1| = |\sigma_2|$ increases. The membrane is then a minimal surface. In reality, one expects some repulsive interactions stabilizing the system at some finite value of the curvature and nonzero value of $G < 0$, e.g., the sponge phase, due to other factors like Helfrich interactions (see Section 14.2.2).

(d) The case $\kappa + \frac{1}{2}\bar{\kappa} < 0$, $\bar{\kappa} > 0$ leads to a contradiction.

Problem 5.7. Show that the contribution of the divergence term to the total energy of the thermal fluctuations vanishes, employing the method of Fourier components of the text. Why was that result expected, in the nematic as well as in the smectic cases?

Problem 5.8. Show that the Frederiks transition in an SmA phase with strong planar anchoring, magnetic field **B** parallel to the layers, is hardly visible ("ghost" transition).[26]

Answers: Let us express the free energy density

$$f = \frac{1}{2}K(\mathrm{div}\,\mathbf{n})^2 + \frac{1}{2}B\left(\frac{d-d_0}{d_0}\right)^2 - \frac{1}{2}\mu_0^{-1}\chi_a(\mathbf{B}\cdot\mathbf{n})^2$$

as a function of the angle θ between the normal to the deformed layer and the unperturbed direction. We remind one that B and $\mathbf{B}$ (or $\mathcal{B} = |\mathbf{B}|$) define two very different quantities, the compressibility modulus and the magnetic induction, respectively. One gets

$$f = \frac{1}{2}K\cos^2\theta\left(\frac{d\theta}{dx}\right)^2 + \frac{1}{2}B(1-\cos\theta)^2 - \frac{1}{2}\mu_0^{-1}\chi_a\mathcal{B}^2\sin^2\theta.$$

The x-axis is along the layers, i.e., perpendicular to the boundaries of the cell, and the z-axis is perpendicular to the unperturbed layers.

(a) assuming that the deviation is small and expanding f to the second order in θ, the free energy reduces to

$$f = \frac{1}{2}K\left(\frac{d\theta}{dx}\right)^2 - \frac{1}{2}\mu_0^{-1}\chi_a\mathcal{B}^2\theta^2,$$

[26]A. Rapini, J. de Phys. (Paris) **33**, 237 (1972).

which yields the Euler–Lagrange equation

$$\frac{d^2\theta}{dx^2} + \frac{1}{\xi^2}\theta = 0$$

($\xi = \frac{1}{B}\sqrt{\frac{K}{\mu_0^{-1}\chi_a}}$ being a characteristic length), and a solution of the form

$$\theta = \theta_M \cos\frac{x}{\xi} \quad \text{for} \quad B > B_c = \frac{\pi}{L}\sqrt{\frac{K}{\mu_0^{-1}\chi_a}}.$$

(Hint: L is the thickness of the cell; at $B = B_c$, the strong anchoring boundary conditions are satisfied; θ_M is the value of θ in the middle plane of the cell). B_c is the critical field for the Frederiks transition. This solution applies to the nematic phase as well.

(b) We take into account the term of compression in order to estimate θ_M. The first integral of the Euler–Lagrange equation that minimizes f is

$$2K\cos^2\theta\left(\frac{d\theta}{dx}\right)^2 = 2B(\cos\theta_M - \cos\theta)$$

$$+ (\mu_0^{-1}\chi_a B^2 + B)(\sin^2\theta_M - \sin^2\theta).$$

This first integral satisfies $\left.\frac{d\theta}{dx}\right|_M = 0$. The right-hand member of the first integral is positive, because the left-hand part is positive. Because $0 < \cos\theta_M \le \cos\theta$, one eventually gets

$$\cos\theta_M > \left(1 + \frac{\mu_0^{-1}\chi_a B^2}{B}\right)^{-1}.$$

This can also be written as

$$\cos\theta_M > \left(1 + \frac{\lambda^2}{\xi^2}\right)^{-1} \cong 1 - \frac{\lambda^2}{\xi^2},$$

where λ is a microscopic length (de Gennes penetration length) and ξ is a macroscopic one. The change in orientation is hardly visible.

Problem 5.9. Find the equilibrium director configuration in a hybrid-aligned film (5.51) with an infinitely strong anchoring (5.50) when $K_1 \ne K_3$.

Answers: The free energy writes

$$f = \frac{1}{2}(K_{11}\sin^2\theta + K_{33}\cos^2\theta)(\theta')^2 = \frac{1}{2}K(\theta)(\theta')^2$$

and does not depend on z explicitly. The first integral of the Euler–Lagrange equation is, thus $K(\theta)(\theta')^2 = c^2$, where c is a constant, defined from the boundary conditions (5.50),

$$\int_{\bar{\theta}_0}^{\bar{\theta}_d} \sqrt{K(\chi)}\, d\chi = \int_0^d c\, dz.$$

The equilibrium director tilt across the nematic slab is then

$$zc = \int_{\bar{\theta}_0}^{\theta(z)} \sqrt{K(\chi)}\, d\chi.$$

Problem 5.10. In the consideration of the hybrid-aligned nematic film with strong anchoring (Section 5.4.2 and Problem 5.9), we assumed that the director is confined to the vertical plane and there are no in-plane distortions (a "homogeneous hybrid film"). Using the one-constant Frank–Oseen elastic energy, show that in-plane director distortions might reduce the elastic energy of the hybrid film.

Answers: One of the simplest ways is to compare elastic energies of the films restricted in the horizontal planes by coaxial cylinders with radii $R > r$. Consider first a homogeneous hybrid film with 2D distortions, $(n_x, n_y, n_z) = [\sin \alpha z, 0, \cos \alpha z]$, where α is the constant defined by the difference in the polar angles at the opposite plates and the thickness of the film, say, $\alpha = \frac{\pi}{2d}$. The elastic energy of such a film is $F_{\text{FO}}^{\text{homo}} = \frac{\pi^3 K}{8d}(R^2 - r^2)$. Now take an ansatz $(n_r, n_\varphi, n_z) = [\sin \alpha z, 0, \cos \alpha z]$, in the cylindrical coordinates (r, φ, z), that describes a 3D director field distorted in the plane of the film. The elastic energy of this distorted film is

$$F_{\text{FO}}^{\text{dist}} = F_{\text{FO}}^{\text{homo}} + \frac{\pi K}{2}\left[-\pi(R-r) + d\ln\frac{R}{r}\right];$$

i.e., $F_{\text{FO}}^{\text{dist}}$ might indeed be smaller than $F_{\text{FO}}^{\text{homo}}$ of the homogeneous film.

Further Reading

P.G. de Gennes and J. Prost, The Physics of Liquid Crystals, 2nd Edition, Clarendon Press, Oxford, 1993.

G. Vertogen and W. de Jeu, Thermotropic Liquid Crystals, Fundamentals, Edited by V. Goldanskii, F. Shafer, and J. Toennies, Springer Series in Chemical Physics, vol. 45, Springer-Verlag, Berlin, 1988.

L. M. Blinov and V. G. Chigrinov, Electrooptic Effects in Liquid Crystal Materials, Springer, New York 1996.

F. C. Frank, Discuss. Faraday Soc. **25**, 19 (1958).

G. Barbero and L.R. Evangelista, An Elementary Course on the Continuum Theory for Nematic Liquid Crystals. Series on Liquid Crystals, Edited by H.C. Ong, World Scientific, Singapore, 2001.

CHAPTER 6

Dynamics of Isotropic and Anisotropic Fluids

Static continuum theory allows one to find the equilibrium configurations of the order parameter field by minimizing an appropriate free-energy functional. If the system is slightly disturbed, it usually relaxes back to this equilibrium. Most of the perturbations relax quickly, over a characteristic time of collisions between the molecules (which is of the order of 10^{-10}–10^{-14} s for classic fluids). However some variables, called hydrodynamic variables, relax slowly. A classic example is the mass density: The frequency ω (inverse time) of the sound waves vanishes when their wavelength becomes infinite, $L \to \infty$, according to the dispersion law $\omega \sim c/L$, where c is the sound velocity. In general, densities of conserved variables (mass, linear momentum, and energy) are hydrodynamic variables. In media with continuous broken symmetries, such as nematic liquid crystals, there is a second class of hydrodynamic variables, namely, the degeneracy parameters (or phases of the order parameter), which is discussed in Chapter 3.

Imagine, for example, a reorientation wave of the nematic director as a perturbation of the uniform state $\mathbf{n} = $ const. If $L \to \infty$, this perturbation is just another uniform state with a different orientation $\mathbf{n'} = $ const. It will practically never relax back to the original state because both have the same energy. If L is finite but still macroscopic, the relaxation time is finite, but still larger than the molecular collision times. The director, which is the phase variable of the order parameter of the nematic, is an independent hydrodynamic variable. In contrast, the amplitude of the order parameter is not a hydrodynamic variable: Even if the wavelength of perturbation is infinitely long, $L \to \infty$, it has to relax to the local equilibrium state.

Hydrodynamics studies processes that occur over the space and time intervals much larger than the characteristic molecular scales: There are many acts of molecular collisions over the hydrodynamics' elementary time dt and space $d\mathbf{r}$ intervals that equilibrate the system locally. This allows one to operate with *field* variables (such as director, velocity, density, etc.), averaged over many particles to characterize the state of an elementary volume located at time t at a space point $\mathbf{r}$. The position $\mathbf{r}$ can be specified in a fixed laboratory reference frame through, e.g., Cartesian coordinates (x_1, x_2, x_3). This approach is called

184

Eulerian. In the alternative Lagrangian approach, one follows a selected particle, moving the coordinate frame with this selected particle.

A moving one-component *isotropic* fluid is described by the following five fields: the three components of local velocity $\mathbf{v}(\mathbf{r}, t)$ and two thermodynamic variables, generally the mass density $\rho(\mathbf{r}, t)$ and the energy density. Because the processes are slow, one can use an appropriate thermodynamic equation of state to find any other thermodynamic quantity from these two. We start our consideration with the hydrodynamics of such a simple one-component fluid. The rest of the chapter is devoted to hydrodynamics of the nematic phase, for which there are two additional variables, namely, two director components (the third one is eliminated by virtue of $\mathbf{n}^2 = 1$).

6.1. Velocity Field and Stress Tensor

6.1.1. Material Derivatives and Components of Fluid Motion

6.1.1.1. Material Derivative

Suppose that some scalar parameter a of a moving fluid (for example, density or a component of velocity) changes in space and time. The changes in an observable $a(\mathbf{r}, t)$ are different for a stationary observer at the point $\mathbf{r}$ and for an observer that moves with a selected material particle. For the stationary observer, the rate of change $\partial a / \partial t$ is caused exclusively by the *local modifications of the observable*. For the moving observer, there are additional changes: During a time interval dt, the element of fluid is carried to a new position $\mathbf{r} + d\mathbf{r}$, where the value of a is generally different from that at $\mathbf{r}, t$:

$$a(\mathbf{r} + d\mathbf{r}, t + dt) - a(\mathbf{r}, t) = dt \left(\frac{\partial a}{\partial t} + \frac{\partial a}{\partial x_i} \frac{dx_i}{dt} \right) = dt \left(\frac{\partial a}{\partial t} + v_i \frac{\partial a}{\partial x_i} \right); \qquad (6.1)$$

$v_i = dx_i / dt$ are the components of the velocity of the particle present at $\mathbf{r}$ at time t. Here and henceforth, we use the Einstein rule of summation over indices that appear twice in a term. Dividing both parts of (6.1) by dt, one obtains an important definition of the *material derivative*,

$$\frac{da}{dt} = \frac{\partial a}{\partial t} + \mathbf{v} \cdot \nabla a. \qquad (6.2)$$

The term $\mathbf{v} \cdot \nabla a$ brings essential *nonlinearity* to the equations of hydrodynamics: The fluid property a is transported by the velocity $\mathbf{v}$, which itself is a fluid property.

6.1.1.2. Components of Motion

Local fluid motion can be decomposed into four distinctive components: (1) pure translation, (2) solid-like rotary flow that involves no deformations, (3) shear deformations, and (4) extensional deformations or dilation.

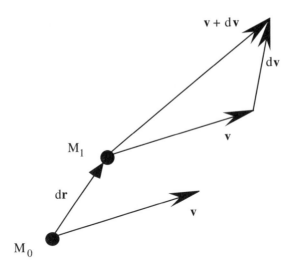

Figure 6.1. Relative motion of two close fluid particles.

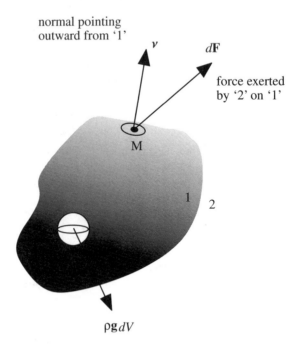

Figure 6.2. A body force $\rho \mathbf{g} \, dV$ acting on a small volume element and a surface force $d\mathbf{F}$ acting on a small surface element with unit normal $\boldsymbol{\nu}$.

Consider two close fluid particles at points M_0 and M_1. At some instant t, these particles are separated by a distance $d\mathbf{r}$ and move with velocities $\mathbf{v}$ and $\mathbf{v} + d\mathbf{v}$ (Fig. 6.1). The velocity of M_1 *relative* to M_0 is the increment $dv_j = \frac{\partial v_j}{\partial x_i} dx_i$, $i, j = 1, 2, 3$, that combines the effects (2)–(4) of rotations and deformations; pure translation (1) is given by $\mathbf{v}$. To separate the solid-like rotations (2) from deformations (3) and (4), the tensor of velocity gradients $\partial v_j / \partial x_i$ is presented as a sum $A_{ij} + W_{ij}$ of symmetric A_{ij} and antisymmetric W_{ij} parts:

$$A_{ij} = A_{ji} = \frac{1}{2}\left(\frac{\partial v_i}{\partial x_j} + \frac{\partial v_j}{\partial x_i}\right), \tag{6.3}$$

$$W_{ij} = -W_{ji} = \frac{1}{2}\left(\frac{\partial v_i}{\partial x_j} - \frac{\partial v_j}{\partial x_i}\right). \tag{6.4}$$

A_{ij} is called the *rate-of-deformation (or strain-rate) tensor*. The diagonal elements A_{ii} describe dilation (4): if, for example, the particles M_1 and M_0 are separated along the axis x_1, then $A_{11} = dv_1/dx_1$ is the extension rate per unit separation distance. The off-diagonal elements A_{ij} are proportional to the shearing velocity in the direction i for two particles M_1 and M_0 separated in the direction j and thus represent the shear motion (3). Finally, W_{ij} describes solid-like rotations (2) of M_1 around M_0 and is often expressed through the angular velocity $\varpi = \frac{1}{2}\nabla \times \mathbf{v}$ as $W_{ij} = -\varepsilon_{ijk}\varpi_k$, where ε_{ijk} are the components of the Levi-Civita tensor equal 1 if $ijk = 123, 231, 312$, (-1) if $ijk = 321, 213, 132$, or 0 if any two indices are alike. As already indicated in Chapter 5, the components of a vector product are $[\mathbf{a} \times \mathbf{b}]_i = \varepsilon_{ijk}a_j b_k$; hence, $[\nabla \times \mathbf{a}]_i = \varepsilon_{ijk}a_{k,j}$, with $a_{k,j} = \partial a_k/\partial x_j$. Note also a useful formula $\varepsilon_{ipq}\varepsilon_{ikl} = \delta_{pk}\delta_{ql} - \delta_{pl}\delta_{qk}$.

6.1.2. Body and Surface Forces. Stress Tensor

A change in the state of fluid motion can be caused by two types of forces, called body (or volume) forces and surface forces (Fig. 6.2). Some forces (e.g., of electrostatic origin) can be represented in either form.

6.1.2.1. Body Forces

Body forces penetrate the whole volume of the system and act on all its elements. An obvious example is gravity. A body force acting on an elementary volume dV is proportional to this volume and can be written as $\rho \mathbf{g}\, dV$, where $\mathbf{g}$ is the force per unit mass.

6.1.2.2. Surface Forces

Surface forces are caused by direct molecular interactions, including interactions between neighboring regions of the same medium. The surface nature of these forces is caused by a very small range of molecular interactions compared with characteristic scales of

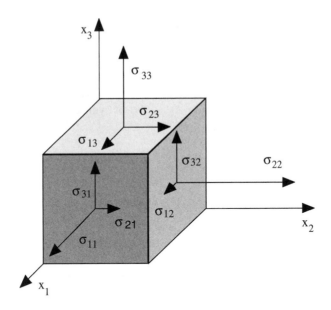

Figure 6.3. A cube with normal and tangenial stresses acting at its faces.

perturbations in a continuum model. The surface forces are conveniently represented by the components σ_{ij} of a *stress tensor*: σ_{ij} is the i-component of an infinitesimal surface force acting on an infinitesimal surface area dA_j perpendicular to the coordinate axis j:

$$dF_i = \sigma_{ij}\, dA_j = \sigma_{ij} v_j\, dA; \tag{6.5}$$

v_j is the component of a unit normal $\boldsymbol{v}$ to the surface; $\boldsymbol{v}$ is directed from the part 1 toward part 2, (Fig. 6.3). The total force acting on a volume V with surface A is then

$$F_i = \oiint_A \sigma_{ij} v_j\, dA = \iiint_V \frac{\partial \sigma_{ij}}{\partial x_j}\, dV, \tag{6.6}$$

where the surface integral is substituted by the volume integral by virtue of the Gauss divergence theorem.

The three diagonal elements of σ_{ij} are *normal stresses*, whereas the six off-diagonal elements are *shear stresses*. The last concept helps to distinguish fluids from solids in hydrodynamic description: A medium is fluid if the shear stresses, no matter how small, cause unlimited large deformations if applied over unlimited time. Fluids do not support shear stresses while at rest.

6.2. Isotropic Fluid in Motion

An ultimate problem of fluid dynamics is to find density, three velocity components and energy density as functions of time and spatial coordinates. Five dynamic equations for these five unknowns follow from the conservation laws for mass, momentum, and energy.

6.2.1. Conservation of Mass: Continuity Equation

Consider a fluid volume V_0 bounded by a surface A_0. The subscript "0" means that the volume is fixed in space. The mass of fluid in this volume at any moment of time is the volume integral over its density, $\iiint_{V_0} \rho(\mathbf{r}, t)\, dV$. The rate of outflow of mass through the closed surface A_0 is $\oiint_{A_0} \rho \mathbf{v} \cdot \boldsymbol{\nu}\, dA$. The rate at which mass is decreasing inside the volume is $-\frac{\partial}{\partial t} \iiint_{V_0} \rho\, dV$. Because mass cannot be created at will,

$$\frac{\partial}{\partial t} \iiint_{V_0} \rho\, dV + \oiint_{A_0} \rho \mathbf{v} \cdot \boldsymbol{\nu}\, dA = \iiint_{V_0} \left(\frac{\partial \rho}{\partial t} + \operatorname{div} \rho \mathbf{v} \right) dV = 0. \tag{6.7}$$

The last equation should hold for any fluid volume V_0. Therefore,

$$\frac{\partial \rho}{\partial t} = -\nabla \cdot \rho \mathbf{v}, \tag{6.8}$$

which is the *law of conservation of mass*, also known as the *continuity equation*. To cast it in the Lagrangian form, i.e., while following the fluid particle, one differentiates the second term in (6.8) and applies the definition of the material derivative (6.2). The result reads as

$$\frac{d\rho}{dt} = -\rho \nabla \cdot \mathbf{v}; \tag{6.9}$$

i.e., the rate $d\rho/dt$ of the density change equals the product of density ρ and the volume expansion rate $\nabla \cdot \mathbf{v}$, taken with an opposite sign. For an incompressible fluid, $\rho(\mathbf{r}, t) = \text{const}$, the continuity equation is simply

$$\nabla \cdot \mathbf{v} \equiv \partial v_k / \partial x_k = 0. \tag{6.10}$$

6.2.2. Linear Momentum Equation

The Lagrangian approach gives a straightforward way to derive the dynamic equations for momentum and energy. One selects a material volume V bounded by a surface A that moves with the fluid (so that both V and A might depend on time) and applies the basic laws of mechanics to this moving volume. The linear momentum equation follows from the principle analogous to the Newton's second law of mechanics: The rate of change of momentum $\frac{d}{dt}\iiint_V \rho \mathbf{v}\, dV$ of a particle equals the net force acting on the particle:

$$\frac{d}{dt}\iiint_V \rho v_i\, dV = \iiint_V \left(\rho g_i + \frac{\partial \sigma_{ij}}{\partial x_j}\right) dV. \tag{6.11}$$

To obtain the balance of momentum in a differential form, one has to bring the operation of differentiation in the left-hand side of (6.11) under the integral. This is not a permissible operation because the volume of integration is time dependent. Observe, however, that one can replace integration over the volume with integration over the mass, $\rho\, dV = dm$. When this is done, the region of integration (mass of the moving particle) is constant and switching of the order of operations is permissible:

$$\frac{d}{dt}\iiint_V \rho \mathbf{v}\, dV = \frac{d}{dt}\int \mathbf{v}\, dm = \int \frac{d\mathbf{v}}{dt}\, dm = \iiint_V \rho \frac{d\mathbf{v}}{dt}\, dV. \tag{6.12}$$

Because V is arbitrary, the Lagrangian form of the momentum equation reads as

$$\rho \frac{dv_i}{dt} = \frac{\partial \sigma_{ij}}{\partial x_j} + \rho g_i. \tag{6.13a}$$

In the Euclidian form,

$$\frac{\partial \rho v_i}{\partial t} = -\frac{\partial}{\partial x_j}(-\sigma_{ij} + \rho v_i v_j) + \rho g_i, \tag{6.13b}$$

where ρv_i is the *density of momentum* and $(-\sigma_{ij}+\rho v_i v_j)$ is the tensorial *flux of momentum*, also called the *momentum current* tensor.

Note first that the source term is the same in both equations of (6.13); this remark proves to be true for all laws of conservation. Second, there is an analogy between (6.13b) and the law of conservation of mass (6.8). The right-hand part of (6.13b) contains the divergence $\frac{\partial}{\partial x_j}$ of the *flux* $(-\sigma_{ij} + \rho v_i v_j)$ and the *source of momentum* ρg_i, which has no equivalent in (6.8), where for the sake of simplicity, we have not introduced any *source of matter* provided by *diffusivity*. Thus, both (6.13b) and (6.8) can be considered as particular examples of the "balance equation"

$$\frac{\partial(\text{density})}{\partial t} = -\text{div (flux)} + (\text{sources}), \tag{6.14}$$

that illustrates the universal structure of all laws of conservation. The derivation of the balance equation can be understood from Fig. 6.4. Let x be the mass density of some

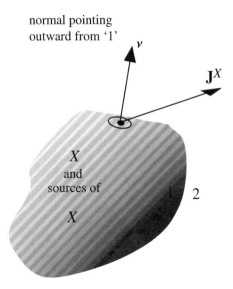

normal pointing
outward from '1'

Figure 6.4. Amount of quantity X in the volume V_0 changes due to the flow of X through the bounding surface A_0 and due to the production of X within V_0.

quantity X (e.g., total momentum, kinetic energy, etc.) specified in the volume V_0 bounded by the fixed surface A_0. The change $\iiint_{V_0} \frac{\partial (x\rho)}{\partial t}\, dV$ of X is caused by two factors: By flow $\oiint_{A_0}(-\mathbf{J}^X \cdot \boldsymbol{\nu})\, dA = -\iiint_{V_0}(\nabla \cdot \mathbf{J}^X)\, dV$ of X through A_0 (the flux $\mathbf{J}^X$ is calculated per unit area normal to $\mathbf{J}^X$ and per unit time), and by the production (or destruction) of X within V_0, which is the volume integral of all the "sources." Equation (6.14) is the differential form of this balance.

6.2.3. Energy Balance Equation

Consider the total energy of a fluid composed of the kinetic energy of its macroscopic motion and the specific internal energy of microscopic motions that does not vanish even for a fluid at rest. In the sequel, we shall note ε the total energy and u the internal energy, both calculated per unit mass, so that $\varepsilon = \frac{v^2}{2} + u$. The change of the total energy, with the rate (Problem 6.1)

$$\frac{d}{dt} \iiint_V \rho\varepsilon\, dV = \iiint_V \rho \frac{d}{dt}\left(\frac{v^2}{2} + u\right) dV,$$

is caused by the work performed on the system and by the flow of heat. The work per unit time (power) is that of volume and surface forces, $\iiint_V \rho(\mathbf{g} \cdot \mathbf{v})\, dV + \oiint_A(\mathbf{F} \cdot \mathbf{v})\, dA$. The heat received per unit time is $\oint_A (-\mathbf{J}^q \cdot \boldsymbol{v}) \, dA$, where $\mathbf{J}^q$ is the heat flux. Hence,

$$\iiint_V \rho \frac{d\varepsilon}{dt} \, dV = \iiint_V \rho(\mathbf{g} \cdot \mathbf{v}) \, dV + \oint_A (\mathbf{F} \cdot \mathbf{v}) \, dA + \oint_A (-\mathbf{J}^q \cdot \boldsymbol{v}) \, dA, \qquad (6.15)$$

or in local (Lagrangian) form

$$\rho \frac{d\varepsilon}{dt} = -\frac{\partial}{\partial x_j}(-v_i \sigma_{ij} + J_j^q) + \rho g_i v_i. \qquad (6.16a)$$

Again, the fluxes of energy and the sources can be written in the Eulerian picture [compare with (6.14)]:

$$\frac{\partial}{\partial t}\rho\varepsilon = -\frac{\partial}{\partial x_j}[v_i(-\sigma_{ij} + \rho\varepsilon \, \delta_{ij}) + J_j^q] + \rho g_i v_i. \qquad (6.16b)$$

It is instructive to separate the kinetic and internal energy changes. Writing the kinetic energy rate as $\rho \frac{d}{dt}(\frac{v^2}{2}) = v_i(\rho \frac{dv_i}{dt})$, where $\rho \frac{dv_i}{dt}$ is expressed through the momentum equation (6.13a), and extracting this term from (6.16a), one gets an equation for the internal energy:

$$\rho \frac{du}{dt} = \sigma_{ij} \frac{\partial v_i}{\partial x_j} - \frac{\partial J_j^q}{\partial x_j}. \qquad (6.17)$$

The specific internal energy is then controlled by the work of the surface forces and the flux of heat; the work of the body forces change only the kinetic energy.

6.2.4. Entropy Production Equation

The fact that the gradients of velocity (the term $\sigma_{ij}\partial v_i/\partial x_j$) cause internal energy changes seems natural: Molecular interaction and transport between adjacent fluid domains that move with different velocities cause intrinsic friction and, thus, dissipate the energy. This phenomenon is cast in a precise form by the second principle of thermodynamics, on the one hand, and the Onsager equations, on the other.

6.2.4.1. Second Principle

In an irreversible process, the entropy variation of a system

$$dS = dS^{(r)} + dS^{(d)} \qquad (6.18)$$

comes from two contributions, a reversible one $dS^{(r)}$ and an irreversible (dissipative) one $dS^{(d)}$. If the system is closed (exchanges heat but not matter with the surrounding), then $dS^{(r)} = \frac{dQ}{T}$, where dQ is the total heat provided to the system. $dS^{(r)}$ might contain additional terms if the system is open and exchanges matter as well. In its most general form,[1] the second principle of thermodynamics claims that for all systems, open, closed, or isolated, the change $dS^{(d)}$ is always positive:

$$dS^{(d)} > 0. \tag{6.19}$$

Even when the system is divided into smaller parts, the entropy production in each of these parts is still positive. This allows one to write the second principle in the local form, as the balance equation

$$\frac{\partial \rho s}{\partial t} + \nabla \cdot \mathbf{J}^s = \sigma \equiv \frac{2R}{T} > 0. \tag{6.20}$$

Here, we introduce the local (coordinate- and time-dependent) quantities: The entropy per unit *mass* $s = s(\mathbf{r}, t)$, so that $S = \iiint_{V_0} \rho s \, dV$ and $ds = ds^{(r)} + ds^{(d)}; ds^{(d)} > 0$; and the entropy flux $\mathbf{J}^s$ (assumed vectorial), associated with the reversible processes, $dS^{(r)}/dt = \oiint_{A_0} (-\mathbf{J}^s \cdot \boldsymbol{\nu}) \, dA$. The entropy source σ (called also entropy production) per unit *volume* is often expressed by an equivalent notation $2R/T$, where R is the *dissipation function*. To find $2R/T$ explicitly in terms of parameters characterizing viscous friction and, for example, heat conduction, we calculate $\frac{\partial \rho s}{\partial t}$ from the first principle of thermodynamics.

The differential of the internal energy ρu per unit volume (which implies $dV \equiv 0$), can indeed be written as

$$d(\rho u) = \mu d\rho + T d(\rho s), \tag{6.21}$$

where the temperature and the chemical potential per unit mass are defined in a standard way as $T = (\partial u/\partial s)_\rho$ and $\mu = (\partial \rho u/\partial \rho)_{\rho s}$, respectively. Dividing the last equation by a time element dt (sufficiently large compared with molecular times, henceforth preserving the validity of the hydrodynamic approach), we obtain the rate equation for the entropy ρs:

$$T \frac{d(\rho s)}{dt} = \rho \frac{du}{dt} - (\mu - u) \frac{d\rho}{dt}, \tag{6.22a}$$

where we have substituted $d(\rho u)/dt = \rho du/dt + u d\rho/dt$. Transforming

$$d(\rho s)/dt = \partial(\rho s)/\partial t + \mathbf{v} \cdot \nabla(\rho s) = \partial(\rho s)/\partial t + \mathrm{div}\,(\rho s \mathbf{v}) - \rho s\, \mathrm{div}\, \mathbf{v}$$

---

[1] D. Kondepudi and I. Prigogine, Modern Thermodynamics, John Wiley & Sons, Chichester, 1998.

with the help of (6.2) and using the continuity equation (6.9), one rewrites (6.22a) as

$$T\frac{\partial(\rho s)}{\partial t} + T\,\mathrm{div}\,(\rho s\mathbf{v}) = \rho\frac{du}{dt} + p\,\mathrm{div}\,\mathbf{v}, \tag{6.22b}$$

where $p = -\rho u + \rho\mu + T\rho s$ is the pressure. With the result (6.17) for $\rho du/dt$, and the relationship $\frac{1}{T}\nabla\cdot\mathbf{J}^q = \nabla\cdot(\frac{\mathbf{J}^q}{T}) - \mathbf{J}^q\nabla(\frac{1}{T})$, (6.22b) acquires the form

$$\frac{\partial(\rho s)}{\partial t} + \nabla\cdot\left(\rho s\mathbf{v} + \frac{\mathbf{J}^q}{T}\right) = \frac{\sigma_{ij} + p\,\delta_{ij}}{T}\frac{\partial v_i}{\partial x_j} + \mathbf{J}^q\nabla\left(\frac{1}{T}\right). \tag{6.23}$$

Comparing it with (6.20), one finds the local entropy production

$$\frac{2R}{T} = \frac{\sigma_{ij} + p\,\delta_{ij}}{T}\frac{\partial v_i}{\partial x_j} - \frac{J_j^q}{T^2}\frac{\partial T}{\partial x_j}, \tag{6.24}$$

caused by viscous friction (the first term) and the irreversible heat flow (the second term). Other processes, such as chemical reactions or diffusion, can contribute to the entropy production.

6.2.4.2. Onsager's Reciprocity Relations

Notice that the terms responsible for entropy production (6.24) have a certain structure: They are products of some flux $\mathbf{J}^\alpha$ (such as momentum flux or heat flux $\mathbf{J}^q$) characterizing an irreversible process and some "thermodynamic force" $\mathbf{F}^\alpha$ that drives the flow; in our example (6.24), the "forces" are gradients of velocity and temperature. We can write phenomenologically

$$\frac{2R}{T} = \sum_a \mathbf{F}^a \cdot \mathbf{J}^a. \tag{6.25}$$

In equilibrium, $R = 0$, both fluxes and forces vanish. If the system is not in equilibrium, but close to it, so that all processes are of small amplitude, one might assume that the fluxes $\mathbf{J}^\alpha$ are linear functions of the forces $\mathbf{F}^\alpha$:

$$\mathbf{J}^\alpha = \sum_\beta L_{\alpha\beta}\mathbf{F}^\beta. \tag{6.26}$$

Equations (6.26) are called *phenomenological* (also *kinetic*) equations; the well-known examples are Ohm's law of electric conduction, Fick's law of diffusion, and Fourier's law of heat conduction.

Certain restrictions are imposed on the values of the phenomenological coefficients $L_{\alpha\beta}$. First, note that (6.26) leads to a quadratic expression $2R/T = \sum_{\alpha\beta} L_{\alpha\beta}\mathbf{F}_\alpha\mathbf{F}_\beta$ for the entropy production (6.25). Thus, the coefficients $L_{\alpha\beta}$ should be such that the second principle $R > 0$ is satisfied for all values of forces $\mathbf{F}^\alpha$; i.e., R is positive definite. For example, $L_{\alpha\alpha}$ should be positive. Second,

$$L_{\alpha\beta} = L_{\beta\alpha}. \tag{6.27}$$

Relations (6.27) are the celebrated Onsager's reciprocity relations that take origin in the time-reversal properties of the processes at microscopic scales. If the system is submitted to a magnetic field, or suffers an angular rotation, then

$$L_{\alpha\beta}(\mathbf{B}) = L_{\beta\alpha}(-\mathbf{B}), \tag{6.28a}$$

$$L_{\alpha\beta}(\Omega) = L_{\beta\alpha}(-\Omega). \tag{6.28b}$$

Heat and stress entropy productions do not couple in (6.24), because the respective fluxes do not have the same tensorial character. One can write the heat flux in an isotropic medium as $\mathbf{J}^q = -\kappa\nabla T$; it is caused by the force $\mathbf{F}^q = \nabla(\frac{1}{T})$. The phenomenon is that of molecular thermoconductivity, where κ is the thermoconductivity coefficient.

6.2.5. Viscous Stress Tensor

If there is no dissipation due to the fluid motion (inviscid fluid), $dS^{(d)}/dt = 0$, we must have, according to (6.24), $\sigma_{ij} = -p\,\delta_{ij}$. In a more general case of a dissipative process, we represent the stress tensor as the sum of the reversible part $\sigma_{ij}^{(r)} = -p\,\delta_{ij}$ and the irreversible part $\sigma_{ij}^{(d)}$, i.e., $\sigma_{ij} = \sigma_{ij}^{(r)} + \sigma_{ij}^{(d)}$. The irreversible part $\sigma_{ij}^{(d)}$ is called the *viscous stress tensor*. When the viscous stresses are assumed to be linear functions of velocity gradients (small perturbations from equilibrium), the corresponding dissipation function is

$$2R = \sigma_{ij}^{(d)}\frac{\partial v_i}{\partial x_j} = \eta_{ijkl}\frac{\partial v_i}{\partial x_j}\frac{\partial v_k}{\partial x_l}. \tag{6.29}$$

There are *a priori* 54 coefficients η_{ijkl}. The following considerations reduce this number to 2 in the case of an isotropic fluid. Note first that $\sigma_{ij}^{(d)}$ should vanish when there is no relative motion (and thus no friction) between the fluid elements, i.e., when $\mathbf{v} = \text{const}$ [this is automatically achieved in (6.29)], or when the fluid rotates as a rigid body with a constant angular velocity ϖ. An example of such a rotation about, say, the z-axis is the velocity field $(v_x, v_y) = (-\varpi y, \varpi x)$. Obviously, for this flow, $\partial v_y/\partial x = -\partial v_x/\partial y$. Therefore, among the four components of motion considered at the beginning of this chapter, only *dilations* (3) and *shears* (4) characterized by the symmetric tensor A_{ij} can cause intrinsic friction. Accordingly, these are the only combinations of the velocity gradients that should appear in the Onsager equations, which thus can be written as $\sigma_{ij}^{(d)} = \eta_{ijkl} A_{kl}$. Because of the symmetry with respect to the interchanges $i \leftrightarrow j$, $k \leftrightarrow l$, and $ij \leftrightarrow kl$ (Onsager's relations), an isotropic fluid is characterized by only two independent viscous coefficients:

$$\sigma_{ij}^{(d)} = \varsigma \, \delta_{ij} \frac{\partial v_k}{\partial x_k} + \eta \left(\frac{\partial v_j}{\partial x_i} + \frac{\partial v_i}{\partial x_j} - \frac{2}{3} \delta_{ij} \frac{\partial v_k}{\partial x_k} \right)$$

$$\equiv \varsigma \, \delta_{ij} A_{kk} + 2\eta \left(A_{ij} - \frac{1}{3} \delta_{ij} A_{kk} \right). \tag{6.30}$$

The scalars ς and η are, respectively, the *dynamic bulk viscosity* and the *dynamic shear viscosity*. To comply with the second law of thermodynamics, both must be positive, $\eta > 0$ and $\varsigma > 0$.

The bulk viscosity vanishes in two important limiting cases: In an ideal gas and when the fluid is *incompressible*, i.e., when the flow causes no dilation (6.10). A fluid can be considered as incompressible if the velocities are smaller than the velocity of sound (about 1.5 km/sec in water at 15°C).

Fluids that obey the linear relationship (6.30) are called Newtonian, after the Newton's law of viscosity, deduced for a simple 1D shear flow. For the velocity field $[0, v_y(z), 0]$, this law reads as

$$\sigma_{yz}^{(d)} = \eta \frac{dv_y}{dz}. \tag{6.31}$$

In a Newtonian fluid, such as water and air in a broad range of velocity gradients, the viscosity does not depend on the fluid shear stress. Fluids of long elongated molecules (polymers) and emulsions (e.g., blood), are usually non-Newtonian: The relationship between the shear stresses and the velocity gradients is not linear.

6.2.6. Navier–Stokes Equations. Reynolds Number. Laminar and Turbulent Flow

Substituting (6.30) in the momentum equations (6.13), one recovers the celebrated *Navier–Stokes equations*

$$\frac{\partial \mathbf{v}}{\partial t} + (\mathbf{v} \cdot \nabla)\mathbf{v} = \mathbf{g} - \frac{1}{\rho} \nabla p + \frac{\eta}{\rho} \nabla^2 \mathbf{v}, \tag{6.32}$$

which are valid when the fluid is incompressible, $\varsigma = 0$, and the dynamical viscosity η does not vary in space (despite possible temperature and pressure gradients). Equation (6.32) should be supplemented by the continuity equation (6.10) and appropriate boundary conditions to get a complete set of equations with one unique solution. The appropriate boundary condition is the so-called *no-slip condition*, which expresses the experimental fact that the fluid velocity at a solid boundary is equal to the velocity of the boundary.

Despite all of the simplifications mentioned above, the Navier–Stokes equations are markedly nonlinear because of the term $(\mathbf{v} \cdot \nabla)\mathbf{v}$ and, thus, are extremely hard to solve. For every concrete problem, the useful first step is to analyze the relative importance of the terms entering (6.32). In most cases, one can identify the largest speed U in the system and some characteristic length L. The quantities U and L together with the kinematic viscosity η/ρ form a dimensionless combination, referred to as the Reynolds number,

$$\mathrm{Re} = \frac{\rho U L}{\eta}, \tag{6.33}$$

which indicates the relative importance of the inertia term $(\mathbf{v} \cdot \nabla)\mathbf{v}$ and the viscous term $(\eta/\rho)\nabla^2\mathbf{v}$ in (6.32). If these terms are estimated as $(\mathbf{v} \cdot \nabla)\mathbf{v} \sim U^2/L$ and $(\eta/\rho)\nabla^2\mathbf{v} \sim (\eta/\rho)U/L^2$, respectively, then their ratio is Re. Flows at low and high Re are different and often referred to as the laminar and turbulent regimes, respectively.

If the Reynolds number is small (because of large viscosity, slow flow or small length scales), the term $(\mathbf{v} \cdot \nabla)\mathbf{v}$ is negligible compared with the viscous term. If, in addition, one considers the steady (time-independent) solutions, then $\partial\mathbf{v}/\partial t = 0$ and the whole inertia term $\sim d\mathbf{v}/dt$ can be dropped out. The Navier–Stokes equations become (in the absence of body forces) much simpler,

$$\eta\nabla^2\mathbf{v} = \nabla p, \tag{6.34}$$

which allows for a number of analytical solutions. One of them is the Stokes formula for the drag force $\mathbf{F}_{\mathrm{drag}}$ exerted by a laminar flow with a constant velocity $\mathbf{U}$, on a sphere of a radius a: $\mathbf{F}_{\mathrm{drag}} = 6\pi\eta a\mathbf{U}$. The derivation is not short, and it can be found elsewhere; see, e.g., books by Landau and Lifshitz and by Batchelor.

Steady flow at low Re is usually able to sustain small perturbations. If such perturbations occur, viscous damping brings the system back to the original state. However, as Re increases, the viscous damping might become unable to overcome the perturbations amplified by the nonlinear terms. The fluid motion becomes turbulent with numerous eddy-like regions.

Transition from the laminar to turbulent regime and fluid behavior when the turbulence is well-developed form two topics of formidable difficulty (just because of the nonlinear character of the Navier–Stokes equations). Turbulence has long been referred to as the only remaining great problem of classic physics. Attempts to understand this phenomenon lead to the formulation of basic principles of nonlinear science that emerged during the past few decades.

6.3. Nematodynamics in Ericksen–Leslie Model

Let us look back at the general scheme of deriving the hydrodynamic equations for the isotropic fluid. We defined five conserved quantities (mass, energy, momentum) and wrote five conservation laws for their densities. We then considered thermodynamics of the fluid slightly perturbed from its equilibrium state and calculated the entropy production as a sum of terms, each of which was a product of a "flux" and a "thermodynamic force" that causes this flux. Then, we assumed that the fluxes are linear functions of the forces, the proportionality coefficients being the viscosities or the thermoconductivity coefficient. Symmetry consideration established which viscosities were independent parameters.

A similar scheme can be applied to the hydrodynamic theory of any system, including the nematic liquid crystals. An important difference is that the nematic fluid has broken orientational symmetry, and thus, the director **n** appears as a new hydrodynamic variable (two independent components). Historically, the hydrodynamics of a nematic fluid has been formulated using two different approaches.

1. The macroscopic Ericksen–Leslie (EL) hydrodynamic theory[2] is based on classic mechanics. The director dynamic equation is derived from the angular momentum conservation. The approach has served as a guideline for many experiments and is used in most textbooks. However, it is not clear how the EL model can be transfered to other ordered systems, for example, SmA, where the broken-symmetry variable is the displacement of the layers rather than the director rotations.

2. The Harvard (H) model[3] is based on the general idea that the equations describing the macroscopic dynamics should result from averaging the microscopic (molecular) interactions of the system. The hydrodynamic equations are rigorously derived by identifying first the hydrodynamic variables, both those conserved and those related to the type of broken symmetry. Each hydrodynamic vartiable satisfies the "balance equation" (6.14), which represents either the conservation law (for densities of mass, energy, momentum) or the dynamic equation for broken-symmetry variables. The thermodynamic potentials depend on all hydrodynamic variables. For example, the director distortions relate to the Frank–Oseen elastic energy terms. The entropy production equation now contains the director-dependent terms and the number of independent viscosity and thermoconductivity coefficients increases.

The two theories are not exactly equivalent. For example, there is a subtle issue of the number and nature of independent viscosities; see Pleiner and Brand[4] for a detailed discussion. Here, we treat both, starting with the EL model.

---

[2]J.L. Ericksen, Trans. Soc. Rheol, **5**, 22 (1961); F.M. Leslie, Arch. Ration. Mech. Analysis **28**, 265 (1968) and Continuum Mech. Thermodyn. **4**, 167 (1992).

[3]D. Forster, T.C. Lubensky, P.C. Martin, J. Swift, and P.S. Pershan, Phys. Rev. Lett. **26**, 1016 (1971).

[4]H. Pleiner and H.R. Brand, in Pattern Formation in Liquid Crystals, Edited by A. Buka and L. Kramer, Springer, New York (1995), p. 15.

6.3.1. Angular Momentum Equation

The new variable is the director $\mathbf{n}$ (two independent components; the scalar order parameter takes its static value $s(T) = $ const). The two director dynamics equations are derived from the conservation of the angular momentum. As for the remaining variables, the continuity equation for mass remains the same as (6.9) for an isotropic fluid; the linear momentum equation is the same as (6.13), but the stress tensor should now account for orientational order; the orientational order also brings new terms to the energy equation (6.16).

The angular momentum of an element of volume dV in a nematic is considered as a sum of a macroscopic "external" contribution $d\mathbf{M}_{\text{ext}} = \mathbf{r} \times (\rho \mathbf{v})\, dV$, where $\mathbf{r}$ is the distance of the element from some origin, and a microscopic "intrinsic" contribution related to the director rotations, which when averaged over a large ensemble of molecules, gives a net angular momentum at the continuum level, $d\mathbf{M}_{\text{int}} = I\mathbf{\Omega}\, dV$. Here, I is the moment of inertia per unit volume and $\mathbf{\Omega} = [\mathbf{n} \times \frac{d\mathbf{n}}{dt}]$ is the local angular velocity of the director. In order of magnitude, the moment of inertia of a single molecule is $m_1 a^2$, where m_1 is a molecular mass and a is a molecular size. If N molecules are aligned perfectly parallel to each other, then their total moment of inertia is $Nm_1 a^2$. Thus, the moment of inertia per unit volume is small, $I \sim \rho a^2$ (Problem 6.4). The inertia terms can be neglected for low-frequency motions, but they are kept in the equations for the sake of completeness.

The law of conservation of the angular momentum $\mathbf{M}$ states that the rate of change $d\mathbf{M}/dt$ is equal to the sum of body and surface torques. *Body* torques are due to the field of gravity and external fields capable to reorient the director. For an element dV, the gravity field yields a torque $[\mathbf{r} \times \rho \mathbf{g}]\, dV$; we shall dispose of it by placing the origin at the center of gravity of the system. The remaining contribution is $[\mathbf{n} \times \mathbf{G}]\, dV$, where the force $\mathbf{G}$ is capable to reorient the director due to the material anisotropy. Here, we will restrict ourselves to the diamagnetic effect only, $G_k = \mu_0^{-1} \chi_a n_j \mathcal{B}_j \mathcal{B}_k$, where $\mathcal{B}_i$'s are the components of the induction $\mathbf{B}$ of the applied magnetic field and χ_a is the anisotropy of the magnetic susceptibility. Inclusion of the electric field is a more subtle issue because of flexoelectricity, ions, and so on.[4]

There are two types of *surface* torques. First, there is the usual torque $[\mathbf{r} \times d\mathbf{F}]$ due to the stresses σ_{kp} acting on an element of area $\boldsymbol{\nu}\, dA$, with components $[\mathbf{r} \times d\mathbf{F}]_i = \rho \varepsilon_{ijk} x_j \sigma_{kp} \nu_p\, dA$. Such a torque would not vanish even in the isotropic phase. The second type of surface torque is exclusively due to orientational interactions. It is related to the surface reorientation of the director and can be written either through the "momentum stress tensor" l_{ij} as $dl_i = l_{ij} \nu_j\, dA$, or through a "surface director stress" tensor τ_{kp} as $[\mathbf{n} \times d\mathbf{t}]_i = \varepsilon_{ijk} n_j \tau_{kp} \nu_p\, dA$. We will use the second presentation because it explicitly includes the director. The relation between τ_{kp} and director distortions will be clarified later on.

Thus, the angular momentum law in the integral form is

$$\frac{d}{dt} \iiint_V \{[\mathbf{r} \times \rho \mathbf{v}] + I[\mathbf{n} \times \dot{\mathbf{n}}]\}\, dV = \iiint_V [\mathbf{n} \times \mathbf{G}]\, dV + \oiint_A \{[\mathbf{r} \times d\mathbf{F}] + [\mathbf{n} \times d\mathbf{t}]\}, \quad (6.35)$$

(the dot is another notation for a material derivative, $\dot{\mathbf{n}} = d\mathbf{n}/dt$) and in the differential form

$$\rho \varepsilon_{ijk} x_j \dot{v}_k + I \varepsilon_{ijk} n_j \ddot{n}_k = \varepsilon_{ijk} n_j G_k + \varepsilon_{ijk} \frac{\partial}{\partial x_p} (x_j \sigma_{kp} + n_j \tau_{kp}). \qquad (6.36)$$

Differentiating the last term and noticing that the combination $(\rho \dot{v}_k - \partial \sigma_{kp}/\partial x_p)$ must vanish by virtue of the conservation of linear momentum (6.13), one gets an equation for the internal angular momentum, also called the *director dynamics equation*,

$$I \varepsilon_{ijk} n_j \ddot{n}_k = \varepsilon_{ijk} n_j G_k - \varepsilon_{ijk} \sigma_{jk} + \varepsilon_{ijk} \frac{\partial}{\partial x_p} (n_j \tau_{kp}). \qquad (6.37)$$

6.3.2. Energy Balance Equation

Energy balance equation for a nematic fluid is obtained from the "isotropic" (6.16a) by adding terms related to the orientational order:

$$\rho \frac{d}{dt} \left(\frac{v^2}{2} + u \right) + \frac{1}{2} \frac{d}{dt} (I \Omega^2) = \rho g_i v_i + G_i \dot{n}_i + \frac{\partial}{\partial x_j} (v_i \sigma_{ij} + \dot{n}_i \tau_{ij} - J_j^q), \qquad (6.38)$$

where $\frac{1}{2} I \Omega^2$ is the rotational energy density. The internal energy density u should now contain contribution from the nematic energy density (5.35), $f = f_{FO} - \frac{1}{2} \mu_0^{-1} \chi_a (\mathbf{B} \cdot \mathbf{n})^2$, where f_{FO} is the elastic Frank–Oseen term specified by (5.2). The right-hand side in (6.38) is supplemented by the power of orienting forces (second term) and that one of the surface director tensor. Finally, in an anisotropic medium, the temperature gradient can cause heat flow in a direction that is different from the direction of the gradient, $J_j^q = -\kappa_{js} \partial T/\partial x_s$, and the thermoconductivity coefficient κ_{js} becomes a tensor of the second rank. In a uniaxial nematic, κ_{js} has two independent components $\kappa_{\parallel}$ and $\kappa_{\perp}$, so that $\kappa_{js} = \kappa_{\parallel} n_j n_s + \kappa_{\perp} \delta_{js}^{\perp}$, where $\delta_{js}^{\perp} = \delta_{js} - n_j n_s$ is the transverse Kronecker delta. At the beginning of the next section, we temporarily drop the thermoconductivity term, as the goal is to clarify the structure of σ_{ij} and τ_{ij}.

6.3.3. Entropy Production Equation

At this moment, we have no information about the possible structure of the tensors σ_{ij} and τ_{ij}. To find it, we consider the entropy production equation, as it was done for the isotropic fluid.

We again single out the internal energy contribution in (6.38). Multiplying the linear momentum equation (6.13) by v_i and the internal angular momentum equation (6.37) by

Ω_i, one obtains two equations that, when added, produce

$$\rho\frac{d}{dt}\frac{v^2}{2} + I\frac{d}{dt}\frac{\Omega^2}{2} = \rho g_i v_i + G_i \dot{n}_i + \frac{\partial}{\partial x_j}(v_i \sigma_{ij} + \dot{n}_i \tau_{ij})$$

$$- \sigma_{ij}\frac{\partial v_i}{\partial x_j} - \Omega_{jk}\sigma_{jk} - n_j \tau_{kp}\frac{\partial \Omega_{jk}}{\partial x_p}, \tag{6.39}$$

where $\Omega_{jk} = \varepsilon_{ijk}\Omega_i$ is the antisymmetric tensor of rotation. Substituting (6.39) into (6.38), one gets

$$\rho\frac{du}{dt} = \sigma_{ij}\frac{\partial v_i}{\partial x_j} + \Omega_{jk}\sigma_{jk} + n_j \tau_{kp}\frac{\partial \Omega_{jk}}{\partial x_p}. \tag{6.40}$$

On the other hand, $\rho du/dt$ can be found from thermodynamic considerations similar to those for an isotropic fluid. An important difference is that the internal energy per unit volume should also include the Frank–Oseen *internal* energy density f (whose Frank moduli are at constant volume and entropy) because the director might be distorted. Hence, $d(\rho u) = \mu d\rho + T d(\rho s) + df$ and the rate of energy change $\rho du/dt = d(\rho u)/dt - u d\rho/dt$ is

$$\rho\frac{du}{dt} = (\mu - u)\frac{d\rho}{dt} + T\frac{d}{dt}(\rho s) + \frac{df}{dt}, \tag{6.41}$$

where $\mu = (\partial \rho u/\partial \rho)_{\rho s,\mathbf{n}}$ is the chemical potential per unit mass and $T = (\partial u/\partial s)_{\rho,\mathbf{n}}$. Transforming $d(\rho s)/dt$ through the definition of the material derivative (6.2) and employing the continuity equation (6.9), one obtains the dissipation function

$$2R = \rho\frac{du}{dt} + p\,\mathrm{div}\,\mathbf{v} - \frac{df}{dt}. \tag{6.42}$$

The rate of change of the nematic energy density f is

$$\frac{df}{dt} = \frac{\partial f}{\partial n_i}\frac{dn_i}{dt} + \frac{\partial f}{\partial(\partial n_i/\partial x_j)}\frac{d}{dt}\frac{\partial n_i}{\partial x_j} \equiv \phi_i\frac{dn_i}{dt} + \pi_{ij}\frac{d}{dt}\frac{\partial n_i}{\partial x_j}, \tag{6.43}$$

where we introduce two new notations, $\phi_i = \frac{\partial f}{\partial n_i} = \frac{\partial f_{\mathrm{FO}}}{\partial n_i} - \mu_0^{-1}\chi_a(n_j\mathcal{B}_j)\mathcal{B}_j$ and $\pi_{ij} = \frac{\partial f}{\partial(\partial n_i/\partial x_j)}$. It is useful to represent df/dt in terms of Ω_{ij} and its derivatives, by transforming $\frac{d}{dt}\frac{\partial n_i}{\partial x_j}$ with the help of (6.2) into $\frac{\partial}{\partial x_j}\frac{dn_i}{dt} - \frac{\partial v_p}{\partial x_j}\frac{\partial n_i}{\partial x_p}$ and by using $dn_i/dt = -\Omega_{ik}n_k$,

$$\frac{df}{dt} = -\phi_i\Omega_{ij}n_j - \pi_{ij}\frac{\partial}{\partial x_j}(\Omega_{ik}n_k) - \pi_{ij}\frac{\partial v_p}{\partial x_j}\frac{\partial n_i}{\partial x_p}. \tag{6.44}$$

Substituting (6.40) and (6.44) in (6.42), the dissipation function can be written as

$$2R = \frac{\partial v_j}{\partial x_k}\left(\sigma_{jk} + p\,\delta_{jk} + \pi_{pk}\frac{\partial n_p}{\partial x_j}\right) + \Omega_{jk}\left(\sigma_{jk} + \phi_j n_k + \pi_{jp}\frac{\partial n_k}{\partial x_p}\right)$$
$$+ \frac{\partial \Omega_{jk}}{\partial x_p}(n_j\tau_{kp} - n_j\pi_{kp}). \tag{6.45}$$

In the presence of temperature gradients, a term $\left(-\frac{J_j^q}{T}\right)\frac{\partial T}{\partial x_j}$ contributes to the right-hand side of (6.45), with $J_j^q = -\kappa_{js}\partial T/\partial x_s$, as discussed in Section 6.3.2.

6.3.4. Nondissipative Dynamics

Let us first analyze (6.45) for a nondissipative process. All three expressions in brackets must vanish, in order to guarantee that $R = 0$ for any $\partial v_j/\partial x_k$, Ω_{jk}, and $\partial \Omega_{jk}/\partial x_p$:

$$\sigma_{jk}^{(r)} = -p\,\delta_{jk} - \pi_{pk}\frac{\partial n_p}{\partial x_j}, \tag{6.46a}$$

$$\varepsilon_{ijk}\sigma_{jk}^{(r)} = -\varepsilon_{ijk}\phi_j n_k - \varepsilon_{ijk}\pi_{jp}\frac{\partial n_k}{\partial x_p}, \tag{6.46b}$$

$$\varepsilon_{ijk}n_j\tau_{kp} = \varepsilon_{ijk}n_j\pi_{kp}. \tag{6.46c}$$

We use the Levi-Civita tensor to stress that the relevant information from the second and the third terms in (6.45) is only about the antisymmetric parts of the expressions in brackets: Because Ω_{jk} is an antisymmetric tensor, any symmetric part within the brackets vanishes when multiplied by Ω_{jk}.

6.3.5. Dissipative Dynamics

In (6.46c) and in the sequel, we assume that the surface director is fixed (strong anchoring) so that the irreversible part of the "surface director stress" tensor τ_{kp} is zero, $\tau_{kp}^{(r)} \equiv \tau_{kp}$.

Equation (6.46a) invites us to define the *Ericksen stress tensor*

$$\sigma_{jk}^{(r)} = -p\,\delta_{jk} - \pi_{pk}\frac{\partial n_p}{\partial x_j}, \tag{6.47}$$

composed of two contributions: the stress of an inviscid liquid and the stress due to director distortions (no diamagnetic contribution). The Ericksen stress is independent of dissipative effects and is usually smaller than the viscous stresses.

Substituting (6.46) in (6.45), the dissipation function can be written as

$$2R = A_{ij}\overline{\sigma}_{ij}^{(s)} + (\Omega_{ij} + W_{ij})\overline{\sigma}_{ij}^{(a)} = A_{ij}\overline{\sigma}_{ij}^{(s)} + \overline{\overline{\Omega}}_{ij}\overline{\sigma}_{ij}^{(a)}. \tag{6.48}$$

Here, we represent the velocity gradients as $\partial v_i/\partial x_j \equiv A_{ij} + W_{ij}$ and introduce the stress tensor $\overline{\sigma}_{ij} = \sigma_{ij} - \sigma_{ij}^{(r)} = \overline{\sigma}_{ij}^{(s)} + \overline{\sigma}_{ij}^{(a)}$ with a symmetric part $\overline{\sigma}_{ij}^{(s)} = \overline{\sigma}_{ji}^{(s)}$ and an antisymmetric part $\overline{\sigma}_{ij}^{(a)} = -\overline{\sigma}_{ji}^{(a)}$. Finally, we denote $\overline{\overline{\Omega}}_{ij} = \Omega_{ij} + W_{ij}$.

The stress tensor $\overline{\sigma}_{ij}$ includes dissipative effects. The first term $A_{ij}\overline{\sigma}_{ij}^{(s)}$ in (6.48) is the dissipation due to the shear flow. In the second term, the antisymmetric tensor W_{ij} specifies a solid-like angular rotation of system as a whole. If $\boldsymbol{\varpi} = \frac{1}{2}\nabla \times \mathbf{v}$ is the angular velocity of such rotation, then $W_{ij} = -\varepsilon_{kij}\varpi_k$. Thus, the second term in (6.48) equals $\varepsilon_{kij}(\Omega_k - \varpi_k)\overline{\sigma}_{ij}^{(a)}$ and is the dissipation caused by the relative director rotation.

The phenomenological equations (6.26) take the form

$$\overline{\sigma}_{ij}^{(s)} = \eta_{ijkl}^{11} A_{kl} + \eta_{ijkl}^{12}\overline{\overline{\Omega}}_{kl}, \quad \overline{\sigma}_{ij}^{(a)} = \eta_{ijkl}^{21} A_{kl} + \eta_{ijkl}^{22}\overline{\overline{\Omega}}_{kl}. \tag{6.49}$$

The viscosities η_{ijkl}'s can depend on the director components. Because the nematic is a centrosymmetric medium, η_{ijkl}'s must be invariant under the transformation $\mathbf{n} = -\mathbf{n}$. The coefficients η_{ijkl}^{11}'s must be symmetric with respect to the operations $i \leftrightarrow j$ and $k \leftrightarrow l$, whereas η_{ijkl}^{12}'s must be symmetric with respect to $i \leftrightarrow j$ and antisymmetric with respect to $k \leftrightarrow l$, and so on, in order to reflect the symmetry of the tensors $\overline{\sigma}_{ij}^{(s)}$, $\overline{\sigma}_{ij}^{(a)}$, A_{kl}, and $\overline{\overline{\Omega}}_{kl}$. Furthermore, if the nematic is incompressible, then the EL model requires that any term containing the trace A_{pp} or the product $\delta_{kl}A_{ij}$ in the stress tensor should vanish (see also Ref. [4] for a critical discussion). The $\delta_{kl}A_{ij}$-terms do not contribute to the entropy production, because $\delta_{kl}A_{ij}A_{kl} = A_{ij}A_{kk} = 0$. All of these requirements are satisfied for the following forms of $\overline{\sigma}_{jk}^{(s)}$ and $\overline{\sigma}_{jk}^{(a)}$:

$$\overline{\sigma}_{ij}^{(s)} = \alpha_1 n_i n_j n_k n_l A_{kl} + \alpha_4 \delta_{ik}\delta_{jl}A_{kl} + \frac{\alpha_6 + \alpha_5}{2}(\delta_{il}n_j n_k + \delta_{jl}n_i n_k)A_{kl}$$
$$+ \frac{\alpha_3 + \alpha_2}{2}(\delta_{il}n_j n_k - \delta_{kj}n_i n_l)\overline{\overline{\Omega}}_{kl},$$
$$\overline{\sigma}_{ij}^{(a)} = \frac{\alpha_5 - \alpha_6}{2}(\delta_{li}n_j n_k - \delta_{kj}n_i n_l)A_{kl} + \frac{\alpha_2 - \alpha_3}{2}(\delta_{li}n_j n_k + \delta_{kj}n_i n_l)\overline{\overline{\Omega}}_{kl}. \tag{6.50}$$

The six coefficients α_i are called the Leslie viscosity coefficients. Their dimension is [kg · m^{-1}·s^{-1}] in SI and [Poise] in cgs; 1 Poise = 1 g·cm^{-1}·s^{-1} = 0.1 kg·m^{-1}·s^{-1}. Only five out of six are independent. According to the Onsager's reciprocal relation, $\eta_{ijkl}^{12} = \eta_{klij}^{21}$, and

$$\alpha_2 + \alpha_3 = \alpha_6 - \alpha_5, \tag{6.51}$$

which is known as the Parodi's equation.[5] Finally, the positive definiteness of the entropy production imposes additional restrictions, such as $\alpha_3 > \alpha_2$ (see Problem 6.6). In the EL model, an incompressible nematic fluid is thus characterized by five independent viscosity coefficients.

The dissipative stress tensor $\sigma_{ij}^{(d)}$ and director dynamics are often written in terms of the vector $\mathbf{N}$ describing the relative rotation rate of the director

$$\mathbf{N} = \dot{\mathbf{n}} - \varpi \times \mathbf{n} \quad \text{or} \quad N_j = -n_k \overline{\overline{\Omega}}_{jk} = n_k \overline{\overline{\Omega}}_{kj}. \tag{6.52}$$

The stress tensor $\overline{\sigma}_{ij}$ expressed through $\mathbf{N}$ and A_{ij} is

$$\overline{\sigma}_{ij} = \alpha_1 n_i n_j n_k n_l A_{kl} + \alpha_2 n_j N_i + \alpha_3 n_i N_j + \alpha_4 A_{ij}$$
$$+ \alpha_5 n_j n_p A_{pi} + \alpha_6 n_i n_p A_{pj}, \tag{6.53}$$

whereas the director equations (6.37) transform with the help of (6.52) and (6.46) into

$$I \frac{d}{dt} [\mathbf{n} \times \dot{\mathbf{n}}] = [\mathbf{n} \times \mathbf{h}] + \mathbf{\Gamma}. \tag{6.54}$$

Here,

$$h_i = -\phi_i + \frac{\partial \pi_{ij}}{\partial x_j} \equiv -\frac{\partial f_{\text{FO}}}{\partial n_i} + \mu_0^{-1} \chi_a (n_j \mathcal{B}_j) \mathcal{B}_i + \frac{\partial}{\partial x_j} \left(\frac{\partial f_{\text{FO}}}{\partial (\partial n_i / \partial x_j)} \right) \tag{6.55}$$

is the molecular field that includes both pure elastic (f_{FO}) and diamagnetic effects, and

$$\mathbf{\Gamma} = (\alpha_2 - \alpha_3)[\mathbf{n} \times \mathbf{N}] + (\alpha_5 - \alpha_6)[\mathbf{n} \times \mathbf{A} \cdot \mathbf{n}] \tag{6.56}$$

is the viscous torque. Note that both (6.37) and (6.54) imply that the internal angular momentum $I \varepsilon_{ijk} n_j \ddot{n}_k$ is not conserved. The quantity $(-\varepsilon_{ijk}\sigma_{jk})$ is a source of the internal angular momentum. Of course, changes in $I\varepsilon_{ijk}n_j\ddot{n}_k$ are possible only at the expense of the external angular momentum $\rho \varepsilon_{ijk} x_j \dot{v}_k$: The same term $\varepsilon_{ijk}\sigma_{jk}$ but with an opposite sign enters an equation for $\rho\varepsilon_{ijk}x_j\dot{v}_k$ [which is found by subtracting (6.37) from (6.36)]. Therefore, the total angular momentum is conserved, as it should be.

At low-frequency excitations, the inertia term can be neglected and (6.54) reduce to

$$[\mathbf{n} \times \mathbf{h}] + \mathbf{\Gamma} = 0, \quad \text{or} \quad [\mathbf{n} \times \mathbf{h}] - [\mathbf{n} \times (\gamma_1 \mathbf{N} + \gamma_2 \mathbf{A} \cdot \mathbf{n})] = 0, \tag{6.57}$$

where we introduce two combinations of viscosities, useful for the discussion in the next sections,

---

[5]O. Parodi, J. Phys. (Paris) **31**, 581 (1970).

$$\gamma_1 = \alpha_3 - \alpha_2 > 0 \quad \text{and} \quad \gamma_2 = \alpha_6 - \alpha_5 = \alpha_2 + \alpha_3. \tag{6.58}$$

In a stationary case, when there are no time derivatives, the director equation (6.57) is simply the Euler–Lagrange equation $[\mathbf{n} \times \mathbf{h}] = 0$ for the equilibrium director field $\mathbf{n}(\mathbf{r})$.

6.4. Nematodynamics in Harvard Theory

According to the general scheme oulined at the beginning of Section 6.3, we identify seven independent hydrodynamic variables: density ρ, momentum density $\rho\mathbf{v}$, energy density $\rho\varepsilon$, and the two components of $\mathbf{n}$. The conservation laws for the first five variables retain the same form (6.9), (6.13), and (6.16) as for the isotropic fluid. The angular momentum is not given a special consideration. The conservation of angular momentum is always guaranteed if the stress tensor σ_{ij} is symmetric. In fact, the requirement for σ_{ij} is softer than that. In the conservation law (6.13) for the momentum density, the observable quantity is not the stress tensor, but its tensorial divergence $\frac{\partial \sigma_{ij}}{\partial x_j}$. The required symmetry can be restored by adding to a nonsymmetric σ_{ij} a term $\frac{\partial \chi_{ijk}}{\partial x_k}$, where χ_{ijk} is any tensor that is antisymmetric with respect to its two last indices j and k; i.e., $\chi_{ijk} = -\chi_{ikj}$. The new symmetric stress tensor $\tilde{\sigma}_{ij} = \sigma_{ij} + \frac{\partial \chi_{ijk}}{\partial x_k}$ yields $\frac{\partial \tilde{\sigma}_{ij}}{\partial x_j} = \frac{\partial \sigma_{ij}}{\partial x_j}$, so that both the linear momentum and the angular momentum are conserved.[6]

6.4.1. Director Dynamics and Dissipative Stress Tensor

The lacking "balance equation" for the director should relate the time derivative $\dot{\mathbf{n}}$ to the velocity gradients A_{ij}, W_{ij} and to the molecular field h_i specified by (6.55). The derivation of the director dynamics equations is helped by observing that the normalization $\mathbf{n}^2 = 1$ leads to $\mathbf{n}\dot{\mathbf{n}} = 0$.

First, recall that the molecular field aligns $\mathbf{n}$ parallel to itself in the equilibrium. Therefore, the quantity of interest in the molecular field (6.55) is the vector $\mathbf{m} = \mathbf{h} - \mathbf{n}(\mathbf{n}\mathbf{h})$ *perpendicular* to $\mathbf{n}$, $\mathbf{n}\mathbf{m} = 0$. The components of $\mathbf{m}$ can be written with the help of the transverse Kroneker delta as $m_i = \delta_{ij}^{\perp} h_j$. The dependence of $\dot{n}_i$ on h_j in the dynamic equation is thus described by the term $(\frac{1}{\gamma_1} \delta_{ij}^{\perp} h_j)$, where the scalar γ_1 has the dimension of the viscosity.

Second, the relationship between $\dot{n}_i$ and the velocity gradients $\frac{\partial v_j}{\partial x_k}$ should be through a third-rank tensor λ_{ijk}, namely, $\dot{n}_i = \lambda_{ijk} \frac{\partial v_j}{\partial x_k}$. Because $n_i \dot{n}_i = 0$, there should be only two independent components of λ_{ijk}. We separate λ_{ijk} into the symmetric and antisymmetric parts, so that $\dot{n}_i = \lambda_{ijk}^{(s)} A_{jk} + \lambda_{ijk}^{(a)} W_{jk}$. Because rotations of the nematic as a whole with the angular velocity $\boldsymbol{\varpi} = \frac{1}{2}$ curl $\mathbf{v}$ reorient the whole director field (with no energy dissipation), the antisymmetric contribution $\lambda_{ijk}^{(a)} W_{jk}$ should be simply $\varepsilon_{ijk}\varpi_j n_k \equiv W_{ik}n_k$.

---

[6]P.C. Martin, O. Parodi, and P.S. Pershan, Phys. Rev. **A6**, 2401 (1972).

The symmetric part can be written with the coefficient $\lambda_{ijk}^{(s)} = \lambda\,\delta_{ij}^{\perp}n_k$, where λ is some dimensionless parameter.

Summarizing, the director dynamics equations are

$$\frac{dn_i}{dt} = W_{ik}n_k + \lambda\,\delta_{ij}^{\perp}A_{jk}n_k + \frac{1}{\gamma_1}\delta_{ij}^{\perp}h_j. \tag{6.59}$$

Comparison with (6.56) and (6.57) justifies the notations introduced earlier in (6.58). As in (6.57), the diamagnetic effect is included in the molecular field $\mathbf{h}$. There is no ∇T terms on the right-hand side of (6.59) because of the symmetry $\mathbf{n} = -\mathbf{n}$.

The entropy production is calculated in a way similar to that in the previous sections. It equals (see Problem 6.5)

$$2R = \sigma_{ij}^{(d,s)}A_{ij} + N_i h_i, \tag{6.60}$$

i.e., coincides with (6.48) when the inertia term is ignored (and it should be ignored in the hydrodynamic limit). As already mentioned, the stress tensor can be made symmetric without altering the linear momentum conservation law (and guaranteeing the angular momentum conservation).[6]

The quantities $\mathbf{N} = \dot{\mathbf{n}} - \boldsymbol{\varpi}\times\mathbf{n}$ and $\sigma_{ij}^{(d,s)}$ in (6.60) are considered fluxes, whereas h_i and A_{ij} are forces. As usual, we suppose that the fluxes are linear functions of the forces. Then we can write $\mathbf{N} = \frac{1}{\gamma_1}\mathbf{h}$, i.e., there is only one scalar dissipative coefficient γ_1 associated with the relative director rotations $\mathbf{N}$. For the dissipative stress tensor, $\sigma_{ij}^{(d,s)} = \eta_{ijkl}A_{kl}$, the viscosity coefficients form a tensor η_{ijkl} that depends on the director components n_i and δ_{ij}. Considering the symmetries of the tensors $\sigma_{ij}^{(d,s)}$ and A_{kl} and the Onsager's symmetry $\eta_{ijkl} = \eta_{klij}$, one finds five independent combinations of n_i and δ_{ij}: $\delta_{ij}\,\delta_{kl}$, $\delta_{ik}\,\delta_{jl} + \delta_{jk}\,\delta_{il}$, $n_i n_j\,\delta_{kl} + n_k n_l\,\delta_{ij}$, $n_i n_k\,\delta_{jl} + n_j n_k\,\delta_{il} + n_i n_l\,\delta_{jk} + n_j n_l\,\delta_{ik}$, and $n_i n_j n_k n_l$. As the result,

$$\sigma_{ij}^{(d,s)} = 2\eta_2 A_{ij} + 2(\eta_3 - \eta_2)(n_i n_k A_{jk} + n_k n_j A_{ik}) + (\eta_4 - \eta_2)\,\delta_{ij}A_{kk}$$
$$+ 2(\eta_1 + \eta_2 - 2\eta_3)n_i n_j n_k n_l A_{kl} + (\eta_5 - \eta_4 + \eta_2)(n_i n_j A_{kk}$$
$$+ n_k n_l\,\delta_{ij}A_{kl}). \tag{6.61}$$

The restrictions on the values of η's are listed in Problem 6.6.

Note that so far we *did not* use the incompressibility condition. For an incompressible nematic, $A_{kk} = 0$, one obtains $\eta_4 = \eta_2$ and $\eta_5 = 0$,[3] and there remain only *three* viscosities.

6.4.2. Summary of Nematodynamics

For convenience, below we summarize all of the relevant equations used by the EL and H models to describe a nematic in motion. The bulk forces include the diamagnetic effect in the molecular molecular field **h**, but not the gravity.

There are seven unknown variables: (1) mass density $\rho(\mathbf{r}, t)$, (2) three components of the velocity field $\mathbf{v}(\mathbf{r}, t)$ or the momentum density $\rho\mathbf{v}(\mathbf{r}, t)$, (3) energy density, and (4) two components of the director field $\mathbf{n}(\mathbf{r}, t)$. These variables are found from seven equations
(1) conservation of mass (6.10):

$$\frac{\partial \rho}{\partial t} = -\frac{\partial \rho v_i}{\partial x_i} \tag{6.62}$$

(2) three equations for the conserved components of the linear momentum:

$$\frac{\partial \rho v_i}{\partial t} = -\frac{\partial}{\partial x_j}(-\sigma_{ij} + \rho v_i v_j). \tag{6.63}$$

The EL model presents the reversible part of the stress tensor as the elastic Ericksen stress tensor $\sigma_{ij}^{(r)}$, (6.47), and the dissipative part through (6.50) or (6.53); in the H-model, the dissipative part $\sigma_{ij}^{(d,s)}$ is given by (6.61).
(3) entropy balance equation:

$$\frac{\partial \rho s}{\partial t} + \frac{\partial}{\partial x_i}\left(\rho s v_i + \frac{J_i^q}{T}\right) = \frac{2R}{T}, \tag{6.64a}$$

where the entropy increase is either zero for reversible process or positive,

$$2R = \sigma_{ij}^{(d,s)} A_{ij} + N_i h_i - \frac{J_i^q}{T}\frac{\partial T}{\partial x_i} > 0, \tag{6.64b}$$

for irreversible ones; the heat flux is $J_i^q = -(\kappa_\| n_i n_s + \kappa_\perp \delta_{is}^\perp)\partial T/\partial x_s$.
(4) director dynamics equations, either in the form of (6.57) or (6.59):

$$[\mathbf{n} \times \mathbf{h}] - [\mathbf{n} \times (\gamma_1 \mathbf{N} + \gamma_2 \mathbf{A} \cdot \mathbf{n})] = 0, \tag{6.65a}$$

$$\mathbf{N} = \lambda[\mathbf{A} \cdot \mathbf{n} - \mathbf{n}(\mathbf{n} \cdot \mathbf{A} \cdot \mathbf{n})] + \frac{1}{\gamma_1}[\mathbf{h} - \mathbf{n}(\mathbf{h} \cdot \mathbf{n})]. \tag{6.65b}$$

The notations imply $[\mathbf{A} \cdot \mathbf{n} - \mathbf{n}(\mathbf{n} \cdot \mathbf{A} \cdot \mathbf{n})]_i = A_{ij}n_j - n_i n_j A_{jk}n_k$ and $\mathbf{N} = \dot{\mathbf{n}} - \varpi \times \mathbf{n}$.

If one compares the dissipative tensors $\overline{\sigma}_{ij}$ in (6.53) and $\sigma_{ij}^{(d,s)}$ in (6.61), the Leslie viscosities are related to the Harvard viscosities of the incompressible nematic in the following way:

$$\alpha_1 = 2(\eta_1 + \eta_2 - 2\eta_3) - \gamma_1 \lambda^2, \quad \alpha_2 = -\gamma_1(1+\lambda)/2,$$

$$\alpha_3 = \gamma_1(1-\lambda)/2, \quad \alpha_4 = 2\eta_2, \quad \alpha_5 = 2(\eta_3 - \eta_2) + \gamma_1\lambda(\lambda+1)/2,$$

$$\alpha_6 = 2(\eta_3 - \eta_2) + \gamma_1\lambda(\lambda-1)/2. \tag{6.66}$$

The Parodi relationship (6.51) is satisfied, $\alpha_6 - \alpha_5 = \alpha_3 + \alpha_2 = \gamma_2$. The dimensionless parameter λ in (6.65b) of the H-model *does not* contribute to the entropy production (6.60) and is thus *nondissipative*, despite the fact that it is the ratio of two dissipative quantities in the EL model:

$$\lambda = -\frac{\gamma_2}{\gamma_1} \equiv \frac{\alpha_2 + \alpha_3}{\alpha_2 - \alpha_3}. \tag{6.67}$$

It is possible to show that the number of viscosities in EL model reduces to three as well, when the director distortions are small.

When the director distortions are small, we can omit the term ϕ_i in the definition (6.55) of the molecular field and write $h_i = \partial \pi_{ij}/\partial x_j$. The tensor $\overline{\sigma}_{ij}$ that appears in the EL model in (6.53) rewrites

$$\overline{\sigma}_{ij} = \alpha_1 n_i n_j n_k n_l A_{kl} + \alpha_4 A_{ij} + \frac{\alpha_3 \alpha_5 - \alpha_2 \alpha_6}{\gamma_1}(n_j n_p A_{pi} + n_i n_p A_{pj})$$

$$+ \frac{\alpha_2}{\gamma_1} n_j \frac{\partial \pi_{ik}}{\partial x_k} + \frac{\alpha_3}{\gamma_1} n_i \frac{\partial \pi_{jk}}{\partial x_k}, \tag{6.68}$$

if one eliminates $N_i = \frac{dn_i}{dt} - W_{ik}n_k$ using (6.59) and the notations (6.58). The tensor $\overline{\sigma}_{ij}$, thus, splits into a truly dissipative part

$$\overline{\sigma}_{ij}^{(d)} = \alpha_1 n_i n_j n_k n_l A_{kl} + \alpha_4 A_{ij} + \frac{\alpha_3 \alpha_5 - \alpha_2 \alpha_6}{\gamma_1}(n_j n_p A_{pi} + n_i n_p A_{pj}), \tag{6.69}$$

with three independent viscosities and a nondissipative part

$$\overline{\sigma}_{ij}^{(r)} = -\frac{\lambda}{2}\left(n_j \frac{\partial \pi_{ik}}{\partial x_k} + n_i \frac{\partial \pi_{jk}}{\partial x_k}\right) - \frac{1}{2}n_j \frac{\partial \pi_{ik}}{\partial x_k} + \frac{1}{2}n_i \frac{\partial \pi_{jk}}{\partial x_k}. \tag{6.70}$$

The dissipative part is symmetric, while the non-dissipative part is not symmetric. However, as already indicated, $\overline{\sigma}_{ij}^{(r)}$ can be transformed into a symmetric tensor, for example, in the form

$$
\overline{\sigma}_{ij}^{(r,s)} = -\frac{\lambda}{2}\left(n_j\frac{\partial \pi_{ik}}{\partial x_k} + n_i\frac{\partial \pi_{jk}}{\partial x_k}\right) - \frac{1}{2}n_k\frac{\partial}{\partial x_k}(\pi_{ij} + \pi_{ji})
$$
$$
+ \frac{1}{2}n_i\frac{\partial}{\partial x_k}(\pi_{kj} + \pi_{jk}) \tag{6.71}
$$

that do not change the physically important quantity $\dfrac{\partial\overline{\sigma}_{ij}^{(r,s)}}{\partial x_j} = \dfrac{\partial\overline{\sigma}_{ij}^{(r)}}{\partial x_j}$ in the conservation law for the linear momentum. The advantage is that the symmetric form of the stress tensor makes the angular momentum automatically conserved.

Therefore, in the limit of small distortions, both models give similar results, although exact correspondence is still lacking. For example, comparing (6.69) to the H-model of the incompressible nematic, (6.61), one finds $\alpha_1 = 2(\eta_1 + \eta_2 - 2\eta_3)$, whereas according to (6.66), $\alpha_1 = 2(\eta_1 + \eta_2 - 2\eta_3) - \gamma_1\lambda^2$. As argued by Pleiner and Brand [4], the discrepancies in the definitions of the viscous coefficients in the two models can be caused by the incompleteness of the stress tensor in the EL model. In the next section, we will partially use the language of Leslie coefficients [referring only to the relationships (6.66)], because it simplifies the notations in the balance of torques.

We conclude the summary with the problems that are common for both hydrodynamic models.

If the frequencies are high, then the whole theory should be modified to include the dynamics of the scalar part of the order parameters (see the review by Beris and Edwards[7]). The relative importance is set by the *Deborah number*

$$
\mathrm{De} = \dot{\gamma}\tau, \tag{6.72}
$$

where $\dot{\gamma}$ is the characteristic shear rate and τ is the characteristic molecular relaxation time. In small-molecular weight liquid crystals, $\mathrm{De} \ll 1$. For polymeric liquid crystals, τ might be large, and De might become of the order of unity; in this regime, the flow changes the scalar order parameter. This chapter deals with $\mathrm{De} \ll 1$.

Both the EL and H-theories are limited to flows without any topological defects, such as hedgehogs, disclinations, and so on. If such defects are present, the director distortions span the whole degeneracy space (see Chapter 12), i.e., averaging the director field over distances larger than the distances between defects gives a zero result; the director is not a hydrodynamic variable any more. Motion with topological defects can be considered using the formalism of Poisson "hydrodynamic" brackets as suggested by Dzyaloshinskii and Volovik.[8] However, simple cases can be studied with a less sophisticated formalism,

---

[7]A.N. Beris and B.J. Edwards, Thermodynamics of Flowing Systems with Internal Microstructure, Oxford University Press, New York, 1994.

[8]I.E. Dzyaloshinskii and G.E. Volovik, Ann. Phys. (N.Y.) **125**, 67 (1980); G.E.Volovik and E.I. Kats, Sov. Phys. JETP **54**, 122 (1981).

inspired by the treatment of the mobility of defects in solids (see Chapter 11). Finally, note that the H-model of nematodynamics is just one example of the unified hydrodynamic theory of systems with broken symmetry.[6] Kats and Lebedev (1994) have reviewed its applications to smectic and columnar phases, freely suspended films, Langmuir monolayers, and membranes.

6.5. Applications of Nematodynamics

6.5.1. Nematic Viscosimetry

The anisotropy of the viscous properties of a nematic fluid can be illustrated by measuring effective viscosities for different director orientations with respect to the flow direction. Consider a plane Couette flow of a nematic confined between two plates $z = 0$ and $z = d$ (Fig. 6.5). The top plate moves with a constant velocity $\mathbf{U}$ along the y-axis, and the bottom

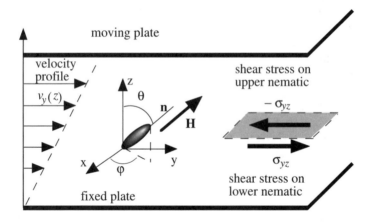

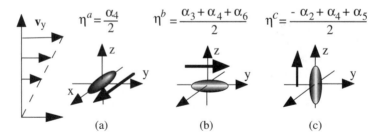

Figure 6.5. Plane Couette flow of a nematic fluid with the director fixed by a strong magnetic field. Miezowicz geometries: (a) $\mathbf{n}$ parallel to the vorticity direction, (b) $\mathbf{n}$ parallel to the flow, (c) $\mathbf{n}$ parallel to the velocity gradient.

plate is fixed; the shear flow is specified as $[0, v_y(z), 0]$. Fluids adhere to solid substrates so that $v_y|_{z=d} = U$ and $v_y|_{z=0} = 0$. Suppose that the director is fixed in space *and* time by a strong magnetic field, $\mathbf{n}(\mathbf{r}, t) = \text{const}$ (Fig. 6.5), and that the nematic satisfies the Newton's law of viscosity (6.31), $\sigma_{yz}^{(d)} = \eta \frac{dv_y}{dz}$, where $\sigma_{yz}^{(d)}$ is a constant shear stress transmitted through the nematic fluid, and η is a (constant) effective viscosity coefficient. Integrating the last equation and employing the boundary conditions, one finds that the velocity is a linear function of the coordinate z: $v_y = \dot{\gamma}z$, where $\dot{\gamma} = U/d$ is the constant shear rate with the dimension $1/s$. Let us express η as functions of the Leslie coefficients α_i's and Harvard viscosities η_i's for the following three most practical geometries of director orientation versus shear flow, known as *Miezowicz geometries*.

1. $\mathbf{n}$ perpendicular to the flow direction and to the velocity gradient, $\mathbf{n} = (1, 0, 0)$. Then, $n_i A_{ij} = 0$, $N_i = n_k W_{ki} = 0$, and the stress tensor reduces to $\overline{\sigma}_{yz} = \alpha_4 A_{yz} = \alpha_4 \dot{\gamma}/2$; i.e., $\eta^a = \alpha_4/2 = \eta_2$. The viscous torque (6.56) vanishes, $\Gamma = 0$; i.e., the nematic behaves as an isotropic fluid with an effective viscosity $\alpha_4/2 = \eta_2$.

2. $\mathbf{n}$ parallel to the flow, $\mathbf{n} = (0, 1, 0)$. The components of $n_i A_{ij}$ and $N_i = n_j W_{ji}$ are $(0, 0, \dot{\gamma}/2)$; $\overline{\sigma}_{yz} = \alpha_3 N_z + (\alpha_4 + \alpha_6)A_{yz}$. Therefore, $\eta^b = (\alpha_3 + \alpha_4 + \alpha_6)/2 = \eta_3 + \gamma_1(1 - \lambda)^2/4$.

3. $\mathbf{n}$ parallel to the velocity gradient $\mathbf{n} = (0, 0, 1)$. The components of $n_i A_{ij}$ and $N_i = n_j W_{ji}$ are $(0, \dot{\gamma}/2, 0)$ and $(0, -\dot{\gamma}/2, 0)$, respectively. Thus, $\overline{\sigma}_{yz} = \alpha_2 N_y + (\alpha_4 + \alpha_5)A_{yz}$ and $\eta^c = (-\alpha_2 + \alpha_4 + \alpha_5)/2 = \eta_3 + \gamma_1(1 + \lambda)^2/4$.

Thus, experiments with shear flows in a strong magnetic field allow for a determination of some combinations of the viscosities. In conjunction with other techniques, such as light scattering, dynamics of Frederiks transitions, rotating and oscillating fields, ultrasound attenuation, and so on, one can find all viscous coefficients. So far, experiments have been performed mainly for small-molecular weight nematics composed of elongated molecules, such as 5CB (see Table 6.1 and Fig. 6.6). The table data for 5CB or MBBA show that the ratio η^b/η^c is small, ≈ 0.18, which is a reasonable result if friction correlates with the cross section of the molecules seen by the flow. For polymeric lyotropic liquid crystals such as poly-γ-benzyl-glutamate (PBG) dissolved in a mixture of methylene chloride and dioxane,[9] this ratio is even smaller, $\eta^b/\eta^c \approx 0.005$. Little is known about discotic materials. The basic difference between them and the usual calamitic materials is in value of λ; $\lambda \sim -1$ for disk-like molecules and $\lambda \sim 1$ for rod-like molecules.[10] Suppose that the director in equilibrium is aligned along the z-axis. Then, in a calamitic nematic, reorientation of the rod-like molecules along $\mathbf{n}$ is caused primarily by the velocity gradients $\partial v_x/\partial z$ and $\partial v_y/\partial z$. Alternatively, reorientation of discs perpendicular to $\mathbf{n}$ is caused by $\partial v_z/\partial x$ and $\partial v_z/\partial y$. The data for the calamitic nematics confirm that $\lambda \sim 1$, being normally slightly larger than 1 (see Table 6.1 and Fig. 6.6).

[9]G. Srayer, S. Fraden, and R.B. Meyer, Phys. Rev. A **39**, 4828 (1989).

[10]G.E. Volovik, Pis'ma Zh. Eksper. Teor. Fiz. **31**, 297 (1980)/JETP Lett. **31**, 273 (1980).

Table 6.1. Typical viscosities for pentylcyanobiphenyl (5CB) (from L.M. Blinov and V.G. Chigrinov, Electrooptic Effects in Liquid Crystal Materials, Springer, New York, 1996, 464 p); MBBA at 25°C (from W.H. de Jeu, Physical Properties of Liquid Crystalline Materials, Gordon and Breach Science, New York, 1980), and PBG, from Ref. [9].

Viscosities, 10^{-3} kg $\cdot$ m$^{-1}\cdot$ s^{-1} (or 10^{-2} Poise)	5CB	MBBA	PBG
α_1	-11	-18 ± 6	-3660
α_2	-83	-109 ± 2	-6920
α_3	-2	-1 ± 0.2	18
α_4	75	83 ± 2	348
α_5	102	80 ± 15	6610
α_6	-27	-34 ± 2	-292

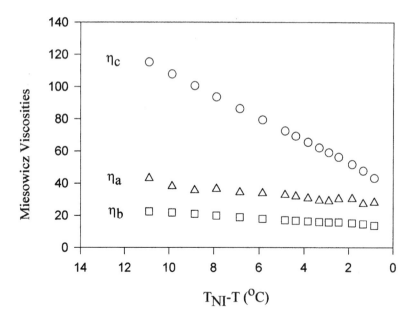

Figure 6.6. Temperature dependencies of the Miezowicz coefficients η^a, η^b, and η^c (units: 10^{-3} kg $\cdot$ m$^{-1}\cdot$ s^{-1}) for 5CB (M. Cui and J. Kelly, Kent State University; see Mol. Cryst. Liq. Cryst. **331**, 49 (1999) for more data and discussion).

6.5.2. Flow-Aligning and Tumbling Nematics with Director in the Shear Plane

The considerations above apply to a uniform director orientation fixed by an external field. If the field is absent, the flow of a confined nematic is determined by a balance of the elastic torque $\sim K/L^2$ and the viscous torque $\sim \eta\dot{\gamma} \sim \eta U/L$. Here, L is a typical scale of deformation, for example, the distance between the cell plates. The relative importance of these torques is expressed by the dimensionless *Ericksen number*

$$\mathrm{Er} = \frac{\eta\dot{\gamma}L^2}{K} = \frac{\eta LU}{K}. \tag{6.73}$$

With the estimates $\eta \sim 10^{-2}\,\mathrm{kg \cdot m^{-1} \cdot s^{-1}}$, $K \sim 10^{-11}\,\mathrm{N}$, $L \sim 10\,\mu\mathrm{m}$ and $\dot{\gamma} \sim 10\,\mathrm{s^{-1}}$, the Ericksen number is around 10^3. The viscous stresses dominate the elastic stresses, at least at high shear rates and until the flow produces spatial director gradients at very small scales L. One might expect strong gradients in the vicinity of the boundary layers.

6.5.2.1. High Shear Rates (Er $\gg$ 1)

To analyze the coupling between the velocity and the orientation at $\mathrm{Er} \gg 1$, we can neglect the elastic torques. Because the EL model gives a more intuitive insight into the balance of torques, we will use the Leslie notations for the viscosities. However, we will also use the parameter λ whenever necessary to stress the nondissipative character of phenomena such as flow alignment.

We are interested in finding an orientation (if it exists) for which the viscous torque vanishes. This orientation is stable if any small deviation causes a viscous torque that drives $\mathbf{n}$ back into the original state.

For $\mathbf{n} = (\sin\theta\cos\varphi, \sin\theta\sin\varphi, \cos\theta)$ and $\mathbf{U} = (0, \dot{\gamma}z, 0)$, $\dot{\gamma} > 0$ (Fig. 6.5), the director equations (6.65) reduce to

$$\gamma_1 \frac{d\theta}{dt} + \dot{\gamma}(\alpha_2\cos^2\theta - \alpha_3\sin^2\theta)\sin\varphi = 0, \tag{6.74a}$$

$$\gamma_1 \frac{d\varphi}{dt}\sin\theta + \alpha_2\dot{\gamma}\cos\theta\cos\varphi = 0, \tag{6.74b}$$

when there is no external field and the elastic terms are neglected. From now on, we make a distinction between flows where the director is in the shear plane (this section) and perpendicular to the shear plane (next section).

As seen from (6.74), a steady state may occur when the director is in the shear plane, $\varphi = \pi/2$. The nonvanishing viscous torque (6.56) is around the x-axis, $\Gamma_x \sim \dot{\gamma}(\alpha_2\cos^2\theta - \alpha_3\sin^2\theta)$ (Fig. 6.7). For the two limiting geometries (b) and (c) of Fig. 6.7, $\Gamma_x^b \sim -\alpha_3\dot{\gamma} \sim \dot{\gamma}\gamma_1(\lambda - 1)$ and $\Gamma_x^c \sim \alpha_2\dot{\gamma} \sim -\dot{\gamma}\gamma_1(\lambda + 1)$. If α_2 and α_3 are of the same sign ($\lambda^2 > 1$),

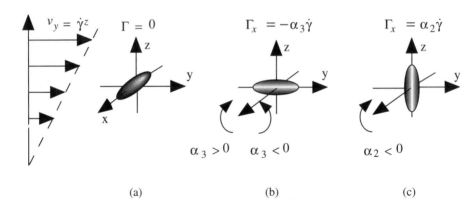

Figure 6.7. Viscous torques acting on the director field in the three Miezowicz geometries. See text.

then the steady state occurs for two angles θ_0 specified by[11]

$$\tan \theta_0 = \pm \sqrt{\frac{\alpha_2}{\alpha_3}} = \pm \sqrt{\frac{\lambda + 1}{\lambda - 1}}. \tag{6.75}$$

Usually, for nematics such as MBBA and 5CB with rod-like molecules $\alpha_2 < 0$, $1.01 < \lambda < 1.1$ (see Table 6.1), and the direction of reorientation in the geometry (c) is the one shown in Fig. 6.7c, which seems to be a reasonable result for elongated molecules placed vertically in a plane Couette flow. The situation is more tricky with α_3, because it can change sign from negative to positive; i.e., λ changes from $\lambda > 1$ to $\lambda < 1$. Such a behavior has been reported close to the N-SmA or N-solid crystal transition. For example,[12] $\lambda = 0.27$ for the nematic 8CB at 34°C. Different signs of the product $\alpha_2\alpha_3$ lead to very different flow behavior.

1. Flow alignment: $\lambda^2 > 1$, $\alpha_2\alpha_3 > 0$.

   If $\lambda > 1$, the *stable* steady solution is that with a "+" sign in (6.75): A director fluctuation $\delta\theta$ will cause a torque $\Gamma_x \sim (-2\dot\gamma\alpha_3\sqrt{\alpha_2/\alpha_3})\,\delta\theta \sim \dot\gamma\gamma_1\sqrt{\lambda^2 - 1}\,\delta\theta$, restoring the angle θ_0 as illustrated in Fig. 6.8. This regime is called the "flow aligning" regime. The director should be closer to the y-axis than to the z-axis, $\pi/4 < \theta_0 < \pi/2$, because the requirement of positive entropy production sets $\alpha_3 - \alpha_2 > 0$ (see Problem 6.6). Indeed, for 5CB and MBBA, the deviations from "perfect" alignment $\theta_0 = \pi/2$ along the flow are small, about 1^0–10^0.

   Alternatively, for discotic molecules $\alpha_3 > \alpha_2 > 0$, $\lambda < -1$, the stable steady solution is the one with a "−" sign in (6.75) (Fig. 6.8); $-\pi/4 < \theta_0 < 0$, and the discs are not expected to deviate much from the xy plane.

[11]F.M. Leslie, Mol. Cryst. Liq. Cryst. **63**, 111 (1981).
[12]H. Kneppe, F. Schneider, and N.K. Sharma, Ber. Bunsenges. Phys. Chem. **85**, 784 (1981).

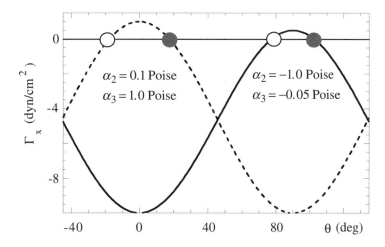

Figure 6.8. Likely angular dependencies of the viscous torque $\Gamma_x = \dot{\gamma}(\alpha_2 \cos^2 \theta - \alpha_3 \sin^2 \theta)$ for flow-aligning calamitic (solid line) and discotic (dashed line) nematics; $\dot{\gamma} = 10 \, \text{s}^{-1}$; the coefficients of viscosity are indicated at the plot. Open circles mark stable, steady director orientations, and closed circles mark unstable orientations.

2. Tumbling: $\lambda^2 < 1, \alpha_2\alpha_3 < 0$.

   Equation (6.75) has no real solutions: **n** rotates in the shear plane, which is called the *tumbling regime*. The period of time needed for the director to rotate by an angle π is calculated by integrating (6.74a): $T = \dfrac{2\pi}{\dot{\gamma}\sqrt{1-\lambda^2}}$, i.e., inversely proportional to the shear rate. Tumbling easily produces disclination lines.

   Tumbling is especially important for LCPs, because these materials generally show $\alpha_2\alpha_3 < 0$. Flow of LCPs shows distinctive features compared with the flow of isotropic polymers. The most striking is the effect of a negative first normal stress difference $N_1 = \sigma_{yy} - \sigma_{zz}$, where σ_{yy} is the normal stress in the direction of the velocity and σ_{zz} that in the shear gradient direction. When the flow of a non-Newtonian fluid occurs in a cone-and-plate rheometer, there is a thrust along the axis of rotation. Usually, for an isotropic polymer, this thrust tends to separate the plate and the cone, which means $N_1 > 0$. Kiss and Porter[13] found that for some shear rates, $N_1 < 0$ in LCP PBG. The mechanism is intimately related to the liquid-crystalline character of PBG and tumbling. In an ordinary isotropic polymer fluid, shear increases the orientational order s of molecules along the flow direction. The elastic response of the fluid is to restore the random ordering, $s = 0$, and thus to push apart the cone and the plate along the z-axis. In the LCP, s is nonzero in the nonperturbed state. Shear might create tumbling and, thus, decrease s. The elastic

---

[13]G. Kiss and R.S. Porter, J. Polym. Sci., Polym. Symp. **65**, 193 (1978).

response would tend to restore a higher value of s and thus to pull the cone and the plate closer together.[14]

6.5.2.2. Low Shear Rates (Er ≪ 1)

Consider now the steady regime at low Ericksen numbers, when the elastic torques resist director reorientations.[15] If the director is in the shear plane and the flow is along the y-axis, the torque equation (6.57) is

$$
(K_1 \sin^2 \theta + K_3 \cos^2 \theta) \frac{\partial^2 \theta}{\partial z^2} + (K_3 - K_1) \sin \theta \cos \theta \left(\frac{\partial \theta}{\partial z} \right)^2
$$

$$
- \frac{\partial v_y}{\partial z} (\alpha_3 \sin^2 \theta - \alpha_2 \cos^2 \theta) = 0, \tag{6.76}
$$

and the linear momentum equation is

$$
\frac{\partial}{\partial z} \left[\tilde{\eta}(\theta) \frac{\partial v_y}{\partial z} \right] = 0, \tag{6.77}
$$

where $-L/2 \leq z \leq L/2$ and

$$
\tilde{\eta}(\theta) = \alpha_1 \sin^2 \theta \cos^2 \theta + \eta^b \sin^2 \theta + \eta^c \cos^2 \theta. \tag{6.78}
$$

The momentum equation integrates to

$$
\tilde{\eta}(\theta) \frac{\partial v_y}{\partial z} = \tilde{\sigma} = \text{const.} \tag{6.79}
$$

The equations (6.76) and (6.79) can be linearized when the shear is small. Suppose also that the surface anchoring at the boundaries sets a "planar" director orientation $\mathbf{n} = (0, 1, 0), \theta = \pi/2$. The linearized momentum equation (6.79)

$$
\eta^b \frac{\partial v_y}{\partial z} = \tilde{\sigma} \tag{6.80}
$$

yields $v_y = \dot{\gamma} z$ with $\dot{\gamma} = \tilde{\sigma}/\eta^b$, whereas the linearized torque equation (6.76)

$$
K_1 \frac{\partial^2 \theta}{\partial z^2} = \alpha_3 \dot{\gamma} \tag{6.81}
$$

---

[14]G. Marrucci and P.L. Maffettone, Mocromolecules **22**, 4076 (1989); R.G. Larson, Ibid. **23**, 3983 (1990).

[15]E. Dubois-Violette and P. Manneville, in Pattern Formation in Liquid Crystals, Edited by A. Buka and L. Kramer, Springer-Verlag, New York, 1995, Chapter 4.

results in the following profile of the polar angle θ across the cell:

$$\theta = \frac{\pi}{2} - \frac{\alpha_3 \dot{\gamma} L^2}{8 K_1} \left[1 - \left(\frac{2z}{L} \right)^2 \right]. \tag{6.82}$$

Comparison of the results for low and high Ericksen numbers shows an apparent non-Newtonian behavior of the nematic fluids. For a planar nematic cell, the apparent viscosity changes from η_1 to $\tilde{\eta}(\theta_0)$ as the shear rate increases. Although locally the relationship between the stresses and strains is taken to be linear, the flow-induced reorientation makes the effective viscosity dependent on the strength of the stress. Not only the flow can cause director distortions, but also the director changes can induce flow. These "backflow effects" have significant implications in nematic display devices.

6.5.3. Instabilities with the Director Field Perpendicular to the Shear Plane

An obvious steady solution of the system (6.74) is $\theta = \pi/2$ and $\varphi = 0$, which corresponds to the geometry (a), i.e., with the director normal to the shear plane (Fig. 6.7a). The viscous torque vanishes when $\mathbf{n} = (1, 0, 0)$; i.e., the nematic behaves as an isotropic fluid with an effective viscosity $\alpha_4/2$ as established above. However, when α_2 and α_3 are of the same sign, the initial orientation $\mathbf{n} = (1, 0, 0)$ is unstable against director fluctuations (Fig. 6.9).[16] Qualitatively, a small deviation $\delta n_z > 0$ leads to a viscous torque $\Gamma_z =$

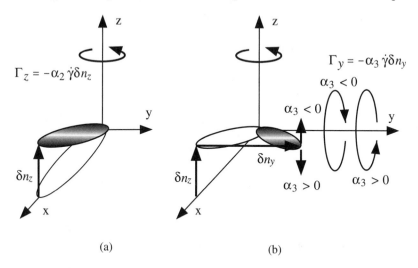

(a)   (b)

Figure 6.9. Pieranski–Guyon instability of the director initially normal to the shear plane. Redrawn from Ref. [16].

[16]P. Pieranski and E. Guyon, Solid State Comm. **13**, 435 (1973).

$-\alpha_2\dot{\gamma}\,\delta n_z > 0$ (Fig. 6.9a). When $\alpha_2 < 0$, this torque rotates the director around the z-axis and produces a component $\delta n_y > 0$. With a nonzero δn_y, there is another torque, $\Gamma_y = \alpha_3\dot{\gamma}\delta n_y$ (Fig. 6.9b). If $\alpha_3 < 0$, this torque amplifies the original deviation $\delta n_z > 0$ and the director turns toward the yz-plane. Of course, if the surface anchoring keeps $\mathbf{n} = (1, 0, 0)$ at the boundaries, the instability occurs only when the shear rate exceeds some threshold, determined by the balance of the viscous and elastic torques (see Problems 6.7 and 6.8).

6.6. Hydrodynamic Modes

The study of the linearized versions of the equations of hydrodynamics provide insight into the nature of the small movements, for small frequencies ($\omega \to 0$) and long wavelengths ($|\mathbf{q}| \to 0$). The number of balance equations ($k = 5$ for an isotropic liquid, $k = 7$ for a nematic) is equal to the number of independent variables. Expanding the fluxes in the balance equations, one expects a similar number of linearized differential equations $\omega = \omega(\mathbf{q})$, of first order in d/dt. Replacing $d/dt \to -i\omega$, one arrives at the determinant equation $\sum_{n=0}^{k} C_n(\mathbf{q})\omega^n = 0$, where $C_n(\mathbf{q})$ are polynomials of order n in $\mathbf{q}$. The roots of this equation, or the "modes," separate into *dissipative modes* (noted D) implying entropy production, and *nondissipative modes*, also called propagative (noted P).

    The same analytical approach allows one to find the response functions (real and imaginary parts of the susceptibilities) and correlation functions. These quantities are directly accessible to experimentation. Ultrasound experiments yield the real susceptibilities, measurements of transport coefficients such as viscosities and diffusivities yield the imaginary parts, and light and neutron scattering yield the correlation functions.

    In *isotropic liquids*, where the number of variables is five (the density ρ, the three components of the velocity $\mathbf{v}$, and the temperature T), the longitudinal (along $\mathbf{v}$) and transversal (perpendicular to $\mathbf{v}$) modes are decoupled. One finds the following five modes:

two P modes:

$\omega = \pm cq - i\gamma q^2$, where the damping term, often neglected, originates in the coupling with shear and thermal diffusion. P modes always exist by pairs, in reason of the invariance under time reversal of the corresponding hydrodynamic equations.

three D modes:

$\omega = -i\kappa_{th}q^2$ thermal diffusion; the dependence in iq^2 is characteristic of a conserved variable that obeys an equation of the diffusion type.

$\omega = -i\frac{\eta}{\rho}q^2$ two shear transverse modes, corresponding to two different polarizations.

    In *nematics*, there are two supplementary D modes, due to the "symmetry-breaking" continuous variables associated with the director, viz., two D modes:

$\omega = -i \frac{K}{\eta} q^2$, where K is a relevant Frank coefficient and η is a rotational viscosity.

It is important to notice that the diffusivity associated to the rotation of the director $D_{\rm rot} = \frac{K}{\eta}$ and the diffusivity associated with the shear of matter $D_{\rm sh} = \frac{\eta}{\rho}$ are not at all of the same order of magnitude: typically, $\frac{D_{\rm rot}}{D_{\rm sh}} = \frac{K\rho}{\eta^2} \approx 10^{-4}$ in thermotropic nematics. The relaxation of vorticity (shear) is a rapid process compared with the relaxation of the director orientation. Light scattering experiments have given a wealth of data on these slow, orientational modes (viscosities, Frank constants).

Problem 6.1. Show that $\frac{d}{dt} \iiint_V a\rho \, dV = \iiint_V \rho \frac{da}{dt} \, dV$, where a is a time- and coordinate-dependent characteristic of a fluid.

Answers: Observe that $\rho \, dV$ is the constant mass of the element of fluid.

Problem 6.2. Reexpress the angular momentum equation for a nematic fluid in the Eulerian form when there are no body forces.

Answers: $\frac{\partial}{\partial t}(I\Omega_i) + \frac{\partial}{\partial x_p}(I\Omega_i v_p - \varepsilon_{ijk} n_j \tau_{kp}) = -\varepsilon_{ijk}\sigma_{jk}$ for the internal part and $\varepsilon_{ijk}\frac{\partial}{\partial t}(\rho x_j v_k) + \varepsilon_{ijk}\frac{\partial}{\partial x_p}(\rho x_j v_k v_p - x_j \sigma_{kp}) = \varepsilon_{ijk}\sigma_{jk}$ for the external part; neither part is conserved by itself because of the "source" terms on the right-hand sides; however, the sources cancel each other when the equations are added, so that the total angular momentum is conserved.

Problem 6.3. Consider the angular momentum conservation law for an isotropic fluid, and show that it leads to the symmetry of the stress tensor $\sigma_{ij} = \sigma_{ji}$.

Answers: The rate of change of the angular momentum of any material element is equal to the sum of the torques created by the body and surface forces: $\rho \frac{d}{dt}(\varepsilon_{ijk} x_j v_k) = \rho\varepsilon_{ijk} x_j g_k + \frac{\partial}{\partial x_p}(\varepsilon_{ijk}\sigma_{kp})$, where x_j is the coordinate of the position of the element with respect to origin; vector products such as $[\mathbf{r} \times \mathbf{v}]$ are expressed through their components $[\mathbf{r} \times \mathbf{v}]_i = \varepsilon_{ijk} x_j v_k$. Because $\partial x_j / \partial x_p = \delta_{jp}$, then $\frac{\partial}{\partial x_p}(x_j \sigma_{kp}) = \sigma_{kj} + x_j \frac{\partial \sigma_{kp}}{\partial x_p}$ and the angular momentum equation can be rearranged as $\varepsilon_{ijk} x_j [\rho \frac{dv_k}{dt} - \rho g_k - \frac{\partial \sigma_{kp}}{\partial x_p}] = \varepsilon_{ijk}\sigma_{kj}$. The linear momentum law implies that the term in the brackets vanishes; thus, $\varepsilon_{ijk}\sigma_{kj} = 0$, i.e., $\sigma_{kp} = \sigma_{pk}$.

Problem 6.4. Calculate the angular momentum of a unit volume of a liquid crystal as the function of the scalar order parameter $s \leq 1$.

Answers:[17] The microscopic density μ of the angular momentum is the sum of angular momenta $\mu_k = I\omega_k$ of individual molecules located at points $\mathbf{r}_k(t)$: $\mu(\mathbf{r}, t) = \sum_k I\omega_k \, \delta(\mathbf{r} - \mathbf{r}_k)$; $I = ma^2$

---

[17]V.I. Sugakov and E.M. Verlan, Hydrodynamics of Liquid Crystals, Kiev State University Publ., Kiev, 1978.

is the moment of inertia of a molecule of mass m and characteristic size a. Let $\mathbf{e}_k$ be a unit vector along the molecular axis. Then, the angular velocity has components $(\omega_k)_i = [\mathbf{e}_k \times \frac{d\mathbf{e}_k}{dt}]_i = [\mathbf{e}_k \times \omega_k \times \mathbf{e}_k]_i = (\omega_k)_j [\delta_{ij} - (e_k)_i (e_k)_j]$, and thus, $\mu_i(\mathbf{r}, t) = \sum_k I(\omega_k)_j [\delta_{ij} - (e_k)_i (e_k)_j] \delta(\mathbf{r} - \mathbf{r}_k)$. Averaging over a small volume gives $\mu_i(\mathbf{r}, t) = \frac{\rho(\mathbf{r},t)}{m_k} \sum_k I_{ij}(\mathbf{r}, t) \Omega_j(\mathbf{r}, t)$, where $\Omega_j(\mathbf{r}, t) = [\mathbf{n} \times \frac{d\mathbf{n}}{dt}]_j$ is the angular velocity of the director, and

$$I_{ij} = I \left\langle \sum_k [\delta_{ij} - (e_k)_i (e_k)_j] \delta(\mathbf{r} - \mathbf{r}_k) \right\rangle$$

$$= \frac{\rho}{m_k} I \left\{ \frac{2}{3} \delta_{ij} - s(\mathbf{r}, t) \left[n_i n_j - \frac{1}{3} \delta_{ij} \right] \right\}$$

is the local moment of inertia; $\langle \ldots \rangle$ denotes an average over a molecular ensemble. Because $\mathbf{\Omega}$ and $\mathbf{n}$ are perpendicular to each other, $n_i n_j \Omega_j = 0$, and, neglecting the dependence of s on space and time coordinates, one arrives at $\mu_i = \rho \frac{I}{m} (\frac{2}{3} + \frac{1}{3} s) \Omega_i$; the moment of inertia per unit volume is $\sim \rho a^2 (\frac{2}{3} + \frac{1}{3} s)$.

Problem 6.5. Reexpress (6.48) $2R = A_{jk} \overline{\sigma}_{jk}^{(s)} + (\Omega_{jk} + W_{jk}) \overline{\sigma}_{jk}^{(a)}$ in terms of the vector $\mathbf{N}$ of the relative director rotation rate (6.52) and the molecular field (6.55).

Answers: Write $(\Omega_{jk} + W_{jk}) \overline{\sigma}_{jk}^{(a)} = \varepsilon_{ijk} (\Omega_i - \varpi_i)(\sigma_{jk} - \sigma_{jk}^{(r)})$, and then use (6.46b) to find $\varepsilon_{ijk} \sigma_{jk}^{(r)}$ and the director equation in the form $I \varepsilon_{ijk} n_j \ddot{n}_k = \varepsilon_{ijk} n_j G_k - \varepsilon_{ijk} \sigma_{jk} + \varepsilon_{ijk} \frac{\partial}{\partial x_p} (n_j \tau_{kp}) = 0$, valid when the *inertia terms are neglected*. The result is $2R = A_{jk} \overline{\sigma}_{jk}^{(s)} + N_j h_j$ [see also (6.60)].

Problem 6.6. From the requirement that the entropy production is positive-definite, which corresponds to the fact that energy is dissipated, find restrictions on the viscosity coefficients in EL and H-models.
Hint. Simplify (6.48) and (6.60) by directing one of the coordinate axis along $\mathbf{n}$.

Answers: Let us choose the x-axis of the Cartesian coordinate frame along the director $\mathbf{n}$. Then, from (6.48), (6.50), and (6.51), one finds the entropy production

$$2R = (\alpha_1 + \alpha_4 + \alpha_5 + \alpha_6) A_{xx}^2 + \alpha_4 A_{ij} A_{ij}$$

$$+ \left[(2\alpha_4 + \alpha_5 + \alpha_6) A_{xi} A_{xi} + 2(\alpha_6 - \alpha_5) A_{xi} \overline{\overline{\Omega}}_{xi} + (\alpha_3 - \alpha_2) \overline{\overline{\Omega}}_{xi} \overline{\overline{\Omega}}_{xi} \right],$$

where i and j take values y and z; there is no summation over the index x. The requirement $2R > 0$ implies that $\alpha_1 + \alpha_4 + \alpha_5 + \alpha_6 > 0$; $\alpha_4 > 0$; $2\alpha_4 + \alpha_5 + \alpha_6 > 0$; $\alpha_3 - \alpha_2 > 0$; and $(2\alpha_4 + \alpha_5 + \alpha_6)(\alpha_3 - \alpha_2) - (\alpha_6 - \alpha_5)^2 > 0$. In a similar way, for the Harvard model, $\eta_2 \geq 0$, $\eta_3 \geq 0$, $\eta_4 \geq 0$, $\eta_4(2\eta_1 + \eta_2) \geq (\eta_5 - \eta_4)^2$, and $2(\eta_1 + \eta_5) - \eta_4 + \eta_2 \geq 0$.

Problem 6.7. Consider a nematic fluid in a planar cell of thickness L in a regime of a plane Couette flow. Find the critical shear rate above which orientation $\mathbf{n} = (1, 0, 0) = \text{const}$ becomes unstable. Suppose that (1) α_2 and α_3 are of the same sign; (2) the surface anchoring is infinitely strong; (3) the flow is of the form $[0, \dot{\gamma}z, 0]$, where $\dot{\gamma} > 0$ is the shear rate.

Answers:[18] For small fluctuations n_y and n_z, the angular momentum balance equations become

$$(\alpha_3 - \alpha_2)\frac{\partial n_z}{\partial t} - K_1 \frac{\partial^2 n_z}{\partial z^2} + \alpha_3 \dot{\gamma} n_y = 0;$$

$$(\alpha_2 - \alpha_3)\frac{\partial n_y}{\partial t} + K_2 \frac{\partial^2 n_y}{\partial z^2} - \alpha_2 \dot{\gamma} n_z = 0.$$

Seeking the solutions in the form of $n_y = n_{y0} \exp(iqz)$ and $n_z = n_{z0} \exp(iqz)$, where $q = \pi/L$, one has $\alpha_3 \dot{\gamma} n_{y0} + K_1 q^2 n_{z0} = 0$ and $K_2 q^2 n_{y0} + \alpha_2 \dot{\gamma} n_{z0} = 0$. From the condition of compatibility of this algebraic linear system of two equations one gets $\dot{\gamma}_{\text{crit}} = \frac{2}{\gamma_1}\sqrt{\frac{K_1 K_2}{\lambda^2 - 1}}(\frac{\pi}{L})^2$. The instability condition can be written $\text{Er} > \pi^2$, where $\text{Er} = \frac{\gamma_1 UL}{2}\sqrt{\frac{\lambda^2 - 1}{K_1 K_2}}$ is the Ericksen number of the problem.

Problem 6.8. Consider the same geometry and conditions as in Problem 6.7, but with an additional magnetic field $\mathbf{B} = (\mathcal{B}, 0, 0)$ stabilizing the initial director orientation $\mathbf{n} = (1, 0, 0) = \text{const}$. Find the critical shear rate above which the director becomes unstable.

Answers: Addition of the magnetic stabilizing torques $\Gamma_{\text{mag}} = \mu_0^{-1}\chi_a(\mathbf{n} \times \mathbf{B})\mathbf{n} \cdot \mathbf{B}$ results in the condition of compatibility

$$\alpha_2 \alpha_3 \dot{\gamma}_c^2 = \left(K_1 \pi^2/L^2 + \mu_0^{-1}\chi_a^2 \mathcal{B}^2\right)\left(K_2 \pi^2/L^2 + \mu_0^{-1}\chi_a^2 \mathcal{B}^2\right),$$

similar to that of Problem 6.7. If the field is weak, a homogeneous distortion of the director appears above $\dot{\gamma}_{\text{crit}} = \frac{2}{\gamma_1}\sqrt{\frac{K_1 K_2}{\lambda^2 - 1}}(\frac{\pi}{L})^2$. Under a strong magnetic field, the threshold is $\dot{\gamma}_{\text{crit}} = \frac{2}{\gamma_1}\frac{\mu_0^{-1}\chi_a^2 \mathcal{B}^2}{\sqrt{\lambda^2 - 1}}$. Above the threshold, a pattern of rolls occurs with the roll axes along the direction of flow (see Dubois-Violette and Manneville[15] and Pieranski and Guyon[18]).

Further Reading

L.D. Landau and E.M. Lifshitz, Fluid Mechanics (Course of Theoretical Physics, Vol.VI, translated from the Russian by J.B. Sykes and W.H. Reid), Pergamon, New York, 1987.

G.K. Batchelor, An Introduction to Fluid Dynamics, Cambridge University Press, 1970.

M.J. Stephen and J.P. Straley, Physics of Liquid Crystals, Rev. Mod. Phys. **46**, 617 (1974).

Pattern Formation in Liquid Crystals, Edited by A. Buka and L. Kramer, Springer-Verlag, New York, 1995.

---

[18]P. Pieranski and E. Guyon, Phys. Rev. A **9**, 404 (1974); P. Manneville and E. Dubois-Violette, J. Phys. (Paris) **37**, 285 (1976).

Hydrodynamics and Nonlinear Instabilities, Edited by C. Godrèche and P. Manneville, Cambridge University Press, 1998.

S.R. de Groot and P. Mazur, Non-Equilibrium Thermodynamics, Dover Publications, Inc., New York, 1984.

D. Forster, Hydrodynamic Fluctuations, Broken Symmetry, and Correlation Functions, W.A. Benjamin, Inc., Reading, MA, 1975.

E.I. Kats and V.V. Lebedev, Fluctuational Effects in Dynamics of Liquid Crystals, Springer-Verlag, New York, 1994.

R.G. Larson, The Structure and Rheology of Complex Fluids, Oxford University Press, 1999.

F. M. Leslie, Adv. Liq. Cryst. **4**, 1 (1979).

G. Marrucci and F. Greco, Flow Behavior of Liquid Crystalline Polymers, Adv. Chem. Phys. LXXXVI, Edited by I. Prigogine and S.A. Rice, John Wiley & Sons, Inc., p. 331 (1993).

P.C. Martin, O. Parodi, and P.S. Pershan, Phys. Rev. **A6**, 2401 (1972).

L.C. Woods, The Thermodynamics of Fluid Systems, Clarendon Press, Oxford, 1975.

CHAPTER 7

Fractals and Growth Phenomena

Euclidian geometry describes the world as a pattern of simple shapes: spheres, triangles, lines, and so on, with an intuitively clear concept of dimension: 0 for a point, 1 for a line, 2 for a plane, and so on. However, this classic picture is not a complete image of Nature, for "Clouds are not spheres, mountains are not cones, coastlines are not circles, and bark is not smooth, nor does lightning travel in a straight line."[1] B. Mandelbrot, who developed the new family of shapes and coined the term fractal, gives one of the possible definitions: "A fractal is a shape made of parts similar to the whole in some way."[2]

In mathematics, the concepts of fractals or "fractal dimension(s)" apply to structures that have a never-ending self-similarity, deterministic or statistical, at all scales of magnification. The fractal dimension serves as an exponent in the power law of the type

$$M(\lambda r) = \lambda^D M(r), \tag{7.1}$$

which shows how the "property" M of the fractal (for example, its mass) changes when the characteristic size in the embedding space is rescaled by a factor λ, $r \to \lambda r$. Note that the exponent is independent of r, which stresses the self-similarity at all scales. Regular homogeneous (called also compact) objects satisfy (7.1) with D being the "usual" integer dimension 1, 2, 3, and so on. The most interesting feature of (7.1) is that there are indeed objects—fractals—fitting (7.1) with D "fractional." Simplified to the limit, interpreting r as a unit length, and $\lambda = L$ as its measure, the scaling of the "property" of fractal with its size L reads as

$$M \propto L^D, \tag{7.2}$$

where, for example, $D = 1.42$ or $D = \ln 3/ \ln 2 = 1.58496\ldots$, indicating that the fractal under consideration is not a line and not a surface. Note that here the symbol "$\propto$" indicates proportionality.

---

[1]B.B. Mandelbrot, The Fractal Geometry of Nature, W.H. Freeman and Company, New York, 1983, p. 1 (first edition: 1977).

[2]B. Mandelbrot, 1987, cited in: J. Feder. Fractals, Plenum Press, New York and London, 1988, p. 11.

In physics, the idea of infinite self-similarity is obviously limited because interactions among particles are different at different scales, say, subatomic and superatomic. Nevertheless, many physical systems do exhibit self-similarity, although of course within some finite range of scales. The list includes colloidal aggregates, polymers (see also Chapter 15), gels, crumpled membranes of surfactants, porous media, Brownian motion, turbulent flows, focal conic domain patterns in smectic liquid crystals (Chapter 10 and Section 13.2.6.), and even bacterial colonies. The geometry of these systems, often based on random processes such as Brownian motion, is complicated; the concept of fractal dimension(s) helps to express, model, and comprehend both the geometrical complexity and its physical consequences. Furthermore, fractal concepts and power laws such as (7.1) establish similarities between growth phenomena (pattern formation) in a variety of equilibrium (such as percolation) and far-from-equilibrium (such as diffusion-limited aggregation and viscous fingering) processes. This connection is of heuristic significance, because presently there is no first-principle theory to describe, for example, diffusion-limited aggregation, which is a markedly far-from-equilibrium and nonlocal process.

We start our consideration with classic fractals (e.g., Koch curve) or "thought-to-be-fractals" (e.g., coastlines) and then consider a number of physical phenomena: percolation, random walks, diffusion-limited aggregation, and viscous fingering in Hele–Shaw cells. In some of the listed cases, the fractals do not necessarily form: Viscous fingering is not always of fractal structure. There is another unifying theme that puts fractal-like aggregates and viscous fingering in one chapter: In both cases, one deals with instabilities of the growth front, when small perturbations of the front (interface) start to grow much faster than do the neighboring regions. The physical description of front instabilities is based on the Laplace equation with appropriate boundary conditions.

7.1. Basic Fractal Concepts

Many natural objects can be approximated by simple Euclidean figures, such as 0D points, 1D straight lines, 2D planes, or 3D spheres. Some objects do not fit this scheme. Their dimension D is different from the Euclidean dimension of the space in which they are embedded and is different from what might be called a "topological dimension." Take a curved line on a plane. The minimum Euclidean dimension D_E of the embedding space is 2. The topological dimension D_t is 1. However, neither D_E nor D_t tell us what is the length L of the line. How do we measure L?

7.1.1. Length of a Line

To measure L, one might divide the line into N smaller parts (using, e.g., a set of dividers) and then approximate each part by a straight segment of length r (opening length of the

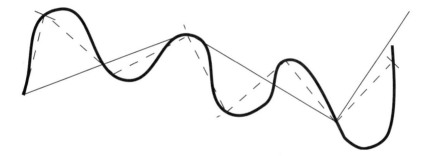

Figure 7.1. Measuring the length of a curved line by different opening lengths of a set of dividers.

dividers). If the line is curved, the measured length depends on r:

$$L(r) = rN(r). \tag{7.3}$$

If we deal with a straight line of length L_0, the dependence $N(r)$ is clearly

$$N(r) = L_0 r^{-1}, \tag{7.4}$$

where the exponent "-1" is obviously related to the topological dimension $D_t = 1$. However, if the line is curved, then reduction of the opening length of the dividers would most probably result in the increase of the length $L(r)$ (Fig. 7.1). The number $N(r)$ would grow faster than r^{-1} when r decreases, for example, N can follow the power law with D *larger* than the topological dimension $D_t = 1$:

$$N(r) \propto r^{-D}, \tag{7.5}$$

at least, within some range of r. However, even if the line is curved, but smooth and differentiable, in the limit $r \to 0$, one would still recover $D = D_t = 1$. Mathematically, such a line is not a fractal, and one should think of more sophisticated examples for which $D > D_t = 1$ even when $r \to 0$. One of them is the triadic Koch curve.

7.1.2. Koch Curve

To construct the Koch curve, one starts with an initiator, a straight line of, say, length L_0. The middle third of the initiator is replaced by the upright sides of an equilateral triangle (Fig. 7.2). The result is the so-called generator, composed of $N = 4$ line segments each of length $r_1 = L_0/3$, so that the total length is $L_1 = 4r_1$. In the next step, each one of the four segments is considered as a base and replaced by a generator scaled down by a

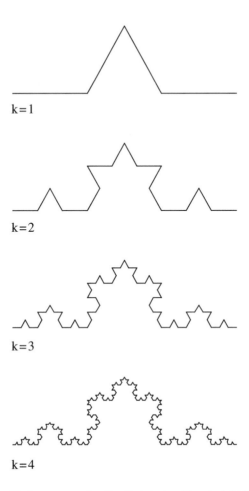

k=1

k=2

k=3

k=4

Figure 7.2. Stages of the Koch curve construction that starts with a straight line of length 1. At stage $k + 1$, one replaces each segment of stage k by a generator scaled down by the factor $(1/3)^k$.

factor of 1/3. The total length becomes $L_2 = 16r_2$, where $r_2 = L_0/9$ is the length of the new straight segment; it can also be considered as an adjustable opening of dividers. The process is carried out ad infinitum.

At the step k, the length of each straight segment is $r_k = L_0/3^k$, and the total length is $L_k = 4^k r_k = L_0(4/3)^k$. Substituting $k = (\ln L_0 - \ln r_k)/\ln 3$, one finds

$$L(r) = L_0^D r^{1-D}, \tag{7.6}$$

Figure 7.3. Koch curves with (a) base angle 0.157 rad and $D \approx 1.0045$; (b) 1.5 rad and $D \approx 1.8205$.

and recovers $N(r) \propto r^{-D}$, i.e., (7.5), now valid for all r's. The exponent

$$D = \lim_{r \to 0} \frac{\ln N(r)}{\ln(1/r)} \tag{7.7}$$

is a constant, $D = \frac{\ln 4}{\ln 3} \approx 1.26186$, *larger* than the topological dimension $D_t = 1$. This *fractal dimension* D of the Koch curve is "fractional," hence, the term "fractal."

If the iterations interrupt at some finite step k, the resulting structure is sometimes called "prefractal." In this sense, physical fractals are prefractals. Some curved line of the type considered in Section 7.1.1 might not be fractal in the strict mathematical sense; however, if such a line is of physical interest (a trajectory, for example) and follows the scaling law of the type (7.5) or (7.6) over decades of lengths, then one might suspect a certain physical reason for such behavior and, thus, employ the fractal description to get a better insight into the problem.

The value of D indicates how far the Koch line is meandering away from a straight line and how effectively it fills the plane. To illustrate this point, one can modify the rules of construction, namely, change the angle at the base of the triangle. Figure 7.3 shows two curves (at $k = 5$) with $D \approx 1.0045$ and $D \approx 1.8205$. The closer D is to 1, the closer the Koch curve in appearance to a straight line, and the closer D is to 2, the more dense it becomes in a 2D plane (see Problem 7.1).

7.1.3. Self-Similarity

The fractal nature of the Koch curve is related to its *self-similarity*. At each step $k + 1$ in Fig. 7.2, one replaces the curve of step k with its $N = 4$ exact copies, each downscaled by a factor of $\lambda = 1/3$. We can also use even smaller copies, those obtained at the step $k + n$, where $n > 1$. Then, to reconstruct the k-step configuration, one would need $N = 4^n$ small copies, each downscaled by $\lambda = (1/3)^n$. The scale factor can thus be written as

$$\lambda = (1/N)^{1/D}, \tag{7.8}$$

where the exponent D

$$D = -\ln N / \ln \lambda(N) \tag{7.9}$$

is the "similarity dimension," in our case, equal to the fractal dimension. In the same way, a straight line is self-similar with the "similarity dimension" $D = 1$, because it can be reconstructed from N self-similar parts each scaled by the factor $\lambda = (1/N)^{1/1} = 1/N$.

7.1.4. Estimating Fractal Dimensions

Structures with fractal properties often occur as a result of aggregation. Aggregation is a process of forming macroscopic structures, aggregates, from smaller parts, that might be aggregates themselves or elementary particles driven by attractive forces. When the process is irreversible, i.e., the particles cannot part once in contact, the resulting aggregate is different from a compact "Euclidian" structure of, say, a crystalline body that forms in equilibrium conditions. Namely, the aggregates are full of large voids that become less and less accessible for the new particles as the aggregate grows: It is more likely that the particle will be captured by a peripheric branch than that it will avoid any contact to make all of its way through to fill the void. The effect is called "self-screening." As the aggregate grows, the self-screening allows larger and larger voids to form; voids at all scales make the aggregate fractal.

Although natural aggregation proceeds through random processes, it is instructive to start with a deterministic aggregation that leads to a so-called Sierpinski gasket, another classic fractal object in mathematics; we will return to physical examples later in this chapter.

Let the elementary particle of the aggregation process (an "atom") be a black triangle of unit edge length ($L = 1$) and unit mass $M = 1$. Following the rules of growth suggested by Stanley,[3] the three triangles are assembled together into a new equilateral triangle of the edge length $L = 2$ and mass $M = 3$. At $k + 1$-step, three exact copies of the aggregate of the previous step k are assembled into a larger aggregate of linear size 2^{k+1} and mass $M = 3^{k+1}$ (Fig. 7.4). The mass and the length are related by the power law

$$M(L) \propto L^D \tag{7.10}$$

---

[3]H.E. Stanley, in Fractals and Disordered Systems, Edited by A. Bunde and S. Havlin, Springer-Verlag, Berlin, 2nd edition, 1996.

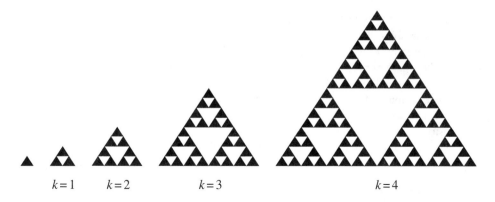

Figure 7.4. Aggregation of small triangles into the Sierpinski gasket.

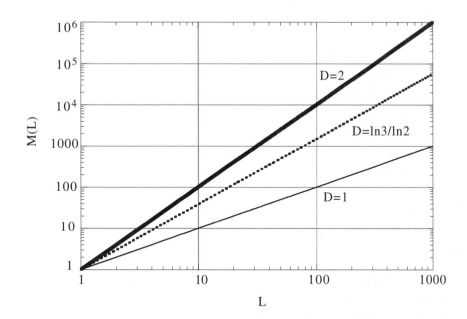

Figure 7.5. Double logarithmic dependence of the mass $M(L)$ on the linear size for a 1D line (thin line), 2D disk (thick line), and an aggregate in the form of the Sierpinski gasket.

with the exponent $D = \ln 3/\ln 2 \approx 1.585$. This exponent is again neither one nor two, but it clearly resembles integer mass dimensions of Euclidean figures. If the growth law was chosen differently, D would change. For example, if the triangles are allowed to contact only by their base vertices along a straight line, then $D = 1$; if the triangles are glued along their sides and the vertices are matched, $D = 2$. If $M(L)$ for the Sierpinski gasket is plotted with a double-logarithmic scale, it produces a straight line with a slope $D \approx 1.585$ intermediate between 1 and 2 (Fig. 7.5). Interestingly and importantly, (7.10) implies that the density ρ of the fractal decreases with its size. If one draws a sphere of radius r around any point of the fractal, then the average density in this sphere scales as

$$\rho(r) \propto r^{D-D_E}. \tag{7.11}$$

Self-similarity of the Sierpinski gasket shows up as the presence of holes (white areas in Fig. 7.4) of every length scale. It is exactly the self-similar and hollow structure of fractals that leads to the power law with fractal dimension in (7.10) and (7.11).

7.1.5. Deterministic and Stochastic Fractals

Fractal Koch lines and Sierpinski gasket considered so far are deterministic: One has well-defined rules to construct the objects, and all points of the fractal can be specified unambiguously. The self-similarity of deterministic fractals can be checked directly and expressed through D. Natural objects are self-similar only in a statistical sense.

A statistical analog of the deterministic fractal Koch curve is a coastline. According to actual measurements, the length of some coastlines satisfies (7.6) over a large range of scales. Applying (7.6) for the scales of "yardstick" between a few and thousands of kilometers, one finds $D = 1.3$ for the west coast of Britain and $D = 1.52$ for the southern coast of Norway, as documented in the classic books on fractals by Mandelbrot and Feder. As long as $D = $ const, the irregularities of the coastline look statistically self-similar at different degrees of magnification. It means that it is impossible to sort out differently magnified pictures of the coastline (of coarse, no vacationing tourists are allowed in the pictures because those have nonfractal shapes).

The fractal dimensions that characterize properties of stochastic fractals such as mass, number of particles, volume, area, and so on, can be determined by the *box counting technique*. The D_E-dimensional Euclidean space in which the fractal is embedded is divided into a grid of boxes of linear size ε. One counts the number of boxes that are nonempty and then repeats the procedure for different values of ε, different orientations, and different origins of the boxes. Each nonempty box has a volume ε^{D_E} and a mass $M(\varepsilon) = \rho\varepsilon^{D_E}$, where ρ is a constant characterizing the material from which the object is made. Within the distance r from any nonempty box, one finds $N = (r/\varepsilon)^D$ boxes covering the object. Their total mass (volume) is $M(r) = NM(\varepsilon)$ or

$$M(r) = (r/\varepsilon)^D M(\varepsilon). \tag{7.12}$$

If we scale the length r by a factor λ, the mass of the object within the sphere of radius $\lambda r < L_0$ is

$$M(\lambda r) = \lambda^D M(r). \tag{7.13}$$

If $D > D_t$ (and usually, $D < D_E$; see Problem 7.2), the object is fractal with the fractal dimension D defined in (7.7). Equation (7.13) tells us that the mass of the fractal within the sphere of radius L scales as $M(L) \propto L^D$, as in (7.10). Returning to the example with the Sierpinski gasket, one can consider $N(\varepsilon) = 3^k$ in (7.7) as the number of triangles of linear size $\varepsilon = (1/2)^k$ needed to cover the gasket at iteration level k; the result is $D = \ln 3/\ln 2$ as already found.

Fractals in Fig. 7.6 clearly illustrate the difference and similarities of deterministic (a) and stochastic (b) fractals that might form during aggregation processes. According to the construction scheme, one divides an original square into nine equal, smaller squares and then throws away four of them. These four can be selected in a deterministic way (Fig. 7.6a) or randomly (Fig. 7.6b). Despite the difference in appearance, both have the same fractal dimension $D = \ln 5/\ln 3 \approx 1.465$, because one needs the same number of squares to cover two objects of the same Euclidean area.

Figure 7.6. (a) Deterministic and (b) stochastic clusters (adapted from T. Vicsek, Fractal Growth Phenomena, 2nd edition, World Scientific, Singapore, 1992).

7.1.6. Brownian Motion and Random Walks

The Brownian motion of a small (micron-size) particle suspended in an isotropic solvent is one of the simplest examples of stochastic fractals. The Brownian particle is in uninterrupted and irregular motion with a zigzag trajectory (Fig. 7.7) due to the fluctuative movement of the solvent molecules and their collisions with the particle. Because the particle jumps in apparently random directions and because each jump has some characteristic mean length $a = \sqrt{\langle r_n^2 \rangle}$, the Brownian motion is often called *random walk*. Let us show that the relationship between the number of jumps N and the mean distance R traveled by the Brownian particle is fractal.

The displacement accumulated over N jumps is the sum $\mathbf{R}_N = \sum_{n=1}^{N} \mathbf{r}_n$. The mean total displacement $R = \sqrt{\langle \mathbf{R}_N^2 \rangle}$ grows linearly with the number of jumps, because any two different jumps are uncorrelated:

$$R^2 = \sum_{n,m}^{N} \langle \mathbf{r}_n \mathbf{r}_m \rangle = Na^2 + 2 \sum_{n>m}^{N} \langle \mathbf{r}_n \mathbf{r}_m \rangle = Na^2, \quad \text{or} \quad R = a\sqrt{N}. \tag{7.14}$$

The number ("mass")-radius relation, thus, writes as

$$N = (R/a)^2, \tag{7.15}$$

which means that the Brownian motion has the fractal dimension $D = 2$. This conclusion is obviously valid for any dimension of the embedding space, as soon as $D_E \geq 2$.

The model of random walk can be used as a very idealized model of a linear polymer in good solvent. This model would consider each step in the random walk as the monomer of

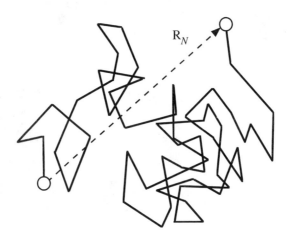

Figure 7.7. Trajectory of a Brownian particle in a 2D plane.

the polymer chain and would assume that any two neighboring links can point in arbitrary directions; moreover, the polymer is allowed to intersect, as the Brownian trajectory does. A more realistic model of a polymer is that of a *self-avoiding random walk* that prohibits self-intersections. Obviously, if the self-avoiding random walk is fractal, then its fractal dimension should be smaller than $D = 2$ calculated above. Indeed, the mean field calculations yield the Flory's formula $D = \frac{D_E+2}{3}$, i.e. $D = 4/3$ for $D_E = 2$ and $D = 5/3$ for $D_E = 3$; for more details, see Gouyet (1996) and Chapter 15.

To conclude this section, note that the concept of random walks can be expanded to objects with $D_t = 2$ that describe other soft-matter systems, namely, membranes. Thermal fluctuations tend to crumple the membrane, and the latter adopts geometry of a "self-avoiding" random surface. For isolated membranes in $D_E = 3$, one finds $D \approx 2.5$.

7.1.7. Pair Correlation Function

One can apply (7.13) to an experimentally observed fractal structure if there is a good (micro)photograph of a fractal. However, in many cases, such photographs are not easy to get; besides, the resolution is obviously limited. Statistical self-similarity of natural fractals can be directly verified in coherent scattering experiments that use probes such as light, X-ray, and neutrons. These experiments provide an insight into the structural properties over many decades of the length scales, from 1 Å to 1 μm, and allow one to verify the scaling symmetry. The measured quantities are directly related to the so-called pair correlation function.

The pair correlation function written for a cluster (an aggregate) of N identical particles

$$C(\mathbf{r}) = \frac{1}{N} \sum_i \rho(\mathbf{r} + \mathbf{r}_i)\rho(\mathbf{r}_i) = \frac{\langle \rho(\mathbf{r} + \mathbf{r}')\rho(\mathbf{r}') \rangle}{\langle \rho(\mathbf{r}') \rangle} \qquad (7.16)$$

is the probability that the point a distance $\mathbf{r}$ away from the occupied site $\mathbf{r}'$ belongs to the same cluster. Here, ρ is the local normalized density; e.g., $\rho(\mathbf{r}) = 1$ if the position $\mathbf{r}$ is occupied and $\rho(\mathbf{r}) = 0$ otherwise. For correlations that depend only on the distance and not on the direction, one can replace $\mathbf{r} \to r$. The total mass M or the number of particles N are integrals over the correlation function, $M(r) \propto \int_0^r C(x)\, d^{D_E}x$. If the object is fractal, $M(r) \propto r^D$, then the correlation function should be of the form

$$C(r) \propto r^{-\alpha}, \qquad (7.17)$$

where α is a noninteger positive number smaller than D_E:

$$\alpha = D_E - D; \qquad (7.18)$$

compare this to (7.11) for density.

The power dependence (7.17) implies that the correlation function is invariant (up to some constant) to a change of length scale by an arbitrary factor λ:

$$C(\lambda r) = \lambda^{-\alpha} C(r). \tag{7.19}$$

In other words, there is no characteristic length scale. This is in striking contrast with the correlation function of a disordered medium with short range order, e.g., a polycrystal made of small disoriented microcrystallites, or an amorphous solid, which show up "correlation" bumps for the values of r corresponding to the distances between nearest or next nearest neighbors, say.

In scattering experiments, one probes the sample with a collimated monochromatic beam and detects scattering caused by variations in properties such as refractive indices (light scattering) or electron densities (X-ray) or nuclei densities (neutrons). For X-ray and neutron scattering, the scales of interest are much larger than is the wavelength of the probe, and one is thus interested in small-angle scattering. If the scattering is elastic, the intensity of radiation scattered by some small angle θ is determined by the momentum transfer vector (the difference between the incoming and the outgoing wave vectors) $q = \frac{4\pi}{\lambda_0} \sin \frac{\theta}{2}$, where λ_0 is the wavelength. The size of the particles a is usually small enough, compared with λ_0, to influence the scattering pattern. To take this effect into account, one assumes that the particles are identical spheres and represents $I(q)$ as a product of a form factor $P(q)$ that describes the scattering from uncorrelated scatterers-spheres and the structure factor $S(q)$ that describes correlations of the spheres.

The form factor is practically constant when $qa \ll 1$ and follows the Porod law $P(q) \propto q^{-4}$ in the limit $qa \gg 1$. The structure factor is the Fourier transform of the pair correlation function:

$$S(q) = 4\pi \int_0^\infty C(r) \frac{\sin qr}{qr} r^2 \, dr. \tag{7.20}$$

To calculate $S(q)$ properly, one has to recall again that physically, the scaling (7.17) is valid only in the range $a \ll r \ll R$, limited from above by the size of the cluster R. To account for this cutoff, one renormalizes the correlation function, $C(r) \propto r^{-\alpha} f(r/R)$, by a factor chosen often as an exponent, $f(x) = e^{-x}$, to satisfy the conditions $f(x) \to 0$ for $x \gg 1$ and $f(x) \to 1$ for $x \ll 1$. For the range of interest, $1/R \ll q \ll 1/a$, the structure and the intensity follow the power law (Problem 7.3)

$$I(q) \propto S(q) \simeq \int C(r) \frac{\sin qr}{qr} r^2 \, dr \propto q^{-D}. \tag{7.21}$$

Therefore, the scattering experiment can directly measure the fractal dimension D, from the slope of the dependence $I(q)$ plotted in log-log coordinates (Fig. 7.8). Note that the

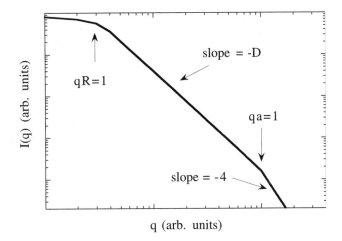

Figure 7.8. Schematic scattering plot revealing fractal behavior for $1/R \ll q \ll 1/a$ and finite size effects at large and small scales.

same expression (7.21) would yield Dirac peaks (in reciprocal space) if the object under consideration is a crystal, or scattering rings in the case of a crystal powder or of an amorphous solid.

7.1.8. Inner and Outer Cutoffs

Fractals formed by aggregated particles have a natural lower bound of scaling properties, which is the size a of a single particle. It might be, for example, a molecular size of monomers in a polymer network. The upper limit is the size of the cluster. In some cases, however, the identification of the inner and outer cutoff limits is not that obvious.

As discussed in Section 10.8, the bases of focal conic domains in smectic polygonal textures form fractal-like patterns similar to those in Apollonian packings of circles. The hierarchy of domains has the scaling properties of the corresponding asymptotic Apollonian fractal gasket, but only within a finite range of scales $r^* < r < L_0$, where L_0 is the outer cutoff (e.g., the overall thickness of the smectic sample) and $r^* \sim K/W$ is the inner cutoff defined by the ratio of the elastic constant to the anisotropy of surface anchoring energy. This lower limit is generally much larger than is the characteristic molecular size.

7.2. Percolation

7.2.1. Geometrical Percolation

Consider a square grid. The cells of the grid are occupied with a probability p (empty with the probability $1 - p$). Neighboring occupied sites (black in Fig. 7.9) with a common edge

(a) p=0.3   (b) p=0.4

(c) p=0.6   (d) p=0.7

Figure 7.9. Percolation networks for different occupation probabilities p.

form a connected cluster. If $p \ll 1$, the clusters are small and isolated, (Fig. 7.9a). When p increases from 0 to 1, so does the mass of the largest clusters (Fig.7.9b). There is a value of $0 < p < 1$, at which a unique cluster appears that connects opposite sides of the grid (Fig. 7.9c). When the size of the grid $L_0 \to \infty$, this *percolating* cluster is infinite; p_c, at which the infinite cluster appears, is called the *percolation threshold* or *critical probability*. Numerical calculations (performed, of course, on finite grids) allow one to conclude that $p_c \approx 0.59275$ for clusters formed by neighboring sites on a 2D square lattice as in Fig. 7.9; they also show that the clusters are fractal distributions of occupied cells.

The model depicted in Fig. 7.9 is only one of numerous variations of geometrical percolation (that differ in grid types or algorithms to identify connected clusters). One also distinguishes percolation of sites, as above, and percolation of bonds between sites. The first prototye of the (bond) percolation theory was the theory of polymerization, developed in the 1940's by Flory.

The percolation models allow one to get insight into the physics of many phenomena, such as insulator-conductor and sol-gel transitions, polymerization, spread of epidemic, behavior of diluted magnetic and porous systems, and so on. For example, one can imagine the empty sites in Fig. 7.9 as an isolating "matrix" and the occupied (black) sites as conducting grains in that matrix. For small concentration of grains, $p < p_c$, the system is an insulator, whereas for $p \geq p_c$, the system is a conductor. Thus, p_c is the position of *the phase transition*. Examples closer to the domain of soft-matter physics are the sol-gel transition (gelation) and the polymerization process. A sol is a system of individual particles dispersed in some liquid medium. If the particles keep apart, the whole system behaves as a liquid. Gelation consists in formation of bonds (chemical, electrostatic, van der Waals, or of other origin) between these particles. When the connected network of bonds is large enough to percolate the whole dispersion medium, the system transforms into a gel and shows some elasticity. The boiling of an egg, and the vulcanization of rubber are the illustrations.

The examples above show that to analyze the percolation-type phenomena, it is important to know the linear size of the *finite* clusters. The finite clusters exist at $p < p_c$ and at $p > p_c$, occupying holes left by the infinite percolating cluster. Two parameters (one can introduce more) serve the purpose: the *radius of gyration R_s* and the *connectedness length ξ*.

R_s is the root mean square radius of the cluster composed of s sites, measured from the center of mass $\mathbf{r}_0$ of this cluster. If $\mathbf{r}_i$ is the position of site i, then $\mathbf{r}_0 = \frac{1}{s}\sum_{i=1}^{s}\mathbf{r}_i$ and

$$R_s^2 = \frac{1}{s}\sum_{i=1}^{s}(\mathbf{r}_i - \mathbf{r}_0)^2 = \frac{1}{2s^2}\sum_{i,j=1}^{s}(\mathbf{r}_i - \mathbf{r}_j)^2. \tag{7.22}$$

The second part of (7.22) expresses R_s through the distances between any two cluster sites (Problem 7.4). The quantity $2R_s^2$ is the averaged squared distance between two cluster sites. The connectedness ξ (or the correlation length) is also the root mean square distance between the pairs of sites that belong to the same cluster, averaged over all finite clusters of size s:

$$\xi^2 = \frac{2\sum_s R_s^2 s^2 n_s}{\sum_s s^2 n_s}, \tag{7.23}$$

where n_s is the number of clusters of size s calculated *per lattice site*. In other words, n_s is the number of clusters of size s on the lattice $L_0 \times L_0$, divided by L_0^2. Equation (7.23) takes into account that any randomly selected site of the lattice has probability $w(s) = sn_s$ to belong to a cluster of size s, because there are s different ways for the site to be incorporated within the cluster.

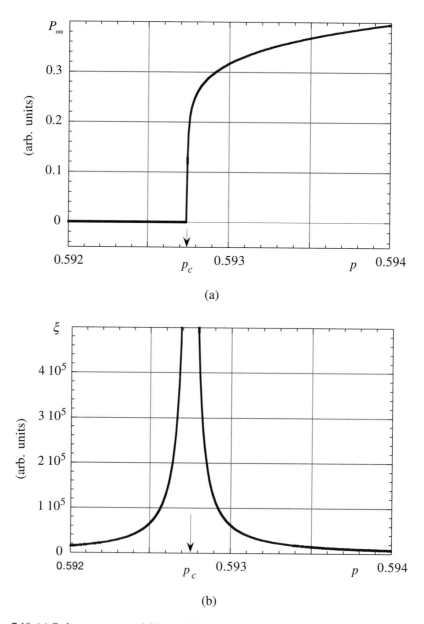

Figure 7.10. (a) Order parameter and (b) correlation length in the vicinity of the percolation threshold.

As p approaches p_c, the finite clusters increase in size; ξ, being the radius of clusters that contribute most to this increase, diverges to infinity at p_c. When ξ diverges, there is no characteristic length to scale the length-dependent physical properties of the system. As fractal structures in Section 7.1, the system looks the same at different magnifications. The properties of the system become nonsensitive to many local details, such as small changes in interactions of particles, lattice structure, and so on, which do not influence the large-scale behavior. This feature results in the *universality* of the critical exponents that describe diverging parameters near p_c. These universal exponents depend on the model under consideration and the dimensionality of the system but not on the details of the local structure.

7.2.2. Percolation and Second-Order Phase Transitions

The discussion above shows that the percolation model has features that unite it with both the fractals and the thermal phase transitions.

Near the transition point $p = p_c$, geometrical percolation can be described in the same terms as the thermal second-order phase transition, say, a transition from a paramagnetic state at high temperatures and a ferromagnetic state at low temperatures. The analog of temperature T is the occupation probability p of one site; the analog of the order parameter, say, the magnetization $M(T)$, is the probability $P_\infty(p)$ that a randomly chosen site belongs to an infinite cluster:

$$p - p_c \leftrightarrow T_c - T, \tag{7.24}$$

$$P_\infty(p) \leftrightarrow M(T). \tag{7.25}$$

In magnetic materials, the magnetization vanishes at the critical temperature T_c, according to the power law $M(T) \propto (T_c - T)^\beta$ with the *critical exponent* β. Immediately above the percolation threshold, $0 < p - p_c \ll 1$, the order parameter $P_\infty(p)$ behaves in a similar way (Fig. 7.10a):

$$P_\infty(p) \propto (p - p_c)^\beta. \tag{7.26}$$

Of course, $P_\infty(p < p_c) = 0$, because only finite clusters exist at $p < p_c$.

The values of the critical exponent β are different for magnets and percolation. For example, mean-field theories predict that in 2D, $\beta = 1/8$ for magnets, whereas $\beta = 5/36$ for percolation; β is also different for 2D and 3D systems. Nevertheless, once the system (magnets versus percolation; 2D versus 3D) is chosen, the critical exponent β is a more universal quantity than is the percolation threshold p_c (see Table 7.1). For example, β remains constant if one switches from site to bond percolation or changes the local structure of the network, say, from triangular to square. In contrast, p_c is different for all of the cases listed above (see Table 7.1). Although p_c and β can always be estimated in computer

Table 7.1. Percolation thresholds and critical exponents for the order parameter P_∞ (β) and the correlation length ξ (ν) for some common percolation networks; astericks mark exact results. Compiled from Bunde and Havlin.[†]

Type	Structure	D_E	p_c	β	ν
site	triangular	2	1/2*	5/36*	4/3*
site	square	2	0.592760	5/36*	4/3*
site	honeycomb	2	0.6962	5/36*	4/3*
site	simple cubic	3	0.31161	0.417	0.875
site	body centered cubic	3	0.245	0.417	0.875
site	face centered cubic	3	0.198	0.417	0.875
bond	triangular	2	$2\sin(\pi/18)$*	5/36*	4/3*
bond	square	2	1/2*	5/36*	4/3*
bond	honeycomb	2	$1 - 2\sin(\pi/18)$*	5/36*	4/3*
bond	simple cubic	3	0.248814	0.417	0.875
bond	body centered cubic	3	0.1803	0.417	0.875
bond	face centered cubic	3	0.119	0.417	0.875

[†]A. Bunde and S. Havlin, in Fractals and Disordered Systems, Edited by A. Bunde and S. Havlin, 2nd edition, Springer-Verlag, Berlin, 1996.

simulations, analytical results are available only for a few special cases. One of these cases is the Bethe lattice, or the Cayley tree, which was used by Flory to develop the theory of gelification in polymers; we consider this model in Section 7.2.5.

The correlation length ξ also diverges when p approaches p_c (both from below and from above) (Fig. 7.10b), with a new critical exponent ν,

$$\xi \propto |p - p_c|^{-\nu}. \tag{7.27}$$

The behavior (7.27) resembles divergence of the correlation length near critical points for thermal phase transitions. Both critical exponents β and ν are universal, because they depend on the dimensions of the system but not on the local details. The aim of the theory is to calculate these exponents from the first principles and to find relationships among them. In the next section, we illustrate relationships between the critical exponents and the fractal characteristics of the percolation networks; the techniques of calculating the values of critical exponents are discussed later.

7.2.3. Finite Clusters at the Percolation Threshold

Let us consider the fractal properties of finite clusters by determining how the mass $M_s(L)$ of an s-cluster depends on the linear size L on which it is considered. The only characteristic length scale of the finite cluster is the radius of gyration R_s. Let us first assume

$L \gg 2R_s$; the mass is then equal to s, $M_s(L) \propto s$, and does not depend on L. On a length smaller than its size, $L \ll R_s$ the cluster appears as a self-similar object, $M_s(L) \propto L^D$, as in (7.10). Because R_s is the only length characterizing the cluster, one might expect that M_s depends on L only through the dimensionless ratio L/R_s and write for both cases above a scaling relationship

$$M_s(L) \propto R_s^D m(L/R_s), \tag{7.28}$$

where the crossover function $m(x)$ is a constant when $x \gg 1$ and scales as $m(x) \propto x^D$ when $x \ll 1$. Because $s \propto M$ when $L \gg 2R_s$, then the number of sites and the linear size of the finite cluster yield typical fractal relationships

$$s \propto R_s^D \quad \text{or} \quad R_s \propto s^{1/D}, \tag{7.29}$$

which has been confirmed by many computer simulations.

The probability $w(s)$ that a *randomly selected* site belongs to a cluster of size s also has the form of a power law. Consider a $L_0 \times L_0$ grid at the percolation threshold, and scale it down by a factor b to a grid $(L_0/b) \times (L_0/b)$. The gyration radius of a cluster is reduced from $R_s(s)$ to $R_s(s') = R_s(s)/b$, where s' is the number of sites of the "new" cluster. According to (7.29), $s' = s/b^D$. The probability $w(s)$ for the original grid can be expressed through its counterpart $w(s')$ of the scaled grid as

$$w(s) = b^{-2}w(s') = b^{-2}w(s/b^D). \tag{7.30}$$

The factor b^{-2}, where the exponent 2 is the Euclidean dimension of the grid, $D_E = 2$, is caused by the fact that each site on the grid $(L_0/b) \times (L_0/b)$ corresponds to b^2 sites on the original grid $L_0 \times L_0$. To satisfy (7.30), $w(s)$ and s should be related by a power law

$$w(s) = sn_s \propto s^{1-\tau}, \tag{7.31}$$

where the critical exponent is determined entirely by the dimensions D and D_E:

$$\tau = D_E/D + 1. \tag{7.32}$$

We replaced the exponent 2 by D_E because the result holds for grids with $D_E < 6$. For dimensions $D_E \geq 6$, the situation is special and surprisingly simple: All percolation models yield the same critical exponents that can be found exactly by the mean-field theory of the Bethe lattice (see Section 7.2.5). Through τ, the dimensions D and D_E determine all other exponents at $p = p_c$.

Let us now calculate the *average* size S of the finite clusters that can be expressed as the sum $S = \sum_{s \geq 1}^{s_{\max}} s\overline{w}(s)$, where $s_{\max}$ is the maximum size of the cluster and $\overline{w}(s)$ is the probability that an *occupied* site belongs to the cluster of s sites. Apparently, at $p \approx p_c$, the cluster maximum size on a finite lattice $L_0 \times L_0$ is $s_{\max} \propto L_0^D$. Because any site is occupied with the probability $p \approx p_c$, we can relate $\overline{w}(s)$ to $w(s) = sn_s$ considered above: $p_c\overline{w} \approx w$. Hence,

$$S \cong \frac{1}{p_c} \sum_{s=1}^{s_{\max}} sw_s. \qquad (7.33)$$

Replacing summation by integration, and employing (7.31), one finds

$$S \propto \int_1^{s_{\max}} s^{2-\tau} ds \simeq s_{\max}^{3-\tau} \propto L_0^{2D-D_E} \qquad (7.34)$$

(notice that $3 - \tau > 0$, because D cannot be smaller than 1 for 2D percolation pattern). The result is different from the scaling law for the largest cluster, $s_{\max} \propto L_0^D$.

7.2.4. Fractal Dimension of the Percolation Cluster

At percolation threshold p_c, the infinite percolating cluster contains holes of all possible sizes because the correlation length ξ diverges (7.27). Above p_c, the length ξ is finite (Fig. 7.10b) and corresponds to the linear size of the largest "holes" left by the percolating cluster. It means that at $p > p_c$, the percolating cluster is self-similar only on length scales $L < \xi$ and homogeneous at larger scales $L > \xi$. An example of such a structure is shown in Fig. 7.11. At $L < \xi$ and $L > \xi$, the mass of the infinite cluster scales differently:

$$M(L) \propto L^D; \quad L < \xi, \qquad (7.35a)$$

$$M(L) \propto P_\infty L^{D_E}, \quad L > \xi. \qquad (7.35b)$$

At $L = \xi$, the two last expressions should recover the same mass: $P_\infty L^{D_E} \simeq (p - p_c)^\beta \xi^{D_E} = \xi^D$. But according to (7.27), $\xi \propto |p - p_c|^{-\nu}$; hence,

$$D = D_E - \frac{\beta}{\nu}, \qquad (7.36)$$

which relates the fractal dimension of the percolation cluster to the exponents β and ν. The exponents β and ν are universal constants in the sense discussed above; therefore, D is universal as well. With $\beta = 5/36$ and $\nu = 4/3$, one gets $D = 91/48 \approx 1.8958$.

The last two sections show that the critical exponents are related, but do not tell how the numerical values of these exponents can be obtained. In the next two sections, we consider two theoretical approaches that allow one to obtain analytical results; for more details, see Stauffer and Aharony (1992) and Gouyet (1996).

Figure 7.11. Periodic lattice formed by Sierpinski gasket units of size $\sim \xi$ (adapted from A. Bunde and S. Havlin, in Fractals and Disordered Systems, Edited by A. Bunde and S. Havlin, 2nd edition, Springer-Verlag, Berlin, 1996).

7.2.5. Percolation on Bethe Lattice

As already indicated, percolation threshold p_c and the critical exponent β can be calculated exactly when percolation takes place at the Bethe lattice, also called the Cayley tree. To build the Cayley tree, one starts with a site and attaches to it z bonds of equal length. The end of each bond is a new site to which new z-bonds are attached, and so on (Fig. 7.12). Any two sites are connected by only one path composed by bonds. The Cayley tree looks like a real tree in the sense that the branches do not form loops and, thus, remain statistically independent; it is exactly this property that allows one to conduct exact calculations.

Consider the Cayley tree as a percolation network: Each site is occupied with probability p. Let us find the probability of finding an (infinite) path of occupied sites that would lead from some "origin" site to the periphery (Fig. 7.12). Of course, all sites are equivalent in the Cayley tree, except those at the very periphery; we exclude these peripheral sites from the consideration. When one walks along the branches of the tree, at a given site, there are always $(z-1)$ bonds leading to $(z-1)$ neighboring sites (zth bond is the one that has led to the site). Because the sites are occupied with probability p, the probability to form a connected path is $(z-1)p$. Each step multiplies the total probability of finding an infinite path by a factor of $(z-1)p$. If $(z-1)p < 1$, there will be no chance to find an infinite path. Therefore, the percolation threshold for the Bethe lattice is defined from the condition $(z-1)p_c = 1$ as

$$p_c = 1/(z-1). \tag{7.37}$$

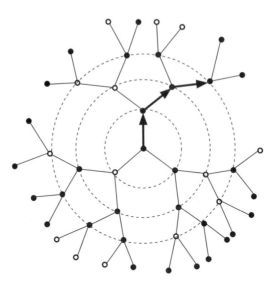

Figure 7.12. Bethe lattice with $z = 3$ at each site (except the surface sites). Each new generation of neighbors is shown on the same shell; the chosen center of the concentric shells does not have a special physical significance because all sites in the interior of the lattice are equivalent. Empty sites are open circles; occupied sites are filled. Arrows show one of the possible paths connecting occupied sites.

For the Cayley tree with $z = 3$, as in Fig. 7.12, $p_c = 1/2$; the same value characterizes site percolation on a triangular lattice.

Let us now calculate the exponent β describing the critical behavior of $P_\infty(p)$, that is the probability of a site to belong to an infinite cluster of connected occupied sites (7.26). One employs a mean-field approach, considering the probabilities of different states in which a chosen site and its neighborhood can find themselves. Consider some site, empty or occupied, with z emanating bonds that lead to neighboring sites. Denote by Q the probability that a chosen bond *does not* connect to an infinite cluster; $0 \leq Q \leq 1$. The original site *does not* belong to the infinite cluster (probability $1 - P_\infty$) when (a) this site is empty (probability $1 - p$), or (b) this site is occupied (probability p), but all of its z-bonds fail to lead to the infinite cluster (probability pQ^z; because there are no loops, the bonds are statistically independent and the probabilities Q multiply each other). Therefore,

$$1 - P_\infty = (1 - p) + pQ^z. \tag{7.38}$$

One can find another expression for Q, considering what happens at the neighboring sites. The bond would not be incorporated into the infinite cluster (probability Q) when either of two things happen. First, the neighboring site (the end of the chosen bond) is empty;

the probability of this is $(1 - p)$. Second, even when the neighboring site is occupied (probability p), the other $(z-1)$ bonds emanating from the neighbor might be disconnected from the infinite cluster (probability Q^{z-1}). The probability of the second event is pQ^{z-1}. All together, Q writes as the sum

$$Q = (1 - p) + pQ^{z-1}. \tag{7.39}$$

If $z = 3$, the solution for $p < p_c = 1/2$ is $Q = 1$ and $P_\infty = 0$; for $p > p_c = 1/2$, it is $Q = (1 - p)/p$ and

$$P_\infty = p - \frac{(1 - p)^3}{p^2}. \tag{7.40}$$

Using Taylor expansion near the threshold $p_c = 1/2$ in powers of $(p - 1/2)$, one gets

$$P_\infty = 6(p - p_c); \tag{7.41}$$

i.e.,

$$\beta = 1. \tag{7.42}$$

It turns out that neither $\beta = 1$ nor the other critical exponents depend on the number of nearest neighbors z in the Cayley tree, hence, the universality of the critical exponents. In contrast, the percolation threshold (7.37) is markedly z-dependent.

Can the results obtained for the model Cayley tree be extrapolated to regular percolations networks? An essential simplifying element in the calculation above was the absence of loops. Effectively, the Cayley tree is of infinite dimension (to see this, calculate the ratio of the surface sites to the interior sites in the Cayley tree of large size). In low-dimensional embedding space, the loops in percolation networks are frequent because there is not much "free space" for different branches to avoid each other. However, if the dimension D_E increases, the probability of loop formation decreases and even becomes zero at some *critical dimension* $D_{E,c}$. For $D_{E,c} \geq 6$, the exact mean field results on the Bethe lattice can thus be assigned to any percolation model. Although the number 6 might seem to be too high to be of any practical interest, some theoretical conclusions can be made by expansion around $D_{E,c} = 6$.

7.2.6. Percolation and the Renormalization Group

The Bethe lattice is one of very few models producing exact results on the critical exponents. In many other cases, a powerful tool to estimate the critical exponents and sometimes even to get exact results is the renormalization group technique. It was developed to treat critical phenomena in thermal phase transitions[4] but can be applied to percolation phenomena as well, because the renormalization concept is based on self-similarity and scaling relationships such as (7.28). Below, we estimate the critical exponent v of the connectedness ξ (7.27), for percolation at a triangular lattice (Fig.7.13).

The percolation threshold for the triangular lattice is known to be exactly $p_c = 1/2$ (Table 7.1). Let us renormalize the lattice by replacing each three neighboring sites (forming a triangle) by a single supersite (Fig. 7.13). The supersite is considered to be occupied if it replaces a triangle that is "connected," i.e., if all three or at least two original sites are occupied. The probability that all three vertices are occupied is p^3. The probability that a given two out of three vortices are occupied is $p^2(1 - p)$. Taking into account that there are three such pairs, one finds that the probability p' for the supersite to be occupied is

$$p' = p^3 + 3p^2(1 - p). \tag{7.43}$$

The renormalization can be repeated, each time increasing the length unit by the same factor b (in our example, $b = \sqrt{3}$). The maximum value of b is limited by the requirement that the renormalized lattice unit length should be much smaller than ξ if one wishes to use self-silimarity and scaling arguments. Recall that ξ sets the upper length limit for self-similarity of the percolation clusters (Fig.7.11).

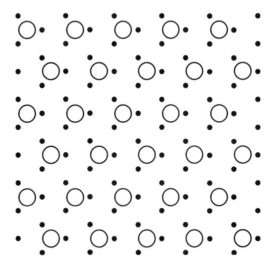

Figure 7.13. Real space renormalization of a triangular lattice. The large circles are supersites that replace three original sites. The supersites are occupied if at least two original sites are occupied.

---

[4]K.G. Wilson and J.B. Kogut, Phys. Rep. **12C**, 75 (1974). For a review of scaling and universality in statistical physics, see L.P. Kadanoff, Physica **A163**, 1 (1990).

Generally, p' for the renormalized lattice does not have to coincide with p of the original lattice. It is only at the threshold where one should definitely have $p' = p = p_c$. The transformation (7.43) preserves only three values of p, which are called *fixed points:* $p^* = 0, 1/2, 1$, as seen by requiring $p' = p$ in (7.43). The limiting values 0 and 1 are trivial, but the value $p^* = 1/2$ is not, because it coincides with the exact value of $p_c = 1/2$ for the triangular lattice. In principle, one should not expect an exact result from the model, because the transformation (7.43) is not exact: Renormalization sometimes changes the connectivity of clusters, connecting originally disconnected clusters or breaking the original single-connected cluster.

Let us now turn to the correlation length ξ. If the self-similarity of both the initial and renormalized structures is preserved, then

$$b\xi(p') = \xi(p), \tag{7.44}$$

which can be also rewritten by using (7.27) as

$$b(p' - p_c)^{-\nu} = (p - p_c)^{-\nu}. \tag{7.45}$$

Equation (7.44) is satisfied when ξ vanishes (which is the case at $p^* = 0, 1$) or $\xi \to \infty$ (which is the case at $p^* = 1/2$). In the vicinity of p_c, we can linearize (7.43), $p' = p^* + A(p - p^*) + \cdots$, where $A = dp'/dp$ is calculated from (7.43). Furthermore, taking the logarithm of both sides of (7.45), one can express the critical exponent ν near p_c as

$$\frac{1}{\nu} = \frac{\ln A}{\ln b}. \tag{7.46}$$

With $A = 6p - 6p^2 = 3/2$ and $b = \sqrt{3}$, one gets $\nu = \frac{1}{2}\ln 3/(\ln 3 - \ln 2) = 1.355$, very close to the exact value $\nu = 4/3$.

7.3. Aggregation

Aggregation of particles produces different results depending on whether it occurs in equilibrium or far from equilibrium, as discussed in Section 7.1.4. If the particles added to the growing aggregate can readjust their position and find the most suitable site to minimize the surface energy, the result would be a compact, nonfractal object. When the added particles stick irreversibly at that location where they first hit the surface of the growing aggregate, then the aggregate might develop a highly irregular shape with numerous voids. The aggregation is also a nonlocal process, unlike the percolation. Because of the self-screening effect, the probability of finding a particle at some point $\mathbf{r}$ depends not only on the position $\mathbf{r}$, but also on the structure of the aggregate as a whole, i.e., on the situation away from $\mathbf{r}$.

Presently, there is no general scheme to describe nonlocal and far-from-equilibrium processes such as aggregation. Available analytical methods can get some insights into the physics of growth but usually cannot capture the whole picture, for example, the values of the fractal dimension. Instead, a growth process can be described by an algorithm, a set of rules. These rules specify:

1. The type of objects added to the growing aggregate; the object might be a (1a) single particle (particle-cluster aggregation) or (1b) another aggregate, usually of a comparable size (cluster-cluster aggregation).

2. The way the objects approach each other. The trajectory of objects might be (2a) Brownian or (2b) linear (ballistic); in addition, the object might rotate in space.

3. Once the two objects encounter each other, there are three more possibilities: (3a) they stick together immediately and forever; there is no repulsion forces to prevent the bond. The rate of aggregation is limited by the time needed for the two objects to find each other; usually, this time is determined by diffusion and the aggregation is called diffusion-limited; (3b) small repulsive forces hinder immediate aggregation, and it takes more than one contact before a permanent bond is formed; the aggregation is called reaction-limited; (3c) the bond is not permanent, and there is a finite probability of disassociation (reversible aggregation).

Combinations of the rules above result in a variety of structures (see Table 7.2).

Table 7.2. Experimental realization and fractal dimension D of different types of aggregation; compiled from Gouyet (1996).

	Particle-Cluster (1a)	Cluster-Cluster (1b)
Diffusion-limited $2a + 3a$	Electrodeposition, dielectric breakdown, growth of bacterial colonies 1.72 ($D_E = 2$) 2.50 ($D_E = 3$)	Colloids and aerosols (screened) 1.44 ($D_E = 2$) 1.75 ($D_E = 3$)
Reaction-limited $2a + 3b$	Epidemics, tumors, forest fires 2.00 ($D_E = 2$) 3.00 ($D_E = 3$)	Colloids and aerosols (partially screened) 1.59 ($D_E = 2$) 2.11 ($D_E = 3$)
Ballistic $2b + 3b$	Sedimentation, deposition 2.00 ($D_E = 2$) 3.00 ($D_E = 3$)	Aerosols in vacuum 1.55 ($D_E = 2$) 1.91 ($D_E = 3$)

7.3.1. Cluster-Cluster Aggregation

Experimental realization of cluster-cluster aggregation can be found in colloidal suspensions of tiny (nanometers and microns) particles suspended in a liquid, say, water. The colloidal particles are subject to two types of forces: van der Waals attraction (Chapter 1) and electrostatic repulsion (Chapter 14). The surface of particles usually carries electric charges, caused by ionized chemical groups or by adsorbed ions. For example, silica particles formed by SiO_2 monomers carry negatively charged OH^- and SiO^{2-} surface groups. A cloud of mobile positive ions (*counter*ions) in the solution screens the surface charges, thus, forming an electric double layer (Chapter 14). Repulsion between the electric double layers prevents a close contact of particles. However, the potential barrier can be reduced, for example, by decreasing the surface charge or by adding a salt to the solution. When the potential barrier is smaller than $k_B T$, the van der Waals forces draw the particles together and they promptly aggregate by forming strong Si–O–Si bonds. First, doublets will form, then quadruplets, and so on. The process is classified as the diffusion-limited cluster-cluster aggregation (DLCA). If the screening is not complete and the potential barrier remains larger than $k_B T$, the aggregation is of the slow reaction-limited (RLCA) type. Both DLCA and RLCA can be realized in the same system, depending on the degree of screening.[5] The RLCA clusters are more compact (higher fractal dimension) than are the DLCA clusters.

The cluster-cluster model is more appropriate than is the particle-cluster model to describe aggregation of colloids or their close counterparts, aerosols (particles in a gaseous medium). However, there are many other processes, not necessarily involving attachment of particles, that are described by the particle-cluster model. The well-known particle-cluster algorithm is that of diffusion-limited aggregation (DLA), suggested by Witten and Sander.[6]

7.3.2. The Witten–Sander Model of Diffusion-Limited Aggregation

The model prescribes the following rules of growth. One starts with a single occupied site. A second particle is released far away from this site and then allowed to exercise a random (Brownian) walk until it either lands on the seed particle or wanders so far away from the seed particle that it can be considered as lost forever. Once the particle hits the seed site, it is immobilized. Emission of random walkers is repeated until the aggregate reaches a desired size.

This simple algorithm results in surprisingly complex and well-organized aggregates; one of them is shown in Fig. 7.14. Statistically, a self-similar and open structure of the aggregates is caused by the self-screening effect: The sites that are farther away from the center have a higher probability of catching new particles. To visualize this feature, particles that arrive at different stages can be marked by different colors or shades of gray, as

[5]M.Y. Lin, H.M. Lindsay, D.A. Weitz, R.C. Ball, R. Klein and P. Meakin, Proc. Roy. Soc. **A 423**, 71 (1989).
[6]T.A. Witten and L.M. Sander, Phys. Rev. Lett. **47**, 1400 (1981).

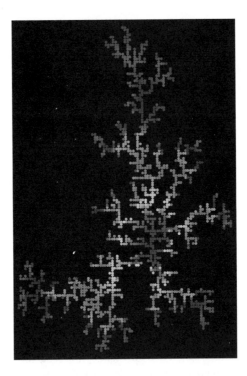

Figure 7.14. Computer-simulated DLA aggregate of 1500 particles. The total number of released particles was 12421. Algorithm used: Gaylord and Wellin (1995).

in Fig. 7.14. Particles released at later stages are seldom seen penetrating the interior of the cluster.

The fractal dimension of the computer-generated DLA clusters can be determined from the correlation function, or from the dependence of the radius of gyration versus the number of particles. Simulations usually reveal that both functions scale according to the power law, $C(r) \propto r^{-\alpha}$, $R_s \propto s^{1/D}$ [see (7.15) and (7.29), respectively]. For planar clusters, $\alpha = D_E - D \approx 0.3$, and thus, $D \approx 1.7$; in 3D space, $D \approx 2.5$ (see Table 7.2). Extensive simulations on very large scales, however, demonstrate that the exponents in the power laws are not exactly constant and become functions of the ratio r/R_s; i.e., $C(r) \propto r^{-\alpha(r/R_s)}$. Physically, it can be understood as a narrowing of the zone of growth compared with the whole size of the cluster when the cluster becomes larger and larger.

7.3.3. Continuum Laplacian Model

Branched patterns, reminiscent of DLA clusters, form in dendritic and snowflake growth, crystallization of thin films, electrodeposition and dielectric breakdown, and even in the growth of bacterial colonies and neuron structures. Many, if not most, of these patterns do not satisfy the definition of fractals over a reasonable scale of lengths.

As already stated, there is no general theoretical scheme to describe far-from-equilibrium growth processes such as DLA and to derive, say, an analytical expression for the fractal dimension from first principles. In some models, the fractal dimension can be given an analytical expression, but these models imply certain simplifications. For example, one can treat the growing tip as a cone[7] and relate D to the angle of the cone. Nevertheless, even when the analytical solution is not available, the behavior of the system can still be understood and predicted if one knows the algorithm of growth. The Witten–Sander model above is one such algorithm. A closer look at it reveals that this algorithm can be approximated by a continuous model, namely, by the Laplace equation with an appropriate set of boundary conditions describing the moving boundary of the growing aggregate. What is even more interesting is that many other growth phenomena are variations of the same unifying theme: a Laplace equation with (different) boundary conditions. Consider the DLA model.[8]

Let $p(\mathbf{r}, t)$ be a probability that a randomly walking particle is at the position $\mathbf{r}$ at time t. This probability depends on the probability $p(\mathbf{r} + \mathbf{a}, t - 1)$ of finding the particle at the neighboring sites $\mathbf{r} + \mathbf{a}$ at the previous time step $t - 1$, averaged over all neighboring z sites ($z = 4$ in a 2D lattice):

$$p(\mathbf{r}, t) = \frac{1}{z} \sum_{\mathbf{a}} p(\mathbf{r} + \mathbf{a}, t - 1). \tag{7.47}$$

When the particles are released steadily, there should be no time dependence: $p(\mathbf{r}, t - 1) = p(\mathbf{r}, t)$. Hence,

$$p(\mathbf{r}) = \frac{1}{z} \sum_{\mathbf{a}} p(\mathbf{r} + \mathbf{a}), \tag{7.48a}$$

which can be wrritten, say, for a 2D case, as

$$p(i, j) = \tfrac{1}{4} \left[p(i - 1, j) + p(i + 1, j) + p(i, j - 1) + p(i, j + 1) \right]. \tag{7.48b}$$

Equations (7.48) are the discretized forms of the Laplace equation (Problem 7.5):

$$\nabla^2 p(\mathbf{r}) = 0 \tag{7.49a}$$

---

[7]L.A. Turkevich and H. Scher, Phys. Rev. **A33**, 786 (1986).
[8]T.A. Witten and L.M. Sander, Phys. Rev. **B27**, 5686 (1983).

or, in 2D,

$$\frac{\partial^2 p}{\partial x^2} + \frac{\partial^2 p}{\partial y^2} = 0. \tag{7.49b}$$

In a similar manner, one can consider the boundary conditions. The growth velocity $v_{\perp}(\mathbf{r})$ at the unoccupied site $\mathbf{r}$ at the surface A of the cluster is proportional to the probability of getting the random walker at that point:

$$v_{\perp}(\mathbf{r}) \propto \frac{1}{z} \sum_{\mathbf{a}} p(\mathbf{r} + \mathbf{a}) \tag{7.50}$$

(the subscript indicates that we are interested in the component of v perpendicular to the surface A). At all sites that belong to the aggregate, $p = 0$: The particles are adsorbed when they reach the aggregate and cannot penetrate inside. On the other hand, far away from the aggregate, $p(\mathbf{r})$ is some constant, $p(\mathbf{r} \rightarrow \infty) = $ const, provided the flux of newly released particles is kept constant. Therefore, the right-hand-side of (7.50) is proportional to the gradient of $p(\mathbf{r})$ and (7.50) can be considered as a discrete approximation of the equation describing the movement of a smooth continuous boundary

$$v_{\perp} = -\mathcal{D}\boldsymbol{\nu}\nabla p|_A, \tag{7.51}$$

where $\mathcal{D}$ is the diffusion constant and $\boldsymbol{\nu}$ is the normal to A.

The fact that the DLA can be approximated by the Laplace equation (7.49) with the boundary condition (7.51) has an important heuristic value, because many growth phenomena can be described by this unifying model. The difference would be in the type of boundary conditions. In the next section, we consider 2D viscous fingering in the Hele–Shaw cell.

7.4. Viscous Fingering in the Hele–Shaw Cell

The Hele–Shaw cell is composed of two horizontal transparent plates of linear size w located at $z = -b/2$ and $z = b/2$. The cell is very thin, $b \ll w$, so that the Reinolds number is small (see Section 6.2). It is filled with a viscous liquid, such as oil or glycerin. Through the opening in the cell, a low-viscosity fluid (air) is injected to replace the high-viscosity fluid. The flow can be considered as 2D laminar flow (or potential flow, with velocity that satisfies the condition $\nabla \times \mathbf{V} = 0$). Numerous experiments, the most celebrated of which have been performed by Saffman and Taylor,[9] show that the moving interface between the

---

[9]P.G. Saffman and G.I. Taylor, Proc. Roy. Soc. **A 245**, 312 (1958).

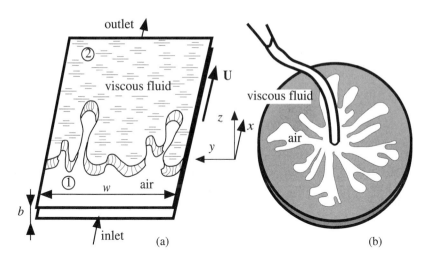

Figure 7.15. Viscous fingering in the Hele–Shaw cell of (a) rectangular and (b) circular shape.

two fluids is unstable. Small fluctuations of the interface (bumps) grow into stable fingers (Fig. 7.15). The width of the fingers changes with the velocity; at high velocities, the finger might branch and split.

The patterns formed in the Hele–Shaw cell vary, depending on a number of factors. Among these factors are intuitively "obvious," such as injection rate, surface tension, viscosity, or shape of the Hele–Shaw cell (e.g., rectangular versus circular). One of the most interesting experimental discoveries[10] is that the patterns strongly depend on the boundary conditions at the interface between the two fluids. Namely, if the Hele–Shaw cell is filled with a porous medium (e.g., randomly packed spheres), the resulting pattern is fractal, strikingly similar to DLA aggregates. When there is no porous medium, the situation is less clear: In some cases, the driving fluid adopts a compact geometry, whereas in other cases, it is fractal, which suggests that there might be all combinations of the two types of behavior. Again, there is no general analytical solution of the Laplace problem capable of predicting a fractal structure and a fractal dimension. Thus, numerial and real experiments remain an indispensible tool for studying the growth forms.

Below, we consider only the very beginning of the viscous fingering instability, assuming that the deviations of the interface from a straight line are small. The aim is to demonstrate the physical mechanism of the growth instability, in which small protrusions of the interface grow much faster than do the neighboring regions.

---

[10]J.D. Chen and D. Wilkinson, Phys. Rev. Lett. **55**, 1892 (1985); K.J. Måløy, J. Feder, and T. Jøssang, ibid., p. 2688.

7.4.1. Flow in Thin Cells

At low Reynolds number, for a 2D velocity field $[v_x(z), v_y(z), 0]$, the Navier-Stokes equations (6.34) reduce to

$$\eta \frac{\partial^2 v_i}{\partial z^2} = \nabla_i p, \quad i = x, y, \tag{7.52}$$

where η is the shear viscosity; p is the pressure, which should not be confused with the probability above (although both satisfy the Laplace equation and therefore play a similar role, as we shall see below); x and y are Cartesian coordinates in the plane of the cell. The gravity effects are absent because the cell is horizontal. Equations (7.52) can be integrated using the no-slip boundary conditions $v_i\big|_{z=\pm b/2} = 0$:

$$v_i = \frac{1}{2\eta} \left(z^2 - \tfrac{1}{4}b^2 \right) \nabla_i p. \tag{7.53}$$

For further analysis, it is helpful to introduce an average (over the z-axis) velocity $\mathbf{U}$ with components $U_i = \frac{1}{b} \int_{-b/2}^{b/2} v_i(z)\, dz$ because $\mathbf{U}$ is directly proportional to the local force:

$$\mathbf{U}(x, y) = -\frac{b^2}{12\eta} \nabla p(x, y). \tag{7.54}$$

(The last equation applies to flows through porous media; in that case, it is often called the Darcy law.) The condition of incompressibility $\nabla \cdot \mathbf{U} = 0$ then indicates that the pressure obeys Laplace equation in two dimensions (compare with (7.49b) for probability):

$$\frac{\partial^2 p}{\partial x^2} + \frac{\partial^2 p}{\partial y^2} = 0. \tag{7.55}$$

7.4.2. Instability of the Interface

We label by the number "1" the less viscous fluid (air) that displaces the more viscous fluid "2" (Fig. 7.15). When the moving interface remains straight in the horizontal plane, its position is given by $x = Ut$. The movement is maintained by the pressure gradients found from (7.54):

$$p_j = p_0 - \frac{12\eta_j U}{b^2}(x - Ut); \quad j = 1, 2; \tag{7.56}$$

p_0 is some constant that does not depend on x. Suppose there is a small perturbation, periodic along the y-axis with a wavenumber q and an amplitude A:

$$x(y) = Ut + A(t) \cos qy. \tag{7.57}$$

When $A \neq 0$, the pressures p_1 and p_2 should suffer periodic perturbations $\sim \cos qy$ similar to that of the interface profile:

$$p_j = p_0 - \frac{12\eta_j U}{b^2}(x - Ut) + B_j(x, t) \cos qy. \tag{7.58}$$

The x-dependence of the amplitudes $B_j(x, t)$ should be in the form of either $\exp(qx)$ or $\exp(-qx)$, because p_1 and p_2 with perturbations $\sim \exp(-iky)$ must satisfy the Laplace equation (7.55). A natural assumption would be that far away from the interface the pressures should remain finite; hence,

$$B_1(x, t) = B_1(t)\exp(qx); \quad B_2(x, t) = B_2(t)\exp(-qx). \tag{7.59}$$

The amplitudes $B_1(x, t)$ and $B_2(x, t)$ can be determined from the conditions at the interface. First, there is a condition of continuity that requires that the normal components of the velocity are equal to each other and to the velocity U_n of the interface:

$$U_n = -\frac{b^2}{12\eta_1}(\nabla p_1)_n = -\frac{b^2}{12\eta_2}(\nabla p_2)_n. \tag{7.60}$$

If $A(t)$ in (7.57) is small, then

$$U_n \approx U + \frac{\partial A}{\partial t}\cos qy. \tag{7.61}$$

From (7.54), (7.58), (7.59), and (7.61), neglecting second-order terms $\sim A^2, \sim AB$, one finds the following approximation to (7.60):

$$\frac{\partial A}{\partial t} = -\frac{qb^2}{12\eta_1}B_1(x, t) = \frac{qb^2}{12\eta_2}B_2(x, t). \tag{7.62}$$

Another boundary condition relates the pressure jump $\Delta p = p_1 - p_2$ across the interface and the mean curvature of the interface (see Chapter 13). Both principal curvatures might be nonzero: The interface is curved in the vertical plane xz with a radius $R_{xz} \approx b/2$, and in the horizontal xy-plane, with some radius R_{xy} that becomes finite when different parts of the interface advance with different velocity. Assuming that the fluid "2" wets the plates of the Hele–Shaw cell while the fluid "1" does not, and neglecting any influence the motion of interface might have on the geometry of wetting, one writes

$$\Delta p = \sigma \left(\frac{2}{b} + \frac{1}{R_{xy}} \right). \tag{7.63}$$

Although the first ("vertical") contribution in (7.63) is often larger than is the second one, it does not depend on x and y and, thus, does not affect the issue of in-plane instability. The second term is important, because it tends to flatten the interface. With $1/R_{xy} \approx -\partial^2 x(y)/\partial y^2$, where $x(y)$ is specified by (7.57), this term produces

$$\Delta p = \sigma q^2 A(t) \cos qy. \tag{7.64}$$

On the other hand, Δp follows from (7.58) as

$$\Delta p = \left[-\frac{12U}{b^2}(\eta_1 - \eta_2)A(t) + B_1(x,t) - B_2(x,t) \right] \cos qy. \tag{7.65}$$

Comparison of (7.64) and (7.65) leads to

$$B_1(x,t) - B_2(x,t) = A(t) \left[\frac{12U}{b^2}(\eta_1 - \eta_2) + \sigma q^2 \right]. \tag{7.66}$$

Finally, eliminating $B_1(x,t)$ and $B_2(x,t)$ from (7.62) and (7.66), one finds

$$\frac{\partial A}{\partial t} = A \frac{12Uq(\eta_2 - \eta_1) - \sigma q^3 b^2}{12(\eta_1 + \eta_2)}. \tag{7.67}$$

When $\eta_1 > \eta_2$ (a more viscous fluid pushes a less viscous one), the interface is stable for all q. However, when $\eta_2 > \eta_1$, any perturbation with a sufficiently large wavelength $\lambda > \lambda_c$,

$$\lambda_c = \pi b \sqrt{\frac{\sigma}{3U(\eta_2 - \eta_1)}}, \tag{7.68}$$

will be unstable: The amplitude A of perturbation grows with time. The *fastest* growing mode is the one with the wavelength $\lambda_f = \sqrt{3}\lambda_c$. As many experiments show, the periodicity of the interfacial instability in Hele–Shaw cells is indeed close to λ_f.

When the (small) viscosity of the driving fluid "1" can be neglected, then

$$\lambda_m \approx \frac{\pi b}{\sqrt{N_{\text{cap}}}}, \tag{7.69}$$

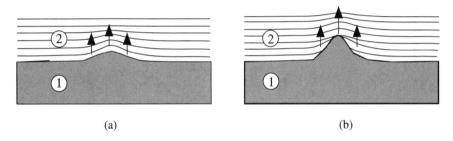

Figure 7.16. (a) Small perturbation of the interface increases field (pressure) gradients, most significantly at the tip of the protrusion; (b) as a result, the interface accelerates.

where $N_{cap} = U\eta_2/\sigma$ is the dimensionless "capillary number." By increasing N_{cap}, one can significantly reduce λ_m, which is an analog of the characteristic size of particles forming DLA clusters.

It is instructive to draw a qualitative physical explanation of the viscous fingering from (7.54) and the geometry of interface (see Fig. 7.16a), where the thin lines represent isobars generated by a perturbation of the interface. Let the pressure difference $p_{1,\text{inlet}} - p_{2,\text{outlet}}$ between the "inlet" and "outlet" be constant. In the low-viscosity fluid "1," the pressure gradients are small; (7.54); we can assume that the pressure within the domain "1," including the tip of the highest finger, is constant. Then the largest pressure gradients $\nabla p \sim (p_{1,\text{inlet}} - p_{2,\text{outlet}})/(x_{\text{outlet}} - x_{\text{tip}})$ in the viscous fluid "2" develop precisely at the tip of the finger. According to (7.54), the tip should move faster than does the rest of the interface. Hence, the initial small perturbation develops into a fast-propagating finger (Fig. 7.16b). The surface tension plays a stabilizing role, tending to reduce the pressure gradients, see (7.63).

The tendency of protruding perturbations to enhance field gradients is characteristic of many pattern-forming systems. Instead of the isobars, the thin lines in Fig. 7.16 might represent isothermal lines in the vicinity of the crystal-melt interface (the Mullins–Sekerka[11] instability), probability (DLA), equipotential lines, and so on.

As already mentioned, viscous fingering in Hele–Shaw cells filled with porous medium produces well-defined fractal structures, whereas "traditional" cells produce well-developed fractals seldomly. The difference might be in the "random" character of boundary conditions set by the porous medium. The randomly varying size of the pores is different from the (constant) characteristic wavelength λ_c set by capillary effects in the traditional cell. The permeability depends on the pore size: A narrow pore is hard to penetrate because of the capillary pressure associated with the pore size.

Problem 7.1. Find analytically how the fractal dimension of the Koch curve depends on the base angle β.

----

[11] W.W. Mullins and R.F. Sekerka, J. Appl. Phys. **34**, 323 (1963).

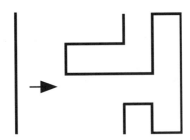

Figure 7.17. See Problem 7.2.

Answers: $D = \frac{\ln 4}{\ln[2(1+\cos\beta)]}$ [see G. Baumann (1996)].

Problem 7.2. Can the fractal dimension be larger than the Euclidean dimension of the embedding space? Give examples.

Answers: Yes. Consider a straight line, and at each iteration step, replace the line segments with, say, 18 segments, as shown in Fig. 7.17. At the next step, the line starts to cross itself. The similarity dimension is larger than 2: $\ln 18/\ln 4 \approx 2.1$.

Problem 7.3. Prove $S(q) \propto q^{-D}$ for $q \gg 1/R$ (7.21).

Answers: $S(q) \propto q^{-D} \Gamma(D-1) \sin[\pi(D-1)2]$.

Problem 7.4. Prove the second part of (7.22).

Answers: The result follows from the transformation

$$\sum_{i,j=1}^{s} (\mathbf{r}_i - \mathbf{r}_j)^2 = \sum_{i,j=1}^{s} [(\mathbf{r}_i - \mathbf{r}_0) - (\mathbf{r}_j - \mathbf{r}_0)]^2$$

$$= s \sum_{i=1}^{s} (\mathbf{r}_i - \mathbf{r}_0)^2 + s \sum_{j=1}^{s} (\mathbf{r}_j - \mathbf{r}_0)^2$$

$$- 2 \sum_{i=1}^{s} (\mathbf{r}_i - \mathbf{r}_0) \sum_{j=1}^{s} (\mathbf{r}_j - \mathbf{r}_0);$$

the last term is zero due to the definition $\mathbf{r}_0 = \frac{1}{s}\sum_{i=1}^{s}\mathbf{r}_i$ of the center of mass.

Problem 7.5. Prove that the Laplace equation (7.49b) can be approximated by the discrete (7.48b) in 2D.

Hint. Approximate the partial derivatives by finite differences, and use the Taylor expansion up to the second order.

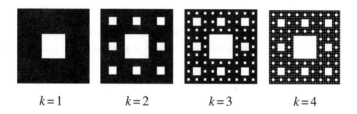

$k = 1$ $\qquad$ $k = 2$ $\qquad$ $k = 3$ $\qquad$ $k = 4$

Figure 7.18. Construction of the Sierpinski carpet.

Problem 7.6.

(a) Find the fractal dimension of the Sierpinski carpet (Fig. 7.18).

(b) The Menger sponge is a fractal constructed in 3D space by rules similar to the construction of the Sierpinski carpet. One starts with a cube of linear size 1. Each face of the cube is divided into nine equal squares. One drills a hole through the central square (from each side). The remaining 20 cubes of linear size $1/3$ are drilled again through the similar process. The sides of the original cube are patterned exactly as the Sierpinski carpet. Calculate the similarity dimension of the Menger sponge.

Answers: (a) $D = \ln 8/\ln 3 \approx 1.9$; (b) $D = \ln 20/\ln 3 \approx 2.7$.

Problem 7.7. Write computer programs to generate (a) deterministic Koch curves and their random counterparts; (b) percolation patterns; (c) DLA clusters.

Further Reading

J. Feder, Fractals, Plenum Press, New York and London, 1988, 283 pp.

Fractals. Selected Reprints, Edited by A.J. Hurd, Americal Association of Physics Teachers, College Park, Maryland, 1989, 139 pp.

J.-F. Gouyet, Physics and Fractal Structures, Springer-Verlag, Berlin, 1996, 234 pp.

Introduction to Nonlinear Physics, Edited by Lui Lam, Springer-Verlag, New York, 1997, 417 pp.

The Fractal Approach to Heterogeneous Chemistry. Surfaces, Colloids, Polymers, Edited by D. Avnir, John Wiley & Sons, Chichester, 1989, 441 pp.

B.B. Mandelbrot, The Fractal Geometry of Nature, W.H. Freeman and Company, New York, 1983.

Solids Far from Equilibrium, Edited by C. Godreche, Cambridge University Press, Cambridge, 1992.

D. Stauffer and A. Aharony, Introduction to Percolation Theory, Taylor & Francis Ltd., London, 1992.

D. Bensimon, L.P. Kadanoff, S. Liang, B. Shraiman, and C. Tang, Rev. Mod. Phys. **58**, 977 (1986).

J.S. Langer, Rev. Mod. Phys. **52**, 1 (1980).

H. Gould and J. Tobochnik, An Introduction to Computer Simulation Methods: Application to Physical Systems, 2nd Edition, Addison-Wesley Publishing Company, Reading, MA, 1996.

R.J. Gaylord and P.R. Wellin, Computer Simulations with Mathematica®: Explorations in Complex Physical and Biological Systems, Springer-Verlag, New York, 1995.

G. Baumann, Mathematica in Theoretical Physics, Springer-Verlag, New York, 1996, Ch. 7, 348 pp.

H. Lauwerier, Fractals: Endlessly Repeated Geometrical Figures, Princeton University Press, Princeton, New Jersey, 1991, 209 pp.

CHAPTER 8

Dislocations in Solids. Plastic Relaxation

Dislocations are responsible for the *plastic deformation* of crystalline materials such as metals, and play a role in a number of other properties of crystals, such as crystal growth, electrical properties of semiconductors, radiation damage through their interaction with point defects, and so on. Their theoretical discovery dates back to the years just before the second World War, and their visualization by various techniques, essentially electron microscopy, to the 1950s. They are an essential ingredient of physical metallurgy. They carry internal stresses and are "topologically" related to the symmetries of the crystal. It is this double character that makes them act as sources of plastic deformation. Their topological properties were fully appreciated with the appearance of a general classification of topological defects in ordered media, which comprehends superfluids, magnetic systems, and liquid crystals. Although defects and their textures in liquid crystals were observed (by optical microscopy) long before defects and their textures in solids, it is only in the last 10 years that investigations on the role of structure and texture on the *rheological properties* have been developed. A good knowledge of the bases of the physics of defects in solids cannot but help to progress in the investigation of rheological properties and instabilities of mesomorphic materials.

8.1. Elasticity of Dislocations

8.1.1. Linear Elasticity; a Summary

Consider a rod of uniform section S, and length ℓ, submitted to a force $\mathbf{F} = \sigma S \hat{\mathbf{z}}$ ($\hat{\mathbf{z}}$ is the unit vector along the rod axis) applied to one of the rod sections, the other one being kept fixed. The rod changes length by an algebraic quantity u, proportional to ℓ (Fig. 8.1). Let $e = u/\ell$; in the new equilibrium state after the force is set up, and as long as $F = |\mathbf{F}|$ is small (this will be qualified later on), we have the linear relationship (Hooke's law):

$$\sigma = Ee; \tag{8.1}$$

261

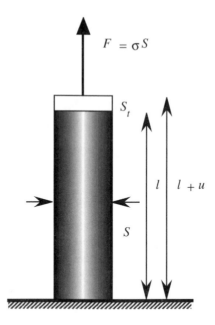

Figure 8.1. Deformation of a rod under a force F applied to one of its extremities.

σ is the tensile stress component acting on the terminal section S_t, and E, the Young modulus, is a material constant. Hence, $u(\ell) = F\ell/SE$.

Let S+, S− be the lips of an imaginary section S anywhere along the length of the rod; each piece of the rod is submitted to a total vanishing force. Hence, σ appears as the force per unit area exerted by the lower part of the rod, on the upper part, this force being applied on S+. There is transmission of the force, and $e(M) = u(M)/\ell$ (M), the relative displacement or *strain*, has the same value $e = u/\ell$ at any point M of the rod. The energy stored in the rod during the *reversible* elastic displacement from $\delta\ell = 0$ to $\delta\ell = u$ is

$$w = S \int_0^1 \sigma(x)\,du(x) = \frac{S\ell}{2}\sigma e$$

(here, $\sigma(x) = \sigma x$, $du(x) = e\ell\,dx$, $0 \le x \le 1$), or per unit volume

$$w/S\ell = \tfrac{1}{2}\sigma e = \tfrac{1}{2}Ee^2. \tag{8.2}$$

The notions above generalize to a stress tensor σ_{ij} and a strain tensor e_{ij}

$$e_{ij} = \tfrac{1}{2}(u_{i,j} + u_{j,i}) \tag{8.3}$$

such that:

1. An infinitesimal surface area dS, with outward pointing normal $\boldsymbol{\nu}$, is the place of application of an infinitesimal force:

$$dF_i = \sigma_{ij}\nu_j\,dS \tag{8.4}$$

acting on the matter outwards (Fig. 6.2). We have used in (8.4) the usual Einstein convention (summation over repeated indices).

The component of dF_i along the surface element, viz. $dF_i(\delta_{ik} - \nu_i\nu_k)$ is called the *shear stress*; the component of dF_i along the normal to the surface element, viz. $(d\mathbf{F}\cdot\boldsymbol{\nu})\nu_i$, is the *tensile stress* (if the scalar product $(d\mathbf{F}\cdot\boldsymbol{\nu})$ is positive) or the *compressive stress* (if the scalar product $(d\mathbf{F}\cdot\boldsymbol{\nu})$ is negative). The equilibrium conditions applied to a small volume of matter yield the following equations:

$$\sigma_{ij,j} = 0, \quad \text{in compact notation: } \nabla\cdot\sigma = 0 \text{ (the total force vanishes)}, \tag{8.5a}$$

$$\sigma_{ij} = \sigma_{ji}, \quad \text{the total torque vanishes.} \tag{8.5b}$$

2. The energy is quadratic in the deformations, as in (8.2); this is a consequence of Hooke's law, which states that a linear relationship between the stresses and the strains holds as long as the forces are small:

$$\sigma_{ij} = C_{ijk\ell}e_{k\ell}. \tag{8.6a}$$

Because σ_{ij} and $e_{k\ell}$ are symmetric tensors, we have $C_{ijk\ell} = C_{jik\ell} = C_{ij\ell k} = C_{ji\ell k}$, so that there are only 36 independent elastic coefficients. This number is reduced by the symmetries of the material, because the elastic energy density $\tfrac{1}{2}\sigma_{ij}e_{ij}$ stored in the material must be invariant in all frames of reference equivalent under the action of these symmetries (this point will be made more precise later on a specific example). As a result, there are only two independent elastic coefficients in an isotropic medium, the two Lamé coefficients λ and μ, so that

$$\sigma_{ij} = 2\mu e_{ij} + \lambda\,\delta_{ij}\,\mathrm{div}\,\mathbf{u}, \tag{8.6b}$$

where $\mathrm{div}\,\mathbf{u} = e_{11} + e_{22} + e_{33} = u_{i,i}$; the Young modulus is $E = \frac{\mu(3\lambda+2\mu)}{\lambda+\mu}$. One also defines the adimensional Poisson ratio $\nu = \frac{\lambda}{2(\lambda+\mu)}$ whose usual value in solids is ≈ 0.3.

In these conditions, we find that (8.5a), expressed in terms of the displacements u_i, takes the form

$$\mu \, \Delta \mathbf{u} + (\lambda + \mu) \, \text{grad div } \mathbf{u} = 0, \tag{8.7}$$

assuming that there are no other applied forces. In a dynamical regime, we add the inertial forces:

$$\mu \, \Delta \mathbf{u} + (\lambda + \mu) \, \text{grad div } \mathbf{u} = \rho \frac{\partial^2 \mathbf{u}}{\partial t^2}, \tag{8.8}$$

where ρ is the mass density.

The approximation of linear elasticity is sufficient to explore the physical properties of dislocations, and we shall stick to this approximation later on.

8.1.2. Applied Stresses and Internal Stresses

Usually, (8.5) are supplemented by conditions on the boundary S_b of the sample that consist either in given applied forces or in given displacements. These two types of conditions might apply to different regions of S_b. There is a general theorem that states that the differential equations in (8.5) have a *unique regular solution*, as soon as the total applied forces $\int_{S_b} \mathbf{f}_{\text{app}} \, dS_b$ and the resulting torque vanish. The continuous stresses that result from the regular solution are *applied stresses*.

However, singular solutions exist, even if $\sigma_{ij} \nu_j \equiv 0$ everywhere, if the medium is not simply connected, i.e., when there are inner boundaries that are not taken into account in the above boundary conditions. An evident case would be a spherical cavity with internal pressure acting on its boundary. But this could be treated by a simple extension of the usual methods, and it is not the case we have in mind. What we have in mind are the *Volterra dislocations*, which correspond to a much less trivial case of inner disconnectedness.

8.2. Volterra Dislocations

8.2.1. Definitions

The Volterra process: The elementary types of dislocations are represented in Fig. 8.2. They are all obtained as follows: Cut the material along a surface Σ (the so-called cut surface) bound by a line L (a loop, or an infinite line); displace the two lips Σ' and Σ'' of the cut surface by a relative *rigid* displacement that can be analyzed as the sum of a translation **b** and a rotation Ω. We shall note Ω: **OM** the transform of vector **OM** under the action of the rotation matrix Ω, where O is a fixed origin, invariant in the rotation, and M is a running point of Σ. The displacement **d**(M) that brings M onto M' (**MM'** = **d**(M) = **b** + Ω:**OM**),

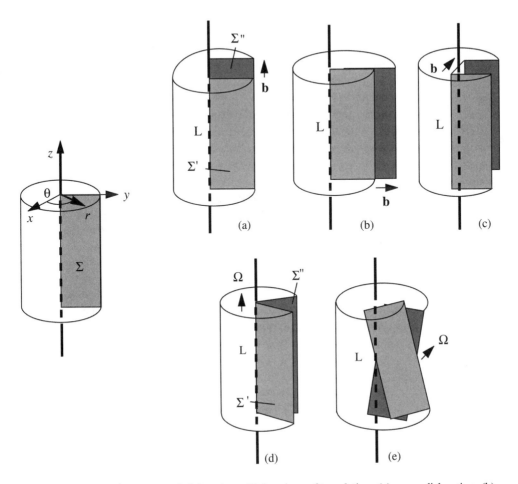

Figure 8.2. Elementary types of dislocations. Dislocations of translation: (a) screw dislocation; (b) and (c) construction of the same edge dislocation. Disclinations: (d) wedge dislocation; (e) twist dislocation.

is *rigid* in the sense that it conserves angles and distances. Now introduce matter in its perfect state (undeformed elastically)—in order to fill the void left by the translation and the rotation—or remove matter in the regions of double covering; glue back the lips Σ' and Σ'' of new matter; let the medium relax elastically. Such a process has, of course, no meaning along L, and one assumes henceforth that a cylindrical region has been removed along L. This "core" is a region where the above process is not valid.

The Volterra process is compatible with a sample-free boundary, but it undoubtedly introduces stresses: They are called *internal stresses*.

1. The core: Clearly, these stresses and strains increase as one goes closer to the core region. The size r_c of the core is such chosen that the stress and strain fields are physically in the elastic range outside r_c, where Hooke's law is valid. They become large near r_c, and their analytical extension is singular somewhere inside the "core" region.

2. Weingarten theorem: The (elastic) stresses and strains are nonsingular on Σ (in fact, Σ' and Σ'') as long as $\mathbf{d}(M)$ is a rigid displacement. Henceforth, the final result does not depend on the exact choice of Σ.

Figures 8.2a, b, and c represent *dislocations of translation*: The displacement is reduced to a pure translation either in the direction of L (8.2a, *screw* dislocation), or perpendicular to L (8.2b, 2c, *edge* dislocations). According to Weingarten theorem, dislocations 8.2b and 8.2c are the same. Figures 8.2d and e represent *dislocations of rotation*, also called *disclinations*. It is clear that the *twist* disclination (8.2e) raises delicate problems of construction, to which we shall come back later. Mixed cases of dislocations (of translation) and disclinations exist.

8.2.2. Elastic Observables Related to Volterra Defects

A general remark: Boundary conditions for the equations of elasticity have to be written on the cut surface Σ. Let $\boldsymbol{\nu}$ be a unit normal vector to Σ (the choice of $\boldsymbol{\nu}$ confers an orientation to the line). The difference between the displacements u^+ and u^- is given by

$$\Delta u = u^+ - u^- = \mathbf{d}(M) = \mathbf{b} + \boldsymbol{\Omega} \times (\mathbf{OM}), \qquad (8.9)$$

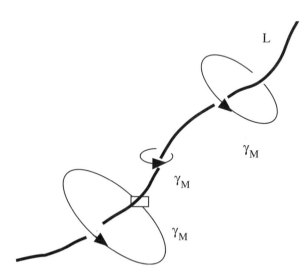

Figure 8.3. Directed circuits γ_M surrounding a dislocation line L, yielding the same value as the integral of (8.11).

where **b** is called the Burgers vector. We have introduced the *rotation vector* $\mathbf{\Omega}$, which is easier to manipulate than is the rotation tensor Ω, and which is a valid representation of a rotation as far as $\mathbf{\Omega} \times (\mathbf{OM})$ is small. Equation (8.9) can also be written in an integral form:

$$\oint_{\gamma_M} d\mathbf{u} = \mathbf{d}(M) \qquad (8.10)$$

on any directed circuit γ_M surrounding L and intersecting the cut surface in M (Fig. 8.3), thanks to Weingarten theorem.

We give hereunder some results concerning classic cases (for details, refer to the textbooks cited at the end of this chapter).

8.2.2.1. Screw Dislocations

We use the cylindrical coordinates of Fig. 8.2. The total stress field in a finite sample (cylinder of radius R) can be split in the sum of a field σ that vanishes at infinity and is singular on the line, and a field σ' that is regular everywhere and opposes the nonvanishing forces due to the singular field σ on the boundary $r = R$; i.e., $(\sigma + \sigma') \cdot \boldsymbol{\nu} = 0$ at any position on the boundary, $\boldsymbol{\nu}$ being a unit vector perpendicular to the boundary. Linear elasticity allows of course for such linear superpositions. Let $\mathbf{u}_{\text{screw}} = \mathbf{u} + \mathbf{u}'$ be the total displacement field. Both **u** and **u'** fields obey (8.8), if one assumes isotropic elasticity, and the singular part obeys, moreover, (8.10), which reduces here to

$$\oint_{\gamma} d\mathbf{u} = \mathbf{b}, \qquad (8.11)$$

where γ is any closed loop encircling the z-axis once. An obvious solution for **u** is

$$\mathbf{u} = \frac{\mathbf{b}}{2\pi}\theta, \quad e_{\theta z} = e_{z\theta} = \frac{b}{4\pi r}, \quad \sigma_{\theta z} = \sigma_{z\theta} = \frac{\mu b}{2\pi r}, \qquad (8.12)$$

all other components vanishing. Note that the deformation field and, consequently, the stresses vanish at infinity, but not on the boundary $r = R$. This is why **u** has to be supplemented by a relaxation field **u'** that brings opposite surface forces on the boundary and is regular everywhere in the bulk. We do not discuss this relaxation field here. The (nonrelaxed) energy of the singular field is, according to the general expression of the elastic free energy of a deformed medium,

$$W = \frac{1}{2}\int \sigma_{ij}e_{ij}\,dV, \qquad (8.13)$$

equal to $\frac{1}{2} \int (\sigma_{\theta z} e_{\theta z} + \sigma_{z\theta} e_{z\theta}) r\, dr\, d\theta = \frac{\mu b^2}{4\pi} \ln \frac{R}{r_c}$ per unit length of line, where the integration is taken in the range $r_c \leq r \leq R$. Adding the relaxation energy that corresponds to stresses $\sigma'_{z\theta} = -\frac{\mu b r}{\pi R^2}$ amounts to adding a term of the same order of magnitude, but necessarily smaller. One gets

$$W_{el} = \frac{\mu b^2}{4\pi} \left(\ln \frac{R}{r_c} - 1 \right). \tag{8.14}$$

Note (1) the dependence of the energy on b, which implies that real dislocations are expected to have Burgers vectors as small as possible; we have indeed $(2b)^2 > 2b^2$; hence, any dislocation of Burgers vector 2**b** tends to split into two dislocations of Burgers vector **b**; (2) the weak logarithmic dependence with the size of the domain influenced by the dislocation; (3) apart of the relaxation terms, expression (8.14) has also to be supplemented by the energy of the "core" region ($r < r_c$). All of these points will be discussed in detail later.

8.2.2.2. Edge Dislocation

Assuming that the medium is infinite, hence, forgetting the relaxation effects, we have (see also Problem 8.1)(Fig. 8.2b):

$$u_x = \frac{b}{2\pi} \left[\theta + \frac{\sin 2\theta}{4(1-v)} \right], \quad u_y = -\frac{b}{2\pi} \left[\frac{1-2v}{2(1-v)} \ell n r + \frac{\cos 2\theta}{4(1-v)} \right],$$

$$u_z = 0, \; W_{el} = \frac{\mu b^2}{4\pi(1-v)} \ln \frac{R}{r_c}. \tag{8.15}$$

8.2.2.3. Mixed Dislocation

Let β be the angle between the line L and the Burgers vector **b**. The calculation or the displacement field **u** proceeds easily from the principle of linear superposition, writing that **b** has a screw component $b \cos \beta$ and an edge component $b \sin \beta$. The energies also add linearly, but this does not follow from the principle of superposition, but from the fact that the two dislocation components do not interact (see Problem 8.2). We find, when the surface relaxation at $r = R$ is included,

$$W_{el} = \frac{\mu b^2}{4\pi K} \left(\ln \frac{R}{r_c} - 1 \right) \quad \text{with} \quad \frac{1}{K} = \cos^2 \beta + \frac{\sin^2 \beta}{1-v}. \tag{8.16}$$

8.2.2.4. Wedge Disclination

The calculation was made by Timoshenko (Fig. 8.2d). For a disclination of (small) angle $\Omega = |\mathbf{\Omega}|$, (8.10) reduces to

$$\oint_{\gamma_M} d\mathbf{u} = \mathbf{\Omega} \times \mathbf{OM},$$

where Ω is along the disclination line, O is an arbitrary origin on the line, M is on the cut surface, and one gets

$$\sigma_{rr} = \frac{\mu\omega}{2\pi(1-\nu)} \left(\ln\frac{r}{R} + \frac{r_c^2}{r^2}\ln\frac{r_c}{R} \right),$$

$$\sigma_{\theta\theta} = \frac{\mu\omega}{2\pi(1-\nu)} \left(\ln\frac{r}{R} + \frac{r_c^2}{r^2}\ln\frac{r_c}{R} + 1 \right),$$

$$\sigma_{zz} = \nu(\sigma_{rr} + \tau_{\theta\theta}), \quad \sigma_{r\theta} = \sigma_{\theta z} = \sigma_{zr} = 0,$$

$$W_{el} = \frac{\mu\omega^2}{16\pi(1-\nu)}(R^2 - r_c^2)\left(1 - \frac{4R^2 r_c^2}{R^2 - r_c^2}\left(\ln\frac{R}{r_c} \right)^2 \right). \tag{8.17}$$

This energy is far larger than for a dislocation of translation. Henceforth, disclinations are not usually met in real solid media, except in the form of pairs of opposite sign. We have mentioned their existence in the Frank and Kasper phases and their possible existence in amorphous media; in both cases they form interlinked networks whose stresses compensate at short distances.

8.3. Simple Topological Characteristics of Dislocations

8.3.1. Equivalent Circuits

It appears evident from the above that a dislocation, which is characterized by its topological invariant the Burgers vector **b**, is a line that must either close on itself, end on another dislocation or on the boundary of the sample, or go to infinity.

The *Burgers circuit* γ_M is a directed circuit that encircles L once and that does not pass through the core region. In that sense, a sample with cylindrical core regions is a *non-connected* medium, because all circuits that can be drawn in the sample are not equivalent under smooth transformations (smooth deformations of the circuits and smooth displacements). But all circuits γ_M deducible one from the other by a smooth transformation yield the same value as in (8.11).

The "node rule" (Fig. 8.4)

$$\sum_i \mathbf{b}_i = 0 \tag{8.18}$$

is a direct consequence of the equivalence of circuits, in the above sense: Note that in (8.18) all dislocations are oriented toward the node or all start at the node, **b** changing sign when the orientation of the line is reversed.

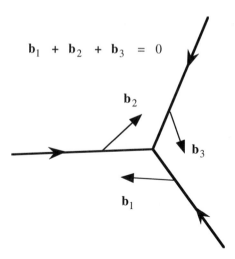

Figure 8.4. To illustrate (8.18).

8.3.2. Dislocations in Crystals

In crystals, the only relevant dislacements $\mathbf{d}(M)$ of the cut surface are those that obey crystalline symmetries, because then the "gluing" step of the Volterra process is equivalent to a reconstruction of the atomic or molecular bonds: The lips of the cut surface do not carry any crystalline singularity if $\mathbf{d}(M)$ splits into a lattice translation and a lattice rotation. Furthermore, Weingarten's theorem applies, and the cut surface does not carry any elastic singularity.

Such a dislocation is "*perfect*." One refers to an "*imperfect*" dislocation when Weingarten's condition is still satisfied, but not the condition on the crystalline symmetries. This is not an unfrequent case.

Figure 8.5 shows the mapping of a deformed crystal (containing a dislocation) onto the perfect crystal. It is visible that the image of the mapping is an open circuit, whose closure failure measures the Burgers vector. The usual sign convention (named FS/RH convention) is as follows: The Burgers circuit γ is traversed clockwise for an observer looking along the positive orientation of the line (right-hand convention); starting from $\mathbf{S}$ and ending at $\mathbf{F} \equiv \mathbf{S}$, the traversal of γ maps on γ' starting from $\mathbf{S}'$ and ending at $\mathbf{F}'$. The resulting Burgers vector is $\mathbf{b} = \mathbf{F}'\mathbf{S}'$.

Figure 8.6 shows a dislocation of rotation angle $\Omega = -\pi/2$ in a tetragonal crystal. This characteristic rotation angle Ω is measured as follows. Let us consider a (oriented) Burgers circuit γ on which one chooses an arbitrary origin M, and let us follow an arbitrary but constant lattice direction $\mathbf{n}$ when traversing γ and parallel-transport $\mathbf{n}$ to an original O. The set of extremities traverses an arc γ' of circle of angle Ω, oriented according to the construction. We call γ' the hodograph of γ. The hodograph of a given crystalline direction

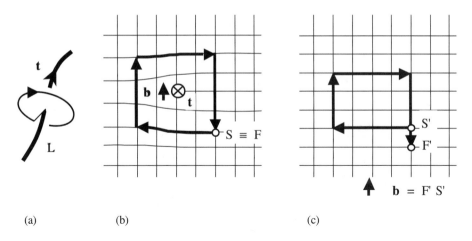

(a)   (b)   (c)

Figure 8.5. Mapping of the Burgers circuit onto a perfect crystal; the FS/RH convention.

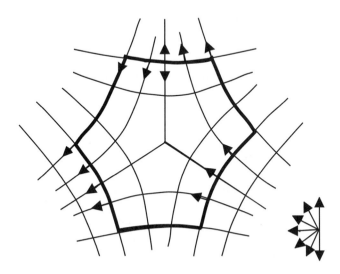

Figure 8.6. Dislocation of rotation angle $\Omega = -\pi/2$ in a tetragonal lattice and construction of the "hodograph" γ' of the rotation circuit γ to measure Ω.

followed along the directed circuit γ yields Ω in magnitude and in sign, independently of the orientation given to γ and of the choice of **n**, for any disclination.

8.3.3. Imperfect Dislocations. Stacking Faults and Twins

Dislocations of small energy have a Burgers vector that is a period of the lattice. We have already indicated that this condition is necessary to have invariance of the distortions with respect to the choice of the cut surface. But there are circumstances in which some nonperiodic Burgers vectors yield low-energy dislocations, although the cut surface maintains its individuality and, thus, carries a nonvanishing surface energy. This is well known in FCC lattices (see Section 2.1), where the correct stacking of (111) planes, i.e, ... ABC⋮ABCA..., can be modified at a small expense of energy, yielding, for example, ... ABC | BCAB.... The slip **b** (A → B, along a [1$\bar{1}$0] direction) is obviously not a periodic translation of the FCC lattice. If such a planar fault is limited to a finite part Σ of the plane of misfit, the boundary between Σ and the rest of the plane is an *imperfect* dislocation line of Burgers vector **b**. A detailed description of the zoology of imperfect dislocations is out of the scope of this textbook, inasmuch as this notion has not yet found application in the physics of soft matter.

The lattices ... ABCABCA... and ... ACBACB... are obviously the same, but to a mirror symmetry. A sequence of stacking faults on parallel adjacent (111) planes brings one lattice onto the other, yielding for example a *twin* ... ABC||BAC.... The plane marked by a double bar in this sequence is the habit plane of the twin. Reciprocally, repeated twinning on two adjacent planes build a stacking fault.

The notion of stacking fault and its relation with twinning extend to other lattices than FCC lattices.

8.4. Some Remarks on the Elastic Energy of a Dislocation

8.4.1. Stability

According to Eqs. (8.14)–(8.16), the elastic energy of a dislocation is of the order of μb^2, because the logarithmic term is of the order of a few units, for a broad range of plausible values of R and r_c.

R can be estimated as being the mean distance between dislocations. Indeed, the internal stresses are believed to fluctuate on the same scale, because neighboring dislocations in a well annealed material form a network (the Frank network), where dislocations of opposite signs, i.e., carrying opposite stresses, are approximately at a mean distance $2R$. Typically, R is of the order of a few tens of nanometers in work-hardened materials, or of the order of a few microns in a well-annealed single crystal metal. We discuss r_c later on.

According to (8.14) or (8.15), the quantity μb^2 is typical of the elastic energy carried by a dislocation line per atomic distance along the line. In Al, an FCC metal, one finds

$\mu b^2 \sim 2.2$ eV ($\mu = 2.7 \times 10^{10}$ N/m^2; $\mathbf{b} = \frac{1}{2}[110]$, $b = 0.29$ nm). In olivine, a rhombohedral mixed silicate of Mg and Fe, which is the most frequent mineral in earth crust, one finds $\mu b^2 \sim 52$ eV ($\mu = 8 \times 10^{10}$ N/m^2, $b = 0.48$ mm). The core energies, as we shall see, are of the same order of magnitude. The total energy of a dislocation line, measured by atomic or molecular length along the line, is therefore considerable, and at least of the same order of magnitude as the binding energy. A dislocation, except in special cases, is therefore an out-of-equilibrium object, because the entropic gain due to the disorder of dislocations is small, except possibly near the melting point.

For a real dislocation, the Burgers vector modulus takes the smallest possible value compatible with crystalline translations, i.e., $\mathbf{b} = \frac{1}{2}[110]$ in FCC, $\mathbf{b} = \frac{1}{2}[111]$ or $[100]$ in BCC, and so on. As mentioned above, the reason is that a dislocation of Burgers vector $2\mathbf{b}$ has an energy four times larger than a dislocation of Burgers vector $\mathbf{b}$, and tends to split into two elementary dislocations, as long as the positive energy of interaction between them does not exceed twice the energy of an elementary dislocation. Similarly, a dislocation $\mathbf{b} = \mathbf{b}_1 + \mathbf{b}_2$, where $\mathbf{b}_1$ and $\mathbf{b}_2$ are two (different) elementary Burgers vectors, tends to split into two dislocations $\mathbf{b}_1$ and $\mathbf{b}_2$, if the scalar product $\mathbf{b}_1 \cdot \mathbf{b}_2$ is positive, because $b^2 = b_1^2 + b_2^2 + 2\mathbf{b}_1 \cdot \mathbf{b}_2$: the dislocations $\mathbf{b}_1$ and $\mathbf{b}_2$ repel. If $\mathbf{b}_1 \cdot \mathbf{b}_2 < 0$, they attract.

8.4.2. Image Forces; Peach and Koehler Forces

8.4.2.1. Image Forces

The relaxation terms due to free boundaries, alluded to in the former section, can be described as due to image dislocations located outside of the sample. The simplest case is that of a screw dislocation $+\mathbf{b}$ parallel to a planar boundary S (Fig. 8.7).

It is easy to check that the screw dislocation $(-\mathbf{b})$ located symmetrically which respect to the midplane S carries a displacement field $\mathbf{u}_{\text{image}} = -\frac{b\theta}{2\pi}$ which yields stresses $\boldsymbol{\sigma}_{\text{im}}$ that

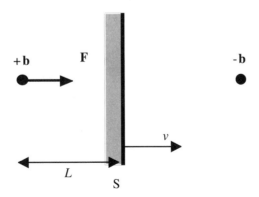

Figure 8.7. Image force for a screw dislocation. The dislocation is attracted to the free boundary.

exactly counterbalance the stresses of the real dislocation on S: $(\boldsymbol{\sigma} + \boldsymbol{\sigma}_{\mathrm{im}}) \cdot \boldsymbol{\nu} = 0$, $\boldsymbol{\nu}$ being the normal to S. Therefore, $\mathbf{u} + \mathbf{u}_{\mathrm{im}} = \frac{\mathbf{b}}{2\pi}(\theta - \theta_{\mathrm{im}})$ is the displacement field of the screw dislocation $+\mathbf{b}$, isolated in its half-space. One gets:

$$W_\ell = \frac{\mu b^2}{4\pi} \ln \frac{2L}{r_c}, \tag{8.19}$$

where L is the distance of the dislocation to S (see Problem 8.3).

Because the core energy obviously does not depend on the position of the line L, one sees that the energy of the screw dislocation line decreases when it approaches the free boundary. This effect can be described by a fictitious force $F = +\frac{\partial W_{el}}{\partial L}$ directed along $\boldsymbol{\nu}$. The plus sign indicates that this force, exerted on the line by the outer medium, is such that the dislocation is attracted toward the free boundary:

$$F = \frac{\mu b^2}{4\pi L}. \tag{8.20}$$

$\mathbf{F}$ can also to be considered as the force due to the presence of the image dislocation which is of an opposite Burgers vector; both dislocations attract each other, as would two electric charges of opposite sign; the analogy is evident.

8.4.2.2. Peach and Koehler Force

Let us now calculate the fictitious force exerted on a dislocation located in the stress field σ_{ij} of another dislocation, or in an applied stress field. We call W the total energy of the system; the spontaneous force to which the dislocation is submitted is $\mathbf{F} = -\nabla W$, with a minus sign now (inner force). We have to analyze the variation dW when the dislocation moves by a quantity $d\mathbf{x}$.

Let us first specialize to an edge dislocation L oriented along the y-axis that moves in the plane (L, $\mathbf{b}$) by a quantity $d\mathbf{x}$ (Fig. 8.8). Take, furthermore, the cut surface as the

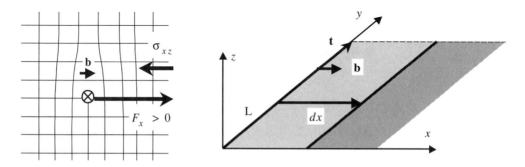

Figure 8.8. To illustrate the Peach and Koehler force.

half-plane (L, **b**) on the right side of L. Moving the dislocation along the x-axis consists in displacing the lips of the cut surface relative one to the other by a quantity $\mathbf{b} = b(1, 0, 0)$. Consider the force per unit area $f_i = \sigma_{ij}\nu_j$, acting on the moving lips of the cut surface, and assume that one of the lips is immobile, the mobile one having the oriented normal $\boldsymbol{\nu} = (0, 0, -1)$. In order to be consistent with the FS/RH convention, the displacement of the mobile lip must be $-\mathbf{b}$. The work dW of the outer stresses is given by

$$dW = -\mathbf{b} \cdot \mathbf{f}\, dx = b\, dx\, \sigma_{xz}.$$

Hence, the fictitious force

$$F_x = \partial_x W = b\sigma_{xz}, \tag{8.21}$$

which drives the line L toward the right if the acting stress σ_{xz} is positive (for another example, see Problem 8.4).

The formula above generalizes to a line of any shape and for any type of displacement, and it can be written in a compact form as

$$\mathbf{F} = (\mathbf{b} \cdot \sigma) \times \mathbf{t}, \tag{8.22}$$

where $\mathbf{t}$ is the unit tangent to the line. In this equation, $\mathbf{b}$ is acting on the first index of σ, i.e. $(\mathbf{b} \cdot \sigma)_j = b_i\sigma_{ij}$, like in (8.21). This remark has its importance when $\sigma_{ij} \neq \sigma_{ji}$ due to the presence of body torques. The full expresion with indices is $F_k = \varepsilon_{kpq}b_i\sigma_{ip}t_q$, but the Peach and Koehler force is often used in the very simple form of (8.21), which describes a shear experiment in the "glide plane" (L, **b**), σ_{xz} being a shear stress acting on this plane.

One will easily deduce from these formulae that two parallel straight dislocations attract if they carry opposite Burgers vectors; they repel if their Burgers vectors are equal.

8.4.2.3. Stability of a Dislocation Near a Surface; Nucleation of Dislocations

Equation (8.20) can be interpreted, using the Peach and Koehler formula, as the product of the image stress by the Burgers vector. Hence, one has $\sigma = \frac{\mu b}{4\pi L}$. Therefore, one must apply stresses of this order to maintain a dislocation inside the material, at a distance L from the boundary. For $L = a$, i.e. an interatomic distance, one gets $\sigma_{\max} = \frac{\mu b}{4\pi a} \approx \frac{\mu}{4\pi}$ (ca. 10^3 kg per mm^2 or 10^{10} N/m^2); reciprocally $\sigma_{\max}$ appears as the stress necessary to nucleate a dislocation at the boundary of a perfect crystal. In fact, plastic deformation of a material does start for stresses that are typically 10^3 times smaller; because plastic deformation requires the presence of a number of fresh dislocations, it is indispensable to understand why existing dislocations can multiply under very small stresses: This is the role of the Frank and Read mechanism (see below).

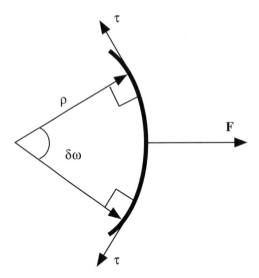

Figure 8.9. To illustrate (8.23).

8.4.3. Line Tension

The energy of a straight dislocation line is proportional to its length; the energy of a curved dislocation line is also proportional to its length, in a first approximation, if the radius of curvature of the line is larger than R, the typical distance between neighboring lines. A curve line has, therefore, a tendency to straighten, in order to decrease its length. This tendency can be described in terms of a line tension τ, which is the ratio of the variation of elastic energy $\delta W = \tau \, \delta \ell$ to the variation in length $\delta \ell$. A reasonable approximation for τ is to take it equal to the line energy, viz. $\tau \sim \mu b^2$. This notion will be developed in some more details in Section 9.1.3. and Problem 9.2.

Let us consider a segment of line anchored in its two extremities, curved under the action of a fictitious force $\mathbf{F}$ (Fig. 8.9). The line takes a curvature ρ^{-1} that is the result of the equilibrium between $\mathbf{F}$ and the line tension. One finds (Problem 8.5)

$$F = \tau/\rho. \tag{8.23}$$

8.4.4. Frank and Read Mechanism

Consider a straight segment of dislocation AB of length 2ℓ anchored in A and B (A and B can be sites of impurities, precipitates, intersections with other dislocation, and so on) (Fig. 8.10), position (a). The segment moves easily in its glide plane under the action of a Peach and Koehler force $F = \sigma b$ per unit length of line. The segment takes a radius of equilibrium $\rho = \frac{\tau}{\mu b} \sim \frac{\mu b}{\sigma}$.

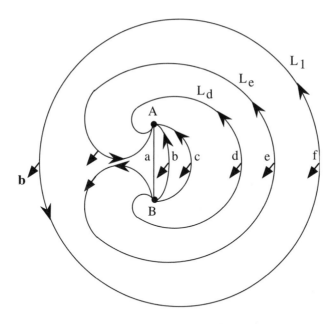

Figure 8.10. Frank and Read mechanism. The arrows show the Burgers vector.

Start with σ small. For $\sigma = 0$, we have $\rho^{-1} = 0$; the radius of curvature first decreases when σ increases, up to a value

$$\rho_c = \frac{\ell}{2} = \frac{\mu b}{\sigma_c} \tag{8.24}$$

reached for $\sigma_c = 2\mu b/\ell$ (Fig. 8.10), position (c). Because of the anchoring, the radius cannot but increase afterward, which requires $\sigma < \sigma_c$. Therefore the position $\sigma = \sigma_c$, $\rho = \ell/2$ is unstable, and the dislocation length increases spontaneously to a new position (d), where segments of opposite signs (still belonging to the same dislocation) are close to one other and, consequently attract each other strongly, see position (e). This process yields the formation of a loop L_1 that is disconnected from the segment AB, position (f). The segment reforms and grows again, pushing away L_1 because it carries a dislocation of the same sign. The mechanism continues, yielding successive loops $L_1, L_2, \ldots$ up to that point where their stresses add to the critical stress σ_c and overcome it. Note that $\sigma_c \sim \sigma \frac{b}{\ell}$ is small in a well-annealed material where there are few dislocations and few anchoring points: Taking $\ell \sim 10^{-6}$ m, $b \sim 10^{-9}$ m, one gets $\sigma_c \sim 10^{-3}\mu$. However σ_c must be larger than the Peierls stress σ_p, which measures the force opposing the displacement of the line and due to friction on the lattice (see analysis of σ_p below).

8.4.5. The Dislocation Core

The problem of the atomic or molecular arrangement in the core has considerably benefited from the powerful computational techniques that are available nowadays. Also, high-
resolution electron microscopy techniques have allowed for a direct visualization of the
core of edge dislocations in covalent structures like Ge. What follows is, therefore, obso-
lete in many respects, but it has the advantage of simplicity.

The core energy W_c is certainly between that of the supercooled liquid at the temperature of the solid, often estimated as $W_s \approx \frac{\mu b^2}{5}$ and that of the elastic solid uniformly
strained ($e_c \approx \frac{b}{4\pi r_c}$, i.e., $W_u \approx \frac{\mu b^2}{8\pi}$). Hence,

$$\frac{\mu b^2}{8\pi} < W_c < \frac{\mu b^2}{5}.$$

With $e_c \approx 0.1$, a typical value beyond which the deformation cannot be considered as
elastic any longer, we expect $r_c \sim \frac{5b}{2\pi} \sim 0.8b$.

One can also put the problem in phenomenological terms. Let f_c be the core energy
density. We have

$$W_{\text{tot}} = \frac{\mu b^2}{4\pi K} \ln \frac{R}{e r_c} + \pi r_c^2 f_c, \tag{8.25}$$

which is minimized for $\frac{\partial W_{\text{tot}}}{\partial r_c} = 0$ (it is a minimum because $\frac{\partial^2 W_{\text{tot}}}{\partial r_c^2} > 0$), i.e., for

$$r_c^2 = \frac{\mu b^2}{8\pi^2 K f_c}, \tag{8.26}$$

which gives a core energy per unit length of dislocation

$$W_c = \frac{\mu b^2}{8\pi K} \tag{8.27}$$

of the same order as the elastic energy outside of the core. Here $K = 1$ for a screw dislocation, and $K = 1 - \nu$ for an edge dislocation. Note that W_c does not depend on f_c. With
this model of the core, the total line energy can be written as

$$W_{\text{tot}} = \frac{\mu b^2}{4\pi K} \left(\ln \frac{R}{e r_c} + \frac{1}{2} \right). \tag{8.28}$$

8.5. Mobility of a Dislocation

8.5.1. Elementary Movements of a Dislocation

There are two types of elementary motions, glide and climb.

8.5.1.1. Glide

It occurs on the glide "plane" (this is the current terminology), i.e., on the surface defined by the dislocation line and the Burgers vector. As shown below, glide occurs on the most compact planes. This movement is conservative, because it does not require anything else than local displacements of atoms, when the core moves through the medium. We mentioned glide when discussing the Peach and Koehler force. Glide is observed at low temperature, in agreement with the fact that it requires a very small activation energy.

8.5.1.2. Climb

Any other movement is called climb. "Pure" climb occurs perpendicularly to the glide plane. Climb requires diffusion of atoms over large distances, and it is nonconservative.

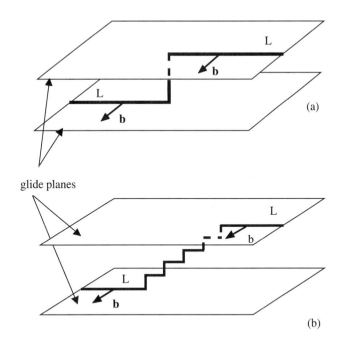

Figure 8.11. (a) An elementary jog; (b) an ensemble of jogs to curve a dislocation out of its glide plane.

This possibility for the dislocation to absorb or emit atoms is facilitated by the presence of "jogs," which are objects of atomic size (Fig. 8.11).

Jogs increase the length of the line, and they add a certain energy of the order of $\mu b^2 \sim 0.1$ eV per jog in a noble metal. Because they also contribute to the free energy by an (entropic) term of disorder, they are present at thermodynamic equilibrium with a concentration

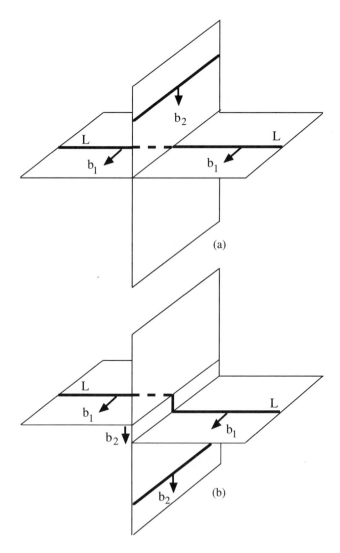

Figure 8.12. Crossing of two dislocations.

$$n = N_0 \exp - \frac{E_c}{k_B T}, \tag{8.29}$$

where N_0 is the largest possible number of jogs (N_0 is of the order of L/b, where L is the total length of dislocation lines) and E_c is the jog energy.

The jogs can also exist out of equilibrium, either in order to curve a dislocation under the action of a Peach and Koehler force or because they are created when two dislocations cross (Fig. 8.12). The sketch of Fig. 8.12 shows how two jogs of length b_1 and b_2, respectively, appear when two dislocations of Burgers vectors $\mathbf{b}_1$ and $\mathbf{b}_2$ cross each other. The reader is urged to figure out how such a process can occur. Hint: Consider the movement of the cut surfaces of the two dislocations.

8.5.2. Glide and Peierls Stress

A tensile test made at low temperature in a single crystal (Fig. 8.13) reveals *glide lines* located in those dense lattice planes that are closest to the planes of maximum shear stress imposed by the applied tensile force, i.e., close to planes making an angle of 45° with the tensile axis. These dense planes are the planes of glide and multiplication of dislocations, an effect that is easily understood. But it is not immediately understandable why the directions of displacement are dense *directions* in these dense planes. Glide starts for a value of the shear stress σ_E called the *yield stress*. The Peierls model explains three facts: (1) A dislocation moves more easily on the plane on which its core is "spread"; these planes are dense planes; (2) The direction of displacement is along a dense direction; (3) the yield stress is the stress necessary to overpass the energy barriers due to the variation of the core energy when the dislocation moves through inequivalent positions in the lattice, i.e., the force of friction.

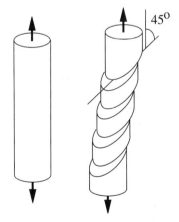

Figure 8.13. Tensile stress.

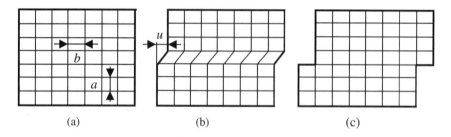

Figure 8.14. Theoretical Frenkel model of glide.

8.5.2.1. Frenkel Model of Critical Shear Stress

This theoretical model refers to a medium void of dislocations (Fig. 8.14). It assumes that the shear σ on the lattice planes is effective in a particular plane P, on which the top part of the crystal, say, *glides* by a quantity u with respect to the bottom part. When u reaches the value b of the lattice parameter, a *glide line* appears at the emerging intersection of P with the boundary of the sample. More specifically, let us first consider $u < b$ small, then we can write

$$\sigma = \mu \frac{u}{a}, \tag{8.30}$$

according to Hooke's law. Because σ is periodic with a period b, we expect to approximate it reasonably well for large u with the expression

$$\sigma = \frac{\mu b}{2\pi a} \sin \frac{2\pi u}{b}, \tag{8.31}$$

which is compatible with (8.30) for small u. The theoretical Frenkel stress is the maximum value of σ, viz. $\sigma_F = \frac{\mu b}{2\pi a}$, which gives the order of magnitude of the stress necessary in this particular glide motion. The value of the yield stress σ_F so obtained is still 10^3 to 10^4 times larger than are the experimental values, but the Frenkel expression has still some merits: First, it tells one that the smaller the ratio b/a is, the smaller σ_F is. The distance a between the atomic planes is maximum when these planes are most densely populated. Therefore, the high-density planes are the best planes for glide. These guesses are in accord with experiments. Furthermore, this theoretical model introduces an interesting concept of *surface energy* $\gamma(u)$ such that

$$\sigma(u)\,\delta u = \delta \gamma \tag{8.32}$$

is the work of the applied stress when the displacement u is incremented by a quantity δu. Integrating (8.32), one gets

$$\gamma(u) = \frac{\mu b^2}{4\pi^2 a}\left(1 - \cos\frac{2\pi u}{b}\right), \tag{8.33}$$

which vanishes for $u = 0$. If the cut surface of an *imperfect* dislocation of Burgers vector β is spread along its glide plane, this cut surface would carry an energy per unit area equal to $\gamma(\beta)$; this is a particular illustration of a *stacking fault*.

8.5.2.2. Continuous Peierls Model

According to the concept of surface energy, it can be reasonably expected that the Burgers vector of a dislocation at rest is spread on its glide plane, creating an extended core. Referring to Fig. 8.15, the analysis of a dislocation whose core is spread can be done in the following way:

- The core energy W_c is the energy of a stacking fault of the same size, assuming in the sense of Frenkel that the mismatch is essentially confined to the glide plane. Hence, $W_c = \int_{-\infty}^{+\infty} \gamma(u)\,dx$, where $u(\pm\infty) = \mp\frac{b}{2}$.
- The stress $\sigma(u)$ at any location x in this plane, in the absence of applied stress, is the sum of the stresses carried by the *infinitesimal dislocations* of Burgers vector $db(x') = -du(x')$ located in x'. Each of these edge dislocations produce at x an infinitesimal shear stress

$$d\sigma(x) = \frac{\mu}{2\pi K}\frac{du(x')}{x - x'},$$

such that in relation with the above discussion of the Frenkel stress, we have at x

$$\frac{d\gamma}{du} = \sigma(x) = \frac{\mu}{2\pi K}\int\frac{du(x')}{x - x'}, \tag{8.34}$$

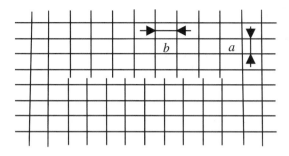

Figure 8.15. Splitting of a dislocation core in its glide plane.

where $d\gamma/du$ takes the value derived from (8.33). The solution of this integrodifferential equation is well-known:

$$\tan \frac{\pi u}{b} = \frac{2K}{a}(x_0 - x),$$ (8.35)

where x_0 is the center of the dislocation. Equation (8.35), which is the distribution of the displacement in the glide plane, tells us that the core is spread over a distance of the order of $L \sim \frac{a}{K}$; this will be soon verified.

Coming now to the total energy, we can see (Problem 8.7) that the core energy diverges; this drawback will disappear in the next section. We adopt provisionally the expression $W_c = \gamma_{max}L$, which still gives a large core energy and scales like the core width, as required. We estimate the elastic energy (outside the core) from (8.16), viz. $W_{el} = \frac{\mu b^2}{4\pi K} \ln \frac{R}{L}$, smaller than in the standard model ($\frac{\mu b^2}{4\pi K} \ln \frac{R}{2b}$), because the core is so large. Therefore, the variation of the energy with respect to the standard model is

$$\Delta W = \gamma_{max}L - \frac{\mu b^2}{4\pi K} \ln \frac{L}{2b} = \frac{\mu b^2}{4\pi^2 a}L - \frac{\mu b^2}{4\pi K} \ln \frac{L}{2b}.$$ (8.36)

Minimizing $W = W_{el} + W_c$ with respect to L, one gets

$$L = \frac{\pi a}{K}, \qquad \Delta W = -\frac{\mu b^2}{4\pi K} \ln \frac{\pi ae}{2bK},$$ (8.37)

which gives a value of L close to the more detailed discrete model below. Note that the value of ΔW in (8.37) favors a large value of a and a small value of b, as in the Frenkel model and as experience shows.

8.5.2.3. Discrete Peierls Model

Apart from the drastic asssumptions already noted, the continuous model does not predict any dependence of the energy with the position x_0 of dislocation, hence, no friction. The model can be improved by evaluating the (core) surface mismatch energy $\int \gamma(a)\,dx$ by a sum $\sum \gamma(u_n)$ on all sites of the stacking fault, assuming that $u_n(x = nb)$ takes the same value as in (8.35). This procedure also gives a finite value for this energy, which we write as $W = W_{el} + W_c + \Delta \Delta W$. The supplementary energy term $\Delta \Delta W$ can be written as

$$\Delta \Delta W = \frac{\mu b^2}{2\pi K} \cos \frac{2\pi x_0}{b} \exp - \frac{2\pi a}{Kb},$$ (8.38)

Figure 8.16. Kinks along a dislocation line.

i.e., the Peierls stress is

$$\sigma_p = \frac{1}{b}\left(\frac{\partial \Delta \, \Delta W}{\partial x_0}\right)_{max} = \frac{\mu}{K}\exp{-\frac{2\pi a}{Kb}}. \tag{8.39}$$

The Peierls stress σ_p decreases exponentially with the ratio a/b and is clearly much smaller than is the Frenkel theoretical stress, to which it has to be compared. For example, in an FCC lattice, a glide on the dense plane {111}, with $\frac{a}{b} = (\frac{2}{3})^{1/2}$, gives $\sigma_p \sim 10^{-3}\mu$ for an edge dislocation ($K = 1 - \nu$) and $\sigma_p \sim 10^{-2}\mu$ for a screw dislocation ($K = 1$).

8.5.2.4. Kinks

A dislocation that is not initially along a potential valley V in which $\Delta \, \Delta W = 0$ follows a sinuous path where it lingers over those directions and crosses the hills H along shorter paths or kinks as in Fig. 8.16, where individual kinks and double kinks are visible.

These kinks play an important role in the deformation processes in which dislocations move under an applied stress. Let us define a few quantities of interest in the analysis of these phenomena. We call microdeformation the kinks displacement along the direction of the line (Fig. 8.16), and macrodeformation the displacement of the line perpendicular to it. At low enough temperatures, the Peierls stress for microdeformation σ_m is smaller than is the Peierls stress σ_p for macrodeformation; microdeformation requires the nucleation of double kinks (energy of nucleation U_{dd}^f) that, when present, decrease the effective yield stress. At higher temperatures, it is the displacement of the kinks that limits the velocity of dislocations. This scheme is valid in FCC metals and covalent materials, but to some differences depending on the structure and the bonds. Low-temperature glide is difficult in BCC metals, ionocovalent structures, and on certain glide systems of hexagonal metals. We shall not develop these concepts in solids.

8.6. Point Defects and Climb

At temperature high enough, typically, $T \geq T_m/2$, where T_m is the melting temperature, the plastic deformation of solids proceeds mostly through nonconservative movements of dislocations, which require diffusion of atoms over large distances. This mechanism in-volves point defects: *vacancies* and *interstitials*. Creep, i.e., deformation developing with time under a constant load, is a typical consequence of easy diffusion at high temperatures.

8.6.1. Vacancies and Interstitials

Figure 2.31 summarizes pictorially how a vacancy and an interstitial are defined in a crys-tal. Under the condition of the constant total mass, the energy U_{vf} of formation of a va-cancy is the energy necessary to bring some atom from inside the crystal to a position of smaller energy on the crystal boundary; U_{vf} is of the order of the energy of sublimation E_S, because in both cases, the same number of bonds have to be cut. The electronic and elastic relaxation around the vacancy contribute to make U_{vf} somewhat smaller that E_S:

$$U_{vf} \sim \frac{1}{10} \mu b^3. \tag{8.40}$$

The energy U_{if} of formation of an interstitial is much higher; it is mostly of an elastic nature (introduction of a sphere of radius b in the matrix):

$$U_{if} \sim \mu b^3. \tag{8.41}$$

There is an equilibrium concentration $c = \frac{n}{N}$ of point defects, which results from a balance between their internal energy U_f and their entropy of translation. An easy calculation for the free energy of an ensemble of N atoms yields

$$F = N\{c(U_f - TS_f) + k_B T[c \ln c + (1-c)\ln(1-c)]\}, \tag{8.42}$$

where S_f is the entropy of formation (changes of the modes of vibration of neighboring atoms, modification of short range order, etc.) which is generally negligible. By minimizing F with respect to concentration c, one gets

$$c = \exp \frac{S_f}{k_B} \cdot \exp -\frac{U_f}{k_B T}, \tag{8.43}$$

where the contribution of S_f is small ($\exp S_f/k_B \cong 1$). Putting numbers in (8.43), one finds that the concentration of vacancies ($U_f = U_{fv}$) is negligible at room temperature ($c \sim 10^{-17}$) but large near the melting point T_m ($c \sim 10^{-4}$). The concentration of intersti-tials is negligible over the whole range of temperatures.

Point defects can be formed out of equilibrium, either by radiation damage (formation of Frenkel pairs, consisting of one vacancy and one interstitial) or by work hardening at low temperature. Vacancies created at high temperature in equilibrium can be quenched at a lower temperature. The vacancies created out of equilibrium subsist, either dispersed or clustered (into dislocation loops, for example): Their energy of migration ($U_{vm} \approx U_{vf}$) is not negligible.

On the other hand, interstitials diffuse easily and disappear, either collapsing with vacancies or on grain-boundaries or the boundaries of the specimen. In FCC metals, interstitials formed under irradiation tend to gather into small, mobile loops.

The notion of point defect, of the vacancy or interstitial sorts, applies equally to liquid crystals with predominant positional order, such as SmB's, where they have been evoked to explain plastic deformation. Their diffusivity is very large, and out-of-equilibrium point defects play no role. Hetero-interstitials are frequent in liquid crystals, whose molecular building blocks are most often thermally unstable or photolysable; it is their diffusivity that is responsible for the Mullins–Sekerka growth instabilities that appear during directional growth of an hexagonal columnar, a SmA, or a nematic in their isotropic phases.

Columns ends in columnar phases, polymer chains ends in liquid crystal polymers, and so on, are point defects characteristic of liquid crystals.

8.6.2. Diffusion of Point Defects and Autodiffusion

8.6.2.1. Diffusion and Random Walks

In a crystal, the diffusion of a vacancy proceeds by jumps from one site to another. In most cases, one will admit that these jumps are elementary jumps, from one site to a neighboring site, along a displacement $\mathbf{r}_o$ that can take z values if there are z-neighbors. We assume that $a^2 = r_o^2$ is the same for these z values. Of course, when we allude to the movement of a vacancy, it is in reality a neighboring atom that shifts to the position of the vacancy, which is, therefore, transported in the place left free by the atom. Let $\mathbf{r}_n$ be the position that is reached by the vacancy after n jumps; we have

$$\mathbf{r}_{n+1} = \mathbf{r}_n + \mathbf{r}_o \tag{8.44}$$

and, taking mean quadratic values

$$\left\langle \mathbf{r}_{n+1}^2 \right\rangle = \left\langle \mathbf{r}_n^2 \right\rangle + 2 \left\langle \mathbf{r}_n . \mathbf{r}_o \right\rangle + \mathbf{r}_o^2. \tag{8.45}$$

This equation simplifies if the jumps are not correlated; this is the case if the point defect is at least as symmetrical as the crystal site on which it sits. Therefore, $\langle \mathbf{r}_n . \mathbf{r}_o \rangle \equiv 0$ and (8.45) gives by recursion

$$\left\langle \mathbf{r}_n^2 \right\rangle = n r_o^2, \tag{8.46}$$

where n, the number of jumps, is also a measure of the time of diffusion t over which the vacancy has gone over a distance $L = \langle \mathbf{r}_n^2 \rangle^{1/2}$. To the foregoing phenomena is related the diffusivity D_v:

$$D_v = \frac{L^2}{t}, \tag{8.47}$$

which is a constant according to (8.46):

$$D_v = \frac{L^2}{t} = \frac{a^2}{\tau}. \tag{8.48}$$

Here, τ is the time characteristic of a unique jump.

Let ν_D be a typical atomic frequency on the lattice, of the order of the Debye frequency ($\nu_D \sim 10^{13}$ s^{-1} in a crystal); the number of effective elementary jumps at a given site is proportional to ν_D and to z, the number of neighboring available sites. Therefore, their frequency goes as

$$\nu_d = Bz\nu_D \exp -\frac{U_{vm}}{k_B T} = \frac{D_v}{a^2}, \tag{8.49}$$

according to (8.48). Here, U_{vm} is the energy barrier that the vacancy has to overcome, and B is a numerical coefficient. Therefore a microscopic value for D_v is

$$D_v = Bza^2\nu_D \exp -\frac{U_{vm}}{k_B T}. \tag{8.50}$$

The coefficient $D_{0v} = Bza^2\nu_D$ is of the order of a few units in cgs units ($D_{0v} \equiv 1$ cm$^2 \cdot$ s^{-1}). Taking $U_{vm} \sim 1$ eV, one finds $D_v \sim 10^{-15}$ cm$^2 \cdot$ s^{-1} in a solid.

8.6.2.2. Einstein Relation for Vacancies

Assume that the point defect is submitted to a force F that helps to overtake the activation energy U_{vm} of the jump, by providing an energy $\frac{1}{2}Fa$. In the direction along $\mathbf{F}$, the effective energy of migration is $U_{vm} - Fa/2$; on the other hand, in the opposite direction, it is $U_{vm} + Fa/2$. The number of effective jumps in the direction of $\mathbf{F}$ is, therefore, per unit time:

$$\delta\nu = \nu_d \left(\exp \frac{1}{2}\frac{aF}{k_B T} - \exp -\frac{1}{2}\frac{aF}{k_B T} \right). \tag{8.51}$$

This expression takes into account only the forward and backward jumps. Assume $aF \ll k_B T$. Expanding the exponentials, one, therefore, gets the drift velocity along $\mathbf{F}$, viz.,

$$v = \delta v a = v_d \frac{a^2 F}{k_B T}, \tag{8.52}$$

which we write in a condensed form:

$$m k_B T = D_v, \tag{8.53}$$

where $m = v/F$ is the *mobility* of the point defect. Equation (8.53) is the celebrated Einstein's relation.

8.6.2.3. Autodiffusion of Atoms in the Matrix

Atoms in the matrix move by exchanging sites with vacancies. We introduce the coefficient of self-diffusion for atoms, which is proportional to the concentration of vacancies c_v:

$$D = c_v D_v. \tag{8.54}$$

D is measured by labeling a certain proportion of atoms in the matrix (radiotracers, for example); their time evolution obeys Fick's second law:

$$\frac{\partial c}{\partial t} = D \nabla^2 c \tag{8.55}$$

(to be proved further), whose solution $c(x, t)$ can be expressed as a function of the adimensional quantity ξ:

$$\xi^2 = \frac{x^2}{4Dt}. \tag{8.56}$$

For example, if at $t = 0$ all labeled atoms are in a plane $x = 0$ $\{c(0, 0) = 1, c(x, 0) = 0\}$, the solution at later times reads as

$$c(\xi) \cong \int_{\xi}^{\infty} \exp -\xi^2 \, d\xi = c_0 \, \mathrm{erfc} \, \xi \tag{8.57}$$

where $\mathrm{erfc}\, z = 1 - \mathrm{erf}\, z$, $\mathrm{erf}\, z$ being the error function (Fig. 8.17).

   In order to establish Fick's law, we first generalize Einstein's relation to the case of autodiffusion. Note first that the labelled atoms practically form an ideal solution with the non-labelled atoms, if their density is small enough and if they do not interact; the cor-

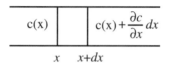

Figure 8.17. Fick's second law.

responding free energy is predominantly entropic; there are no relevant specific chemical interactions between isotopes. Hence

$$F(\text{labeled atoms}) = Nk_B T[c \ln c + (1 - c) \ln(1 - c)], \tag{8.58}$$

which yields a chemical potential per atom:

$$\mu = \frac{1}{N}\frac{\partial F}{\partial c} = k_B T \ln \frac{c}{1 - c} \sim k_B T \ln c \tag{8.59}$$

and a spontaneous force acting on labeled atoms:

$$\mathbf{F} = -\nabla\mu = -\frac{k_B T}{c}\nabla c\,\hat{\mathbf{x}}, \tag{8.60}$$

where $\hat{\mathbf{x}}$ is an unit vector in the direction of $\mathbf{F}$.

The Einstein's relation now reads as

$$\mathbf{v} = -D_v c_v \frac{\nabla c}{c} = -D\frac{\nabla c}{c}, \tag{8.61}$$

where $\mathbf{v}$ is the velocity of a labeled atom and D is its diffusivity. This is *Fick's first law*, which relates the flux of labeled atoms $\mathbf{J} = c\mathbf{v}$ to the gradient of concentration.

Fick's second law immediately follows; consider, for example, the balance of ingoing and outgoing atoms in a slab of thickness dx at the position x. One has

$$\frac{\partial c}{\partial t} = -\frac{\partial}{\partial x}J, \tag{8.62}$$

according to (8.61); one recovers a 1D version of Eq. (8.55). This 1D equation easily generalizes to the full 3D (8.55).

Note that Fick's laws can be used with any type of atom, defect, or "particle," as soon as (8.59) is valid, and using the right coefficient of diffusion. For example, Fick's laws apply to vacancies, with $D = D_v$. Note also that (8.59) does not depend on a precise definition of the concentration, which can be atomic, weight, and so on. For example, taking for c the volume fraction $c = n\Omega$, where Ω is the volume of a particle and $\mathbf{J} = \frac{\mathbf{v}c}{\Omega} = \mathbf{v}n$ is the flux of particles per unit surface and unit time, one gets

$$\mathbf{J} = -\frac{D}{\Omega}\nabla c, \tag{8.63}$$

a most useful formula to be used later on.

8.6.3. Creep

At high temperature, a defect can flow under a constant applied stress $\sigma_0 > \sigma_y$, the yield stress: a typical creep experiment consists in recording the deformation ε as a function of time (Fig. 8.18).

At low temperatures, creep is logarithmic, and the deformation vanishes on long durations $\varepsilon - \varepsilon_0 \propto t^{-1}$; at higher temperatures or high stresses, there are typically three stages: β creep with slowing down of the deformation; κ creep, linear; *tertiary* creep with speeding of the deformation, ending in rupture.

The Herring–Nabarro model of linear creep figures out that the sample is made of grains (typical size L), without dislocations except maybe in the grain boundaries, and that the grain boundaries emit or absorb point defects (Fig. 8.19) under the action of an applied tensile or compressive stress σ: The vacancies migrate from the regions under compression to the regions under tension. The energy of formation of a vacancy in a region under compression is increased by a quantity σb^3 and decreases by the same quantity in a

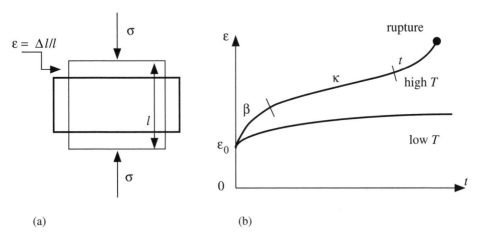

Figure 8.18. Creep experiments (a) principle of the test; (b) logarithmic creep at low temperatures and three stages of creep at high temperatures.

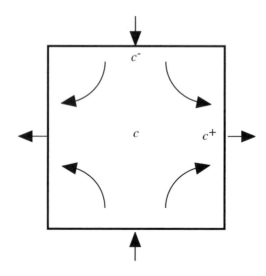

Figure 8.19. Herring–Nabarro creep.

region under tension. Hence, the corresponding concentrations c^+ and c^- in these regions are

$$c^+ = c_0 \exp \frac{\sigma b^3}{k_B T}, \quad c^- = c_0 \exp -\frac{\sigma b^3}{k_B T}, \quad c_0 = \exp -\frac{U_{vf}}{k_B T}. \tag{8.64}$$

Whence a gradient of concentration

$$|\nabla c| \cong \frac{c^+ - c^-}{L} \sim \frac{2}{L} c_0 \frac{\sigma b^3}{k_B T}, \tag{8.65}$$

if $\sigma b^3 \ll k_B T$.

Using (8.63), one gets, for the total flux:

$$J L^2 = -\beta \frac{D_v}{\Omega} \frac{\sigma b^3}{k_B T} L, \tag{8.66}$$

where β is a constant of the order of unity.

The variation of length of the grain per unit time being $\delta \dot{L} = \frac{b^3}{L^2} J L^2$, one gets

$$\dot{\varepsilon} = \frac{b}{dt} \frac{\delta L}{L} = J L^2 \left(\frac{b}{L}\right)^3, \tag{8.67}$$

or, with $\Omega = b^3$,

$$\dot{\varepsilon} = \beta c_0 D_v \frac{\sigma \Omega}{k_B T L^2}.$$ (8.68)

Equation (8.67) has the shape of a flow equation, with a viscosity $\eta = \frac{\sigma}{\dot{\varepsilon}}$:

$$\eta = \beta' \frac{L^2}{D_v \Omega} k_B T$$ (8.69)

(where $\beta' = \frac{1}{c_0 \beta}$) of the order of 10^{13} poises for $L \sim 10 \mu$m, $T = 750°$K, $D = 10^{-10}$ cm$^2 \cdot$ s^{-1}, and $\Omega = 10^{-23}$ cm^2. This is an order of magnitude that is met at the glass transition, for example. Under high stresses, or at high temperatures, the solid behaves like a liquid.

Note, furthermore, that a liquid would correspond to $L^3 \equiv \Omega$, i.e. $\eta \sim \frac{k_B T}{D_v a}$, an expression that yields a correct order of magnitude, with the same value of D as above. We recall also that the viscosity of the movement of a spherical particle of radius a in a liquid of diffusivity D_v is given by the Stokes–Einstein formula (which is exact in the framework of the Navier–Stokes equations) :

$$\eta_{\text{SE}} = \frac{k_B T}{6\pi D_v a},$$ (8.70)

whose analogy with the formula above is clear.

8.7. Ensembles of Dislocations

8.7.1. Frank Network

In a well-annealed single crystal, the density of dislocations can be greatly reduced by various processes of movement of dislocations under their mutual stresses and thermal stresses; dislocation loops may shrink to zero, and dislocations of opposite signs annihilate. A typical density is 10^4 to 10^6 dislocations per cm^2, to be compared with 10^{10} to 10^{12}/cm^2, the density of dislocations in a work-hardened material.

The Frank network of well-annealed crystals is a low-energy configuration made of dislocation segments that merge at nodes where the Kirchhoff condition $\sum_i \mathbf{b}_i = 0$ is automatically satisfied (Fig. 8.20). There are at least three dislocations of different Burgers vectors at a node. Equilibrium of line tensions τ_i at the node, if it is satisfied, brings a supplementary condition:

$$\sum_i \boldsymbol{\tau}_i = 0,$$ (8.71)

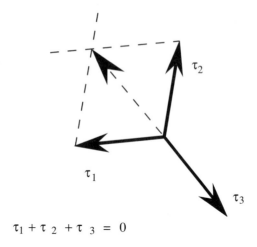

$$\tau_1 + \tau_2 + \tau_3 = 0$$

Figure 8.20. Triple node of dislocations and equilibrium of line tensions.

where $\pmb{\tau}_i = \tau_i \mathbf{t}_i$ is a vector directed along the outer tangent $\mathbf{t}_i$ to the dislocation line L_i at the point of merger and τ_i is the line tension.

Dislocations in excess produced by plastic deformation, for example by Frank–Read mills induced on segments of the Frank network by external stresses, are all of the same sign and cannot gather in a Frank network. They build a "polygonized" structure of sub-boundaries (dislocations walls), located along the walls of the Frank network. These sub-grains boundaries are at the origin of a slight misorientation θ of the order of $\theta \sim \frac{b}{\ell}$, where ℓ is the distance between equal dislocations.

The Frank network and the polygonized structure yield in X-ray diffraction patterns of single crystals a widening of the Laue spots from which this misorientation can be measured (typically, 1 to 30 minutes of arc for blocks of the order of 10^{-3} cm): this block structure is also known as the mosaic structure.

8.7.2. Sub-Boundaries

The misorientation angle due to a wall of dislocations can be calculated as follows. Consider (Fig. 8.21) a closed loop Γ encircling a segment $\delta \mathbf{L}$ of wall, though which passes a set of dislocations whose total Burgers vectors sum to $\delta \mathbf{b}$. Let us map the Burgers loop Γ into the perfect crystal, $\Gamma \rightarrow \gamma$. The image γ is generally an open circuit whose closure failure equals $\delta \mathbf{b}$. But this closure failure is proportional to the length $\delta L = |\delta \mathbf{L}|$, if the same dislocation configuration repeats along the wall. Hence one can write $\delta \mathbf{b} = \pmb{\theta} \times \delta \mathbf{L}$, where $\pmb{\theta}$ is a rotation vector orthogonal to $\delta \mathbf{b}$, which measures the misorientation when traversing any loop encircling a segment of wall parallel to $\delta \mathbf{L}$.

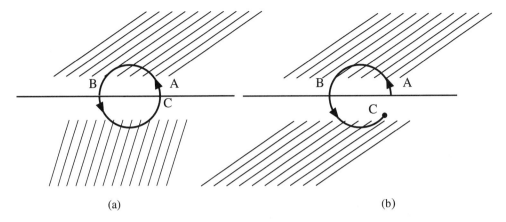

(a)   (b)

Figure 8.21. Burgers circuit for a sub-boundary.

Two simple examples of sub-boundaries are as follows:

1. The *tilt sub-boundary*, whose misorientation $\boldsymbol{\theta}$ is parallel to the plane of the boundary. This can be achieved by a set of parallel edge dislocations, whose Burgers vectors are perpendicular to the boundary. Such tilt boundaries are met in the Grandjean–Cano geometry of Fig. 8.22, either in cholesterics or in smectics. In both cases, the anchoring conditions bring the layering parallel to the boundaries of the sample, these boundaries forming a wedge of angle θ, of the order of 10^{-4}rad in most experiments. This misorientation is relaxed by a tilt subgrain boundary in the midplane of the sample. Hence, the distance ℓ between dislocations of Burgers vector b is $\ell \sim \frac{b}{\theta}$.

   In a solid crystal, a tilt boundary is obtained by the merging of two lattices, limited by a plane at a small angle $\pm\theta/2$ to a common glide plane, otherwise *perfect*, joined together along this plane, and allowed to relax. One therefore expects that the long-range stresses of the dislocations vanish at long distance, as it has been proved rigourously.

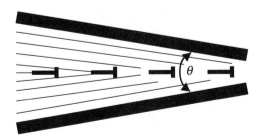

Figure 8.22. Dislocation model of a tilt sub-boundary in a smectic wedge.

The stress remains confined to a region parallel to the boundary whose extent is of the order of the distance between the dislocations, and further decreases exponentially. The energy associated with the dislocations can then be computed as the sum of the individual energies of the dislocations stored in individual cylinders of radius $\ell/2$. The energy per unit area of the wall is, thus,

$$E(\theta) = \frac{\mu b}{4\pi(1-\nu)}\theta \ln \frac{b}{2r_c\theta}. \tag{8.72}$$

This expression of $E(\theta)$ shows a maximum at $\theta = b/2er_c$. This seems to be well verified experimentally.

2. The *twist sub-boundary*, whose misorientation θ is perpendicular to the plane of the boundary. This can be achieved by one or more sets of parallel screw dislocations of different Burgers vectors. In smectics, whose repeat parameters are in only one direction, such twist boundaries are made of a unique set of parallel dislocations. We recall that according to a well-established model, the TGB phases are made of parallel twist boundaries (see Fig. 4.11). One expects three sets of screw dislocations in hexagonal columnar phases in twist boundaries perpendicular to the columns.

In a solid crystal, a twist boundary without long range stresses must contain two sets of screw dislocations with nonparallel Burgers vectors (Frank).

Note that three sub-boundaries with energy per unit area E_i merging along a line must be such that $\mathbf{E}_1 + \mathbf{E}_2 + \mathbf{E}_3 = 0$, where $\mathbf{E}_i = E_i\boldsymbol{\nu}_i$, $\boldsymbol{\nu}_i$ being a unit vector in the plane of sub-boundary i, perpendicular to the line of merger. This equilibrium of "surface tensions" is true as long as the crystal is well annealed.

8.7.3. Large Misorientations, Twin and Epitaxy Dislocations, Martensitic Transformations

One expects that splitting of a misorientation boundary into dislocations has a physical reality as long as the distance ℓ between dislocations is larger than the core diameter (of the order of $2b$). Equation (8.72) is then valid for $\theta \leq 25°$, approximately. Above this value, the wall energy is practically constant, because of atomic rearrangements that take place in a strip of atomic size and do not depend much on the misorientation. However, the wall energy drops drastically for some special values of the misorientation and special wall orientations, corresponding to twinning between the two lattices (cf. Section 8.3.3). It is noticeable that the wall energy increases very sharply for small deviations to the twinning conditions: This is due to the appearance on the wall of twin dislocations, which carry long-range stresses (incoherent twin), and whose density and nature depend on the deviation. These dislocations are akin to imperfect dislocations (cf. Friedel's textbook and the current literature for more details).

Epitaxy is the generic name to conote the easy matching of two different crystal species along a plane for which they have equal (perfect epitaxy) or nearly equal 2D lattice parameters. Coherent twinning is a special case of perfect epitaxy when there is only one crystal species and the two lattices conserve common symmetry elements of the crystal. Any deviation to perfect epitaxy is usually relaxed by the appearance of epitaxial dislocations on the boundary.

Eventually it is worth mentioning that in a phase change inside a crystal with production of nuclei of the second phase by local rearrangements of the atoms, the walls between the two phases can often be analyzed in terms of dislocations, whose movement is important to understand the growth of the second phase. This is called a martensitic transformation.

Problem 8.1. Check (8.15), and calculate the strain and stress fields of an edge dislocation; check that they scale as $\sim 1/r$. Calculate the dilatation and show that it vanishes on the average.

Answers:

$$\sigma_{rr} = \sigma_{\theta\theta} = -\frac{\mu b}{2\pi(1-\nu)}\frac{\sin\theta}{r}, \sigma_{r\theta} = \frac{\mu b}{2\pi(1-\nu)}\frac{\cos\theta}{r}, \sigma_{zz} = -\frac{\mu\nu b}{\pi(1-\nu)}\frac{\sin\theta}{r},$$

$$\sigma_{rr} + \sigma_{\theta\theta} + \sigma_{zz} = -\frac{\mu b}{\pi}\frac{\sin\theta}{r}\frac{1+\nu}{1-\nu},$$

$$\text{dilatation } \delta = \sigma_{ii}/\lambda + 2\mu = -\frac{b}{2\pi}\frac{1-2\nu}{1-\nu}\frac{\sin\theta}{r},$$

$$e_{rr} = e_{\theta\theta} = \frac{\sigma_{rr} - \lambda\delta}{2\mu} = -\frac{b(1-2\nu)}{4\pi(1-\nu)}\frac{\sin\theta}{r}, e_{zz} = 0,$$

$$e_{r\theta} = \frac{b}{\pi(1-\nu)}\frac{\cos\theta}{r}.$$

Problem 8.2. Calculate the energy of interaction of an edge and a screw dislocations which are parallel.

Answers: Let $\sigma_{ij}^{(1)}, \sigma_{ij}^{(2)}$ be the stresses due separately to two dislocation lines L_1, L_2, independently of the presence of the other. When together, the total stresses are the sum of the individual stresses, due to Hooke's law: $\sigma_{ij} = \sigma_{ij}^{(1)} + \sigma_{ij}^{(2)}$. The total energy [see (8.13)] $W = \frac{1}{2}\int(\sigma_{ij}^{(1)} + \sigma_{ij}^{(2)})(e_{ij}^{(1)} + e_{ij}^{(2)})dV$ can then be split into three parts: the proper energies of the 2 lines $W_1 = \frac{1}{2}\int\sigma_{ij}^{(1)}e_{ij}^{(1)}dV$ and $W_2 = \frac{1}{2}\int\sigma_{ij}^{(2)}e_{ij}^{(2)}dV$ and the energy of interaction $W_{\text{int}} = \frac{1}{2}\int(\sigma_{ij}^{(2)}e_{ij}^{(1)} + \sigma_{ij}^{(1)}e_{ij}^{(2)})dV \equiv \int\sigma_{ij}^{(1)}e_{ij}^{(2)}dV$.

The stress-fields of two parallel screw and edge lines are orthogonal (no common components), according to (8.12) and (8.15); furthermore, the dilatation carried by a screw dislocation is vanishing. Therefore, their energy of interaction is vanishing.

Problem 8.3. Prove the equivalence of the two expressions of the energy of interactions between two dislocations of cut surfaces Σ_1 and Σ_2, stresses $\sigma_{ij}^{(1)}$ and $\sigma_{ij}^{(2)}$, $W_{int} = \int \sigma_{ij}^{(1)} e_{ij}^{(2)} \, dV$ and $W_{int} = - \oint_{\Sigma_2} b_{2,i} \sigma_{ij}^{(1)} \, dS_j$.

Answers: $\sigma_{ij}^{(1)}$ and $e_{ij}^{(2)}$ are both symmetric tensors; the contracted product of a symmetric tensor by an antisymmetric tensor is zero. One can therefore replace $e_{ij}^{(2)} = \frac{1}{2}(u_{i,j}^{(2)} + u_{j,i}^{(2)})$ by $e_{ij}^{(2)} + \omega_{ij}^{(2)} = u_{i,j}^{(2)}$ (where $\omega_{ij}^{(2)} = \frac{1}{2}(u_{i,j}^{(2)} - u_{j,i}^{(2)})$ is antisymmetric) in the product $\sigma_{ij}^{(1)} e_{ij}^{(2)}$. The next step is to replace the volume integral by a surface integral (on the two lips of the cut surface Σ_2, which are separated by a constant vector $\mathbf{b}_2$). We have indeed $\sigma_{ij}^{(1)} \frac{\partial u_i^{(2)}}{\partial x_j} = (u_i^{(2)} \sigma_{ij}^{(1)})_{,j} - u_i^{(2)} \sigma_{ij,j}^{(1)}$, where the second term in the right-hand member vanishes identically (8.5a). The rest of the demonstration proceeds by the use of Gauss theorem. For details, in particular the appearance of the minus sign, see Nabarro.

Problem 8.4. An edge dislocation line whose Burgers vector $\mathbf{b}$ is in the z-direction lies along the y-direction, and it is submitted to an applied longitudinal compression along the z-axis; what is the fictitious force acting on the dislocation, and in which direction does it move?

Answers: An applied longitudinal compression along the z-axis can be represented by a stress tensor with a unique component $\sigma_{zz} < 0$. Applying the Peach–Koehler formula, one gets $\mathbf{F} = b(\sigma_{zz}, 0, 0)$. The dislocation moves to the right if $b > 0$. Indeed, such an edge dislocation carries an extra half-layer in the xy-plane ($x > 0$) if the above data satisfy the FS/RH convention.

Problem 8.5. Demonstrate (8.23).

Answers: Let $d\theta$ be the angle under which the arc ds is seen from the center of curvature. The radius of curvature $\rho = ds/d\theta$ is supposed to be large compared with the arc ds. Then, the line tensions at both ends of the arc of line project along the mid-radius of the arc, having a total component $2\tau \sin d\theta/2 \approx \tau \, d\theta$, which opposes the Peach–Koehler force $\mathbf{F} \, ds$, hence, (8.23).

Problem 8.6. Calculate the core radius r_c of a hollow dislocation. Show that the condition $r_c > a$ implies that $b \gg a$.

Answers: The core energy of a hollow dislocation scales in a first approximation as the area of its boundary, say, $W_c = 2\pi r_c \gamma$, where $\gamma \sim \mu a$. Minimizing $W_{el} + W_c$ yields $r_c = \frac{\mu b^2}{8\pi^2 K \gamma} \sim \frac{1}{8\pi^2 K} \frac{b^2}{a}$, which length defines a true hollow core if $r_c > a$; i.e., $b > 2\pi \sqrt{2K} \, a$.

Problem 8.7. Show that the core energy $W_c = \int_{-\infty}^{+\infty} \gamma(u)\, dx$ diverges in the continuous Peierls model of the core.

Answers: The core energy can be written as

$$W_c = \int_{+b/2}^{-b/2} \frac{\mu b^2}{4\pi^2 a} \left(1 - \cos \frac{2\pi u}{b} \right) \frac{dx}{du}\, du,$$

where $\frac{dx}{du} = -\frac{\pi a}{2Kb}(1 + \tan^2 \frac{\pi u}{b})$ can be obtained from (8.35). Letting $t = \tan \frac{\pi u}{b}$, one gets $W_c \propto \int_{+\infty}^{-\infty} \frac{t^2}{1+t^2}\, dt$, which obviously diverges.

Further Reading

L. Landau and I. Lifshitz, Theory of Elasticity, Pergamon, New York, 1995.

J. Friedel, Dislocations, Pergamon, New York, 1964.

J.P. Hirth and J. Lothe, Theory of Dislocations, Second Edition, John Wiley, New York (1982).

F.R.N. Nabarro, Theory of Crystal Dislocations, Oxford University Press, 1967.

Report on the Conference on "Defects in Crystalline Solids" held at the H. H. Wills Laboratory, University of Bristol, The Physical Society, London, 1955.

Dislocations in solids, volumes 1–8, Edited by F.R.N. Nabarro, North-Holland., Amsterdam, 1975.

CHAPTER 9

Dislocations in Smectic and Columnar Phases

Because of their translational symmetries, 1D (smectic) and 2D (columnar) positionally ordered liquid crystals show dislocations of translation of various characteristics. Their elastic and plastic properties, which depend strongly on the Burgers vector, their relationship with disclinations, and so on, make them original with respect to their counterparts in 3D solid crystals. These dislocations are also worth studying in some detail because of their role in complex phases with thermodynamical defects, such as TGB phases (smectics with periodic twist sub-boundaries), which we have already alluded to, or Moiré phases (columnar phases with periodic twist sub-boundaries), whose existence has been recently predicted.

We shall limit ourselves to the simplest cases of 1D or 2D liquid crystals, namely, SmA and columnar phases of hexagonal symmetry. In SmA, the notion of layer is not complicated by supplementary order, as in SmB or SmC, for instance, and the order parameter is restricted to a nematic component (the director $\mathbf{n}$ is identical to the layer normal $\boldsymbol{\nu}$) and to a 1D mass density component, figuring the periodic variation of the mass along Oz. Elementary defects are, therefore, disclinations (singularities of the director field) and dislocations of Burgers vectors $\mathbf{b} = n\, d_o \boldsymbol{\nu}$ (singularities of the 1D solid) and defects involving both. Here, d_0 is the repeat distance of the smectic phase, and n is an integer. For 2D columnar phases with hexagonal symmetry, the free-energy density, written to the second order, has cylindrical symmetry, and is therefore relatively easy to handle.

9.1. Static Dislocations in Smectics

9.1.1. Edge Dislocation of Small and Large Burgers Vectors

9.1.1.1. Small Burgers Vectors

Let us consider a dislocation along the y-axis. If the Burgers vector is small ($n = 1, 2, \ldots$), the elastic field can be treated in the small deformation approximation, in terms of the displacement field $u(x, z)$, which is odd in z.[1] Take the cut surface along the half-plane $z = 0$, $x > 0$, and consider the half-space $z > 0$. The boundary conditions are

$$z = 0; \quad u(x < 0) \equiv 0; \quad u(x > 0) \equiv \frac{b}{2}. \tag{9.1}$$

We look for a solution of the Euler–Lagrange differential equation $K_1 \Delta_\perp^2 u = B \frac{\partial^2 u}{\partial z^2}$, which corresponds to the elastic energy functional in its linear form (5.27). The boundary conditions can be expressed as a Fourier series:

$$u(x, 0) = \frac{b}{4} + \frac{b}{4\pi} \int_{-\infty}^{+\infty} \frac{dq}{iq} \exp iqx, \tag{9.2}$$

where we have used the fact that the integral of the delta function $\delta(x)$ is the step function $H(x) = \frac{1}{2\pi} \int_{-\infty}^{+\infty} \frac{dq}{iq} \exp iqx$, taken such that $H(x < 0) = -\frac{1}{2}$, $H(x > 0) = +\frac{1}{2}$.

We can, therefore, write the general solution $u(x, z)$ as

$$u(x, z) = \frac{b}{4} + \frac{b}{4\pi} \int_{-\infty}^{+\infty} \frac{dq}{iq} g_q(z) \exp iqx, \tag{9.3}$$

Using the Euler–Lagrange equation, one finds $g_q(z) = \exp(-\lambda q^2 z)$, where $\lambda = \sqrt{K_1/B}$, so that

$$u(x, z) = \frac{b}{4} + \frac{b}{4\pi} \int_{-\infty}^{+\infty} \exp(iqx - \lambda q^2 z) \frac{dq}{iq} \equiv \frac{b}{4\sqrt{\pi}} \int_{-\infty}^{x/\sqrt{\lambda z}} \exp(-t^2/4) \, dt; \tag{9.4}$$

hence, the angle $\theta(x)$ between $\boldsymbol{\nu}$, the layers normal, and the z-axis can be written:

$$\theta(x) = \frac{\partial u}{\partial x} = \frac{b}{4\sqrt{\pi \lambda z}} \exp -\frac{x^2}{4\lambda z}. \tag{9.5}$$

This quantity decreases slowly with z (on a typical distance $\frac{x^2}{4\lambda}$), decreases rapidly with x inside the parabola $x^2 = \pm 4z\lambda$, and takes a value quickly close to zero outside of these parabolae (Fig. 9.1). Let ξ be the length over which the core extends along the x-axis; the associated perturbation along the z direction extends over a distance $\frac{\xi^2}{4\lambda}$, according to (9.5).

---

[1]P.G. de Gennes, Comptes Rendus Acad. Sci. (Paris) **275B**, 939 (1972).

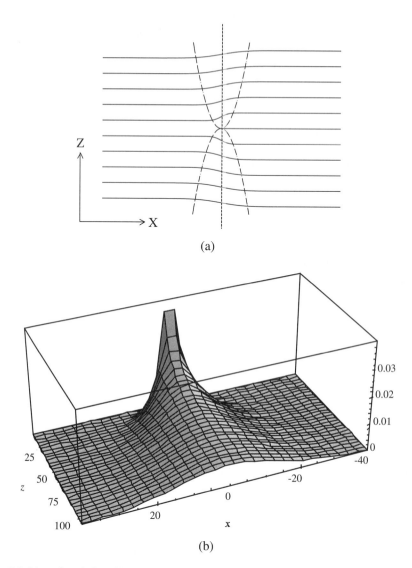

(a)

(b)

Figure 9.1. Distortions induced by an edge dislocation: (a) displacement field $u(x, z)$ (9.4) and (b) the tilt $\theta(x, z)$ (9.5).

The line energy can be calculated from the expression for $u(x, z)$. The curvature contribution, for instance, involves the following sequence of expressions, along the calculation:

$$\frac{\partial^2 u}{\partial x^2} = \frac{ib}{4\pi} \int q \, dq \, \exp iqx \, \exp(-z\lambda q^2);$$

$$\frac{1}{2} K_1 (\operatorname{div} \mathbf{n})^2 = \frac{1}{2} K_1 \frac{\partial^2 u}{\partial x^2} \frac{\partial^2 u^*}{\partial x^2}, \tag{9.6a}$$

where $\frac{\partial^2 u^*}{\partial x^2}$ is the complex conjugate of $\frac{\partial^2 u}{\partial x^2}$;

$$\frac{1}{2} \int K_1 (\operatorname{div} \mathbf{n})^2 \, dx \, dz = \frac{K_1 b^2}{16\pi} \int q^2 \, dq \, \exp -2z\lambda q^2 \, dz$$

$$= \frac{K_1 b^2}{32\pi \lambda} \int dq = \frac{K_1 b^2}{8\lambda \xi}, \tag{9.6b}$$

where we have taken $q_{\max} = \pm \frac{2\pi}{\xi}$. The above expression is the result of an integration in the upper half-plane, and it must be multiplied by 2. The compression contribution $\frac{1}{2} B (\frac{\partial u}{\partial z})^2$ is equal to the curvature contribution for each mode q, and the energies per mode add up linearly. Hence, eventually,

$$W_e = \frac{K_1 b^2}{2\lambda \xi} + W_c \equiv \frac{B}{2} b^2 \frac{\lambda}{\xi} + W_c, \tag{9.7}$$

where W_c is the core energy, at the moment unspecified. We recognize in the last expression a $\frac{B}{2} b^2$ term that is akin to what is found in 3D solids and that suggests that small Burgers vectors are favored.

So far, we have supposed that the nonlinear term $(\partial u/\partial x)^2$ in the elastic energy density (5.28) is small; $\frac{(\partial u/\partial x)^2}{|\partial u/\partial z|} \ll 1$. However, the ratio $\frac{(\partial u/\partial x)^2}{|\partial u/\partial z|}$ is of the order of b/λ and independent of z inside the parabola $x^2 = \pm 4z\lambda$ (Fig. 9.1). Therefore, the linear theory is justified only for $b \ll \lambda$, whereas in a general case $b \sim \lambda$, one has to take the nonlinear term into account. Fortunately, the problem of finding the dislocation profile from the energy density (5.28) with the nonlinear term $(\partial u/\partial x)^2$ preserved is solvable analytically[2] (Problem 9.1):

$$u(x, z) = 2\lambda \ln \left[1 + \frac{\exp(b/4\lambda) - 1}{2\sqrt{\pi}} \int_{-\infty}^{x/\sqrt{\lambda z}} \exp(-t^2/4) \, dt \right]. \tag{9.8}$$

---

[2]E.A. Brener and V.I. Marchenko, Phys. Rev. E **59**, R4752 (1999).

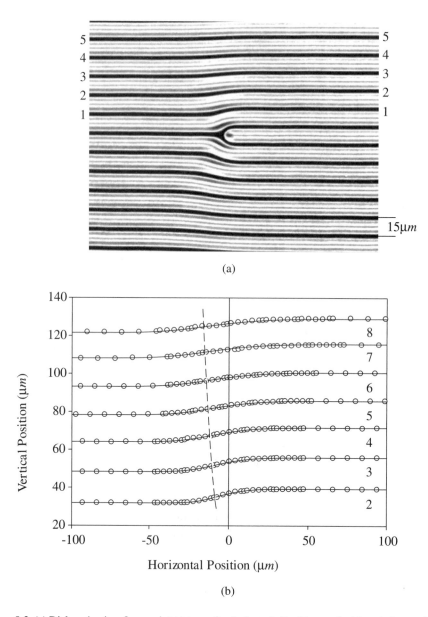

(a)

(b)

Figure 9.2. (a) Dislocation in a fingerprint texture of a cholesteric liquid crystal with a pitch $p = 15\mu$; dislocation profile fitted by nonlinear theory, (9.8), with $\lambda = 2.65\mu$m as a fitting parameter. The dashed line marks inflection points. (See T. Ishikawa and O.D. Lavrentovich, Phys. Rev. E **60**, R5038 (1999).)

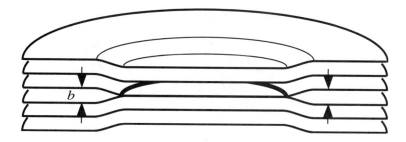

Figure 9.3. Free-standing smectic film with an edge dislocation.

The most notable departure from the linear model could be observed when $\lambda < b$. The condition $\lambda < b$ is naturally obeyed for cholesterics, even for elementary dislocations: In the de Gennes-Lubensky coarse-grained model, $\lambda = \sqrt{K/B} = \frac{p}{2\pi}\sqrt{3K_3/8K_2}$ is normally smaller than the pitch p. A dislocation $b = p$ in the cholesteric fingerprint texture is shown in Fig. 9.2: The displacement of layers is closely fitted by (9.8), which results in experimental determination of λ. Qualitatively, the difference in the linear and nonlinear dislocation profile is manifested by the position of the inflection points $\partial^2 u/\partial x^2 = 0$: In a linear model, the inflection points are located at the axis $x = 0$, whereas in the nonlinear model, they locate at $x < 0$.

Edge dislocations of small Burgers vectors are observed in thin SmA slabs, e.g., in thin free-standing films and in homeotropic wedges. A free-standing film sustained in a hole is in contact with the edge of the hole through a meniscus of matter that acts as a reservoir.[3] The curvature of this meniscus is at the origin of a Laplace pressure drop $\Delta p = \frac{\gamma}{R}$ (γ is the surface tension) between the liquid crystalline matter and the outside, which drains matter from the film toward the reservoir. This draining occurs through the movement of edge dislocation loops that nucleate in the film (Fig. 9.3) more easily at higher temperature because it is an activation barrier, until the film is only a few layers thick.

Homeotropic wedges have been described in Section 8.7.2. The dislocations have Burgers vector unity (n = 1) as long as their distance to the dihedron wedge is smaller than some value r^*, and then there is a transition to a larger Burgers, which increases with r. Consequently, the distance between neighboring dislocations increases with r (see Fig. 9.4 and Nallet and Prost[4]).

9.1.1.2. Large Burgers Vectors

Dislocations with macroscopic Burgers vectors, large enough to be estimated by light polarizing microscopy, are found in thermotropic as well as in lyotropic systems. The explanation is in a specific model of the core that is split into two disclinations of opposite signs

[3]P. Pieranski et al., Physica A **194**, 364 (1993).
[4]F. Nallet and J. Prost, Europhys. Lett. **4**, 307 (1987).

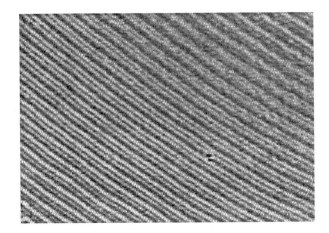

Figure 9.4. Dislocations observed in a homeotropic wedge of lyotropic L_α phase. Courtesy of C. Blanc.

(Fig. 9.5a). The model yields $\xi \sim \frac{b}{2}$. Hence, the elastic energy (9.7) outside of the core scales essentially as b and not as b^2. The stability of the dislocation would be practically independent of the magnitude of the Burgers vector, were this off-core energy the main contribution to the total energy.

It is precisely the core contribution that makes the difference; according to the disclination model of Fig. 9.5a, the contribution of curvature elasticity is increased in the total

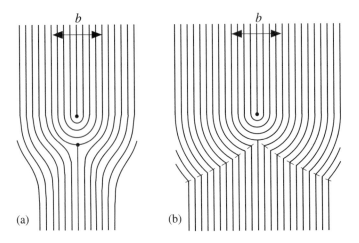

Figure 9.5. (a) Splitting of the core of a large Burgers vector dislocation into two disclinations; (b) a model of an edge dislocation with vanishing solid elasticity; note the appearance of a singular wall.

balance, because now the core energy can be written as:

$$W_c = \frac{\pi K_1}{2} \ln \frac{b}{2d_0} + \tau_c.$$

(9.9)

The core energy τ_c of the two disclinations L_1 and L_2 is of the order of K_1 and does not depend on b. Note that

$$W_c(2b) - 2W_c(b) = \frac{\pi K_1}{2} \ln \frac{4d_0}{b}$$

(9.10)

is negative for $b > 4d_0$; hence, the energy difference (9.10) decreases logarithmically when b increases. The large Burgers vectors are favored.

Because the above calculation of the elastic energy outside of the core is done in the small perturbation approximation, our conclusions should be qualified when b is large, $b \geq \lambda$. And indeed, as we have already argued in Section 5.4.2., as soon as the characteristic length (here, b) is larger than λ (i.e., in practice for any Burgers vector larger than d_0), we expect that the contribution of curvature elasticity prevails over compression elasticity. The complete disappearance of compression elasticity requires that the layers become parallel, which is achieved in Fig. 9.5b, at the expense of a wall singularity. The model in Fig. 9.5b is an extreme case of a nonlinear dislocation profile discussed above for elementary dislocations with $b \geq \lambda$, with the wall defect being an analog of the locus of inflection points in Fig. 9.2b.

The disclination model of the core, which is justified here by energetical considerations, also has a value in cholesterics (Figs. 9.2a and 11.17) and lyotropics, where it allows for a complete separation between the solvent and the bilayers (Fig. 9.6). In that case, it is valid even for small Burgers vectors.

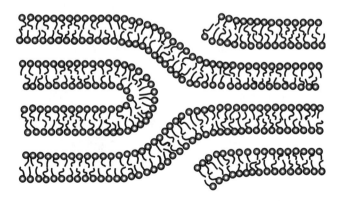

Figure 9.6. Edge dislocation in a lyotropic L_α phase.

The most frequent manifestation of edge dislocations with large Burgers vectors are oily streaks, frequently present in cholesterics (Sections 10.7.1 and 11.2.5.3) and in thick, homeotropic smectic samples. According to G. Friedel's model, oily streaks are made of pairs of dislocations of (large) opposite Burgers vectors. Each element of the pair is probably due to the coalescence of small Burgers vector dislocations of the same sign (as stated above, a large Burgers vector dislocation $b = nd_0$ can be favored with respect to small Burgers vectors dislocations $n_i d_0$, $\sum n_i = n$), produced when the sample is formed. Oily streaks are typical defects of lyotropic phases; the special core topology of edge dislocations in these phases makes the collapse of dislocations of opposite signs difficult, even for small n's, hence, the preferred formations of large n, n' dislocations. Another feature of oily streaks is the frequent occurrence of a longitudinal instability, which sometimes even splits the streak into a series of focal conic domains, to be studied in Chapter 10.

9.1.2. Screw Dislocation

The Euler–Lagrange equation

$$K \, \Delta_\perp^2 u = B \frac{\partial^2 u}{\partial z^2} \tag{9.11}$$

has for solution

$$u(r, \theta) = \frac{b\theta}{2\pi}, \tag{9.12}$$

i.e., a screw dislocation with the same configuration as in a crystalline solid. One can easily show as an exercise that the elastic energy of the line vanishes, when calculated to the second order; a core energy W_c subsists. Screw dislocations have, therefore, a small energy $W_s = W_c$; they have been observed in various water-surfactant systems by electron microscopy of freeze-fractured specimens, which cleave easily along the interfaces of the paraffinic chains. Screw dislocations pierce the cleavage surfaces and carry, in the wake along the direction of cleavage, steps that gather in rivers as in usual crystals (Fig. 9.7). They have also been observed by polarizing light microscopy in homeotropic samples of thermotropic SmA, near the SmC transition,[5] where they pin the edge dislocations. Finally, dislocation lines at a small angle to the Burgers vector direction are most probably made of long segments of a pure screw character linked by small edge segments, as a result of the large line tension of screw dislocations (see Section 9.1.3).

In fact, (9.12) is the exact minimization solution for the elastic energy density (5.12), $f = \frac{1}{2}K_1(\text{div}\mathbf{n})^2 + \frac{1}{2}B\gamma^2$, composed of both curvature and dilation terms.[6] It is a solution

---

[5] R.B. Meyer, B. Stebler, and S.T. Lagerwall, Phys. Rev. Lett. **41**, 1393 (1978).
[6] M. Kleman, Phil. Mag. **34**, 79 (1976).

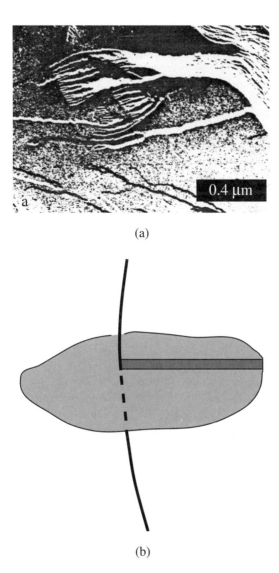

(a)

(b)

Figure 9.7. (a) Screw dislocations, steps, and rivers on the cleavage surface of a lyotropic phase of $C_{12}EO_6$; Courtesy of M. Allain; (b) the cleavage pattern of a lamellar sample, where the screw dislocation emerges, and a step begins and runs as the fracture propagates.

for which div $\mathbf{n} = 0$, i.e., a minimal surface. The displacement u is no longer the relevant variable, and (9.12) takes the form

$$\varphi = -b\theta/2\pi + z, \tag{9.13}$$

which describes an isolated screw dislocation as a ruled helicoid; here, $\varphi = \varphi(\theta)$ is a phase variable, constant on each layer, z is the coordinate along the dislocation, and θ is a polar angle. Taking $\varphi = $ const, one immediately sees that this helicoid is generated by a half straight line perpendicular to the z-axis, leaning on it, and rotating helically with a constant angular velocity and a pitch b (Fig. 9.8).

The (out-of-core) free energy of a dislocation, which is vanishing in the quadratic approximation, takes the following value:

$$W_s = \frac{Bb^4}{128}\left(\frac{1}{r_c^2} - \frac{1}{R^2}\right) + W_c, \tag{9.14}$$

when calculated to next order; here, R is the external radius of the sample. This result is valid in the limit $\lambda = \sqrt{K_1/B} > b$, i.e., when the predominant energy contribution comes

Figure 9.8. A ruled helicoid.

from position elasticity (Section 5.5.2). The inequality $\lambda > b$ also implies that the Burgers vectors under consideration are small, $b = d_0$, typically. The condition $\lambda > d_0$ is well satisfied near a smectic-to-nematic transition, because $B = 0$ in the nematic phase.

The above analysis is valid when the Burgers vector is small enough. As already emphasized (see also Section 5.5.2), as soon as any characteristic length in the sample geometry, such as b, becomes larger than λ, the contribution of curvature elasticity becomes predominant, and the layers prefer to stack parallel to each other. Thus, the condition that the derivative in the displacement along the normal of the layers vanishes, $\partial u / \partial n = 0$, is nearly satisfied, while the condition div $\mathbf{n} = 0$ is not. As we shall see in the next chapter, the condition $\partial u / \partial n \equiv 0$ yields a precise geometry of defects, the focal conic domains, but it can be approximately satisfied in other geometries, and the giant screw dislocation, as opposed to the screw dislocation with a small Burgers vector discussed above, is such a geometry, Figs. 9.9 and 9.10.[7]

Start from a ruled helicoid H_0 whose pitch $p = b = n\, d_0$. This surface figures out one of the layers of the screw dislocation. The other ones H_n are stacked along surfaces, parallel to H_0, in such a way that $\frac{\partial u}{\partial n} \equiv 0$. The H_n surfaces are no longer minimal surfaces, but they are close to minimal surfaces, because H_0 is minimal. In other words, the term of

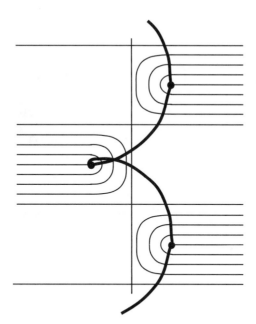

Figure 9.9. Model for the stacking of layers in a giant screw dislocation.

[7]C.E. Williams, Philos. Mag. **32**, 313 (1975).

compressibility in the free energy density vanishes identically, and the term of curvature keeps small.

This stacking can be pursued as long as the layers do not physically meet. More precisely, because the layers H_n are parallel, their normals (i.e., the director field) are straight lines, and they envelop a *focal surface* F made of two sheets. This geometrical problem is in fact formally analogous to the problem met in optical geometry: the normals being light rays, and the layers being surfaces of equal phase. The two sheets are the geometrical locus of the two centers of curvature of the parallel surfaces H_n (all of the intersections of parallel H_n's with the same normal share the same center of curvature). F constitutes the natural boundary of the H_n stacking. In fact, there is one helical cuspidal edge on each sheet of F, say, L_1 and L_2, and these cuspidal edges generate two physical line defects, about which the layers fold beyond F, in the shape of two *disclination lines* of strength $k = \pm\frac{1}{2}$ (Fig. 9.10). This remarkable mode of splitting of a screw dislocation of large Burgers vector is indeed observed in a thermotropic SmA (Fig. 9.9), where the two helical lines L_1 and L_2 are conspicuous. Even more remarkably, such lines can be guessed in the textures of a condensed chromosome of a prokaryotic microscopic alga, where the DNA is cholesteric.[8] The layers are now the cholesteric layers of repeat distance $p/2$; the chromosome is limited by a double helical crease, and the layers do not continue outside of this region, but the DNA filaments fold about the $k = \frac{1}{2}$ lines.[9]

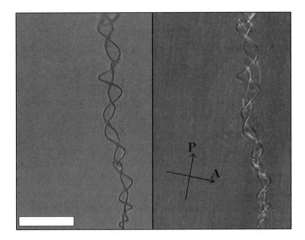

Figure 9.10. Giant screw dislocation in a thermotropic smectic. Courtesy of C. Blanc.

[8]F. Livolant and Y. Bouligand, Chromosoma **80**, 97 (1980).
[9]M. Kleman, Physica Scripta **T19**, 565 (1987).

9.1.3. Line Tension of a Screw Dislocation

Let, as in Section 8.1, τ be the line tension of a dislocation; when taking into account the variation of the line energy $W(\alpha)$ with the orientation α, τ can be written (Problem 9.2) as

$$\tau = \left(W + \frac{\partial^2 W}{\partial \alpha^2} \right)_{\alpha=0}. \tag{9.15}$$

In most solids, even very anisotropic, the $\frac{\partial^2 W}{\partial \alpha^2}$ contribution is small compared with W. This is not the case here for the screw dislocation. One can show that the displacement field carried by a line tilted from the z-axis, by a small angle α, can be written as

$$u(x, y, z) = \frac{b}{2\pi} \tan^{-1} \frac{y - \alpha z}{x}. \tag{9.16}$$

The energy per unit length of line is

$$W(\alpha) = \frac{Bb^2}{8\pi} \alpha^2 \ln \frac{R}{r_c}; \tag{9.17}$$

hence,

$$\tau = \frac{Bb^2}{4\pi} \ln \frac{R}{r_c}. \tag{9.18}$$

The line tension is large; therefore a screw dislocation remains straight along the z-axis when no configurational force F_{conf} is applied on the line. Otherwise, it curves with a radius

$$\rho = \frac{\tau}{F_{\text{conf}}} \tag{9.19}$$

between the pinning points. Here, F_{conf} is a force per unit length of line.

9.1.4. Stresses in a SmA and Peach and Koehler Forces

In order to define a stress field in a smectic phase, we proceed as in the solid case. We restrict to small deformations of the layers; the total free energy F then reads as

$$F = \int f \, dV = \int \left[\frac{1}{2} K_1 (\Delta_\perp u)^2 + \frac{1}{2} B \left(\frac{\partial u}{\partial z} \right)^2 \right] dV. \tag{9.20}$$

Starting from a state of deformation $u(\mathbf{r})$ that obeys the Euler–Lagrange equations and the boundary conditions, let us introduce a small displacement field δu that increases F: $F \rightarrow F + \delta F$, where

$$
\begin{aligned}
\delta F = \int &\left(K_1 \Delta_\perp^2 u - B \frac{\partial^2 u}{\partial z^2} \right) \delta u \, dV \\
&+ \int \left\{ B \frac{\partial u}{\partial z} \, dS_3 - K_1 \frac{\partial}{\partial x} (\Delta_\perp u) \, dS_1 - K_1 \frac{\partial}{\partial y} (\Delta_\perp u) \, dS_2 \right\} \delta u \qquad (9.21) \\
&- \int K_1 \Delta_\perp^2 u \left(\delta \left(\frac{\delta u}{\delta x} \right) \, dS_1 + \delta \left(\frac{\delta u}{\delta y} \right) \, dS_2 \right).
\end{aligned}
$$

This expression contains

- bulk terms, which vanish because of the Euler–Lagrange equations, and
- surface terms of two types, surface forces and surface torques.

9.1.4.1. Surface Forces

The first surface integral in (9.21) represents the work done by the external forces when the boundary is displaced by δu. Because the displacement field δu can be chosen at will, assume that it vanishes in some arbitrary subvolume of the system; in such a case, the external forces are the components of the stress tensor acting on the boundary of the complementary subvolume. Hence, at any point in the bulk, one has

$$
\sigma_{zx} = -K_1 \frac{\partial}{\partial x} (\Delta_\perp u); \quad \sigma_{zy} = -K_1 \frac{\partial}{\partial y} (\Delta_\perp u); \quad \sigma_{zz} = B \frac{\partial u}{\partial z}. \qquad (9.22)
$$

All other components of σ vanish; σ is not symmetric, and its asymmetric part must be interpreted in terms of torques, which appear in the second type of surface terms.

Note that our definition of stresses is such that the forces per unit area exerted on a surface of external normal ν_i can be written as $f_j = \sigma_{ji} \nu_i$. The Euler–Lagrange equation can be written in terms of stresses as

$$
\sigma_{zx,x} + \sigma_{zy,y} + \sigma_{zz,z} = 0. \qquad (9.23)
$$

9.1.4.2. Surface Torques

Because $\mathbf{n} = (-\frac{\partial u}{\partial x}, -\frac{\partial u}{\partial x}, 1)$, we can write the last integral of (9.21) as

$$
\int K_1 \Delta_\perp^2 u \left(\delta n_x \, dS_1 + \delta n_y \, dS_2 \right). \qquad (9.24)
$$

With $\delta\boldsymbol{\omega} = \delta\mathbf{n} \times \mathbf{n}$, we have

$$\delta n_x \, \delta S_1 + \delta n_y \, \delta S_2 = \delta\mathbf{n} \cdot \delta\mathbf{S} = (\mathbf{n} \times d\mathbf{S}) \cdot \delta\boldsymbol{\omega}. \tag{9.25}$$

Hence, the surface torques per unit area applied on the surface at a point of external normal $\boldsymbol{\nu}$ are

$$\mathbf{C} = K_1 \, \Delta_\perp^2 u \mathbf{n} \times \boldsymbol{\nu}. \tag{9.26}$$

The stress field being defined, the Peach and Koehler force follows:

$$\mathbf{F}^{PK} = (\mathbf{b} \cdot \sigma) \times \mathbf{t}, \tag{9.27}$$

where some care should be taken when calculating $\mathbf{b} \cdot \sigma$: $\mathbf{b}$ saturates the first index of the components of the stress tensor; $\mathbf{t}$ is the unit tangent to the line, as in (8.22).

9.2. Dislocations in Columnar Phases

Columnar hexagonal liquid crystals are 2D solids in the planes perpendicular to the columns and possible Burgers vectors belong to this 2D hexagonal lattice. The elastic free-energy density can be written as

$$f = 2\mu \left(u_{xy}^2 - u_{x,x} u_{y,y} \right) + \tfrac{1}{2} B_\perp \left(u_{x,x} + u_{y,y} \right)^2 + \tfrac{1}{2} K_3 \left(u_{x,zz}^2 + u_{y,zz}^2 \right), \tag{9.28}$$

where $\mathbf{u} = (u_x, u_y)$ is the displacement vector in the hexagonal plane, $u_{xy} = \tfrac{1}{2}(u_{x,y} + u_{y,x})$; μ is the shear modulus; $B_\perp = \lambda + 2\mu$ is the modulus of compressibility; and K_3 is the Frank bend modulus. We assume that columns relax like 1D liquids; hence, there is no contribution of the displacement u_z to the free-energy density. We also assume that $\nabla\mathbf{u}$ is small, which yields $\mathbf{n} = (-u_{x,z}, -u_{y,z}, 1)$ for the director along the column.

9.2.1. Longitudinal Edge Dislocations in Columnar Hexagonal Liquid Crystals

Line dislocations parallel to the columns do not involve any bend deformation; therefore, we are left with a free energy density similar to that one of a 2D solid; (8.15) can be applied, and the line energy (surface relaxation and core included) can be written as

$$W_{\text{long}} = \frac{\mu b^2}{4\pi(1 - \nu)} \left(\ln \frac{R}{er_c} + \frac{1}{2} \right), \tag{9.29}$$

where ν is the Poisson ratio. Assuming that the penetration length $(\frac{K_3}{\mu})^{1/2}$ is of the order of the lattice parameter a, we expect $\mu \cong \frac{K_3}{a^2}$, i.e., $W_{\text{long}} \sim K_3(\frac{b}{a})^2$. For $b = a$, this is of the same order as the energy of a disclination, typically, 0.01 eV per molecular length ℓ measured along the line ($\ell \sim 0.2$ nm, $a \sim 3$ nm in a thermotropic phase). This is comparable to $k_B T$ at room temperature, so that one expects important entropy effects, in contrast with usual crystals, and eventually a total energy

$$W_{\text{long}} \cong K_3 - \frac{k_B T}{a}, \tag{9.30}$$

which becomes negative when T is large enough: The phase would melt, by a process akin to the melting of solids, due to a multiplication of defects (see Section 4.5).

The above discussion is of course very crude. It does not take into account the other types of dislocations (screw dislocations, transverse edge dislocations) that are of a very different nature. The line energy of a generic dislocation in a columnar phase varies considerably with the line direction and involves bend deformation, i.e., depends directly (not through μ) on the Frank modulus K_3. If μ is small enough compared with λ, one expects that the longitudinal dislocations have the smaller line energy, and the above arguments might be true; if μ and λ are comparable, the screw dislocations win.

Edge longitudinal dislocations are very much akin to edge dislocations in solids. The situation is more original for edge transversal dislocations and screw dislocations.

9.2.2. Edge Transversal Dislocations

For an edge dislocation running along the x-axis, of Burgers vector b, the only relevant displacement is u_y and $\partial/\partial x = 0$. The free energy density reads as

$$f = \tfrac{1}{2}(\lambda + 2\mu)(u_{y,y})^2 + \tfrac{1}{2}K_3(u_{y,zz})^2, \tag{9.31}$$

and $\int f\, dV$ is minimized for

$$(\lambda + 2\mu)\frac{\partial^2 u}{\partial y^2} = K_3 \frac{\partial^4 u_y}{\partial z^4}. \tag{9.32}$$

This equation is akin to to the equation of an edge dislocation in a smectic. Following the treatment used in this case (Section 9.1.1), and noticing the symmetry $y \rightarrow -y$, we write for, say, $y > 0$:

$$u_y(y, z) = \frac{b}{4} + \frac{b}{4\pi} \int \frac{dq}{iq} g(y, q) \exp iqz \tag{9.33}$$

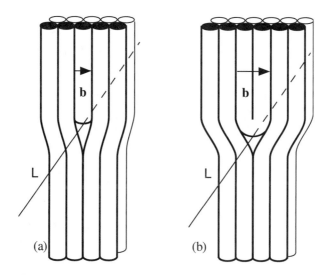

Figure 9.11. Transverse edge dislocation.

with Burgers vector $(0, b, 0)$ and $g(0, q) = 1$,

$$g(y, q) = \exp -q^2 y \Lambda_3, \tag{9.34}$$

where $\Lambda_3 = \sqrt{K_3/(\lambda + 2\mu)}$ is the penetration length attached to bend deformation (compare with $\lambda_1 = \sqrt{K_1/B}$ in smectics). The angular deformation $\theta(z) = u_{y,z}$ of the columns outside the dislocation core (Fig. 9.11) can be written as

$$\theta(z) = \frac{b}{4\pi(\Lambda_3 y)^{1/2}} \exp -\frac{z^2}{4\Lambda_3 y}. \tag{9.35}$$

Note that in this calculation, we have neglected the splay contribution in the free energy density

$$f_{\text{splay}} = \frac{1}{2} K_1 \left(\frac{\partial^2 u_y}{\partial y \partial z} \right)^2. \tag{9.36}$$

This is all the more justified now that the columns keep nearly parallel and present no free ends: One expects $K_1 \gg K_3$ in columnar phases and, therefore, a vanishing div **n**. This is certainly not the case in the core region. Let us introduce f_{splay} in the free energy: Instead of (9.34), one gets

$$g(y, q) = \exp\left[-q^2 y \Lambda_3 (1 + q^2 \Lambda_1^2)^{-1/2}\right], \qquad (9.37)$$

where $\Lambda_1 = \sqrt{K_1/(\lambda + 2\mu)}$ is another penetration length ($\Lambda_1 \gg \Lambda_3$). It is obvious at first sight that outside a region $|q|^{-1} > \Lambda_1$, the divergence contribution is small and the expression (9.35) is valid. Contrarywise, for $|q|^{-1} \ll \Lambda_1$, one gets

$$g(y, q) \cong \exp - qy \frac{\Lambda_3}{\Lambda_1}; \qquad (9.38)$$

hence,

$$u_y = \frac{b}{4} + \frac{b}{2\pi} \tan^{-1} \frac{z}{y} \frac{\Lambda_3}{\Lambda_1}. \qquad (9.39)$$

The core region, where the columns show free ends, is therefore expected to be large, of the order of Λ_1.

In analogy with the smectic case, one expects

$$W_{\text{trans}} = \frac{\lambda + 2\mu}{2} b^2 \frac{\Lambda_3}{\xi_\perp} + W_{\text{core}}, \qquad (9.40)$$

where $\xi_\perp$ is the actual core size in the z-direction.

9.2.3. Screw Dislocations

We consider a screw dislocation aligned along the x-direction. The relevant displacements are u_x and u_y; $\partial/\partial x = 0$. The free energy density reads as

$$f = \tfrac{1}{2}(\lambda + 2\mu)(u_{y,y})^2 + \tfrac{1}{2}\mu(u_{x,y})^2 + \tfrac{1}{2}K_1\left(\frac{\partial^2 u_y}{\partial y \partial z}\right)^2 + \tfrac{1}{2}K_2\left(\frac{\partial^2 u_x}{\partial z \partial y}\right)^2$$

$$+ \tfrac{1}{2}K_3\left[\left(\frac{\partial^2 u_x}{\partial z^2}\right)^2 + \left(\frac{\partial^2 u_y}{\partial z^2}\right)^2\right] \qquad (9.41)$$

One notices at once that u_x and u_y are decoupled. Consider the u_x term. One gets, after minimization,

$$\mu \frac{\partial^2 u_x}{\partial y^2} = K_2 \frac{\partial^4 u_x}{\partial z^2 \partial y^2} + K_3 \frac{\partial^4 u_x}{\partial z^4}. \qquad (9.42)$$

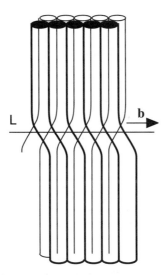

Figure 9.12. Screw dislocation in a columnar phase. No physical singularity on the columns.

Note that one can take $u_y \equiv 0$, because the Burgers vector is in the x-direction: There is no splay deformation, but there is a twist. The calculation goes as above and yields

$$g(y, q) = \exp\left\{-q^2 y \Lambda_3' \left(1 + q^2 \Lambda_2'^2\right)^{1/2}\right\},\tag{9.43}$$

where $\Lambda_2' = \sqrt{K_2/\mu}$ and $\Lambda_3' = \sqrt{K_3/\mu}$. It is reasonable to assume that in columnar systems $K_2 \gg K_3$, because a twist deformation does not allow the columns to keep parallel and equidistant. A discussion of (9.43) as above for (9.37) tells us, therefore, that the core region is a region where twist is concentrated (Fig. 9.12). As will be shown in Chapter 11, a periodic lattice of screw dislocations is a possible model of a twisted hexagonal phase[10] (Fig. 11.14).

A remarkable feature of this core is that there is no necessity to introduce any physical singularity on the columns. Notice also that a jog of elementary length on a screw dislocation introduces one cut column; the appearance of an adequate density of such jogs on screw dislocations can be in practice a way of relaxing plastically applied stresses.

The line energy is of the form

$$W_{\text{screw}} = \frac{1}{2}\mu b^2 \frac{\Lambda_3'}{\xi_\perp} + W_{\text{core}},\tag{9.44}$$

where $\xi_\perp$ is the core size in the z-direction.

---

[10]R.D. Kamien and D.R. Nelson, Phys. Rev. Lett. **74**, 2499 (1995).

9.2.4. Free Fluctuations of Longitudinal Dislocations

A fluctuation of the line is akin to the formation of a double kink. We, therefore, discuss this possibility in the terms of Chapter 8 (Section 8.5.2) for thermally activated glide. Because the dislocations of different types in a columnar phase show up very anisotropic properties, the comparison of U_{dd}^f, the energy of nucleation of a double kink, and U_d^m, the energy of displacement of the kinks, is mostly dependent on this anisotropy. As an example, if μ is small enough, i.e., $\Lambda_3 = (K_3/\mu)^{1/2} \gg a$, where a is a molecular length, we expect $W_{\mathrm{screw}} > W_{\mathrm{long}}$, which most probably yields a large energy U_{dd}^f (kinks are screw dislocations segments) compared with U_d^m. Therefore, the yield stress for glide of longitudinal dislocations is of the "low temperature" type, in terms of the classic analysis for metals. However, the nucleation of double kinks is a thermally activated process, and we expect that they appear in number, due to the coherent thermal molecular motion, and effectively reduce the Peierls–Nabarro friction. The calculation goes as follows.

Consider a double kink of length 2ℓ nucleating under an applied stress $\sigma < \sigma_p$ (Fig. 9.13). The energy required is the difference of energy between the double kink (whose nucleation requires to extend the cut surface by an area of the order of $b\ell$) and a straight dislocation, i.e.,

$$U(\sigma) \approx 2W_{\mathrm{screw}} \frac{b^2}{\ell_\sigma}, \tag{9.45}$$

where $\ell_\sigma = \sqrt{W_{\mathrm{screw}}/(\sigma_p - \sigma)}$.

In these equations σ_p is expected to have an expression similar to that one in a standard solid (8.39), because W_{long} is standard. Putting order of magnitudes, one gets for $\sigma = 0$:

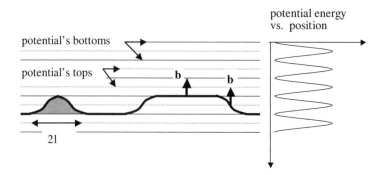

Figure 9.13. Thermally activated glide against a Peierls–Nabarro force.

$$\ell_{\sigma=0}^2 \approx \frac{b^2}{8\pi}\left(\frac{\Lambda_3}{b}\right)^{1/4}\exp\frac{2\pi}{1-\nu}\frac{a}{b},$$

$$U(0) \approx 2\mu b^2\left(\frac{\Lambda_3}{b}\right)^{1/4}\exp-\frac{\pi a}{(1-\nu)b}; \qquad (9.46)$$

i.e., typically $\ell_{\sigma=0} \approx 100$ nm for $b \approx 3$ nm and $\Lambda_3 \approx b$, and $U(0)$ of the order of $k_B T$ for $\mu \le 10^7 \mathrm{N/m^2}$ (10^8 dyne/cm^2). Therefore, free fluctuations of longitudinal dislocations are to be expected in most thermotropic columnar hexagonal liquid crystals.

9.3. Hydrodynamics of a Smectic Phase

The hydrodynamics of a slightly deformed perfect smectic phase (i.e., in the approximation of weak perturbations) is developed in this section. Section 9.4 describes briefly the (small amplitude) lamellar dynamic modes, which it summarizes without entering into details. The hydrodynamical theory is then applied to the movement of edge dislocations of *small* Burgers vectors, for which the precise model of the core is unessential (Section 9.5). A priori, *climb*, which is the displacement of the dislocation in the plane of the layers, perpendicularly to the Burgers vector, is easier than glide, contrarily to the case of solids. This is due to the fact that glide necessitates layer breaking (particularly difficult in lyotropic smectics), whereas climb does not. Finally, Section 9.6 reports on the behavior of macroscopic samples under simple shear or compression; the few experiments that have been performed reveal interesting collective motions of defects, multiplication of dislocations, and instabilities toward the formation of defects of another nature (see Chapter 10).

The theory of the movement of dislocations can be cast either in a language borrowed from metallurgy, and using henceforth concepts developed in solid materials science (such as diffusion for the case of climb of edge dislocations), in the language of hydrodynamics, or in both. It is indeed one of the nice features of the defects dynamics in lamellar phases that they mix concepts borrowed from both theories.

Hydrodynamical properties of layered materials are remarkable. One can distinguish between correlated (Fig. 9.14a, b) and uncorrelated (Fig. 9.14c; 9.15) motions of molecules and layers. In the first group, the layers follow the movement of the molecules, which can be treated as forming an ordinary (anisotropic) fluid. One would therefore expect three different viscosities $\eta_\perp$, η, and $\eta_\parallel$. Let $\mathbf{v} = (v_x, v_y, v_z)$ be the components of the fluid velocity (we assume that the fluid is but slightly perturbed with respect to the perfect smectic; the z-direction is along the unperturbed director), and let $A_{\gamma\delta} = \frac{1}{2}(v_{\gamma,\delta} + v_{\delta,\gamma})$ be the components of the shear rate tensor. The three independent viscosity coefficients appear in the expressions of the viscous stress tensor components:

$$\sigma''_{xx} = 2\eta_\perp A_{xx}, \quad \sigma''_{yy} = 2\eta_\perp A_{yy}, \quad \sigma''_{xy} = 2\eta_\perp A_{xy},$$

$$\sigma''_{zx} = 2\eta A_{xz}, \quad \sigma''_{zy} = 2\eta A_{zy}, \quad \sigma''_{zz} = 2\eta_\parallel A_{zz}. \qquad (9.47)$$

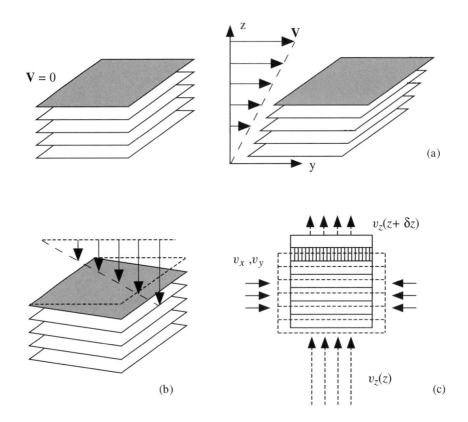

Figure 9.14. Various types of shear in a lamellar phase, (a) velocity field parallel to the layers, viscosity $\eta_\perp$; (b) velocity field with $v_z(x, y)$ component perpendicular to the layers, η; (c) velocity field perpendicular to the layers, $\eta_{||}$.

Generally speaking, one should consider five viscosity coefficients. This number is reduced to three if one assumes (reasonably) that the smectic fluid is incompressible.

Fig. 9.14 represents two types of "correlated" shears. In Fig. 9.14a, the velocity field (v_x, v_y) is parallel to the layers, and the layered media flow much like ordinary liquids. The layers keep globally fixed in space and slide easily past each other; the corresponding viscosity $\eta_\perp$ is expected to be small. In Fig. 9.14b, the layers configuration is supposed to be dragged with the motion, which is governed by η. The case of the viscosity $\eta_{||}$ (Fig. 9.14c) is somewhat more complex: It shows up if $\partial v_z/\partial z \neq 0$; but because of the incompressibility condition div $\mathbf{v} = 0$, it is coupled with $v_x, v_y \neq 0$, i.e., with the other viscosities. Generically, such a flow would considerably modify the layers thickness, if the layers and the molecules move in company. However, a large change of the layers thickness is energetically unfavorable. Therefore, one expects that this type of motion is coupled with a flow of matter through the layers (or, equivalently, a motion of the layers through the fluid) (Fig. 9.14c), which is reminiscent of (thermally activated) vacancy or interstitial diffusion in crystals.

The motion of fluid through the layers is called permeation.[11] Generally, $\eta_{\parallel}$ is expected to be large compared with η and $\eta_{\perp}$.

A very simple "uncorrelated" motion of the molecules and the layers is as follows. Assume that the layers are fixed in space and a pressure gradient $\frac{\partial p}{\partial z}$ is applied perpendicularly to them. A weak flow develops across the layers, with velocity:

$$v_z = -\lambda_p \frac{\partial p}{\partial z},$$ (9.48)

where λ_p is the so-called permeation constant. Dimensionally, $\lambda_p \sim a^2 \eta_p^{-1}$, where η_p should be a (large) viscosity for molecular motion perpendicular to the layers and a is a length comparable to the square root of the cross-section of the molecules with the bilayer surface. In fact, permeation is a slow process, as we shall argue later, and any motion considered on a time $\tau_q < \frac{1}{\lambda_p B q^2}$, where q^{-1} is a characteristic length of the motion, is not of this type (Problem 9.3). In any real motion, correlated and uncorrelated processes should exist, according to the scales and the durations under study.

Let us develop these notions. There are two relevant velocities, as follows:

1. The velocity **v** of the molecules, which satisfies the conservation law:

$$\text{div } \mathbf{v} = 0$$ (9.49)

   in the absence of dilatation.

   The fundamental equation that expresses the conservation of momentum has been derived by Martin et al.[12]; it can be written as

$$\rho \frac{d\mathbf{v}}{dt} = -\nabla p + \text{div } \sigma' + \text{div } \sigma'',$$ (9.50)

   where:

   (a) σ' is the elastic stress tensor carried by the deformed layers (9.22). The quantity div σ', which is the restoring force per unit volume acting on the layers, vanishes at rest

$$\text{div } \sigma' = \left(0, 0, B \frac{\partial^2 u}{\partial z^2} - K \nabla_{\perp}^4 u \right).$$ (9.51)

[11]W. Helfrich, Phys. Rev. Lett. **23**, 372 (1969).
[12]P. Martin, O. Parodi, and P. Pershan, Phys. Rev. **A6**, 2401 (1972).

We shall note

$$B\frac{\partial^2 u}{\partial z^2} - K\nabla_\perp^4 u = g.$$

(b) σ'' is the viscous stress tensor, which can be taken as symmetrical, because the torques are already present in σ'. The vectorial quantity $\mathrm{div}\,\sigma''$ is the friction force per unit volume.

(c) p is the pressure.

2. The velocity of the layers $(0, 0, \dot{u})$, directed along the normal to the unperturbed layers, in the small perturbation picture, which is adopted here. The relative velocity $\dot{u} - v_z$ is conjugate to g; hence, the entropy source can be written as

$$T\sigma_{\mathrm{irr}} = \sigma'' \cdot \mathbf{A} + g(\dot{u} - v_z), \tag{9.52}$$

where for the sake of simplicity, we have omitted transport terms like heat on electric current. One gets

$$\sigma''_{\alpha\beta} = \Lambda_{\alpha\beta\gamma\delta}A_{\gamma\delta}, \quad \dot{u} - v_z = \lambda_p g. \tag{9.53}$$

The $\Lambda_{\alpha\beta\gamma\delta}$ reduce to three independent viscosity coefficients (9.47). The second equation in (9.53) easily yields (9.48) if the layers are kept fixed, the flow stationary, and the viscous stresses neglected. Equation (9.48) is analogous to Darcy's law for *porous media*, which states that the velocity of a fluid is proportional to the pressure drop (see Section 7.4.1): In our case, the smectic phase appears as its own filter.

From a microscopic point of view, the coefficient of permeation is related to the activated passage of the molecules from one layer to the next. Let us apply Einstein law (8.53) to this process. The force acting on a molecule is $g\omega_m$, where $\omega_m = a^2 d_0$ is a molecular volume. Hence,

$$\dot{u} - v_z = \frac{\mathcal{D}_\|}{k_B T}g\omega_m = \lambda_p g \tag{9.54}$$

and

$$\lambda_p = \frac{\mathcal{D}_\|\omega_m}{kT}, \tag{9.55}$$

where $\mathcal{D}_\|$ is the diffusivity of the molecules through the layers. The coefficient of permeation can be measured in relation with the mobility of edge dislocations (see below). The

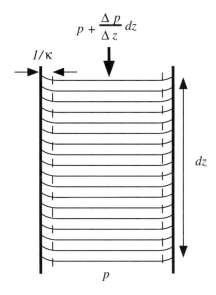

Figure 9.15. Gedanke experiment to measure λ_p.

following *gedanke* experiment, where the layers are strongly anchored perpendicular to the walls of a cylindrical capillary of radius L, and submitted to a constant pressure gradient $p' = \frac{\partial p}{\partial z}$, should in principle also measure the coefficient of permeation (Fig. 9.15). The flow is assumed stationary and the layers fixed; hence, $\dot{u} = 0$. The Navier–Stokes (9.50) can be written, using (9.54), as

$$-\frac{v_z}{\lambda_p} - p' + \eta \frac{\partial^2 v_z}{\partial x^2} = 0. \tag{9.56}$$

This equation is nothing else than a generalization of (9.48) to the case when the layers do not keep planar. We look for a solution independent of the z-coordinate [hence, the term $\eta_\parallel \partial^2 v_z / \partial z^2$ does not appear in (9.56)]. The solution is

$$v_z = -p' \lambda_p \left(1 - \frac{\cosh \kappa x}{\cosh \kappa L} \right), \tag{9.57}$$

where $\kappa^{-2} = \lambda_p \eta$. Note that κ^{-1} appears here as the thickness of a boundary layer (see Fig. 9.15 and Clark[13]).

---

[13]N.A. Clark, Phys. Rev. Lett. **40**, 1663 (1978).

9.4. Dynamic Modes in Smectics

There are six modes, i.e., one mode more than in an isotropic liquid. This mode is the phase of the layers. Large rotations of the director are excluded in this model, which employs the small perturbation approximation for the free energy (see the de Gennes and Prost textbook):

- Four P modes

  $\omega = \pm c_1 q$ first sound, related to the variation of the density ρ; phase velocity $c_1 = \sqrt{\frac{A}{\rho}}$,

  $\omega = \pm c_2 q$ second sound, related to the variation of the layers thickness, at constant density ρ; $c_2 = \sqrt{\frac{B}{\rho}} \sin\phi \cos\phi$, where ϕ is the angle of $\mathbf{q} = (q \cos\phi, q \sin\phi)$ with the layer. A and B are elastic moduli.

- Two D modes:

  $\omega = -i\kappa_{th}q^2$ thermal diffusion,

  $\omega = -i\nu q^2$ transverse shear mode.

The two situations in which $\mathbf{q}$ is either parallel or perpendicular to the layers are special, because the two anisotropic propagative modes vanish for those values of ϕ. The six modes are now replaced by the following modes:

1. $\mathbf{q}$ perpendicular to the director (parallel to the layers):

   (a) Two P modes:

   $\omega = \pm c_1 q_\perp$ first sound.

   (b) Four D modes:

   $\omega = -i\frac{K}{\eta}q_\perp^2$ undulation mode of the layers, reminiscent of the slow mode in nematics, (Problem 9.4)

   $\omega = -i\kappa_{th}q_\perp^2$ thermal diffusion, coupled to the sound mode,

   $\omega = -i\nu q_\perp^2$ transverse shear mode,

   $\omega = -i\nu' q_\perp^2$ longitudinal shear mode coupled to the sound mode ν, $\nu' \propto \frac{\eta}{\rho}$.

2. **q** parallel to the director:

(a) Two P modes:

$\omega = -\pm c_1 q_{||}$ first sound; $c_1 = \sqrt{A/\rho}$.

(b) Four D modes:

$\omega = -i B \lambda_p q_{||}^2$ mode of permeation (Problem 9.3),

$\omega = -i K \lambda_{th} q_{||}^2$ thermal diffusion, coupled to sound mode,

$\omega = -i \nu q_{||}^2$ transverse shear mode,

$\omega = -i \nu' q_{||}^2$ longitudinal shear mode coupled to the sound mode.

9.5. Movement of Isolated Dislocations in an SmA Phase

9.5.1. Edge Dislocation

9.5.1.1. Climb

Let $\sigma = \sigma' + \sigma''$ be the total stress to which an edge dislocation is submitted (excluding its proper stress field). The Peach and Koehler force (9.27) that acts on it is $F^{PK} = b(\sigma_{zy}t_z - \sigma_{zz}t_y, \sigma_{zz}t_x - \sigma_{zx}t_z, \sigma_{zx}t_y - \sigma_{zy}t_x)$. We shall assume that the dislocation lies along the y-axis ($t_x = 0, t_y = 1, t_z = 0$).

Consider, therefore, the following simple model[14] (Fig. 9.16), in which the dislocation is symbolized by a semi-infinite rigid plate of vanishing thickness, moving with velocity $-V$ in a smectic sample at rest at infinity or, equivalently, a plate at rest in a smectic sample moving with velocity $+V$ at infinity. A boundary layer of thickness $\delta(x)$ forms along the plate, and we shall furthermore assume that $\delta(+\infty) = \frac{b}{2}$. The three equations of the problem are Darcy's law, the Navier–Stokes equation, and the equation of continuity. The layers are supposed fixed, and the molecules permeate through them ($\dot{u} = 0$). Therefore,

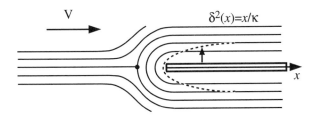

Figure 9.16. Model for the movement of an edge dislocation under climb.

[14]Orsay Group on Liquid Crystals, J. de Physique **36**, C1-305 (1975).

one expects

$$v_z \cong \lambda_p \frac{p}{\delta}, \quad \frac{p}{x} \cong \eta \frac{V}{\delta^2}, \quad v_z x \cong V \delta, \tag{9.58}$$

where we restrict ourselves to the orders of magnitude of the various observables. These equations yield $\delta^2(x) \sim \kappa^{-1}x$; i.e., $4x_{\max} = \kappa b^2$. The total friction experienced by the plate on each side is

$$\frac{1}{2}f = \int \sigma''_{xz}\, dx \sim \eta \int_0^{x_{\max}} \frac{V}{\delta(x)}\, dx = \eta \kappa b V. \tag{9.59}$$

This force is balanced by the Peach and Koehler force $b\sigma_{zz}$; i.e., one finds that the mobility of the dislocation $m = \frac{V}{\sigma_{zz}} = \frac{\kappa^{-1}}{2\eta}$ reads as

$$m \sim \lambda_p \kappa \sim (\lambda_p \eta^{-1})^{1/2} \sim \frac{\mathcal{D}_{\|}\omega_m}{k_B T}\frac{1}{d_0}. \tag{9.60}$$

This expression of the mobility has been used to measure the coefficient of permeation (Problem 9.6), from homeotropic samples in Grandjean–Cano wedges, either measuring the time of annealing of the sample from the most disordered state to the state with a unique grain boundary[15] or measuring the relaxation time τ_c of the grain boundary moving under an applied stress σ'_{zz} (see (9.66) below). A typical value is $\lambda_p \sim 10^{-13}\,\mathrm{cm}^2/\mathrm{poise}$ in a thermotropic SmA, in which $\eta \sim 1$ poise; this value is in agreement with the value of $\mathcal{D}_{\|}$ measured from NMR or neutron quasielastic scattering.[16]

The above expression of the mobility is independent of the Burgers vector b. Note that Darcy's law [the first relation in (9.58)] is obtained from the z-component of the Navier–Stokes equations under the assumption that the contribution of the viscous stresses is small compared with the contribution of the pressure gradient $\frac{\partial p}{\partial z}$. A little algebra shows that this assumption is equivalent to the inequality $\delta < x$; i.e., $x > \kappa^{-1}$. Therefore, this calculation is not valid near the core of the dislocation. But this region is precisely where the diffusion of molecules away from the dislocation takes place, and this is the largest effect to be considered in a metallurgical model of the mobility of the defect. Let v_m be the velocity of the molecules diffusing from one layer to the next. According to Einstein's relation (8.53), v_m can be written as

$$v_m = \frac{\mathcal{D}_{\|}f_m}{k_B T}, \tag{9.61}$$

---

[15]W.K. Chan and W.W. Webb, J. de Physique **42**, 1007 (1981).
[16]G.T. Krüger, Phys. Rep. **8**, 231 (1982).

where f_m is the force exerted on the molecules due to the acting stress σ_{zz}. We have $f_m \sim \sigma_{zz} a^2$. Since the velocity V of the dislocation is related to v_m by the relation of conservation $Vb = v_m a$, one gets:

$$m = \frac{V}{\sigma_{zz}} \cong \frac{a}{b} \frac{\mathcal{D}_{\parallel} a^2}{k_B T} = \frac{a}{b} \frac{\mathcal{D}_{\parallel}}{k_B T} \frac{\omega_m}{d_0}. \tag{9.62}$$

This expression is akin to (9.60), except for the presence of the factor $\frac{a}{b}$, which takes into account the Burgers vector. The difference is unphysical for small Burgers's vectors, but the effect of the core could be considerable for large b's.

9.5.1.2. Glide

As already stated, glide is more difficult than is climb. Elementary edge dislocations involve a change in the nature of the core, from the configuration of Fig. 9.17a to that of Fig. 9.17b, one of them being energetically more favorable. Glide needs breaking of layers, whereas climb requires diffusion or permeation, two processes that are easy in a liquid. The energy involved in the breaking of a layer in a lamellar phase can be estimated in a thermotropic phase by an argument similar to Frenkel's (see Chapter 8, and J. Friedel,

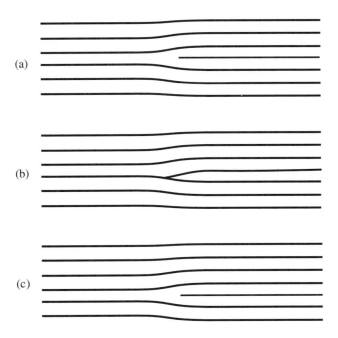

Figure 9.17. (a), (c) Dislocation in a well of potential; (b) intermediary state.

"Dislocations" Chapter 5) for the calculation of the theoretical shear stress in solids. This leads to $\sigma_F \sim K \frac{d_0}{a^3}$ and to a Peierls–Nabarro activation energy of the order of $K \frac{d_0^2}{a^2}$ per unit length of line. The condition for "fast" glide, i.e., without any thermal activation, would require, therefore, a Peach and Koehler force acting on the dislocation that should be at least of the order of F_c, with $F_c = K \frac{d_0}{a^2} \approx B d_0$.

The foregoing considerations assume that the line is moving entirely parallel to itself. But glide of dislocations of small Burgers vector can also proceed by activated jumps between equivalent configurations of lower energy, along finite segments of the dislocation line (kinks of screw character). Edge dislocations with large Burgers vectors, whose core is split into two disclinations, should not be mobile, except under relatively high stresses: Their motion would indeed involve the motion of disclinations, which cannot proceed without the creation and annihilation of elementary edge dislocations, i.e., in the present case, with the exchange of such edge dislocations between the two disclinations. Finally, in lyotropics, glide could be the result of complex interactions between membranes, involving the formation of passages under stress.

9.5.2. Screw Dislocation

The velocity field and the layer structure of a screw dislocation oriented along the z-axis, say, and moving with constant velocity perpendicular to the screw axis have been calculated by solving the hydrodynamic equations.[17] According to Pleiner, such a motion does not involve diffusion of molecules, as long as the dislocation does not oscillate. Furthermore, permeation is negligible: The molecules flow inside the layers, taking a small z-component velocity due to the screw geometry. The motion is opposed by (a) a drag force F_x^{dr} of frictional origin, similar to the force exerted on a solid cylinder of radius r_c (the core radius) in a simple fluid, given by Oseen formula below, and (b) a friction force F_x^{sc} specific to the screw layer structure:

$$F_x^{dr} = \frac{4\pi \eta_\perp V}{\frac{1}{2} - \gamma - \ln \frac{\text{Re}}{4}}, \tag{9.63a}$$

$$F_x^{sc} = \frac{1}{4\pi^2} \eta V \frac{b^2}{r_c^2}, \tag{9.63b}$$

where $\gamma = 0.577\ldots$ is the Euler's constant and $\text{Re} = \rho \frac{r_c V}{\eta_\perp}$ is the Reynolds number of the flow. The Oseen formula (and the whole theory of the motion of a screw dislocation) is valid for small Reynolds number, i.e., for rather small velocities $V < \frac{\eta_\perp r_c}{\rho}$. The hydrodynamic core value r_c is believed to be larger than the core radius at rest.

A screw dislocation in a shear flow $\sigma_{zy}'' = \eta \dot{\gamma}$, with the shear rate $\dot{\gamma}$, experiences a Peach and Koehler force $\mathbf{F}^{PK} = (b \sigma_{zy}'', 0, 0) = (b \eta \dot{\gamma}, 0, 0)$, which is perpendicular to

---

[17]H. Pleiner, Phil. Mag. A **54**, 421 (1986).

the shear plane $0yz$. It, therefore, acquires in the direction x a velocity V obtained by balancing F_x^{PK} and $F_x = F_x^{sc} + F_x^{dr}$. This velocity is small compared with any velocity $v_y \sim \dot{\gamma} d$ related to shear, where d is a characteristic size of the sample. In effect, the dislocation is also dragged along by the shear, while keeping anchored by its extremities to the boundaries of the sample.

9.6. Collective Behavior of Dislocations and Instabilities

9.6.1. General Remarks

A few experiments, some of them detailed below, have put into evidence climb of edge dislocations in SmA's and the role of screw dislocations. Phenomena remain simple as long as (a) one starts from well-oriented samples, and (b) stresses are not too large (compared with $B\frac{d_0}{L}$, say, where L is a characteristic macroscopic size of the sample). At higher fast stresses, instabilities are often observed. For example, the linear undulations studied in Section 5.5.1 transform into edge dislocations, whose motion by climb relaxes the applied stresses.[18] And at larger stresses, focal conics domains appear, sometimes in the shape of regular patterns, most frequently forming irregular arrangements; they will be discussed in the next chapter.

Collective glide of edge dislocations has been observed in a few cases, in which it yields large instabilities and the formation of focal conic domains (see next chapter).

SmB's are 3D-ordered solids, characterized by a weak coupling between layers and henceforth by a strong anisotropy of the viscoelastic constants. Their plasticity properties can be discussed to a large extent in terms of a pure metallurgical concept, at least in what concerns simple geometries. For example, activated vacancies diffusion has been invoked to explain creep under compressive stresses σ_{zz} in the SmB phase of butyloxybenzilidene aniline (also called 40.8). Under a constant shear-stress σ applied parallel to the layers, the same SmB also displays stationary creep[19]

$$\dot{\gamma} = \dot{\gamma}_o \exp{-\frac{U}{k_B T}} \cdot \exp{\frac{\sigma \Omega}{k_B T}}, \qquad (9.64)$$

which has been interpreted as due to the activated glide of the basal dislocation lines (parallel to the layers, and whose Burgers vectors δ_o belong to the 2D hexagonal lattice of the layers—δ_o is the lattice parameter within the SmB layers), crossing the "forest" of the dislocation lines, which are perpendicular to the layers and whose Burgers vector major component is along the normal to the layers. U is the activation energy necessary to cut a "tree," and $\sigma \Omega$ is the work of the stress in one elementary jump of the basal dislocations

[18]R. Bartolino and G. Durand, Phys. Rev. Lett **39**, 1346 (1977).

[19]P. Oswald, J. Physique **46**, 1255 (1985).

from one tree to the next. The activation volume Ω is large ($\sim 10^5$ molecular volumes). It is equal to $\Omega = L\,\delta_0\,d_0$, where L is the average distance between two neighboring trees). However, because stacking faults are frequently observed in SmB's, it may be reasonable to assume that the basal dislocations are split into partials, separated by a distance ℓ_0. All of those considerations are typical of a (2D) solid. Furthermore, the whole topic of plastic deformation in "ordered" smectics is still in its infancy, and we shall not insist on these considerations.

9.6.2. Collective Climb of Dislocations in SmA[20]

We consider a Grandjean–Cano wedge of angle α (Section 8.5) with homeotropic boundary conditions. A tilt boundary sits in the midplane of the wedge made of edge dislocations parallel to the wedge, separated by a distance $\ell = \frac{b}{\alpha}$. We shall assume that the dislocations are elementary ($b = d_0$) and that the wedge is perfect in the sense that no other defects are present. Under the action of a compressive stress $\sigma_0 = \sigma_{33}$, the dislocations climb cooperatively to relax the stress, rightward or leftward, according to the direction along which the layers are removed or added. Let x be the displacement of a dislocation. One has

$$\sigma = \sigma_0 - \frac{B}{L}\alpha x. \tag{9.65}$$

Here, L is the thickness of the sample. Because α is so small (typically, 10^{-3} rad), one can neglect the thickness variation of the sample. Let $v = m\sigma$ be the velocity of the dislocation: $v = \frac{dx}{dt}$. Hence,

$$\sigma(t) = \sigma_0 \exp - t/\tau_c, \quad \tau_c = L/mB\alpha. \tag{9.66}$$

These equations are valid as long as the stresses are small enough and the dislocations move parallel one to another. The experiments yield $B(= \frac{L}{d_0}\sigma_0)$ and m; i.e., either the coefficient of self-diffusivity or the coefficient of permeation. The temperature dependence of $m \sim \exp - U/k_B T$ yields the activation energy for the self-diffusivity.

At high enough stresses, above some yield stress σ_c[21], typically, of the order of $10^{-5}B$, i.e. small, the sample shows an exponentially increasing deformation rate $\dot{\varepsilon} = \dot{\varepsilon}_0 \exp \frac{t}{\tau}$, where τ is no longer equal to τ_c but shows a complex behavior (see Section 9.6.3). The simplest interpretation of the exponential increase of $\dot{\varepsilon}$ with time is that the dislocation density, which is initially $\rho_0 = \alpha/b$, increases with time. One expects indeed, on simple geometrical grounds (Problem 9.7), that the deformation rate $\dot{\varepsilon}$ varies as

$$\dot{\varepsilon} = \rho b v. \tag{9.67}$$

[20]P. Oswald and M. Kleman, J. Physique Lett. **45**, L319 (1984).

[21]First put into evidence by R. Bartolino and G. Durand, Mol. Cryst. Liq. Cryst. **40**, 117 (1977).

This expression, which relates the strain rate to the density of moving dislocations and to their velocity, is known as an Orowan relation. Because $v = m\sigma$ and b is constant, it is only the increase of ρ that can explain the exponential behavior.

There have been no systematic studies of σ_c until now; it can be due to the nucleation of defects along the free boundaries.

9.6.3. Multiplication of Edge Dislocations

Experimentally, the relaxation time τ ($\dot{\varepsilon} = \dot{\varepsilon}_0 \exp t/\tau$) reported in the previous paragraph varies by successive jumps when the applied stress is increased. Let us call p the serial number of the jump. The jumps can be characterized by two quantities:

1. The number of layers $N_p = pN_1$ that are involved at jump p, N_p is proportional to p.

2. The relaxation time $\frac{1}{\tau} = \frac{1}{\tau_c} + \frac{1}{\tau_p}$, where $\tau_p \sim p^{-1}$.

These results have been explained on the basis of an instability of the screw dislocation lines, which take a helical shape under the action of a compressive (or dilative) strain $\gamma = \frac{\delta L}{L}$ (Fig. 9.18). The helical shape corresponds to the removal (in compression) or addition (in dilation) of an extra layer in the area πr^2 bound by the cylinder on which is inscribed the helical line of equation $z = \frac{1}{2\pi}\frac{L}{p}\theta$. The pitch is L/p, and there are p 2π-turns through the thickness L of the sample. The chirality of the helix depends on the signs of the strain γ and of the Burgers vector b. This instability has been analyzed in the static limit.[22]

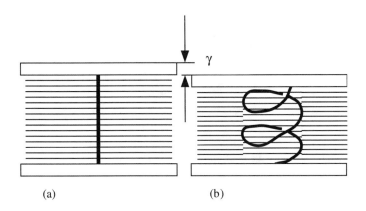

(a)   (b)

Figure 9.18. Helical instability of a screw dislocation line under stress.

[22]P. Oswald and M. Kleman, J. Physique Lett. **45**, L-319 (1984); L. Bourdon, M. Kleman, L. Lejcek, and D. Taupin, J. Physique **42**, 261 (1981).

Problem 9.1. Using the energy density (5.28) with the nonlinear term $(\partial u/\partial x)^2$ preserved, show that (9.8) describes the profile $u(x, z)$ of an edge dislocation of the Burgers vector b.

Answers: (Ref.2): Variation of the free energy

$$\frac{1}{2} \int \left\{ B \left[\partial u/\partial z - (\partial u/\partial x)^2/2 \right]^2 + K \left(\partial^2 u/\partial x^2 \right)^2 \right\}$$

leads to the Euler–Lagrange equation $\lambda \partial_x^4 u - \partial_z^2 u + 2(\partial_x^u)\partial_z \partial_x u + (\partial_x^2 u)\partial_z u - \frac{3}{2}(\partial_x^u)^2 \partial_x^2 u = 0$. Expressing all lengths in units of $\lambda = \sqrt{K/B}$ and assuming that the displacement field $u(x, z)$ depends only on one variable $v = x/\sqrt{z}$, one arrives at the differential equation for the strain field $\varphi = du/dv$:

$$\varphi''' = \frac{v^2}{4}\varphi' + \frac{3v}{4}\varphi + \varphi^2 + \frac{3v}{2}\varphi\varphi' + \frac{3}{2}\varphi^2\varphi'. \tag{9.P1}$$

Noticing that (9.P1) is satisfied when $\varphi' = -\frac{1}{2}(v\varphi + \varphi^2)$; one finds the general solution

$$\varphi = \frac{2\exp(-v^2/4)}{\int_{-\infty}^{v} \exp(-t^2/4)\,dt + C},$$

with the constant of integration C determined from the constraint $u(x \to \infty) - u(x \to -\infty) = \int_{-\infty}^{\infty} \varphi\,dv = b/2$. Restoring the original variables, one recovers (9.8).

Problem 9.2. Prove (9.15).

Answers: Consider a segment of screw dislocation line of length ℓ, of energy $\ell W(\alpha)_{\alpha=0}$, oriented along the z-axis. A fluctuation of this element brings it to a length $\ell + \delta\ell$ and an energy $\int_{z=0}^{z=\ell} W(\alpha)\,ds$, where $ds \approx dz(1 + \frac{1}{2}\alpha^2)$ i.e., $\delta\ell = \frac{1}{2}\int \alpha^2\,dz$. The mean value of the fluctuation is $\langle \alpha \rangle = 0$ i.e., $\int \alpha\,dz = 0$. By definition, the line tension is $\tau\,\delta\ell = \int_{z=0}^{z=\ell} W(\alpha)ds - \ell W(0)$. It suffices to expand $W(\alpha)$ in a Taylor series about $\alpha = 0$; the result

$$\tau = (W(\alpha) + \frac{d^2 W(\alpha)}{d\alpha^2})_{\alpha=0}$$

follows.

Problem 9.3. In a regime of pure permeation (i.e., $v_z = 0$), and assuming that the displacement $u(\mathbf{r}, t)$ of the layers is 1D ($\frac{\partial u}{\partial x} = \frac{\partial u}{\partial y} = 0$), show that $u(\mathbf{n}, t)$ obeys a simple diffusion law, with diffusivity $D_{per} = B\lambda_p$. What is the typical relaxation time attached to the permeation of a fluctuation of wavelength q^{-1}?

Answers: We have $\dot{u} = \lambda_p g$ according to (9.53), and $g = B\frac{\partial^2 u}{\partial z^2}$. Hence, $\dot{u} = \lambda_p B\frac{\partial^2 u}{\partial z^2}$. This expression has the form of an equation of diffusion for the layers displacement, with $D_{per} = B\lambda_p$. The characteristic time attached to permeation is $D_{per}^{-1}q^{-2} = \frac{1}{\lambda_p B q^2}$.

Problem 9.4. Prove that the undulation mode of the layers, reminiscent of the slow mode in nematics, relaxes with a frequency of $\omega = -i\frac{K}{\eta}q_\perp^2$.

Answers: Consider a mode $u = u_0 \exp i(qx - \omega t)$. Equations (9.53) and (9.50) can be written as $g = -Kq^4u = \lambda_p^{-1}(\dot{u} - v_z)$ and $\rho\dot{v}_z = -\eta q^2 v_z + g$. Neglecting the inertial term (small Reynolds number) and the higher powers of q (wavelengths large compared with microscopic scales), one gets, after elimination of v_z: $\eta\dot{u} + Kq^2 u = 0$.

Problem 9.5. Prove (9.66) (relaxation time).

Problem 9.6. Show that the mobility for climb of an edge dislocation of small Burgers vector is controlled by permeation.

Answers: We start from the model of dislocation mobility of Section 9.5.1. According to the result of Problem 9.3, the relaxation frequency for permeation is $\omega_{\mathrm{per}}(x) \sim \lambda_p B\frac{\kappa}{x}$, taking $q_{||}(x) \sim \frac{1}{\delta(x)}$. Now, the matter brought by permeation along the central layer (here symbolized by the plate of vanishing thickness) diffuses along this layer with a relaxation frequency $\omega_{\mathrm{diff}}(x) \sim \frac{\eta}{\rho x^2}$. This equation is obtained from the expression of the transverse shear mode (see Section 9.4), with $q_\perp(x) \propto \frac{1}{x}$. It is easy to see that employing reasonable experimental values of the constants (e.g., $\eta \approx 0.1$ poise, $\kappa^{-1} \approx 5 \times 10^{-7}$ cm, $B \approx 10^7$ dyne/cm^2), one gets $\omega_{\mathrm{diff}}/\omega_{\mathrm{per}} \gg 1$.

Problem 9.7. Prove the Orowan relation[23] (9.67), for glide and for climb.

Answers: Consider a parallelopipedic element of matter submitted to (a) a simple shear deformation $\dot{\varepsilon}_{xz} = V_x/L$ in the xy-plane (Fig. 9.19a) and (b) a compression (or dilation) along the z-direction $\dot{\varepsilon}_z = V_z/D$ (Fig. 9.19b), and reacting to these actions by the movement of dislocations only (plastic deformation),

by *glide* (dislocations of Burgers vector $\mathbf{b} = (b, 0, 0)$ along the y-direction, say). Each time a dislocation moves by a quantity Δx, the mean shear deformation of the element of matter changes

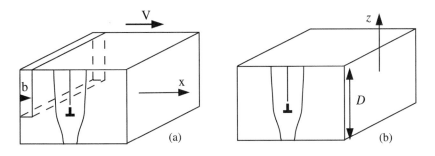

Figure 9.19. See Problem 9.7.

[23]E. Orowan, Nature **149**, 643 (1942).

by a quantity $\Delta\varepsilon_{\text{tot}} = \rho b\,\Delta x$. If $N = \rho DL$, such dislocations (of density ρ) have moved by the same distance during the same time, the mean total shear deformation is $\Delta\varepsilon_{\text{tot}} = \rho b\,\Delta x$, and the associated shear rate is $\dot{\varepsilon}_{xz} = d\,\Delta\varepsilon_{\text{tot}}/dt$. The Orowan formula is obtained in the case when the density of defects varies slowly.

by *climb* (dislocations of Burgers vector $\mathbf{b} = (b, 0, 0)$ along the y-direction, say, moving in the yz-plane). The displacement of the dislocations in the xy-plane requires the diffusion of molecules toward or away from the dislocations; these molecules are provided (or caught) either at long distances (e.g., at the boundaries) or by the opposite climb of another dislocation. The second case is certainly met in 3D solids, whereas it is the first case that is at work in the examples considered in lamellar phases. In both cases, the total quantity of matter is constant and the total dilation vanishes, as it should be in a pure shear deformation ($\dot{\varepsilon}_x + \dot{\varepsilon}_y + \dot{\varepsilon}_z = 0$). One gets eventually $\dot{\varepsilon}_z = \rho b v$ if the dislocations do not multiply.

Further Reading

J. Friedel, Dislocations, Pergamon, New York, 1954; J. de Physique **C3-40**, 45 (1979).

P. S. Pershan, J. Appl. Phys. **45**, 1590 (1974).

R. Holyst and P. Oswald, Int. J. Mod. Phys. **B9**, 1515 (1995).

J.-P. Poirier, Plasticité à haute température des solides cristallins, Eyrolles, Paris, 1976.

CHAPTER 10

Curvature Defects in Smectics and Columnar Phases

Both smectic A and columnar liquid crystals possess quantized symmetry translations (one such translation in SmAs, two in columnar phases). This is the reason why dislocations, the defects that break translational symmetries, are found in these phases. Defects involving other kinds of symmetry breaking might exist, such as disclinations or topological singular points, which break rotations.

Dislocations carry long-range strain fields, whose description (see the two preceding chapters) depends on the exact knowledge of the elastic constants and, thus, differs from one material to the other. However, there are cases for which the long-range distortions are practically material independent, for example when dislocations of the same sign gather into walls; this is well-known in 3D crystals (see Chapter 8), or in situations that imply only curvature. This latter case is all the more important in smectics and columnar phases as these phases are liquid-like (either in two dimensions or in one dimension). The layers of a SmA phase can take any shape at constant density and thickness, at a low cost in energy. Similarly, the columns of, say, a discotic hexagonal phase can bend at constant density. In both cases, we are interested in textures where the local structure is conserved; i.e., the layers or the columns preserve equispacing while curved or bent. Set as such, the problem acquires a pure geometrical nature and has solutions with 0D (points), 1D (lines), or 2D (walls) singularities. The solution for smectics has been known for more than 80 years now and consists in the so-called cofocal domains, whose 1D singularity set is a pair of conjugate conics.[1] Columnar phases have been investigated more recently: The (1D) singularity of the so-called developable domains is the cuspidal edge of a developable surface.

---

[1] G. Friedel, Annales de Physique Fr. **18**, 273 (1922).

10.1. Curvature in Solid Crystals

Curvature, as a typical mode of deformation, was first introduced for solid crystals.[2] This concept implies geometry, mostly. The Bravais lattice of a 3D crystal is built on a local tri-hedron of vectors $\mathbf{a}$, $\mathbf{b}$, $\mathbf{c}$ that are deformable in two ways: by changing the angles between the directions $\mathbf{a}$, $\mathbf{b}$, $\mathbf{c}$, and by allowing their lengths $|\mathbf{a}|$, $|\mathbf{b}|$, $|\mathbf{c}|$ to vary. This descrip-tion of the deformation can be cast into a strain tensor e_{ij}. Let us assume for the sake of simplicity that the local trihedron is rectangular, and then the components of e_{ij} in this tri-hedron are such that the diagonal components e_{aa}, ..., correspond to the relative variations in length of the vector $\mathbf{a}$, ..., and the off-diagonal components e_{ab}, ..., correspond to their angular variations. Such a description of the deformation is local; it also imposes, however, a small rotation from one trihedron to the next. Let us write this rotation as

$$d\omega_i = K_{ji}\, dx_j. \tag{10.1}$$

In this expression, K_{ji}, the so-called *tensor of contortion*, can be easily obtained from the antisymmetric part of the deformation tensor $\beta_{ji} = \partial u_i / \partial x_j$. This antisymmetric part writes indeed $\omega_1 = \frac{1}{2}(\partial u_2/\partial x_3 - \partial u_3/\partial x_2)$, and so on; i.e.,

$$\boldsymbol{\omega} = \tfrac{1}{2}\,\mathrm{curl}\mathbf{u}; \quad \omega_i = \tfrac{1}{2}\varepsilon_{ijk}u_{k,j}, \tag{10.2}$$

an expression that is akin to the vorticity of a fluid in motion (cf. Chapter 6). One has, therefore (Problem 10.1),

$$K_{ji} = \partial\omega_i/\partial x_j = \tfrac{1}{2}\varepsilon_{ik\ell}\beta_{j\ell,k} = \varepsilon_{ik\ell}e_{\ell j,k}. \tag{10.3}$$

We consider now a generalization of the contortion tensor, which can be easily under-stood in the following 2D example, where a family of parallel atomic layers is deformed in a process called single glide (Fig. 10.1). The result appears, at large scales, as a cur-vature of the layers. But at small scales, the rigidity of the lattice and the requirement of small elastic energy yield an array of edge dislocations, whose number (per unit area), is equal to $1/bR$, where R is the radius of curvature, and b is the Burgers vector modulus. In the limit $b \to 0$ for Burgers vector, this number becomes infinite and the strain e_{ij} is vanishing. The quantity $\alpha = 1/R$ is called a *dislocation density*; it is also the only relevant component $K = K_{33} = 1/R$ of the contortion tensor for this process. It is noteworthy that the dislocations are all of the same sign; again, let us stress that the global effect is geometrical and does not imply any knowledge of the material constants. Also, notice that $K = 1/R$ becomes infinite on the focal surface F of the set of parallel layers. This central region features a disclination, because the normals to the atomic layers rotate about F. The whole geometry can also be thought of as an extended grain boundary.

---

[2]J.F. Nye, Acta Met. **1**, 153 (1953).

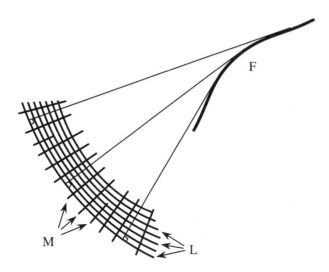

Figure 10.1. The layers (or lattice planes) L are those that suffer curvature; the M layers keep straight but are bounded by dislocations with Burgers vectors parallel to the L layers. F is the singular focal surface.

This 2D analysis can be extended to 3D.[3] The generalized contortion tensor in the presence of a dislocation density reads as

$$K_{ji} = \varepsilon_{ik\ell}e_{\ell j,k} - \alpha_{ij} + \tfrac{1}{2}(\alpha_{11} + \alpha_{22} + \alpha_{33})\delta_{ij}, \tag{10.4}$$

where α_{ij} is the density of dislocations parallel to the i direction, whose Burgers vector is along the j direction. Two sources of lattice curvature appear clearly in (10.4). One is of elastic origin and does not interest us. The second one is a geometric effect due to dislocations and can be called Nye's curvature (Problem 10.2).

10.2. Curvature in Liquid Crystals: Some General Remarks

This discussion of curvature in solids can be extended to liquid crystals. It is useful to distinguish two types of Burgers vectors in a liquid crystal: finite Burgers vectors relating to discrete translational symmetries and infinitesimal Burgers vectors relating to continuous translations.

---

[3]B.A. Bilby, L.R.T. Gardner, and E. Smith, Acta Met. **6**, L9 (1958). E. Kröner, Kontinuums Theorie der Versetzungen und Eigenspannungen, Springer, New York (1958).

As we already know, finite dislocations exist in smectics and columnar phases. Therefore, as in solid crystals, one expects that 3D densities of dislocations of finite strength are unstable with respect to polygonization, i.e., with respect to the formation of tilt (or twist, or mixed) boundaries in which dislocations of the same sign gather and through which the layers rotate abruptly by an angle proportional to the dislocation content of the wall (see Chapter 8 and Section 10.8).

Dislocations of vanishingly small Burgers vectors are proper to liquid crystals. It has been shown[4] that the distortions of the director field in a nematic can be described in terms of a pure contortion tensor K_{ji}, with $e_{lj} \equiv 0$ (the latter condition ensures constant mass density, and no elastic energy attached to the dislocation densities themselves), i.e., with the language of infinitesimal dislocations. Although the recourse to this language is not of absolute necessity, it has the advantage of establishing a link between the distortion field of an anisotropic *directional* liquid and the curvature field in a solid. We shall come back to the case of nematics (and of cholesterics) in the next chapter, in which we shall discuss the relation between infinitesimal dislocations densities and disclinations. In what concerns smectics and columnar phases, the dislocations in question have Burgers vectors along the layers (smectics), or along the rods (columnar phases). Therefore, they are representative of the bending of layers or of rods. Note that fluid relaxation, which results in changes of curvature of the director field at constant mass density, can be described by a variation in the contortion field, i.e., as a movement of infinitesimal dislocations. Pure contortion ($e_{lj} \equiv 0$), which is precisely the subject of this chapter, does not exist in 3D solids.

10.3. Curvature in Smectics

10.3.1. Historical Remarks

At the beginning of this century, Georges Friedel inferred from optical microscope observations of the since-called Smectic-A (SmA) phases that these phases were lamellar. Diffraction methods were not yet invented, and the discovery of the lamellarity was at that time based only on the visible geometrical properties of SmA line singularities. The singularities appear as pairs of conics, an ellipse E and a branch of hyperbola H, situated in two orthogonal planes in such a way that the apices of any one of them is at the foci of the other.[5] G. Friedel's conclusion rested on the remarkable guess that the observed conics were *focal lines*, i.e., that they constituted the geometrical locus of the centers of curvature of a family of parallel surfaces disposed along the physical layers. The (limited) region of space where the bending lamellae are ascribed to the presence of the pair of conics is called a "cofocal domain" (also called a focal conic domain, in short, FCD).

The subject of FCDs, which is still attracting a lot of interest, especially in lyotropic phases, is probably the oldest subject in the physics of defects. Let us remark that one of

---

[4]M. Kléman, Phil. Mag. **27**, 1057 (1973), J. de Phys. **34**, 931 (1973).
[5]G. Friedel and F. Grandjean, Bull. Soc. Franc. Minér. **33**, 192, 409 (1910).

the main characteristics of the theory of defects, namely, that the full knowledge of the defects belonging to an ordered medium is enough to understand the nature (scalar, vectorial, etc.) of its "order parameter," was already visible to its founders; but the importance of this discovery was not really appreciated before the topological theory of defects (see Chapter 12). SmA is certainly not the only medium where the knowledge of FCDs is of great use. Similar geometries occur in other smectic phases, such as SmC, SmC*[6] or even ordered smectics, and in vesicles.[7]

Our subject has been treated a number of times in excellent review articles; apart from the historical and still fundamental article of G. Friedel, it is worth mentioning the Bragg's[8] paper and the paper of Bouligand,[9] who revisited the subject and contributed to its renewal. Since then, there have been a number of experimental and theoretical works; this outburst of a subject of such an old seating, 80 years after their discovery, is not without some special gusto.

Paragraph 10.3.2 is a direct introduction to Section 10.4, which is focused on FCDs. Section 10.3 embraces general properties of bundles of normals, not only the case of normals to families of parallel physical surfaces, which occur in FCDs.

10.3.2. Congruences of Straight Normals and Focal Conic Domains

The normals to a set of parallel surfaces are straight lines and form a *"congruence" of straight normals*; i.e., the normals bundle depends on two parameters, for example, the two coordinates of a point on a fixed, but arbitrarily chosen surface of the set. The only gradient of the director **n** (along the normals) that is different from zero is div **n**. As it is well known, the same situation occurs in geometrical optics: The straight normals are akin to the light rays in a medium of constant refraction index, and the surfaces are akin to the wavefronts—this congruence envelops two *focal surfaces* F_1 and F_2 on which the layer's curvature becomes infinite, and consequently, the associated energy density $f = \frac{1}{2}K(\sigma_1 + \sigma_2)^2 + \overline{K}\sigma_1\sigma_2$ becomes infinite. Here, K and $\overline{K}$ are the splay and the saddle-splay elastic constants, respectively. The energy is certainly decreased if F_1 and F_2 degenerate into *lines*. Now it results from *Dupin theorem* (see Darboux, 1954) that in such a case, one of the focal lines (F_1, say) is an ellipse and the other one is an hyperbola whose foci are the apices of F_1, and that is located in a plane perpendicular to F_1. Conversely, the foci of F_1 are the apices of F_2. The layers are curved into Dupin cyclides, i.e., surfaces whose lines of curvature are circles. The conjugate ellipse and hyperbola are easily observed in polarizing light microscopy (Fig. 10.2), and the domain (the FCD) they carry can be made visible by confocal microscopy.

Experimentally, the eccentricity of the ellipse takes a broad range of values $0 \leq e < 1$, depending on material parameters, boundary conditions, and the geometry of the system.

---

[6]L. Bourdon, M. Kleman and J. Sommeria, J. de Phys. **43**, 77 (1982).

[7]See for example B. Fourcade, M. Mutz, and D. Bensimon, Phys. Rev. Lett. **68**, 2551 (1992).

[8]W. Bragg, Trans. Faraday Soc. **29**, 1056 (1933).

[9]Y. Bouligand, J. Physique **33**, 525 (1972).

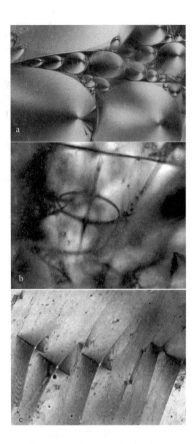

Figure 10.2. Textures of cofocal ellipses and hyperbolae in thermotropic smectic samples observed (a) in the plane of ellipses; (b) in a tilted plane; (c) in the plane of hyperbolas.

A particular case corresponding to an ellipse of zero eccentricity, $e = 0$, is a pair of a circle and a straight line. This domain is called a toric FCD, or TFCD. The focal conics may also form around a pair of parabolae;[10] however, the parabolic FCD (PFCD) is *not* the limiting case $e = 1$ of an ellipse-hyperbola FCD, and it has different elastic features (see Section 10.4.3).

One might wonder why the most usually observed focal domains are not those in which the singularities are still further reduced, i.e., to one point (in such a case, the layers are concentric spheres). These "spherulites," as a special type of focal domain, are indeed observed when the saddle-splay coefficient $\overline{K}$ is of favorable value (see (10.34)).

---

[10]C.S. Rosenblatt, R. Pindak, N.A. Clark and R.B. Meyer, J. Physique **38**, 1105 (1977).

The purpose of this chapter is to give a complete review of the analytical basis of the subject, supported by the description of a number of physical situations that have been recently studied. However, we shall not prove Dupin's theorem nor the cyclide theorem; the interested reader is referred to standard books of geometry (Darboux, Hilbert and Cohn Vossen).

10.3.3. Congruences of Normals, Variations of Perfect Focal Conic Domains

Small variations of perfect cofocal domains, made of a congruence of straight normals, can be analyzed using the concepts of *congruence of normals* and of *congruence of straight lines*. In a congruence of normals, the molecules no longer align along straight lines and the layers are no longer parallel; i.e., $\mathbf{n} \times \text{curl}\mathbf{n} \neq 0, \mathbf{n} \cdot \text{curl}\mathbf{n} = 0$. In a congruence of straight lines, the notion of layers disappear; i.e., $\mathbf{n} \cdot \text{curl}\mathbf{n} \neq 0, \mathbf{n} \times \text{curl}\mathbf{n} = 0$. In both cases, energetical considerations would favor F_1 and F_2 still being lines, but there is no geometrical obstruction for them being surfaces. Mixed situations can prevail.

A congruence of normals is elastically distorted, but this distortion can be relaxed by the appearance of a number of quantized edge dislocations (all of the same sign, Fig. 10.3a). This situation occurs in cholesterics,[11] where it is observed that dislocations frequently polygonize. The variation in shape of F_1 and F_2 with respect to perfect conics allows for an easier space filling than with perfect FCDs (see below).

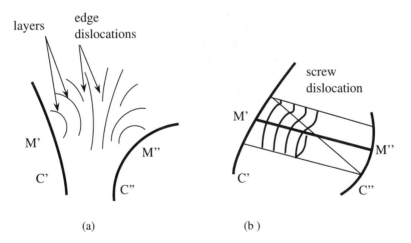

(a)   (b)

Figure 10.3. (a) Edge dislocations in a focal domain; C_1 and C_2 are elements of focal lines. (b) A typical situation that must show screw dislocations.

[11] Y. Bouligand, J. Physique **38**, 1011 (1973).

It is possible to reintroduce the physical notion of layers in a congruence of straight lines: The quantity $\mathbf{n} \cdot \mathrm{curl}\,\mathbf{n}$ can be analyzed as a density of quantized screw dislocations imposed on a set of layers, all of the same sign (Fig. 10.3b). If they polygonize, one gets a situation akin to a twist grain boundary. However, they do not necessarily polygonize, because the layers in their close vicinity are, satisfactorily, curved into minimal surfaces (hence, the curvature energy vanishes) and because their elastic energy is so small.

The giant screw dislocations observed in thermotropic SmAs[12] are a remarkable example of a topology where the layers keep at constant distances one from the other but the focal sets are surfaces. However, the visible singularities are still lines, made of a pair of helices that are the cuspidal edges of the evolutes of a helicoid generated by a straight line (the concept of "virtual surface" is introduced for that purpose; see Chapter 9). All layers are parallel to this helicoid, whose pitch is equal to the Burgers vector of the dislocation (Fig. 9.9). The physical part of the focal sheets is reduced to a line, the rest being "virtual." The phenomenon originates in the fact that some regions of space are multicovered by the surfaces Σ_M parallel to the helicoid and thus, a choice has to be made to eliminate some parts of the Σ_M's that are close to their evolutes. Other focal domains with virtual focal sheets have been imagined more[13] recently, in relation with oblique boundary conditions met in some special cases; the same problem of the boundary conditions could also lead to focal domains with "canal" surfaces[14] (when only one of the focal surfaces is degenerate to a line, and the other one is partly virtual or not virtual).

Darboux theorem is common to congruence of straight lines and congruence of straight normals, but it does not apply to congruence of normals. It states that given a congruence of straight lines D, two planes tangent to the focal surfaces F_1 and F_2 along any line D are orthogonal. G. Friedel noticed that in SmAs, the focal lines pertaining to the same cofocal domain and observed by optical microscopy always cut in projection at right angles and concluded that the molecular alignments obey Darboux theorem. This situation is not well observed in cholesterics, where the perfect geometry of FCDs is considerably distorted.

We have been describing above the (small) variations to cofocal conics in terms of (quantized) dislocation densities (in the sense of Nye). Similarly, one could use (unquantized) dislocation densities in order to describe the curvature in a perfect cofocal domain, by introducing a suitable contortion tensor K_{ij}. However, this procedure is not straightforward here because the curvatures are large, and the above theory is valid only for small curvatures. For an extension of the contortion tensor to large curvatures, see Bilby.[15] Let us remark that our "empirical" description of defects paves the way toward the understanding of more abstract gauge theories, in which dislocation densities are given the status of gauge field densities.[16]

---

[12]C. Williams, Phil. Mag. **32**, 313 (1975).

[13]J.B. Fournier, Phys. Rev. Lett. **70**, 1445 (1993).

[14]J. Sethna and M. Kleman, Phys. Rev. **A26**, 3037 (1982).

[15]B.A. Bilby, Progr. Solid Mech., **1**, 329 (1960).

[16]L. Dzyaloshinski and G. Volovik, J. Physique **39**, 493 (1978). B. Julia and G. Toulouse, J. Physique Lett. **40**, L-395 (1979).

10.4. Focal Conic Domains

10.4.1. The Analytical Approach: Basic Formulae

In an FCD, the layers are folded around two conjugated lines, viz., an ellipse E and one branch H of a hyperbola, in such a way that they are everywhere perpendicular to the straight lines joining any point M′ on the ellipse to any point M″ on the hyperbola (Fig. 10.4). Any point M on the line M′M″ is the orthogonal intersection with this line of a uniquely defined surface (a "layer") Σ_M, perpendicular everywhere to the two-parameter family (the congruence) of lines M′M″. All parallel surfaces Σ_M orthogonal to M′M″ have the same centers of curvature, M′ and M″. The curvatures $|\sigma'| = 1/M''M$ or $|\sigma''| = 1/M''M$ become infinitely large when M approaches either M′ or M″. Correspondingly, the Σ_M's are singular on M′ and M″, where the energy density grows without limit. We shall prove that each layer is the (common) envelope of a set of spheres centered on E or H, and it is characterized by a scalar parameter r.

Let E and H be a set of mutually cofocal ellipse and hyperbola, located in two perpendicular planes, and let the equation of the ellipse be, in standard notations,

$$z = 0, \qquad \frac{x^2}{a^2} + \frac{y^2}{b^2} = 1 \tag{10.5a}$$

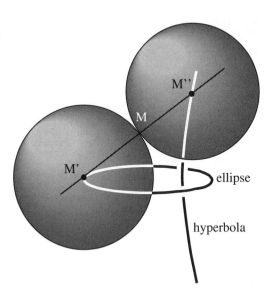

Figure 10.4. General geometrical aspect of an FCD. The line M′M″ is the direction perpendicular to the layer at point M.

(the semiminor axis-length b should not be confused with a Burgers vector). The equation of its cofocal hyperbola H then reads as

$$y = 0, \qquad \frac{x^2}{a^2 - b^2} - \frac{z^2}{b^2} = 1. \tag{10.5b}$$

Let M$'(x', y', 0)$ be a point on E and M$''(x'', 0, z'')$ be a point on H, and let us parameterize the conics in the usual way:

$$\text{M}' \begin{cases} x' = a \cos u, \\ y' = b \sin u, \end{cases} \qquad 0 \le u < 2\pi,$$

$$\text{M}'' \begin{cases} x'' = \pm c \cosh v, \\ z'' = b \sinh v, \end{cases} \qquad -\infty \le v \le \infty,$$

where $c^2 = a^2 - b^2$. The $\pm$ signs refer to one or the other branch of the hyperbola. In the sequel, we adopt the $+$ sign. The length M$'$M$''$ is

$$\text{M}'\text{M}'' = \Delta = |a \cosh v - c \cos u| = |ex' - e^{-1}x''|, \tag{10.6}$$

where $e = c/a$ is the eccentricity of the ellipse.

Let us now consider a point M on the line M$'$M$''$ and introduce the signed lengths $r' = $ MM$'$, $r'' = $ MM$''$. For the sake of clarity, we shall always orient the line from M$'$ to M$''$, i.e. from E to H, so that

$$\Delta = r'' - r'.$$

Now, we introduce the parameter r, defined for M$'$M$''$ oriented as above, as

$$r' = ex' - r; \quad r'' = e^{-1}x'' - r. \tag{10.7}$$

To each value of r is attached a surface $\Sigma(r)$, and it is easy to show that M$'$M$''$ is perpendicular to $\Sigma(r)$ at M. Let indeed

$$S_r(\text{M}') \equiv (X - x')^2 + (Y - y')^2 + Z^2 - r'^2 = 0 \tag{10.8a}$$

be the equation of a sphere centered in M$'$ and passing through M, and similarly,

$$S_r(\text{M}'') \equiv (X - x'')^2 + Y^2 + (Z - z'')^2 - r''^2 = 0 \tag{10.8b}$$

be the equation of a sphere centered in M″ and passing through M. The set of spheres $S_r(M')$ envelop a surface that is precisely $\Sigma(r)$. Indeed, the point of contact of $S_r(M')$ with its envelope belongs at the same time to $S_r(M')$ and to the derived surface:

$$\frac{dS_r(M')}{dM'} \equiv \frac{dS_r}{dx'} = \frac{\partial S_r}{\partial x'} + \frac{\partial y'}{\partial x'}\frac{\partial S_r}{\partial y'} = y'X + (e^2 - 1)x'Y - ey'r = 0, \qquad (10.9)$$

and it is a matter of simple algebra to show that M belongs to the intersection of $S_r(M')$ and $\frac{dS_r(M')}{dM'}$. M obviously also belongs to the intersection of $S_r(M'')$ and $\frac{dS_r(M'')}{dM''}$. Therefore, $\Sigma(r)$ is the locus of the common envelope to $S_r(M')$ and $S_r(M'')$, and M′ and M″ are its centers of curvature at M. $\Sigma(r)$ is a surface that is everywhere perpendicular to the line joining any point M′ on E to any point M″ on H: It is therefore the surface we are looking for. Also, $\Sigma(r)$ is a cyclide (i.e., its lines of curvature are circles), because the complete contact of each $S_r(M')$ [respectively, $S_r(M'')$] with $\Sigma(r)$ is the circle $C_r(M')$ [respectively, $C_r(M'')$] along which $S_r(M')$ [respectively $S_r(M'')$] and $\frac{dS_r(M')}{dx'}$ [respectively $\frac{dS_r(M'')}{dx''}$] intersect.

Note also that M′ is the vertex of a cone of revolution whose basis is $C_r(M')$ and that lies on H; reciprocally, M″ is the vertex of a cone of revolution whose basis is $C_r(M'')$ and that lies on E.

10.4.2. Different Species of Focal Conic Domains

Fig. 10.4 illustrated the case when the physical layer intersects the line M′M″ between M′ and M″. According to the theory of congruences of straight lines, the centers of curvature of the cyclide in M are M′ and M″, which are on the focal lines. In this illustration, the curvatures σ' and σ'' are of opposite signs, and the cyclide is therefore hyperbolic in M (saddle point), see Fig. 5.5b. Depending on the location of the point M and on the Gaussian curvature of layers, one can distinguish different types of FCDs (Fig. 10.5).

10.4.2.1. Focal Conic Domain of the First Species (FCD-I)

If the physical part of the layer is located between M′ and M″, one gets a *focal conic domain of the first species* (FCD-I) (Fig. 10.5a). This geometry yields $\sigma'\sigma'' < 0$. Dupin cyclides in a FCD-I show features varying with the position of M on the segment M′M″; either Σ_M ends on the ellipse on two point singularities (layer marked 1 in Fig. 10.5a), Σ_M is free of singularities and looks like a deformed half-torus (layer 2), or it ends on the hyperbola and has the form of a spheroid limited to its $\sigma'\sigma'' < 0$ part, with two conical indentations along the hyperbola (layer 3). One sees that in an FCD-I, the ellipse and the branch of hyperbola are both line defects and thus both visible.

Different projections of an FCD-I observed in polarizing microscopy obey *Darboux theorem*, which states that the orthogonal projections on any plane of a pair of conjugate

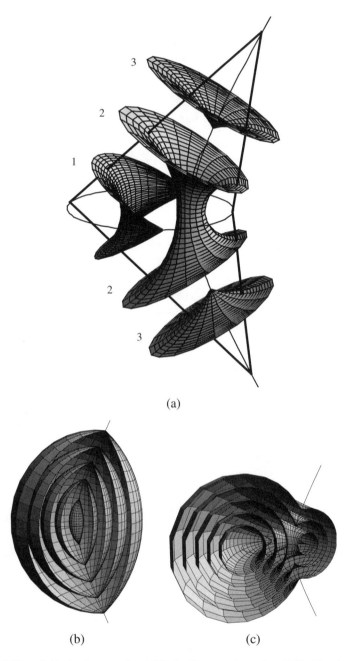

(a)

(b)   (c)

Figure 10.5. FCDs of (a) the first species (FCD-I); (b) second species (FCD-II); (c) third species (FCD-III).

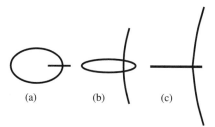

(a)   (b)   (c)

Figure 10.6. Darboux' theorem: projections of FCD-I observed in polarizing microscope: (a) in the plane of the ellipse; (b) in any plane; (c) in the plane of the hyperbola.

ellipse and hyperbola are at right angles; compare Figs. 10.2 and 10.6. The experimental observations of FCDs fit remarkably well with this property and bring, thus, a further proof to the Friedel's geometrical model.

Analytically, for FCD-I, $r' < 0$ and $r'' > 0$. Hence, r obeys the inequality

$$c \cos u < r < a \cosh v. \tag{10.10}$$

The whole set of r-values for the *complete* FCD-I (appearing in Fig. 10.5a) is in the range $[-c, +\infty[$. Special values of r correspond to the following situations:

$r = -c$; the Dupin cyclide is reduced to a point that is the apex of the ellipse opposite to the physical branch of the hyperbola.

$-c < r < +c$; the Dupin cyclide has two singular points on the ellipse (obtained for $r' = 0$) and none on the hyperbola, layer 1 in Fig. 10.5a.

$c < r < a$; the Dupin cyclide has no singular points; the complete cyclide, i.e., made of the $\sigma'\sigma'' < 0$ and $\sigma'\sigma'' > 0$ parts, is homotopic to a torus, layer 2 in Fig. 10.5a. The mean Dupin cyclide $r_0^2 = ac$ is very special: According to a conjecture due to Willmore, the *complete* cyclide r_0 (a toroidal surface) is an absolute minimum of the curvature energy $\frac{1}{2}\kappa \int (\sigma' + \sigma'')^2 \, d\Sigma$.

$r > a$; singular points on the hyperbola only, layer 3 in Fig. 10.5a.

10.4.2.2. Focal Conic Domain of the Second Species (FCD-II)

FCD-IIs are of positive Gaussian curvature, $r'r'' > 0$. A FCD-II is obtained when the physical part of the normal to Σ_M is located along the half-line with origin in M′ (Fig. 10.5b). In that case, the cyclides look like rugby balls, with two conical cusps located on the hyperbola. The ellipse is now virtual, and only the hyperbola is visible under the microscope. The stacking of the cyclides can fill the entire space.

Analytically, we have $r' < 0$ and $r'' < 0$, because the point M is located outside of M′M″ and the directions of MM′ and MM″ are opposite to the direction of M′M″. Therefore, along a given normal (u, v), the following inequality holds:

$$r > a \cosh v, \tag{10.11}$$

and the whole set of r-values for the complete FCD-II domain is in the range $[a, +\infty]$. For $r = a$, the Dupin cyclide is reduced to the apex F of the branch of H, which is physical.

10.4.2.3. Focal Conic Domain of the Third Species (FCD-III)

FCD-IIIs are formed by cyclides with both positive and negative Gaussian curvature (Fig. 10.5c).

10.4.2.4. Useful Formulae

In all three cases, the following formulae for principal curvatures defined through the reciprocal radii of curvature are useful:

$$\sigma' = \frac{1}{r'} = \frac{1}{c \cos u - r}; \quad \sigma'' = \frac{1}{r''} = \frac{1}{a \cosh v - r}. \tag{10.12}$$

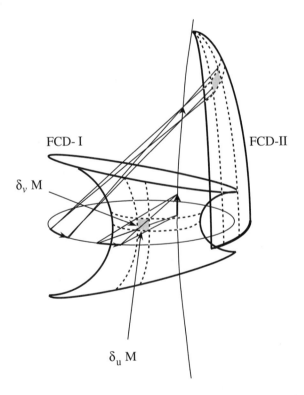

Figure 10.7. Infinitesimal elements of surface $d\Sigma$ of a Dupin cyclide in the case of FCD-I and FCD-II. $d\Sigma = |\delta_v\,||\,\delta_u\,|$.

Note also that the infinitesimal element of surface $d\Sigma(r) = AB\,du\,dv$ of the $\Sigma(r)$ cyclide can be expressed as the function of the principal curvatures (Fig. 10.7). Consider an infinitesimal variation δu of u along the line of principal curvature labeled by a constant v. M is displaced from M to $M + \delta_u M$, and M' from M' to $M' + \delta_u M'$; r' is modified, but M'' and r'' stay fixed. We have

$$A\,du = |\delta_u M| = \left| \delta_u \left(\frac{r'}{\Delta} M'M \right) \right| = \pm \frac{b\sigma''}{\sigma - \sigma''}\,du, \quad B\,dv = \pm \frac{b\sigma'}{\sigma' - \sigma''}\,dv.$$

Hence,

$$d\,\Sigma(r) = \frac{b^2 |\sigma'\sigma''|}{(\sigma' - \sigma'')^2}\,du\,dv. \tag{10.13}$$

10.5. Curvature Energy of FCDs

The curvature energy of an FCD is defined as the integral over the FCDs volume of the energy density f associated with the mean and Gaussian curvatures of layers:

$$f = \tfrac{1}{2}K(\sigma' + \sigma'')^2 + \overline{K}\sigma'\sigma''. \tag{10.14}$$

It is convenient to split the integral into two parts:

$$W = \int f\,d\Sigma\,dr = W_1 + W_2, \tag{10.15}$$

$$W_1 = \mp \tfrac{1}{2}Kb^2 \int \sigma'\sigma''\,du\,dv\,dr$$

$$= \mp \tfrac{1}{2}K(1 - e^2)a \int \frac{du\,dv\,d\rho}{(e\cos u - \rho)(\cosh v - \rho)}, \tag{10.16a}$$

$$W_2 = \mp(\overline{K} + 2K)b^2 \int \frac{\sigma'^2\sigma''^2\,du\,dv\,dr}{(\sigma' - \sigma'')^2}$$

$$= \mp\Lambda(1 - e^2)a \int \frac{du\,dv\,d\rho}{(\cosh v - e\cos u)^2}, \tag{10.16b}$$

where $\Lambda = \overline{K} + 2K$, $\rho = r/a$, e is the eccentricity, and a is the semimajor axis of the ellipse; the upper signs correspond to FCD-I, and the lower signs correspond to FCD-II.

Note that the K- and the $\overline{K}$-term both contribute to the "topology," because they appear in W_2, which is an integral of the Gauss–Bonnet type.

10.5.1. FCD-I: Negative Gaussian Curvature

The W_2 term can be easily integrated in the range $c \cos u < r < a \cosh v, 0 \leq u < 2\pi, -\infty \leq v \leq \infty$ and be given an exact form:

$$W_2 = -4\pi \Lambda a(1 - e^2)\mathbf{K}(e^2), \tag{10.17}$$

where $\mathbf{K}(x) = \int_0^1 \frac{dt}{\sqrt{(1-t^2)(1-xt^2)}}$ is the complete elliptic integral of the first kind. W_2 is negative when Λ is positive, a fact that is always ensured if $\overline{K} > -2K$. Note that for the free energy density (10.14) to be positive-definite for the lamellar phase, $\overline{K}$ must be within the range $-2K \leq \overline{K} < 0$, which also means $0 < \Lambda \leq 2K$ (K is always positive).

The W_1 term is singular near the ellipse and hyperbola, where $r \to c \cos u$ and $r \to a \cosh u$. The phenomenological elastic theory should not be applied in these regions, and one has to restrict the region of integration by a cutoff length, called the core radius. Assume that the core radius does not depend on the layer (i.e., does not depend on r):

$$r_{\text{cutoff}} = a \cosh v - r_c \qquad \text{near the hyperbola,}$$

$$r_{\text{cutoff}} = c \cos u + r_c \qquad \text{near the ellipse.}$$

This assumption is obviously greatly oversimplifying the situation. For example, it does not take into account that the layers that intersect the hyperbola far from the ellipse show practically no singularity. Furthermore, near the defect cores, the layers might suffer dilation; for a critical discussion, see Fournier.[17]

Integration over ρ splits W_1 into the singular $W_{1\text{-sing}}$ and nonsingular $W_{1\text{-non sing}}$ parts:

$$W_{1\text{-sing}} = 4\pi K a(1 - e^2)\mathbf{K}(e^2) \ln \frac{a}{r_c}, \tag{10.18a}$$

$$W_{1\text{-non sing}} = 4\pi K a(1 - e^2)\mathbf{K}(e^2) \ln \left(2\sqrt{1 - e^2}\right), \tag{10.18b}$$

where r_c is typically of the order of the repeat distance of the layers. A specific core energy, which cannot be calculated with (10.14) at hands, should be added: $W_{1\text{-sing}} \to W_{1\text{-sing}} + W_{\text{core}}$, where W_{core} is proportional to the length of the defect. We omit W_{core}. In some cases, this omission can be justified by the fact that the parameter r_c can be renormalized to adsorb W_{core} into $W_{1\text{-sing}}$.

---

[17] J.B. Fournier, Phys. Rev. **E 50**, 2868 (1993).

The total FCD-I curvature energy $W = W_{1\text{-non sing}} + W_{1\text{-sing}} + W_2$, expressed as the function of the semimajor axis a and the eccentricity $0 \le e < 1$, adopts a compact form:[18]

$$W = 4\pi a(1 - e^2)\mathbf{K}(e^2)\left[K \ln \frac{2a\sqrt{1 - e^2}}{r_c} - \Lambda\right]; \qquad (10.19)$$

notice that $a\sqrt{1 - e^2} = b$, where b is the semiminor axis; $\Lambda = 2K + \overline{K}$.

The dependence $W(\overline{K})$ is clear: The larger $\overline{K}$ is, the smaller the energy is; the reason is simply the negative sign of the Gaussian curvature of Dupin cyclides in an FCD-I. A further remark concerns the sum of the two nonsingular terms $W_{1\text{-nonsing}} + W_2$ at $a = \text{const}$. When Λ increases, the coordinate of the minimum of the sum shifts from $e \to 1$ to $e \to 0$ (Fig. 10.8). The tendency of $W_{1\text{-nonsing}} + W_2$ to reach a minimum at small eccentricity $e \to 0$ is, of course, in competition with the increase of $W_{1\text{-sing}}$ at $e \to 0$. Thus, the minimum of curvature energy can be achieved at e substantially different from 1 only when the domains are extremely small, $a/r_c \sim 10$, and when the saddle-splay constant is close to its upper limit $\overline{K} = 0$ set by the requirement of a positive-definite value of f in (10.14). Generally, for a reasonably large domain, $a/r_c > 10$, the curvature energy becomes minimum only at $e \to 1$. However, it would be a mistake to conclude that an FCD-I tends to increase its eccentricity as much as possible on the grounds of (10.19). In real samples, the FCDs are rarely isolated; their elastic energy is only a part of the total energy that includes the energy of surface anchoring, dislocations, layers compressions, and so on, as discussed below.

First, note that the plots in Fig. 10.8 correspond to $a = \text{const}$. The volumes of FCD-Is with identical a's but different e's are obviously different; an increase of e means a decrease of the semiminor axis b. Thus, the curvature energies of two FCDs with different e's should be compared under additional geometrical constrains. These constrains in concrete experimental situations involve the finite size of the system and, thus, require one to consider surface anchoring energies that are usually large in smectic phases.

The second reason that limits e relates to the fact that any FCD-I should match the layers in the adjacent regions. Because of their peculiar shape, FCD-Is cannot fill a bounded piece of space as isolated objects. They have to be embedded in the surrounding matrix of smectic layers, which might be flat or curved. As understood from Fig. 10.5a, when an FCD-I of small eccentricity is embedded into a system of flat layers, the tilt of smectic layers *inside* the FCD-I (with respect to the horizontal plane) requires a matching dislocation set outside of the FCD-I. The total Burgers vector of this set equals $2ae$. Thus, although the trend $e \to 1$ is favored by the curvature of layers *inside* the FCD-I, an opposite trend $e \to 0$ is favored by the line tension $\sim \sqrt{BK}ae$ of dislocations *outside* of the FCD-I. The interplay between FCDs and dislocations plays an important role in the structure of FCDs arrays, such as oily streaks and tilt grain boundaries, discussed later in this chapter. Overall, the problem of finding an equilibrium e requires consideration of dislocations, compressibility, and anchoring terms in addition to the curvature energy.

[18]M. Kleman and O.D. Lavrentovich, Phys. Rev. **E61**, 1574 (2000).

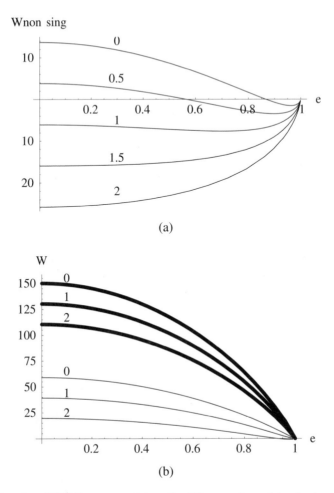

Figure 10.8. Energies of FCD-Is vs eccentricity e for different values of the elastic parameter Λ/K (indicated by numbers above the curves); normalized units with $K = 1$ and $a = 1$. (a) Nonsingular terms $W_{\text{non sing}} = W_{1-\text{non sing}} + W_2$ vs. e. (b) Total energy $W = W_{1-\text{sing}} + W_{1-\text{non sing}} + W_2$ vs. e for FCD-Is of a large ($a/r_c = 1000$, thick lines) and a small size ($a/r_c = 10$, thin lines).

10.5.2. Toric FCD with Negative Gaussian Curvature

The curvature energy (10.19) takes a very simple form for the toric FCD-I, or TFCD, in which the ellipse degenerates into a circle and the hyperbola is a straight line, $e = 0$:

$$W = 2\pi^2 a \left(K \ln \frac{a}{r_c} + K \ln 2 - \Lambda \right). \tag{10.20}$$

Notice that for $\Lambda/K = \ln 2 \approx 0.693$, the energy is reduced to its singular term. The core radius is invariant along the circle. For $\Lambda/K > 1 + \ln 2$, it is easy to see that W takes some minimum value for a particular solution of $\frac{\partial W}{\partial a} = 0$:

$$a \sim r_c \exp\left(\frac{\Lambda}{K} - \ln 2 - 1\right). \tag{10.21}$$

For $\Lambda/K < 1 + \ln 2$, the energy $W(a)$ does not have a physical minimum—because $\ln \frac{a}{r_c}$ cannot be negative—and the TFCDs are unstable defects, little mobile because of the lattice friction. This is probably the most general case, whereas the case $\Lambda/K > 1 + \ln 2$ should probably be indicative of a precritical or critical regime, near a phase transition.

10.5.3. Parabolic FCD with Negative Gaussian Curvature

This case has to be treated apart, because it does not follow analytically from the general case. We summarize the results.

Let P and Q be two running points on two cofocal parabolae (Fig. 10.9):

$$\mathrm{P}\begin{cases} x = 2f\alpha, \\ y = 0, \\ z = -\dfrac{f}{2} + f\alpha^2, \end{cases} \qquad \mathrm{Q}\begin{cases} x = 0, \\ y = 2f\beta, \\ z = \dfrac{f}{2} - f\beta^2. \end{cases} \tag{10.22}$$

Here, f is the semiparameter of the parabola; we hope there is no confusion with the same notation f for the curvature energy density. The distance PQ reads as

$$\mathrm{PQ} = f(1 + \alpha^2 + \beta^2), \tag{10.23}$$

and the radii of curvature (measured along the line oriented from P to Q) are

$$R_\alpha = \mathrm{PM} = \frac{1}{\sigma_\alpha} = f\alpha^2 + \frac{f}{2} - r,$$

$$R_\beta = -\mathrm{MQ} = \frac{1}{\sigma_\beta} = -f\beta^2 - \frac{f}{2} - r, \tag{10.24}$$

where $-(\beta^2 + \frac{1}{2}) < \frac{r}{f} < \alpha^2 + \frac{1}{2}$ for the PFCD of the first species (PFCD-I) with $\sigma_\alpha \sigma_\beta < 0$. Equalities are forbidden because they yield an infinite curvature; the excluded regions near P and Q are core regions.

Rosenblatt et al.,[10] who first observed PFCD-Is, have shown in a series of pictograms that an PFCD-I extends over all space: The layers are practically planar and perpendicular to the common axis of the parabolae, at long distance from the foci ($|r/f| \gg 1$). When

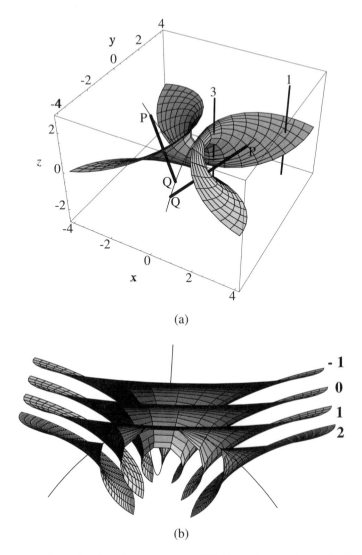

(a)

(b)

Figure 10.9. PFCD of negative Gaussian curvature (PFCD-I). (a) A part of a nonsingular symmetric layer ($r = 0$) wrapped around two parabolae. Lines PQ are normals to the layer. Line 1 crosses the layer once, whereas line 3 crosses the very same layer three times. (b) Four equidistant layers, including the symmetric layer labeled by "0."

$|r/f| > 1/2$, $\Sigma(r)$ cuts one of the parabolae in two points, where the curvature increases without limit. The layers with $|r/f| < 1/2$ have no point singularity: They do not cut the parabolae, but rather envelop them, forming handles; the Dupin cyclide $r = 0$ is symmetric with respect to the plane $z = 0$, with a twist of $\frac{\pi}{2}$ when passing from the region $z > 0$ to the region $z < 0$. Using the definition of the cyclides as envelopes of spheres (see (10.8) and (10.9) above for the general case), we find that the cyclides can be expressed analytically by the parametric equations:

$$\begin{cases} x = \alpha(p + r - z), \\ y = \beta(p - r + z), \\ z = \alpha x - \beta y - r. \end{cases} \tag{10.25}$$

The element of area $d\Sigma(r, \alpha, \beta)$ reads as

$$d\Sigma = \frac{4}{|\sigma_\alpha \sigma_\beta|} \frac{d\alpha \, d\beta}{(1 + \alpha^2 + \beta^2)^2}, \tag{10.26}$$

and the energy of an PFCD-I is of the form:

$$W = -4\Lambda \int \frac{d\alpha \, d\beta \, dr}{(1 + \alpha^2 + \beta^2)^2} + 2K \int d\alpha \, d\beta p^2 |\sigma_\alpha \sigma_\beta| \, dr. \tag{10.27}$$

The full expression of W will not be developed here. It suffices at this point to remark that:

- The topological term sums up to $-4\pi \Lambda p$ in the continuous limit ($p \gg d$) for the layers $|r/f| < 1/2$: Each layer has a topological contribution $\propto (-4\pi)$. For $|r/f| \geq 1/2$, the topological contribution diverges logarithmically with the size $2L$ of the PFCD-I measured along the z-axis, which scales (for L larger than f) like the maximum value of r. We eventually have

$$W_{\text{topol}} \approx -4\pi \Lambda f \left(1 + 2\ln \frac{2L}{f}\right). \tag{10.28}$$

- The K term yields a nonsingular, finite, contribution for $|r/f| < 1/2$, viz.:

$$W_{K-\text{non sing}} = 2\pi^3 K f, \tag{10.29}$$

and a singular contribution for $|r/f| \geq 1/2$. Because the effects of curvature decrease when r increases, the curvature of those surfaces is certainly not of a large order of magnitude in the region outside of the core r_c. Taking r_c constant on segments of parabolae of the order of f, assuming that the singularities are negligible outside of these segments, one expects

$$W_{K\text{-non sing}} \approx K \ln^2 \frac{fL}{r_c^2}. \tag{10.30}$$

Again, the sum of the contributions in (10.28), (10.29), and (10.30) does not take into account the core energy. A more complete discussion of all those energies is still wanting.

10.5.4. FCD-II: Positive Gaussian Curvature

Let us choose the range of variation $r > a \cosh v$, and introduce also a lower cutoff r_c, related to the maximal value of v on each $\Sigma(r)$ by the relation $r = a \cosh v_{\max} + r_c$ and an upper cutoff $r_{\max} = +R$. This latter condition means that the FCD-II is bounded by the cyclide parameterized by $r = R$. Because the layers have positive Gaussian curvature, this closed surface is of genus zero, with two singular points at opposite points on the hyperbola. Outside of the FCD-II, the layers adapt to the boundary conditions in various possible ways (see Problem 10.3).

Let us first calculate the energy of the layer parameterized by r. The ranges of variation for v and u

$$1 < \cosh v < r/a; \quad 0 < u < 2\pi. \tag{10.31}$$

The elastic energy of the layer is split similarly to (10.15):

$$w_r = \int f \, d\Sigma = w_{1,r} + w_{2,r}, \tag{10.32}$$

with

$$w_{1,r} = \frac{2\pi \kappa a^2 (1 - e^2)}{\sqrt{r^2 - e^2 a^2} \sqrt{r^2 - a^2}} \operatorname{arctanh} \left\{ \sqrt{\frac{r+a}{r-a}} \tanh \left[\frac{1}{2} \operatorname{arccosh} \left(\frac{r - r_c}{a} \right) \right] \right\}$$

$$w_{2,r} = 4\pi (2\kappa + \bar{\kappa}) \sqrt{\frac{r^2 - a^2}{r^2 - e^2 a^2}}. \tag{10.33}$$

In these equations, $\kappa = K d_0$, $\bar{\kappa} = \bar{K} d_0$, and d_0 is the repeat distance of the layers. Expressions in (10.33) have to be integrated in the range $r_c < r < R$, where r_c depends on r; but the total energy $\int_a^R dr(w_{1,r} + w_{2,r})$ can be suitably evaluated by taking $r_c = \text{const}$. The special simpler case $e = 0$ is proposed in Problem 10.3. The main result, easy to prove directly, is that the elastic energy stored in a spherulite (also called an onion) of radius R is

$$W = 4\pi \Lambda R \equiv 4\pi (2K + \bar{K}) R. \tag{10.34}$$

10.6. Curvature Defects in Columnar Phases

10.6.1. General Considerations

Call $\mathbf{t}(\mathbf{r})$ the unit vector along the physical rod, $\mathbf{r}$ being the position. We assume that rods in the distorted state do not break. This implies, by virtue of the conservation of rod flux,

$$\operatorname{div} \mathbf{t} = 0. \tag{10.35}$$

Because we are considering a situation of pure curvature, the rods are parallel in the distorted state. We can visualize this property of parallelism on the surfaces Σ's generated by a one-parameter subset of rods. Such surfaces can be safely defined in the continuous limit, because the distance a between rods is small compared with their length. On any of those surfaces, the orthogonal trajectories of a rod are *geodesic lines*, because these trajectories cut orthogonally a set of parallel rods (this result refers to the property of frontality of geodesic lines; for a simple and illustrative review of this geometrical concept as well as of others appearing here and elsewhere in this chapter, see Hilbert and Cohn-Vossen).

Two different surfaces Σ_i and Σ_j intersect along a rod and make a constant angle on this intersection; this angle is equal to the angle in the undistorted state, as required in a situation of pure curvature. Here, we use the following:

> *Joachimstahl theorem*, stating that if two surfaces cut at a constant angle along their intersection, then the intersection is either a line of curvature on both surfaces or is on none.

If the intersection is not a line of curvature, it is not possible to avoid a certain quantity of twist $q = -\mathbf{t} \cdot \operatorname{curl} \mathbf{t}$, which is necessarily attended by elastic distortions, as a detailed analysis can show. But this is precisely what one wants to avoid. Therefore, the only solution is when the rod is along a line of curvature on any Σ; this implies that the orthogonal trajectories of the rods are at the same time geodesic lines and lines of curvature. Therefore, they are planar curves, because being lines of curvature means that their geodesic torsion must vanish, and being geodesic lines means that their geodesic torsion is equal to their natural torsion.[19] Hence, the rods are perpendicular to a family of planes Σ in a situation of pure curvature, which also means that each surface Σ is generated by a planar curve invariable in shape (the geodesic line), moving in such a way that the velocities of all its points are normal to the planes Π, which contains this planar curve. Σ surfaces deserve to be called Monge's surfaces, by the name of the famous geometer who studied them in detail for the first time.[20]

Figure 10.10 represents two infinitesimally close Π planes; D is their straight line of intersection, which envelops a space curve L when the plane Π moves. The infinitesimal

---

[19]See Darboux or Hilbert and Cohen-Vossen.

[20]G. Monge, Applications de l'analyse à la géométrie, Paris, 1807.

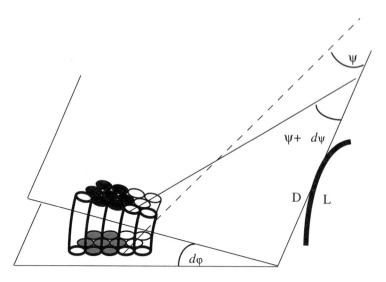

Figure 10.10. Disposition of rods between two infinitesimally close Π planes; on each plane, there is a perfect hexagonal lattice. See text for other details.

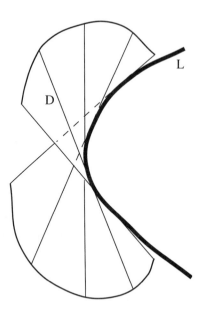

Figure 10.11. Scheme of a developable surface generated by the tangents to a space curve L, the so-called cuspidal edge of the developable.

motion of Π can be divided in a motion of pure rotation about D, viz. $d\phi$, and a motion about a normal to Π passing through the point of contact of D with L, viz. $d\psi$.

Consider now any column:[21] Its center of curvature is on D, and its osculating plane perpendicular to D. Its curvature is $\frac{1}{R} = |\mathbf{t} \cdot \mathrm{curl}\, \mathbf{t}| = |d\phi/ds|$, where s is the curvilinear abscissa along the column. In what concerns $d\psi$, it is related to the curvature of L at its point M of contact with D; i.e. $1/\rho = |d\psi/d\sigma|$, where σ is the curvilinear abscissa along L. The osculating plane of L in M is Π. Properties of reciprocity exist between L and any column, viz.

$$\frac{\rho}{\tau} = -\frac{T}{R}, \tag{10.36}$$

where T and τ are, respectively, the torsions of the rod and of L, oriented by the motion of the Π planes. The curvature $1/R$ of a rod becomes infinite on D, whose locus is, consequently, the curvature defect of the system. This locus is a developable surface that is the boundary of the domain of existence of rods (Fig. 10.11). We call it a *developable domain*.[21]

10.6.2. Developable Domains

First, let us consider the case when the developable is a circular cylinder of radius a; the curve L is reduced to a point at infinity in the direction of the generatrices of the cylinder. Consider a circular section of this cylinder; the planes Π cut this section along the tangents to the circle of radius a, and the rods are evolutes of this circle (Fig. 10.12). Clearly, such a configuration is that one of a disclination line of strength $k = +1$, of core radius r_c necessarily larger than a so that we expect an empty core, or at least a core in which the rods take an orientation entirely different from the orientation outside, for example, parallel to the generatrix of the cylinder. Such objects are observed in very thin samples of discotic phases in which rods orient parallel to the surface (Fig. 10.13). The $k = 1/2$ lines are more numerous than the $k = 1$ lines, and it can be shown experimentally that their geometry is that of a half-core. There is an example of a $k = +1$ line in Fig. 10.13, but the core is extended (Problem 10.4); for a more detailed analysis, see Oswald and Kleman.[22]

Among the other simple developable domains, let us quote that one in which the cuspidal edge L is a point at finite distance and the developable is a cone of revolution. It is possible to show that the columns are spherical curves (the spheres centered on L are Monge's surfaces of the problem), which are orthogonal trajectories of a set of great circles (which are here the geodesic lines of the problem) drawn on the sphere and bitangent to the intersection of the cone with the sphere (Fig. 10.14). The columns are easily obtained by equispacing, and symmetry about the axis of the core. Notice that the solution

---

[21] For a detailed analytical treatment, see M. Kleman, J. de Phys. **41**, 737 (1980).

[22] P. Oswald and M. Kleman, J. Phys. **42**, 1461 (1981).

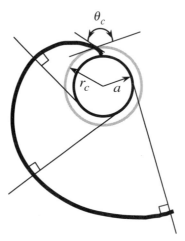

Figure 10.12. Cross section of a developable domain equivalent to a disclination of strength unity. a is the radius of the developable, and r_c is the radius of the core. The rods are evolutes of the circle.

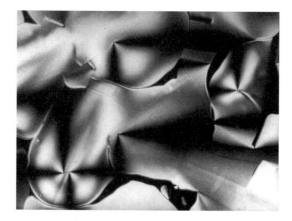

Figure 10.13. A thin droplet of a discotic hexagonal phase (C_5 hexaalkoxy derivative of triphenylene) spread on a glass plate shows the appearance of numerous disclination lines of strength of half unity, which have the geometry of half-developable domains (the columns are partly along evolutes, partly straight).

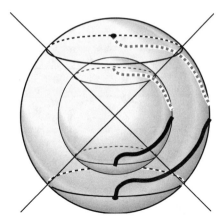

Figure 10.14. A singular point as a developable domain. Columns are along spherical curves joining one point on a circle of intersection of the sphere with a fixed cone to another point on the other circle of intersection. Other rods on the same sphere are obtained by axisymmetry. Then, concentric spheres are to be considered.

can be either right-handed (as in Fig.10.14) or left-handed, which means that it is possible to define a chirality for a set of columns without twist.

10.6.3. Classification of Developable Domains

By definition, a developable surface D can be developed on the plane P by pure bending without stretching. In such a mapping, the cuspidal edge becomes a line $\overline{L}$ in the plane and the straight lines D (see Fig. 10.11) map on one or the other of the half-infinite segments $\overline{D_1}$ and $\overline{D_2}$ of $\overline{D}$ bounded by the point of contact $\overline{M}$ on $\overline{L}$.

Lengths and angles are preserved, which implies that the curvatures of L and of $\overline{L}$ are the same at corresponding points. Hence, L and $\overline{L}$ differ only by a torsion function $\tau(\sigma)$, which can be chosen at will.

Because any torsion $\tau(\sigma)$ can be chosen to build L, it is clear that another way of mapping the developable domain on the plane P is to untwist it (rather than unbend it), i.e., decrease $\tau(\sigma)$ in a continuous manner without changing the curvature $\rho(\sigma)$ until $\tau(\sigma)$ vanishes. Any intermediary situation in the course of such a process is a developable domain that possesses the same representation in the plane P. In particular, the final state in P is a developable domain in which all planes tangent to S are in P. This also means that the hexagonal pattern of rods of any plane Π maps on the same pattern in P (Fig. 10.15), whereas each physical rod becomes a straight line perpendicular to P.

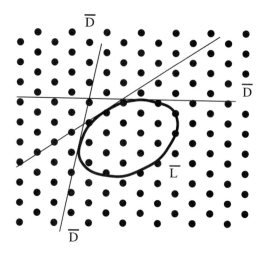

Figure 10.15. Representation of a developable domain on P. $\overline{L}$ is the image of L; each plane Π maps on P. Hexagonal patterns carried by planes Π map on a unique hexagonal pattern on P independent of the choice of Π.

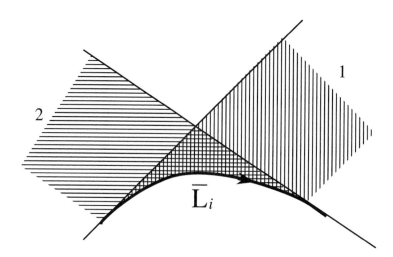

Figure 10.16. The convex arc $\overline{L}_i$, and its forward and reverse end tangents divide the exterior part in various sectors. Sector 1 is spanned by the forward half-tangents and lifts to one sheet of S; similarly, for sector 2 (reverse half-tangents), which lifts to another sheet of S.

Notice that the reality of the analytical process of untwisting does not mean that it is possible to decrease continuously to zero the energy of a developable domain by a physical process of that sort, because untwisting is not a conservative process.

Figure 10.15 is drawn for a convex closed curve $\overline{L}$. The lifting to the two sheets of the corresponding developable in space is easy to imagine. The situation is a bit more involved if $\overline{L}$ contains asymptotes, points of inflexion, termini. Then, it is necessary to divide $\overline{L}$ in simple convex arcs $\overline{L}_i$ and lift each of them. This is represented in Fig. 10.16.

10.7. FCDs in Lyotropic Lamellar Phases: Oily Streaks and Spherulites

Lyotropic lamellar phases and thermotropic SmAs show different textures of defects, based in both cases on arrangements of FCDs. Because most of the examples that can be found in the literature are devoted to thermotropic SmAs, we specialize this section to some special textures of FCDs more frequently met in lyotropic systems than in thermotropic ones.

10.7.1. Oily Streaks

Oily streaks are the most usual structural defects in lamellar phases, especially of a lyotropic type (Fig. 10.17). They appear as long bands with a complex inner structure and

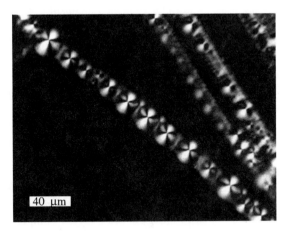

Figure 10.17. Oily streak that contains FCD-Is. The oily streak is a chain of FCD-Is. The gap between two domains is filled with edge dislocations. CpCl/hexanol/Brine = 14.8/15.2/70. Sample thickness 130 μm. Courtesy P. Boltenhagen.[23]

---

[23]P. Boltenhagen, O.D. Lavrentovich, and M. Kleman, J. Phys. II France **1**, 1233 (1991).

subdivide homogeneous regions with flat layers parallel to the boundaries of the preparation (so-called homeotropic alignment in case of SmA).

These oily streaks, first described by G. Friedel, are edge dislocations of large Burgers vectors, comparable to the thickness of the sample. Dislocations of small Burgers vector appear in the sample as a result of the capillary process of introduction of the chemical between the glass plates, and gather together after a few hours, to form large Burgers vector dislocations. As already discussed in Section 9.1.1, the current model assumes that the oily streak consists of two dislocations with large (sometimes of the order of the sample thickness) Burgers vectors of opposite signs, see Figs. 11.18 and 11.19. In lyotropic phases, a large activation energy of a topological nature (two fluid media) prevents the opposite dislocations from collapse.

The elastic energy stored by the oily streaks is often relaxed by the splitting of the streak into a series of FCDs; see Colliex et al.[24] and Rault[25] for other modes of instabilities of oily streaks. The ellipses are in the plane of the observations, with their long axes transverse to the oily streaks. The topological relations between FCDs and dislocations, which allow for the splitting in question, will not be discussed at length in this book. We shall content ourselves with the remark that the singularities in FCDs are of the disclination type and, as such, are susceptible to interact topologically with dislocations, according to the rules that are discussed in Chapters 9 and 12. The interested reader may consult Bourdon[26] for a detailed treatment of this question.

The intervals between the FCDs are filled either with the segments of dislocations (Fig. 10.17) or with FCDs of smaller size. The largest FCDs have geometrical features that are directly related to the Burgers vector β (here, $\beta = n\, d_0$ is a multiple of the L_α phase repeat distance, not to be confused with the parameter β in PFCDs; most generally, $\beta = b_1 - b_2$, meaning that the dislocation is the sum of two dislocations of opposite signs b_1, $-b_2$, whose Burgers vectors scale with the size of the sample). The largest FCDs have a major axis $a = (b_1 + b_2)/2$, (i.e., scan the whole dislocation inner region), a minor axis $\sqrt{b_1 b_2}$, and an eccentricity $e = (b_1 + b_2)/|b_1 - b_2|$.

Figure 10.18 illustrates these characters. The plane that contains the ellipse is always parallel to the layers belonging to the homeotropic region. By refocusing the microscope, it is possible to follow a line singularity, which lies in the vertical plane containing the major axis of the ellipse and passes through the focus of the ellipse. This line is one branch of hyperbola. Between crossed polarizers, the focus of the ellipse is the center of dark brushes (Fig. 10.17). The domains are FCDs of the first species.

The stability of the FCD splitting of the edge dislocation has been theoretically studied in Boltenhagen et al.[23] There are two major elastic contributions to this stability: (1) Formation of FCD-Is decreases the elastic energy because layers' compression is reduced, and (2) FCD-Is contain layers with negative Gaussian curvature, which decreases the energy if

[24]C. Colliex, M. Kleman, and M. Veyssié, Eighth Int. Congress on Electron Microscopy, Camberra, **1**, 718 (1974).

[25]J. Rault, Comptes Rendus Acad. Sci. (Paris) **B280**, 417 (1975).

[26]L. Bourdon, M. Kleman, and J. Sommeria, J. de Phys. **43**, 77 (1982).

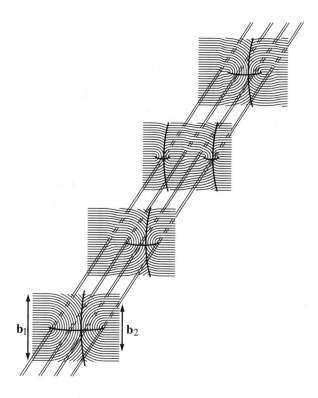

Figure 10.18. General structure of an oily streak formed by a chain of FCD-Is joined by edge dislocations.

$\Lambda = 2K + \overline{K}$ is positive. An important element in this stability can be in the case illustrated in Fig. 10.17 the proximity of the sponge phase, the only region of the phase diagram where FCD-Is are visible. Surface anchoring (Chapter 13) also influences the stability, as large FCD-Is require large tilt of layers at the boundaries, see Fig. 10.18.

10.7.2. Spherulites

This is the case when the lamellar phase formed, for example, by CpCl/brine/hexanol mixture, is near the micellar phase, which is made of globular or rod-shaped micelles. Observations with the polarizing microscope show textures that differ drastically from typical textures ever reported for smectic-like phases. One observes birefringent, elongated domains, located at different levels of the surounding matrix, which is also birefringent (Fig. 10.19). The meridian cut of the boundary of each of those domains, has the shape of the intersection of two circles with the same radius R. In these domains the optical axis

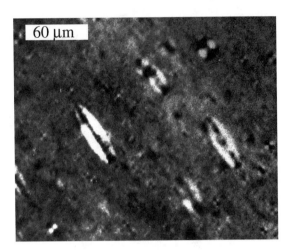

Figure 10.19. FCDs of the second species CpCl/hexanol/Brine = 12.5/7.5/80. Courtesy P. Boltenhagen (See P. Boltenhagen, O.D. Lavrentovich, and M. Kleman, Phys. Rev. A **46**, R1743 (1992).)

is perpendicular to the domain boundary. These observations lead to the conclusion that these domains are FCDs of the second species (Fig. 10.5b), in which the layers are folded around two conjugated lines, viz., a circle and a straight line. The circle is virtual, and in this case, only the straight line is visible under a microscope.

These domains are not stable, and as time elapses, they relax into spherical ones. During this process, the virtual circle is reduced to a point and the number of layers constituting the domain is conserved.

Contrary to the case of FCD-I, layers in the FCD-IIs have positive Gaussian curvature; this is probably related to a negative value of $\overline{K}$, which has been estimated by several methods. But it has also been argued (see below) that this is related to the method of preparation of the sample, which always involves shearing: FCD-IIs being the result of an hydrodynamic instability. These two origins are not contradictory: In SmAs, shear instabilities yield PFCD-Is (see Section 10.8).

10.8. Grain Boundaries and Space Filling with FCDs

FCDs have a nontrivial shape that does not tile space. Because they are observed (light microscopy) to assemble in large scale clusters, the question occurs which rules they obey when doing so, and how the layers are organized in-between, i.e., at smaller scales. Here, we consider examples with usual FCD-Is and FCD-IIs.[27] The case of TFCD-Is is simple

---

[27] M. Kleman, J. de Phys. **38**, 1151 (1977).

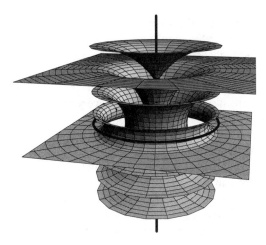

Figure 10.20. FCD with a circular base continuously embedded into a set of horizontal flat layers.

because a 2D periodic network of equal TFCDs can fill the space with gaps where the layers are strictly parallel and planar, which match smoothly with the negative Gaussian curvature parts of the tori (Fig. 10.20). The case of PFCD-Is has been considered experimentally and theoretically;[10] each PFCD-I fills all space, and the layers are pratically flat (see above) at some distance of the axis of the parabolae; there is therefore a practically smooth matching from one PFCD-I to another, as soon as the axes are parallel.

10.8.1. Focal Conic Domains of the First Species

10.8.1.1. Friedel Laws of Association

Usually in a polarizing microscope, one observes not isolated FCDs, but entire ensembles of domains in contact. These ensembles have been described in great detail by G. Friedel, who has established the geometrical laws that they obey:[1]

A domain is said to be *complete* if all of the layers with negative Gaussian curvature attached to a pair of conjugated conics have physical existence. A complete FCD is bounded by two semi-infinite cylinders of revolution that have the ellipse as a common basis, and whose generatrices are parallel to the asymptotes of the hyperbola. Fig. 10.5a shows only those parts of Dupin cyclides that are within these cylinders. *Incomplete* domains can be either *closed* or *fragmented*. A closed domain is bounded by two cones of revolution that have the ellipse as a common basis, and whose vertices are on the physical branch of the hyperbola, one on each side of the plane of the ellipse usually (e.g., point M'' in Fig. 10.5a might be one of the vertices). The largest possible closed domain is obtained by taking the vertices to infinity on the asymptotes; i.e., it is a complete domain. A domain will be called fragmented if the ellipse is not entirely physical. It is then bounded by two sheets of cone of revolution whose vertices are on the ellipse.

The laws of association that follow have a simple geometric content. As one will realize, they correspond physically to the tendency to fill space by ensuring the *continuity of the layers* from one domain to the other, and by preserving a *constant thickness* of layers. Of course, these requirements cannot be fulfilled everywhere, and numerous discrepancies are observed. For example, as already stated, PFCD-Is do not stack according to these laws. But on the whole, they allow for a correct analysis of the stacking of the layers at large scales, particularly in thermotropic specimens.

The most frequent experimental situation in thermotropic systems is that two FCDs are tangent to each other along two generatrices, and that the conics carried by these domains touch each other in pairs; i.e., the pairs of conics either interrupt each other or are tangent to each other. This observation due to Friedel means that the contact between two FCDs obeys the "law of corresponding cones," described below. Alternatively, if two FCDs are tangent along only one generatrix, it means that the conics they carry do not touch each other in pairs. Friedel summarizes his findings as follows.

> *Law of Impenetrability:* Two FCDs cannot penetrate each other; if they are in tangential contact at a point M, they are tangent to each other along at least one generatrix common to two of the bounding cones.

This "law" of impenetrability is of an experimental nature; physically, its meaning is that the interactions between FCDs are of a steric nature.

> *Law of Corresponding Cones* (LCC): When two conics C_1 and C_2 belonging to two different FCDs (FCD$_1$ and FCD$_2$) are in contact at a point M, the two cones of revolution with common vertex M, which rest on the two other conics C'_1 and C'_2 of the two FCDs, coincide. Therefore, C'_1 and C'_2 have two points of intersection P and Q on the common cone, and the straight lines PM and QM are two generatrices along which the two FCDs are in contact.

This law says infinitely more than the former one; it has a physical content, discussed in the next section, and a geometrical content, which we emphasize now. If two FCDs are in contact at two points like M and P (to simplify matters, the reader may visualize C_1 and C_2 as two ellipses, and C'_1 and C'_2 as two hyperbolae (see also Fig. 10.21), but the result does not depend on such a specialization), then indeed the cones of revolution with common vertex M and resting, respectively, on C'_1 and C'_2 are identical, because they have the same axis (the tangent to C_1 and C_2 at M), and a point in common, point P. Because they coincide, C'_1 and C'_2, being on the same cone, have two intersections, P and Q. In other words, two FCDs that have one common generatrix and two contacts on their generating conics, obey LCC.

LCC leads to the geometrical construction of Fig. 10.21. If two conics are *coplanar*, the triples F_1, M, F'_2, and F'_1, M, F_2, are aligned; F_1, F_2, F'_1, F'_2, are the foci of C_1 and C_2

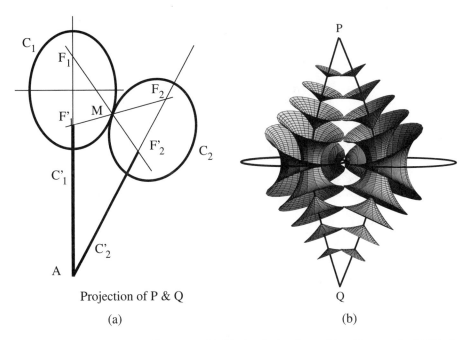

Figure 10.21. Law of corresponding cones: (a) 2D view in the plane of the ellipses and a (b) 3D view.

(Problem 10.5). Point A in Fig. 10.21a is the common projection of the two intersections P and Q of C'_1 and C'_2.

10.8.1.2. Grain Boundaries; Apollonian Packing

If Friedel's laws are obeyed, and because the layers are everywhere perpendicular to the generators, they cross the lines of contact between FCDs without singularities. A particular simple type of grouping of FCDs occurs in *grain boundaries* (Figs. 10.22 and 23c,d): Instead of dislocations, as in Fig. 10.23b, the boundary is filled with ellipses of equal eccentricities and parallel long axes. The hyperbolae have consequently parallel asymptotes. The angle between the asymptotes measures the disorientation between the two grains, because the layers are perpendicular to the asymptotic directions. The "common apex" is at infinity on both sides, and the domains are complete (Fig. 10.22 and 23c,d).

But many questions subsist. How are the ellipses distributed in size? It is clear that in order to reduce the area of the boundary that is not occupied by the ellipses, i.e., which does not belong to any FCD and, consequently, whose energy is not curvature only, one has to introduce smaller ellipses between the larger ones, in contact with them. This "iterative filling" has to yield a minimum of the energy of the grain boundary, which requires that the

Figure 10.22. A grain boundary split into FCDs; hyperbolae are in the plane of the figure and ellipses are perpendicular to it (Courtesy: Claire Meyer).

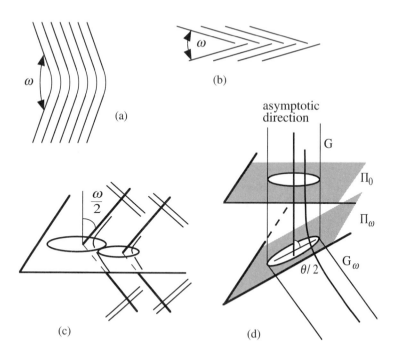

Figure 10.23. Tilt grain boundaries: (a) curvature wall; (b) dislocations wall; (c) a grain boundary split into FCDs obeying Friedel's laws of association; (d) its construction; $\theta = \pi - \omega$.

local contacts obey Friedel's laws; the typical size of the non-FCD remaining interstices is also fixed by the *global* energy minimization.

This question is discussed in Bidaux et al.[28] for a particular simple case of grain boundary: a boundary of vanishing disorientation. The hyperbolae are degenerate to straight lines, the ellipses are degenerate to circles, and the FCDs are toric. TFCDs built on coplanar tangent circles trivially obey Friedel's laws. The problem is then reduced to the two following steps: iteration of a compact packing of circles in a plane (known in mathematics as Apollonius's problem; at each step of the iteration, the radius of the newly introduced circles, tangent to the circles that are already present, decreases), calculation of the size of the remaining gaps such that the global energy is minimized. The unknowns are: (a) the number of circles of radius $R > b$, when the iteration reaches circles of radius b: Let $g(b)$ be this number; (b) the total perimeter of these circles, $P(b)$; (c) the residual uncovered surface area, $\Sigma(b)$. Let L be the size of the largest circles; one expects that all relevant quantities scale algebraically with the dimensionless quantity L/b. We can write:

$$g(b) = \text{const.} \left(\frac{L}{b}\right)^n ; \quad P(b) = \text{const.} \, b \left(\frac{L}{b}\right)^n ; \quad \Sigma(b) = \text{const.} \, b^2 \left(\frac{L}{b}\right)^n .$$

$$(10.37)$$

Numerical calculations[29] indicate that the exponent n is approximately 1.306. The energy of the grain boundary can be estimated as follows.

The energy of a TFCD of radius b scales as Kb; hence the contribution to the total energy of the circles of radius $R > b$ scales as $W_{\text{line}} \sim K P(b) = \text{const.} \, Kb(L/b)^n$. The residual regions are elastically deformed over a distance from the plane of the boundary of order $\lambda = \sqrt{K/B}$, the penetration length; hence, we have $W_{\text{resid}} \sim B\lambda\Sigma(b) = \text{const.} \, Bb^2\lambda(L/b)^n$; B is the compression modulus. After minimization of $W_{\text{line}} + W_{\text{resid}}$ with respect to b, the value of b at the final iteration is $b^* \sim \sqrt{K/B} = \lambda$. The energy per unit area of the grain boundary is $\sigma_{\text{FCD}} \sim L^{-2}(L/b^*)^n B\lambda^2 b^*$. We leave as an exercise to the reader to compare this energy with the energy of a curvature wall and the energy of a wall made of dislocations. We now discuss the applicability of this result to a grain boundary of nonvanishing disorientation.[30]

Consider the set of complete FCDs carried by the Apollonius tiling. Its boundary is made of a set of parallel cylinders of revolution C tangent along common generatrices G, Fig. 10.23d. Cut this set by a plane Π_ω at an angle $\omega/2$ with the plane Π_0 of the Apollonius tiling. It is easy to show that the set of ellipses E_ω of eccentricity $e_\omega = \sin\omega/2$ at the intersection of Π_0 and Π_ω is a valid set of LCC conics. In effect, any cone of revolution lying on an ellipse has its vertex on the conjugate hyperbola; a cylinder C lying on a circle belonging to Π_0 is such a cone; therefore, the asymptotic directions are along

[28]R. Bidaux, N. Boccara, G. Sarma, L. de Sèze, P.G. de Gennes, and O. Parodi, J. Physique **34**, 661 (1973).

[29]P.B. Thomas and D. Dhar, J. Phys. A: Math. Gen. **27**, 2257 (1994).

[30]M. Kleman and O.D. Lavrentovich, Eur. Phys. J. E **2**, 47 (2000).

G and the direction G_ω symmetric to G with respect to Π_ω taken as a mirror plane. The half-cylinders of revolution with generatrices G and G_ω and resting on the ellipses E_ω are tangent along the same generatrices G as the set of complete FCDs carried by the Apollonius tiling for the upper half-cylinders and the mirror generatrices G_ω for the lower half-cylinders. Therefore, *each complete FCD of the ω-tilt grain boundary is in contact along two generatrices with its neighboring complete FCDs (Fig. 10.23c) and is in one-to-one relationship with a complete toric domain belonging to the Apollonius filling.*
Equations (10.37) are modified as follows:

$$g(b) = \text{const.} \left(\frac{L}{b}\right)^n ; \quad P(b) = \text{const. } \mathbf{E}(e^2)(1 - e^2)^{-1/2} b \left(\frac{L}{b}\right)^n ;$$

$$\Sigma(b) = \text{const.}(1 - e^2)^{-1/2} b^2 \left(\frac{L}{b}\right)^n , \tag{10.38}$$

where b is the half-length of the minor axis (the radius of a circle becomes a minor axis in the mapping); $\mathbf{E}(x) = \int_0^1 (1 - t^2)^{-1/2}(1 - xt^2)^{1/2}\, dt$ is the complete elliptic integral of the second kind; the perimeter of the ellipse is $4a\mathbf{E}(e^2) = 4b\mathbf{E}(e^2)/\cos\frac{\omega}{2}$; $e^2 = \sin^2\frac{\omega}{2}$; and the ratio of the areas in the mapping is $\sqrt{1 - e^2} = \cos\frac{\omega}{2}$.

To calculate the wall energy, one needs the expression (10.19) for the curvature energy W of an isolated FCD-I and an estimate of the core energies. Assume that the core radius is constant ($r_c \approx \lambda = \sqrt{K/B}$) along the ellipse and that the core energy is equal to K per unit of line along the ellipse, whereas the total core energy is twice the ellipse core energy; then the core energy of one confocal pair is roughly $8a\,K\mathbf{E}(e^2)$. Therefore, the wall energy comprises the following three terms:

$$W_{\text{bulk}} = - \int_{x=b}^{L} dg(x) W(x)$$

$$\sim \alpha_b K (1 - e^2)^{1/2} \int_b^L n \left(\frac{L}{x}\right)^n dx \mathbf{K}(e^2) \left[\ln 2\frac{x}{\lambda} - 2\right], \tag{10.39a}$$

$$W_{\text{core}} = - \int_{x=b}^{L} dg(x) w_{\text{core}}(x)$$

$$\sim \alpha_c K (1 - e^2)^{-1/2} \mathbf{E}(e^2) \int_b^L n \left(\frac{L}{x}\right)^n dx, \tag{10.39b}$$

$$W_{\text{resid}}^{\text{disl}} \sim eB\lambda\Sigma(b) = Bb^2\lambda e(1-e^2)^{-1/2}\left(\frac{L}{b}\right)^n, \tag{10.39c}$$

where α_b and α_c are numerical constants.

Although the establishing of (10.39a,b) is straightforward, (10.39c) needs some comment. The residual regions are pieces of grain boundaries and henceforth must adopt either a model of a curvature wall (Fig. 10.23a) (when the disorientations ω are large) or of a dislocation wall (Fig. 10.23b) (when the disorientations are small). Below, we consider only one scenario, when the energy is relaxed by edge dislocations; the second scenario is discussed in Problem 10.6.

The very possibility of relaxation by dislocations results from the fact that the singular lines in a FCD are indeed bent (not straight) disclinations, which according to a general analysis[31] are terminations of dislocations. As shown in Boltenhagen et al.,[23] the dislocations that are attached to the ellipse of an isolated FCD characterized by the quantities a and e have a total Burgers vector $2ae$. Assume now that the core of each segment of dislocation of Burgers vector $2ae$ is extended over a region of similar size $r_c = 2a$. Such an assumption is all the more true as the Burgers vector is large, and has the advantage to yield a (small) dislocation line tension w that scales as the Burgers vector,[32] not as the square of the Burgers vector; i.e., $w \sim B\lambda ae = B\lambda be/\sqrt{1-e^2}$. We have to calculate the total length of such dislocations, which we do as follows. The area of each connected residual element, at stage "b" of the iterative filling, between FCDs, scales as $\sim ab$. One can convince oneself of this scaling relation by the consideration of four equal (a, e) ellipses in symmetrical contact two by two; they enclose a "residual" area equal to $(4-\pi)ab$. Therefore, the number of connected residual elements scales as $\sim \Sigma(b)/ab = (1-e^2)^{-1/2}b^2(L/b)^n/ab = (L/b)^n$. The length of a segment of dislocation in a connected element scales as $\sim b$, hence, a total residual energy as in (10.39c). Note that this expression yields a vanishing W_{resid} for $\omega = 0$, as expected in a realistic model.

Now, minimizing the total energy $W = W_{\text{bulk}} + W_{\text{core}} + W_{\text{resid}}^{\text{disl}}$ with respect to b, one gets for the minimal b:

$$\frac{b_{\text{disl}}^*}{\lambda} \sim \frac{1}{e}\frac{n}{2-n}\left\{\alpha_b(1-e^2)\mathbf{K}(e^2)\left[\ln\frac{2b_{\text{disl}}^*}{\lambda} - 2\right] + \alpha_c\mathbf{E}(e^2)\right\}, \tag{10.40}$$

and for the energy per unit area of wall (with dislocation-relaxed residual areas),

$$\sigma_{\text{FCD}}^{\text{disl}}(e) \sim \frac{1}{(n-1)L^2}\left(\frac{L}{\lambda}\right)^n B\lambda^3\left(\frac{\lambda}{b_{\text{disl}}^*}\right)^{n-1}$$

$$\times (1-e^2)^{-1/2}\left[e\frac{b_{\text{disl}}^*}{\lambda} + \alpha_b\frac{n}{n-1}(1-e^2)\mathbf{K}(e^2)\right], \tag{10.41}$$

---

[31] J. Friedel and M. Kleman, J. de Phys. **30**, C4:43 (1969).

[32] M. Kleman, Points, lines and walls, John Wiley & Sons, Chichester, 1982.

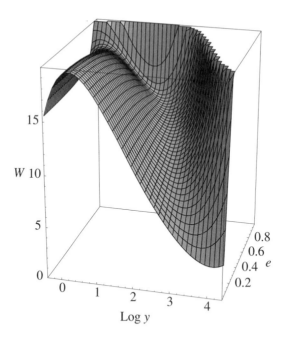

Figure 10.24. Elastic energy W (arb. units) of a grain boundary formed by FCD-Is vs. e and $\log_{10} y$ for $\alpha_b = 4\pi$, $\alpha_c = 8$, $L = 50\,\mu$m, and $\lambda = 2$ nm; $y = b/\lambda$.

which obtains from the integration of (10.39a,b), and the insertion of (10.40) in the sum $W_{\text{bulk}} + W_{\text{core}} + W_{\text{resid}}^{\text{disl}}$. The value of b^* for $\omega = 0$ is infinite, and the energy vanishes, as expected.

Figure 10.24 shows the total energy W for $\alpha_b = 4\pi$, $\alpha_c = 8$, as a function of the eccentricity e and the dimensionless variable $y = b/\lambda$. It is visible that for each value of e, i.e., of the disorientation ω, there are two solutions in b/λ that make the derivative to vanish. The solution with the smaller, microscopic b/λ is a maximum, whereas the other one is a minimum. This is the valid solution; it varies quickly with e (or ω). The residual area is large for small disorientations ω, and the energy is small; at $\omega = 0$, the "residual" area invades the whole boundary. It is only for very large disorientations that the residual area becomes microscopic, as in the Apollonian model. We leave it as an exercise to compare the energy of the grain boundary formed by FCD-Is to the energy of dislocation grain boundary (at small ω) and curvature grain boundary (at large ω) (Problem 10.6).

The FCD model above explains the well-known experimental feature that the residual areas in smectic grain boundaries are of micron sizes, i.e., macroscopic. The same feature of large residual areas is met in other types of packing, e.g., at interfaces between a smectic phase and another phase[33] (which might be the isotropic phase of the same smectic com-

---

[33]O.D. Lavrentovich, Sov. Phys. JETP **64**, 984 (1986); Mol. Cryst. Liq. Cryst. **151**, 417 (1987).

pound[34]). These interfaces are often observed to be tiled with conics, because of anchoring properties of the interface. We leave the discussion to Section 13.2.6, where the concept of surface anchoring is discussed.

10.8.2. Focal Conic Domains of the Second Species

Dense packings of "onions" have been observed in lyotropic L_α phases under shear.[35] They subsist after shear ceasing; the size of the onions depends critically on the shear rate. Consequently, this is not an iterative filling.

Although it is not clear whether iterative space filling of a smectic phase with spherulites has been observed—various electron microscopy studies of the lyotropic phases have revealed the presence of packings with spherulites that have different sizes ranging from macroscopic (a few micrometers) to microscopic (tens and hundreds Angström); e.g., in the vicinity of the $L_\alpha - L_1$ phase transition[36] or after shaking of a very dilute L_α phase, it appears interesting to consider such a possibility.

There are some important differences in packing of space with FCD-Is and FCD-IIs. First of all, an FCD-I with conical shape has volume $L \times a^2$ and extends through the whole sample (length L). A FCD-I often has its base located at the sample surface. This is why the anisotropy of the surface energy is so important for the scenario with FCD-Is (Section 13.2.6). In contrast, the FCD-II has a closed shape with characteristic volume a^3, where a might be much smaller than L. Thus, the physical limit for the iteration process is defined solely by the bulk properties of the lamellar phase.

The largest spherulites have a macroscopic size $R \gg \lambda$ (defined, e.g., by the sample size or by the shear rate) and distort the lamellar matrix. The energy W_B of the layer compressibility outside the spherulites of radius R scales like $B\lambda R^2$ if a mean separation is larger than R^2/λ and like BR^3 if the separation is smaller than R (in the latter case, the dilatation ϵ of the layers is of the order of unity). These distortions can be relaxed by smaller FCD-IIs lying in between the large FCD-IIs because the geometry of the FCD-II implies only curvature deformations and energy $\sim \Lambda a$. The iteration process will interrupt at scales a^* that do no provide a sufficient energy gain when substituting curvature by dilation. To define a^*, let us introduce the number g of spheres of radius $R \geq a$ packed in a volume L^3, $g \sim (L/a)^\gamma$. The residual volume (that is not occupied by FCD-IIs) is $V(a) \sim L^\gamma a^{3-\gamma}$. Then, the total free energy scales as

$$W(\rho) \sim \Lambda L^\gamma a^{1-\gamma} + BL^\gamma a^{3-\gamma}. \tag{10.42}$$

Here, γ, contrarily to the Appolonius exponent that is met in the problem of 2D circles filling, is not a universal constant. But, in some experimental cases (concrete), a relevant

[34]J.B. Fournier and G. Durand, J. Phys. II France **1**, 845 (1991).

[35]P. Sierro and D. Roux, Phys. Rev. Lett. **78**, 1496 (1997).

[36]P. Boltenhagen, M. Kleman, and O. Lavrentovich, J. de Physique II (France) **4**, 1439 (1994).

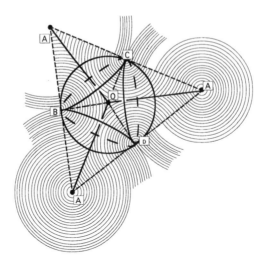

Figure 10.25. Fusion of the FCD-IIs centered at points A's via the FCD-Is ABOC, ACOD, ABOD.

value[37] of γ is approximately 2.8. This value yields indeed a minimum of $F(\rho)$ when $\Lambda > 0$ and a physical limit of iterations:

$$a^* \sim \sqrt{(\gamma - 1)\Lambda/(3 - \gamma)B}. \qquad (10.43)$$

Thus, the iterations go down to a very small scale $\sim \lambda$ as in the Bidaux et al. model for FCD-Is filling. It is difficult to imagine some other cutoffs of the iteration because in balance of the curvature and dilatation elastic terms, the only characteristic length is that defined by λ.

The space-filling spherulitic state is a (meta)stable one if $\Lambda = 2K + \overline{K}$ is negative and large enough. The fusion of two neighboring spherulites is hindered by the necessity of creating an energetically unfavorable "passage" with negative Gaussian curvature between them (Fig. 10.25).

10.9. Rheophysics of FCDs

Rheological properties are related to the *microstructure* only when the sample is perfect, i.e., remains in its ground state of orientational order when put in motion. For example, if the velocity gradient is perpendicular to the layers (the velocity being in the plane of the layers), a sheared specimen sees the microstructural viscosity related to the interlayer

---

[37]R. Omnès, J. Physique France **46**, 139 (1985).

friction. But even such a simple flow can present instabilities even at very low shear rates. The nonlinear behavior that then occurs can be related to *defects* of the related structure, like dislocations and FCDs. The theory of the onset of these instabilities and of their growth is a developing field.[38,39] Therefore, this section will be mostly devoted to experimental results. Lyotropic and thermotropic systems show similar behaviors from that point of view at low frequencies (or low shear rates), but display different textures of defects at high shear rates.

10.9.1. Global Viscoelastic Behavior and Alignment Under Shear

It is well known that in a linear viscoelastic fluid, the measurement of the low-frequency dependence of the complex modulus $G^* = G' + iG''$, yields $G' \approx \eta\tau\omega^2$, $G'' \approx \eta\omega$, where τ is the longest relaxation time of the material and is related to the molecular relaxation processes. Such a behavior is, for example, documented in high temperature copolymers, when the typical relaxation frequencies τ^{-1} are of the order of a few reciprocal seconds or more, comparable to frequencies feasible with rheometers. This behavior disappears when the material is cooled into a macroscopic ordered phase: The relaxation times now depend on collective modes, and there is no longest relaxation time. $G'(\omega)$, $G''(\omega)$ plots should then be related to the structure of the ordered phase.

As shown by Larson et al.,[40] this is so for a number of lamellar phases of very different molecular structures, like that of blockcopolymers PS-PI and thermotropic 8CB. In both cases, $G'(\omega)$ and $G''(\omega)$ follow the same way the effect of temperature changes, shear variations, or duration of shear application. This is interpreted as meaning that different lamellar samples *anneal* in similar ways, under a temperature increase or under a long-continued shear. In other words, whatever the type of specimen, the density of defects is the leading factor in the viscoelastic properties.

Another global effect of the changes of temperature and shear-rate or shear duration is the orientation of the sample. According to most authors,[41] the lamellar phases align with the director perpendicular to both the flow and shear-gradient directions (alignment "a"), at sufficiently low continued shear rates (or low frequencies) and sufficiently high temperatures. After an intermediary range of shear rates, at which the director is in the direction of the shear gradient (alignment "c"; the layers are then parallel to the plates), at higher shear rates, there is again a transition to the "a" orientation. These effects have been related to the nonlinear interaction of shear with fluctuations.[42,43]

---

[38] S.I. Ben-Abraham and P. Oswald, Mol. Cryst. Liq. Cryst. **94**, 383 (1983).

[39] R. Ribotta and G. Durand, J. Physique **38**, 179 (1977).

[40] R.G. Larson, K.I. Winey, S.S. Patel, H. Watanabe and R. Bruinsma, Rheol Acta **32**, 245 (1993).

[41] see for example K.A. Koppi et al., J. Phys. II (France) **2**, 1941 (1992); C.R. Safinya et al. Science **261**, 588 (1993); P. Panizza et al., J. Phys. II, France **5**, 303 (1995).

[42] R.F. Bruinsma and C.R. Safinya, Phys. Rev. Lett. **A43**, 5377 (1991).

[43] M. Goulian and S.T. Milner, Phys. Rev. Lett. **74**, 775 (1995).

10.9.2. Textures

Textures of lyotropic and thermotropic lamellar phases show relatively different aspects under shear, even if the principles at the origin of the coupling between shear and fluctuations are the same. This is because, as repeatedly stated, the defects, although of the same topological nature, have different types of physical realizations. Also, the anchoring conditions can be of a different nature and oppose strongly or not the bulk effects.

10.9.2.1. Thermotropic Specimens

An homeotropic SmA under shear develops instabilities that transform into a set of PFCD-Is. For short times of shearing, the texture is disordered, but it eventually aligns in the direction of shear, forming a regular square lattice, whose period ℓ increases with the sample thickness. Defect textures can be characterized by the intensity I of light transmitted through the sample between crossed polarizers; $I = 0$ for the homogeneous state (director is along the wave vector). Figure 10.26 reproduces experimental results[44] in 8CB: In the steady state, the relation between stress and shear rate is linear and of the so-called

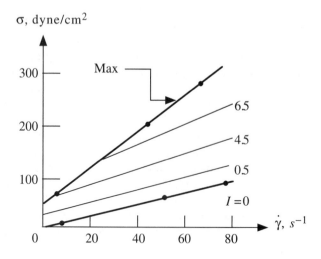

Figure 10.26. Shear-stress plotted against shear rate for various intensities of transmitted light (indicated by numbers, arbitrary units). Quasi-homogeneous sample $I = 0$; $I = 6.5$ corresponds to about 80% of the sample containing a disordered texture of PFCDs; curve "max" is the maximum shear stress measured in function of the applied shear rate. For clarity, most of the actual experimental data are shown by lines.

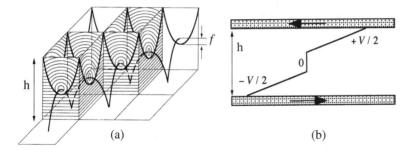

Figure 10.27. Model for the flow in a specimen where a network of PFCD-Is has developed under shear in the central region.

Bingham type; i.e., with a yield stress σ_Y,

$$\sigma = \sigma_Y + A\dot{\gamma}, \qquad (10.44)$$

where $\dot{\gamma}$ is the shear rate and A is a viscosity-like coefficient.

The yield stress σ_Y can be written tentatively, on dimensional grounds, as K/ℓ^2. This expression provides the right order of magnitude, which indicates that σ_Y measures the resistance of the PFCDs lattice to the flow. In fact, it has been shown by the observation of dust particles in the flow that the central region of the sample, which is the region where the layers are most disturbed by the PFCDs, is practically immobile: Most of the flow takes place near the boundaries[45] (Fig. 10.27).

A few experiments carried on parallel anchored samples[46] also show that the defects nucleate at small shear rates or shear frequencies. The essential effect is the nucleation of two Grandjean-Cano walls of opposite signs, parallel to the boundaries of the sample, and that relax the disorientation introduced by the shear (Fig. 10.28). These walls are split into FCDs of the same eccentricity. The continuous shear is presumably matched by a relative glide of the two sets of FCDs along common generatrices.

10.9.2.2. Lyotropic Specimens

The most spectacular difference between lyotropic and thermotropic textures under shear is the appearance of spherulites (onions) in the former. Nucleation of these FCD-IIs has been observed in dilute samples.[47] They form dense, regular packings, whose characteristic length (size R of the onions) is related to the shear rate $R \propto 1/\sqrt{\dot{\gamma}}$. The fact that some dilute lyotropic materials prefer FCD-IIs (onions) rather than PFCD-Is above the onset of the shear instability, which is probably the same in both cases, is not well understood.

---

[45]P. Oswald et al., Philos. Mag. **A46**, 899 (1982).

[46]C.E. Williams and M. Kleman, J. Physique **36**, C1-315 (1975). J. Marignan, G. Malet, and O. Parodi, Ann. Physique **3**, 221 (1978).

[47]O. Diat and D. Roux, Langmuir **11**, 1392 (1995).

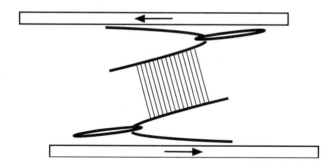

Figure 10.28. Under shear, a planar sample shows three regions of different orientations [after J. Marignan et al., Ann. Phys. France **3**, 221 (1978)].

Problem 10.1. Prove (10.3).

Answers: Because ε_{ikl} is antisymmetric with respect to any pair of indices, a contracted product of the form $\varepsilon_{ikl} a_{kl} \equiv 0$, if a_{kl} is symmetric ($a_{kl} = a_{lk}$). Hence, $\varepsilon_{ikl} u_{j,lk} \equiv 0$. One can therefore write $\frac{1}{2}\varepsilon_{ikl}\beta_{lj,k} = \frac{1}{2}\varepsilon_{ikl}(\beta_{lj,k} + \beta_{jl,k})$ and identify $\frac{1}{2}(\beta_{lj} + \beta_{jl})$ with e_{lj}.

Problem 10.2. Prove the relation between the contortion tensor and the density of dislocations (10.4).

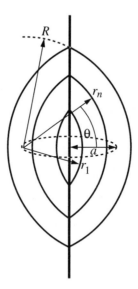

Figure 10.29. FCD-II, case $e = 0$. The meridian cut of the domain is the intersection of two circles with radius R; θ is the angle between the normal to the layer and the plane containing the circle.

Answers: The demonstration is due to Nye[2]; see also F.R.N. Nabarro, Theory of Crystal Dislocations, Oxford, 1967, p. 45.

Problem 10.3. Section 10.5.4. is a set of results that require the reader to take a careful look, because
most of the formulae are given with only a few hints for the demonstrations. We propose here an
exercise that completes the list of (observed) cases studied in this section:
  Consider a FCD-II of rotational symmetry ($e = 0, a = b, a < r < R$) (Fig. 10.29).

(a) Show that the principal curvatures and the infinitesimal element of surface can be written as
    $\sigma' = \frac{1}{r}, \sigma'' = \frac{\cos\theta}{r\cos\theta - a}, d\Sigma = 2\pi r(r\cos\theta - a)\,d\theta$. Deduce the nonsingular curvature energy
    $w_{2,r}$ of the r-layer and the corresponding nonsingular contribution to the total energy of the
    FCD-II. Evaluate the core energy in the model of Fig. 10.30. Show that the core energy vanishes
    when $a \to 0$. Calculate the total energy of the FCD-II (an "onion") when $a \to 0$.

(b) Estimate the energy of deformation of the matrix, corresponding to the introduction of the FCD-II
    in the homeotropic region (Fig.10.31).

Answers:

(a) The nonsingular part elastic energy of the r-layer is obtained from (10.33) by setting $e = 0$ as
    $w_{2,r} = 4\pi(2\kappa + \bar{\kappa})\sqrt{1 - \frac{a^2}{r^2}}$, where $r = a/|\cos\theta_{\max}|$ (Fig. 10.29); integration over r yields

$$W_2 = \int_a^R \frac{w_{2,r}}{d_0}\,dr = 4\pi\Lambda R\left(\sqrt{1 - \frac{a^2}{R^2}} - \frac{a}{R}\arccos\frac{a}{R}\right).$$

The total energy of the FCD-II reads as $W = \int_a^R \frac{w_{1,r}+w_{2,r}}{d_0}\,dr + W_c$. By analogy with a nematic
phase with splay, the energy density of the core is of the order of $K\sin^2\omega$, where ω is the angle
between the normal to the layer and the defect line. The core energy of the line defect reads as

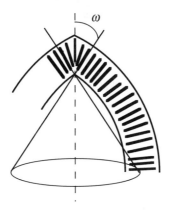

Figure 10.30. FCD-II core model. In the cone with angle ω, the deformation is of a pure divergence
type.

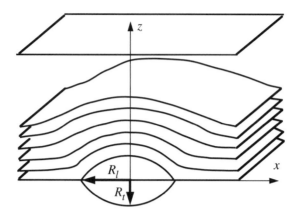

Figure 10.31. Deformation of the homeotropic region due to the introduction of a rotationally symmetric FCD-II.

$$W_c \approx K \int_{a+d_0/2}^{R} \frac{a^2}{r^2} \, dr = K a^2 \frac{R - a - d_0/2}{R(a + d_0/2)}.$$

When $a \to 0$, $W \to 4\pi \Lambda R$. This is by far the most often experimentally observed case.

(b) The energy density of this deformation is of the form $w_B = \frac{1}{2} B (\frac{\partial u}{\partial z})^2$, where u is the perpendicular displacement (in the z direction) of the layers from their initial position corresponding to the flat layers. In a very simplified model, u can be taken as $u = R_t \cos qx \cos hy \exp(-\frac{z}{L})$, with $q = \frac{\pi}{2R_l}$ and $h = \frac{\pi}{2R_t}$. One finds $1/L = \lambda(q^2 + h^2)$, where $\lambda = \sqrt{K/B}$.

The integration of the energy density relating to the deformation of the matrix over the ranges $-R_l < x < R_l$; $-R_t < y < R_t$; $-\infty < z < \infty$ has the form

$$W_B = \frac{\pi^2}{8} B \lambda R_t^2 \left(\frac{R_t}{R_l} + \frac{R_l}{R_t} \right).$$

W_B has a minimum when $R_l = R_t$, i.e, for a spherical domain (onion).

Problem 10.4.

(a) Calculate the energy of a disclination of strength $k = 1$ in the columnar hexagonal phase, in the three following cases: (1) hollow core, (2) isotropic core, (3) full core. Hint: the only relevant bulk energy term is the term of curvature of the columns. The notations below are those of Fig. 10.12.

(b) Calculate the energy of a disclination of strength $k = 1/2$ assuming a full core.

(c) Discuss the stability of the $k = 1$ line relative to its splitting into two $k = 1/2$ lines (see Oswald and Kleman[22]).

Answers:

(a) $k = 1$. Let γ_{hc} be the surface energy density for a hollow core. One finds

$$(1) \qquad\qquad W_{hc} = \frac{\pi K_3}{2} \ln \frac{R^2 - a^2}{r_c^2 - a^2} + 2\pi \gamma_{hc} r_c,$$

which must be minimized with respect to r_c and a, and must satisfy $r_c > a$.

$$(2) \qquad\qquad W_{hc} = \frac{\pi K_3}{2} \ln \frac{R^2 - a^2}{r_c^2 - a^2} + 2\pi \gamma_i r_c + w_c,$$

with $w_c \approx \frac{\Delta H \Delta T}{T_c}$, where ΔH is the enthalpy of transition, T_c is the temperature of transition to the isotropic liquid, and $\Delta T = T_c - T$. One gets

$$a = 0, \quad r_c = \frac{1}{2 w_c} \left(\sqrt{\gamma_i^2 + 2 w_c K_3} - \gamma_i \right).$$

(3) $\gamma = \gamma_0 + \Delta\gamma \sin^2 \theta_c$; there are two solutions: $a = 0, r_c = \frac{K_3}{2\gamma_0}$ and

$$r_c = \frac{K_3}{4(\gamma_0 + \Delta\gamma)}, \quad a = r_c \sqrt{\frac{\gamma_0 + 2\Delta\gamma}{\Delta\gamma}}.$$

(b) $k = 1/2$. There are 2 solutions:

$$a = 0, \quad r_c = \frac{K_3}{\gamma_0(1 + 2/\pi) + 2\Delta\gamma/\pi}$$

and

$$r_c = \frac{K_3}{8[\gamma_0(1/2 + 2/\pi) + \Delta\gamma/2]}, \quad a = r_c \sqrt{\frac{\gamma_0(1/2 + 2/\pi) + \Delta\gamma(1 - 1/\pi)}{\Delta\gamma(1/2 - 1/\pi)}}.$$

Problem 10.5. (a) Prove the assertion of Section 10.8.1.1 related to the case when the conics C_1 and C_2 are coplanar; (b) Prove that if two FCDs have only one common generatrix, they are not in contact along two conics.

Answers: (a) M is the vertex of the common cone of revolution resting on the conics C'_1 and C'_2, Fig. 10.21. But C'_1 (respectively, C'_2), being conjugate to C_1 (respectively, C_2), pass through the foci of C_1 (respectively, C_2). Therefore, $F_1 M F_2^!$ and $F_1^! M F_2$ are generatrices of the common cone. Note: The tangent common to C_1 and C_2 at M is the axis of this cone.

Problem 10.6. Compare the energy of the grain boundary formed by FCD-Is to the energy of (a) dislocation grain boundary (small ω) and (b) curvature grain boundary (large ω) (Fig. 10.23).

Answers:

(a) At small disorientation, the dislocation model, which is observed in Grandjean-Cano wedges, has an energy per unit area of the order of $\sigma_{\text{disl}} \approx B\lambda e$, practically independent of the Burgers vector (assumed small) of the dislocations. Comparing to

$$\sigma_{\text{FCD}}^{\text{disl}} \approx \frac{B\lambda^3}{L^2}(\frac{L}{\lambda})^n(\frac{\lambda}{b*})^{n-1},$$

where $\frac{\lambda}{b*} \sim e$, see (10.40), one gets

$$\frac{\sigma_{\text{disl}}}{\sigma_{\text{FCD}}^{\text{disl}}} \sim \left(\frac{Le}{\lambda}\right)^{2-n}.$$

Except for very large samples, the dislocation model is favored at small e.

(b) At large disorientations ω (or small $\theta = \pi - \omega$), the residual regions are filled with curvature walls. The corresponding contribution is calculated the same way as $W_{\text{resid}}^{\text{disl}}$ in (10.39c); it reads as

$$W_{\text{resid}}^{\text{curv}} \sim 2Bb^2\lambda\left[1 - \frac{e\,\text{Arc cos }e}{\sqrt{1-e^2}}\right]\left(\frac{L}{b}\right)^n.$$

Here, we have used the following exact expression of the curvature wall energy,[48] valid for all angles:

$$\sigma_{\text{curv}} = 2B\lambda\left(\tan\frac{\theta}{2} - \frac{\theta}{2}\right)\cos\frac{\theta}{2},$$

where $\theta = \pi - \omega$. Minimizing the total energy $W = W_{\text{bulk}} + W_{\text{core}} + W_{\text{resid}}^{\text{curv}}$ with respect to b, one gets the minimal b

$$\frac{b_{\text{curv}}^*}{\lambda} \sim \frac{1}{2\left(\sqrt{1-e^2} - e\,\text{Arc cos }e\right)}\frac{n}{2-n}$$

$$\times \left\{\alpha_b(1-e^2)\mathbf{K}(e^2)\left[\ln\frac{2b_{\text{curv}}^*}{\lambda} - 2\right] + \alpha_c\mathbf{E}(e^2)\right\},$$

and the energy per unit area of the FCD-split wall in which the residual areas are relaxed by the layers' curvatures:

$$\sigma_{\text{FCD}}^{\text{curv}}(e) \sim \frac{1}{(n-1)L^2}\left(\frac{L}{\lambda}\right)^n B\lambda^3\left(\frac{\lambda}{b_{\text{curv}}^*}\right)^{n-1}\left[\frac{2b_{\text{curv}}^*}{\lambda}\left(1 - \frac{e\,\text{Arc cos }e}{\sqrt{1-e^2}}\right)\right.$$

$$\left. + \alpha_b\frac{n}{n-1}\sqrt{1-e^2}\mathbf{K}(e^2)\right].$$

---

[48] C. Blanc and M. Kleman, Eur. Phys. Journ. **B10**, 53 (1999).

The dependence W on the eccentricity and the ratio b/λ reveals that the residual areas are of macroscopic size. The dislocation and curvature models of the residual areas in the FCD-split grain boundary are complementary, the first being favorable at smaller e and the second at larger e. Finaly, one can compare the energy of the pure curvature wall $\sigma_{\text{curv}} \approx B\lambda\theta^3 \approx B\lambda(1-e^2)^{3/2}$ to the energy of the FCD-split boundary

$$\sigma_{\text{FCD}}^{\text{curv}} \approx \frac{B\lambda^3}{L^2}(\frac{L}{\lambda})^n(\frac{\lambda}{b*})^{n-1}\frac{1}{\theta},$$

with

$$\frac{b*}{\lambda} \rightarrow \frac{12\alpha_c n}{(2-n)\theta^3},$$

to see that

$$\frac{\sigma_{\text{curv}}}{\sigma_{\text{FCD}}^{\text{curv}}} \sim \left(\frac{L}{\lambda}\right)^{2-n}\theta^{7-3n};$$

i.e., except for very large samples, the curvature model is favored at large e.

Further Reading

G. Darboux, Théorie Générale des Surfaces, Chelsea Publishing, New York, 1954.

J. D. Ferry, Viscoelastic Properties of Polymers, 3rd edition, John Wiley & Sons, New York, 1980.

G. Friedel, Annales de Physique (Paris) **18**, 273 (1922).

D. Hilbert and S. Cohn-Vossen, Geometry and the Imagination, Chelsea Publishing, New York, 1952.

M. Kléman, Points, Lines and Walls, John Wiley & Sons, Chichester, 1983.

CHAPTER 11

Disclinations and Topological Point Defects. Fluid Relaxation.

Dislocations in crystalline solids are rather simple examples of *topological* singularities: The symmetry of translation is broken along a line, and the Burgers vectors of merging dislocations add up linearly; on the other hand, they are complex *physical* objects, whose most important properties relate to irreversible phenomena that attend *plastic deformation*.

Liquid crystals, which are endowed with continuous symmetries and physical prevalence of correlations of orientation over correlations of position, show an amazingly greater and more complex variety of topological singularities. As in the case of dislocations, the role of these defects in irreversible phenomena is of physical importance; however, flow deformation (*rheology*) is far less understood than is plastic deformation. Note furthermore that singularities in liquid crystals and, more generally, anisotropic liquids, magnetic systems, and spin glasses are defects of various dimensionalities, not only line defects, but also points, walls, and "configurations" (walls, topological solitons).

Dislocations of rotation, more usually called *disclinations*, are line defects along which a symmetry of rotation is broken. These defects are most common in uniaxial nematics (Fig. 11.1) and in all fluid media with directors (like cholesterics, biaxial nematics, etc.). Other topological defects in uniaxial nematics are *point defects*. Unlike their namesakes in solid crystals (which are not topological defects), they distort order in the whole volume of the system. Disclinations can in principle be classified by the elements of rotational symmetry of the ordered medium. Keeping with the definition and construction of defects inspired by the Volterra process for dislocations (Chapter 8), it is easy to figure out the type of distortion they carry when the rotation vector $\mathbf{\Omega}$ (the analog of the Burgers vector; $\mathbf{\Omega} = \Omega \boldsymbol{\nu}$; $\boldsymbol{\nu}^2 = 1$) is along the line.[1] These lines are wedge disclinations in nematics (Section 11.1.1). In contrast, the construction of disclinations whose lines are not parallel to $\mathbf{\Omega}$ (e.g., twist disclinations in nematics; Section 11.1.3) does not result from a "bare" Volterra process. Excellent drawings of the Volterra process for wedge and twist disclinations in solids that can be found in the literature[2] illustrate this difficulty. A discussion of

---

[1] F.C. Frank, Disc. Faraday Soc. **25**, 19 (1958).

[2] W.F. Harris, Disclinations, Scientific American **237**, 130 (1977).

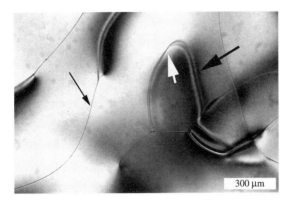

Figure 11.1. Thin and thick disclination lines and singular point (white arrow) as seen in the bulk of a nematic slab under the polarizing microscope.

this question can be skipped in the first reading, and we delay it to Section 11.3. The notion of continuous dislocation density, introduced in Chapter 10, sheds here some light. Other methods, of a very unifying *topological* nature, are required for a thorough understanding of disclinations with a nonsingular core (Sections 11.1.2 and 11.2.3) and for singular points; these methods encompass defects of *all* dimensionalities and are applicable to *all* ordered media. They will be studied in detail in Chapter 12.

11.1. Lines and Points in Uniaxial Nematics: Static Properties

11.1.1. Wedge Disclinations in Nematics

A wedge disclination in a *solid* can be constructed with the help of the Volterra process, *wedge* meaning here that the axis of rotation is along the disclination line (Fig. 11.2). Although such an object is conceptually possible, there is little chance to find an isolated wedge disclination experimentally because of the considerable elastic energy that it would carry.

Twist disclinations, whose axis of rotation is perpendicular to the line, are even more difficult to construct, even conceptually, as already mentioned. Henceforth, disclinations in solids exist only in the form of pairs of opposite sign interacting at short distances. We have mentioned the existence of such pairs in the Frank and Kasper phases and their possible existence in amorphous media. In both cases, the disclinations of opposite signs form interlinked networks.

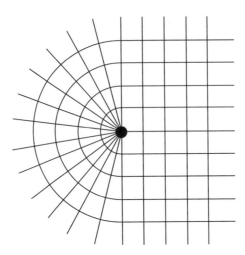

Figure 11.2. A π-wedge disclination in a solid.

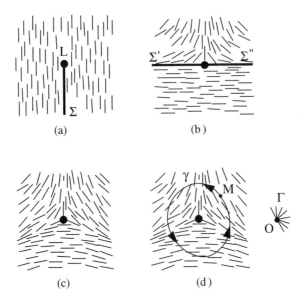

Figure 11.3. Volterra process and Burgers circuit for a wedge disclination $k = -1/2$ in a nematic: (a) cut surface Σ at a line L, (b) opening of the lips of Σ by an angle π and introduction of extra matter, (c) elastic relaxation, (d) the Burgers circuit γ and its hodograph Γ.

In liquid crystals, disclinations are standard objects. Figure 11.3 illustrates the Volterra construction of a wedge line of angle π in a nematic. One recognizes the cut surface Σ, the angle π between its opened lips, the introduction of perfect matter in the dihedral angle, and eventually the final elastic relaxation. The *strength k* of a wedge disclination line is the ratio $k = \alpha/2\pi$, where α is the angle by which the director rotates after the traversal of the Burgers circuit surrounding the line (Fig. 11.3d). It is also, with sign, the angle by which the lips of Σ have been opened. The sign of α is determined by the comparison of the traveling directions along the Burgers circuit and the hodograph Γ of the directors (Fig. 11.3d). Let us consider a (oriented) Burgers circuit γ on which one chooses an arbitrary origin M; each director $\mathbf{n}$ that is met when traversing γ is now parallel-transported to an origin O. The set of extremities traverses a circular arc of angle α, oriented according to the construction. This is the hodograph Γ (see also Section 8.3 and Fig. 8.6). In the example of Fig. 11.3, Γ and γ are oriented in opposite directions, hence, the minus sign attached to the strength k: Although the orientation of the Burgers circuit γ is arbitrary, the sign of k is not. Thus, Fig.11.3 shows construction of a disclination of strength $k = -\frac{1}{2}$. The same result holds if one takes opposite vectors in Fig. 11.3d. Notice that a wedge of matter has been *introduced* during the Volterra process. Contrarywise, the line $k = +\frac{1}{2}$ is built by *removing* a wedge of matter.

Examples of wedge disclinations are pictured in Fig. 11.4; all strengths are a multiple of $\pm\frac{1}{2}$. The reader will convince oneself as an exercise that the strengths of all these lines can be obtained by the construction of the hodographs of the directors. Notice also addition rules: two parallel lines $k = +\frac{1}{2}$ and $k = -\frac{1}{2}$ add, at a distance, to $k = 0$ (use the hodograph method). The hodograph method always provides a way of measuring the strength k of wedge lines whatever it may be, whereas the Volterra process does not have a clear meaning for lines of high strength $|k| \geq 1$.

The defect configurations in Fig. 11.4 are 2D ($\mathbf{n}$ is confined to planes perpendicular to the disclination axis) and twistless. This *planar model* has been used by Frank[1] to calculate the elastic energies of disclinations, under the assumption $K_1 = K_3 = K$. The director components write

$$n_x = \cos\varphi, \quad n_y = \sin\varphi, \quad n_z = 0, \tag{11.1}$$

where φ is a function of the Cartesian coordinates x, y, or the polar coordinates r, θ.

The Euler–Lagrange equation coincides with the 2D Laplace's equation of electrostatics:

$$\Delta\varphi = \frac{\partial^2\varphi}{\partial x^2} + \frac{\partial^2\varphi}{\partial y^2} = 0, \tag{11.2a}$$

which writes in polar coordinates (r, θ) as

$$\frac{1}{r}\frac{\partial}{\partial r}\left(r\frac{\partial\varphi}{\partial r}\right) + \frac{1}{r^2}\frac{\partial^2\varphi}{\partial\theta^2} = 0. \tag{11.2b}$$

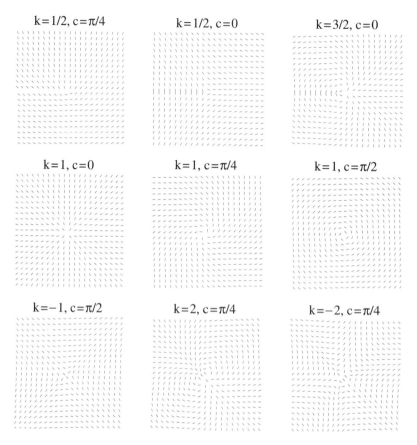

Figure 11.4. Director configurations $n_x = \cos(k\theta + c)$, $n_y = \sin(k\theta + c)$ for wedge disclinations of different strength k and constant c; $\theta = \arctan(y/x)$.

We look for its singular solutions, which are of two types:

$$\varphi = A\theta + B(r, \theta), \quad \varphi = C \ln r + D(r, \theta), \tag{11.3}$$

where B and D are harmonic functions; A and C are constants. We are interested in the first type of solution, which yields dislocations of rotation. Note that the second one describes a rotation of the director infinitely repeated when traversing a direction $\theta = $ const. We must have, by virtue of the hodograph rule,

$$\oint d\varphi = 2\pi k, \tag{11.4}$$

where the integral is taken along any circuit surrounding the origin. Hence,

$$A \equiv k, \tag{11.5}$$

where k is an odd or even multiple of $\pm\frac{1}{2}$. The reader will easily check that the solution

$$\varphi = k\theta + B(\theta, r) \tag{11.6}$$

fulfills all of the topological requirements for a disclination at the origin; it is also required that $B(\theta, r)$ is a harmonic regular function at the origin. The energy per unit length of line $\frac{1}{2} K \int (\nabla \varphi)^2 \, dx \, dy$ is

$$W = \pi K k^2 \ln \frac{R}{r_c} + W_c, \tag{11.7}$$

where r_c and W_c are the radius and the energy of the disclination "core." This expression is totally similar to the expression for the energy of a screw dislocation in solids, and it calls for the same comments. The 2D anisotropic case ($K_1 \neq K_3$) should yield a similar expression, with $K = K(K_1, K_3)$ being a function of K_1 and K_3. Note that the K_{24} and K_{13} energies (5.4) of the planar lines are identically zero. The k^2 coefficient indicates that the out-of-the-core energy of a $k = \pm1$ line is twice as large as the sum of the energies of two $k = \pm\frac{1}{2}$ lines; therefore, a $k = \pm1$ line in the Frank model is expected to split into a pair of $k = \pm\frac{1}{2}$ lines.

The core contribution cannot be studied within the Frank–Oseen elastic approach, as the director gradients are too large. In the simplest model, one can assume that the core preserves uniaxial order and that the amplitude s of the order parameter decreases to 0 over a correlation length $\xi \sim r_c$. A complete calculation of the variation of s is possible, in the framework of the Ginzburg–Landau theory, for instance. The limitation of such types of calculation is that the Ginzburg–Landau model is valid essentially when the typical gradients $\nabla s < s/\xi$. In the case of stronger gradients, ab initio calculations are preferable.

In the model of an isotropic core, $s(r \to 0) \to 0$, the radius r_c can be roughly estimated as follows. We take $k_B(T_{NI} - T)$ as an energy estimate per degree of freedom, T_{NI} being the clearing point. Because the number of degrees of freedom is proportional to the number of molecules, W_c scales as r_c^2: $W_c \sim k_B(T_{NI} - T)\pi r_c^2 \rho N_A/M$, where ρ is the mass density of the nematic, M is the molecular mass, and N_A is the Avogadro's number. Minimization of the total energy (11.7) yields

$$r_c \sim k \sqrt{\frac{MK}{\rho N_A k_B(T_{NI} - T)}}, \quad W_c \sim \pi k^2 K. \tag{11.8}$$

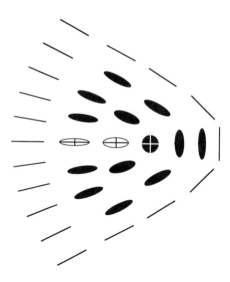

Figure 11.5. Biaxial core structure of the core of a disclination $k = 1/2$. Redrawn from Lyuksyutov.[3]

Therefore, r_c is typically of the order of ξ, i.e., a few molecular lengths in the nematic phase; when $T \to T_{NI}$, r_c diverges.

The second model of the disclination core lifts the restriction on uniaxiality and allows for a biaxial order. The order parameter of the nematic liquid crystal, as discussed in Chapter 3, is a symmetric traceless second-order tensor $\overline{\overline{Q}}$. Uniaxial order corresponds to the case when the two eigenvalues of $\overline{\overline{Q}}$ are degenerate. Generally, all three are different and

$$Q_{ij} = s \left(n_i n_j - \tfrac{1}{3}\delta_{ij} \right) + p(l_i l_j - m_i m_j), \qquad (11.9)$$

where $\mathbf{n}, \mathbf{l}, \mathbf{m} = \mathbf{n} \times \mathbf{l}$ is the set of orthogonal unit directors and p is the "biaxiality parameter": $p = 0$ in the uniaxial phase; see (3.3). Strong distortions at the core may relax by varying both s and p instead of keeping $s = 0$ and $p = 0$. The biaxial core model is shown in Fig. 11.5. Far away from the core, the nematic is uniaxial, $s \neq 0$, $p = 0$. The order is also uniaxial right at the center, with the local optical axis along the line. The two uniaxial regions are separated by a biaxial ring.[3]

11.1.2. Nonsingular Disclinations

The result above according to which $k = \pm\frac{1}{2}$ lines are less energetic than are any other lines, and the k^2 behavior of the elastic energy, suggests that "half-integer" disclination

---

[3]I.F. Lyuksyutov, Zh. Eksp. Teor. Fiz. **75**, 358 (1978)/Sov. Phys. JETP **48**, 178 (1978); N. Schopol and T.J. Sluckin, J. Phys. (Fr.) **49**, 1097 (1988).

lines should be frequent and $k = \pm 1$ should be rare in uniaxial nematics. The experimental situation is more subtle, as follows from the observation of the Schlieren textures. We first discuss the case of SMLCs.

The Schlieren textures are formed when the nematic slab is sandwiched between two flat glass plates and the director is tangential or slightly tilted at the surfaces. In-plane director distortions lead to extinction brushes (see Chapter 3). In tangentially anchored samples, one observes defects as centers from which two or four brushes emerge. One might suggest that the centers represent the ends of planar disclinations seen from above. Recalling that an extinction brush marks the region where $\mathbf{n}$ is either parallel or perpendicular to the polarization of the incident light, Section 3.3.3, one easily finds a relationship between $|k|$ and the number N of brushes pinned at the center:

$$|k| = N/4. \tag{11.10}$$

Thus, centers with four brushes can be identified as $k = \pm 1$ defects and with two brushes as $k = \pm\frac{1}{2}$ disclinations.

The first unexpected feature is that slabs with tangential boundary conditions show comparable numbers of defects $k = \pm 1$ and $k = \pm\frac{1}{2}$. Moreover, when the director at the bounding plates becomes slightly tilted (as a result of temperature-induced "anchoring" transition), the $k = \pm\frac{1}{2}$ lines cease to exist as isolated objects and bring about wall defects (Fig. 11.6b,c). In contrast, the $k = \pm 1$ defects survive. The difference stems from the fact that for tilted surface alignment, the director projection $|\mathbf{n}_s| < 1$ onto the surface does not have the property $\mathbf{n}_s = -\mathbf{n}_s$ anymore. Across the wall, $\mathbf{n}_s$ reorients into $-\mathbf{n}_s$, through $|\mathbf{n}_s| = 1$ at the center of the wall (dotted area in Fig. 11.6c). For $k = \pm 1$, the fact $\mathbf{n}_s \neq -\mathbf{n}_s$ does not mean much, since $\mathbf{n}_s \to \mathbf{n}_s$ when going once around the defect.

The second striking difference between $k = \pm 1$ and $k = \pm\frac{1}{2}$ lines, which is also not foreseen by the planar Frank model, is revealed when one of the plates is gently shifted (Fig. 11.7). Upon separation in the plane of observation, a pair of centers with $N = 2$ is always seen as connected by a thin thread (the disclinations core) that strongly scatters light. In contrast, pairs with $N = 4$ brushes separate without leaving any visible singularity between them: The distortions along the imaginary line joining the two centers are fuzzy and extended in the bulk. The only singularities seem to be two surface point defects. The shear experiment helps to understand the nature of lines in Fig. 11.1: The "thins" are $k = \pm\frac{1}{2}$ ($N = 2$) lines, whereas the "thicks" are $k = \pm 1$ ($N = 4$) lines. We shall see now that the thick $k = \pm 1$ lines do not fit the planar model at all, as the director configuration is not planar, at least in SMLC uniaxial nematics.

The wedge disclination $k = 1$ with a radial $\mathbf{n}$ (Fig. 11.8a) can lose its core singularity by an "escape in the third dimension":[4] At the central axis, $\mathbf{n}$ is parallel to the disclination line, but preserves its radial orientation at the periphery (Fig. 11.8b). This geometry can be easily observed in a capillary, where perpendicular (= radial) anchoring ensures the pres-

---

[4]P.E. Cladis and M. Kleman, J. Physique **33**, 591 (1972); R. Meyer, Phil. Mag. **27**, 405 (1973).

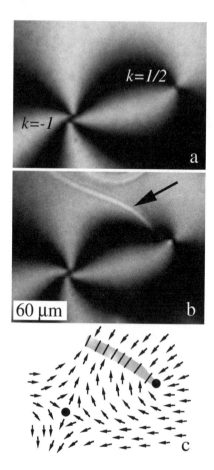

Figure 11.6. Nematic Schlieren texture with $k = -1$ and $k = 1/2$ defects. Crossed polarizers. Director surface orientation is tangential in (a) and slightly tilted in (b). Note the wall defect attached to the $k = 1/2$ line, marked by an arrow. Scheme (c) shows surface director field for texture (b); small arrows indicate tilted director. Photos courtesy Yu. A. Nastyshyn.

ence of a $k = 1$ disclination; the central line is diffuse, and often shows *point singularities*, corresponding to transition regions from an escape up to an escape down (Fig. 11.8c).

The escape of the director along the disclination line can remove singularity from the core of any line with an integer k. In Fig. 11.8, the escape replaces (a) pure splay with (b) splay-bend. If a disclination were of a circular type ($k = 1, c = \pi/2$ in Fig. 11.4), it could escape by replacing pure bend with twist-bend. In contrast, *half-integer lines* are necessarily singular. This result will be made clearer in Chapter 12.

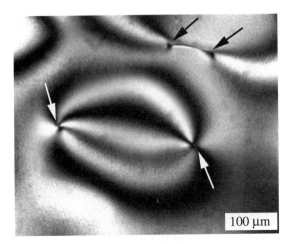

Figure 11.7. Shear of the Schlieren texture reveals that the centers with two brushes (black arrows) are the ends of singular thin disclinations, whereas centers with four brushes (white arrows) are just singular points with no singularity in the bulk.

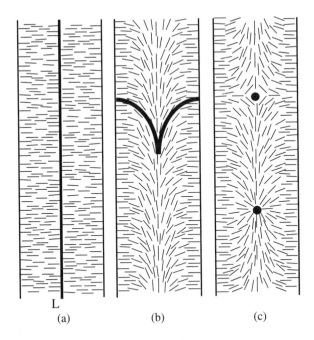

Figure 11.8. Meridional cuts of a cylinder with a singular (a) and nonsingular escaped $k = 1$ disclination (b); opposite directions of escape produce singular points, a hyperbolic and radial hedgehogs.

The elastic energy per unit length of an integer line with an escaped core varies as $K|k|$, and it does not depend on the lateral size R of the system (Problem 11.1):

$$W = 3\pi K'|k|, \tag{11.11}$$

where K' is some function of K_1, K_2, and K_3. It is clear that the escape phenomenon is generally favored (compare 11.11 and 11.7), but might be prevented if the elastic constants differ too much. In SMLCs, where K_1, K_2, and K_3 are of the same order of magnitude, the escape is favored, except if R is not much larger than r_c (multiplication of lines with isotropic core near the clearing point).

In LCPs, however, K_1, K_2, and K_3 can be much different, and the nature of the distortion (i.e., the predominance of splay, twist, or bend) plays a major role. When $K_3 > K_1$, which is true for some LCPs, especially lyotropic solutions of hard rod particles,[5] and many SMLCs, the energy of the radial escaped configuration writes (Problem 11.1) as

$$W = \pi K_1 \left(2 + \frac{\arcsin \beta}{\beta\sqrt{1 - \beta^2}} \right), \quad \beta = \sqrt{1 - K_1/K_3}, \tag{11.12}$$

and is smaller than the energy of the radial singular line $W = \pi K_1 \ln(R/r_c) + W_c$ for reasonably large R/r_c and not very large K_3/K_1 (Fig. 11.9). But in the nematic phase of rigid polymers, if $K_3/K_1 \gg 1$, a singular integer line can be favored over an escaped one.

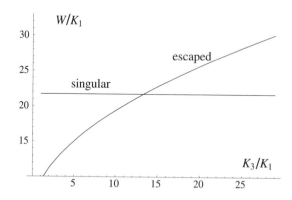

Figure 11.9. Line tensions of a radial $k = 1$ disclinations vs. K_3/K_1 for the singular planar configuration, $W/K_1 = \pi \ln(R/r_c)$ (straight line, $R/r_c = 10^3$) and for the escaped configuration (11.12).

[5]S.D. Lee and R.B. Meyer, J. Chem. Phys. **84**, 3443 (1986).

Because singular $k = \pm 1$ lines are unstable against splitting into $k = \pm\frac{1}{2}$ disclinations, one expects to observe most frequently those later, which is verified.

11.1.3. Twist Disclinations

Twist disclinations occur when following the director about a Burgers circuit enclosing the line, **n** is viewed as continuously rotating about a direction $\mathbf{\Omega}$ perpendicular to the line. In principle, this definition restricts twist geometries to configurations in which the director remains in a plane orthogonal to the rotation direction (Fig. 11.10) or at a constant angle to its normal. Clearly, twist disclinations are favored when the twist modulus K_2 is smaller than K_1 and K_3 (Problem 11.2). Since this is often the case in SMLCs, the consideration of twist disclinations is of some physical importance.

However, there is no clearcut distinction in topological status between a twist line and a wedge line, and it is possible to pass continuously from one to the other (Fig. 12.15). This points already to the insufficiencies of the hodograph method in measuring the strength of a line when it is not a planar configuration. A generalization is as follows (Bouligand): Attach to each point of the circuit the director that meets it at that point, and consider the ribbon formed by the circuit and those attached directors. This is a Möbius ribbon (which has only one side) when $|k| = 1/2$; i.e., when the total rotation is $\pm\pi$. Note that because the mirror image of a Möbius ribbon is not equal to the original, the chirality of the ribbon reflects a nonscalar property of the disclination (Fig. 11.10). More will be said in Chapter 12 about these topics.

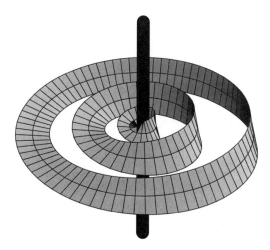

Figure 11.10. Disclination line with a π-twist of the director. Invariance of the Möbius strip $k = 1/2$ with the position of the strip.

11.1.4. Defect Lines in LCPs

The differences between disclination lines in LCPs and in SMLCs alluded to above extend to the core structure. Clear experimental data for the core structure of twist $|k| = \frac{1}{2}$ lines and wedge $k = -\frac{1}{2}$ lines have been obtained for colloidal solutions of tobacco mosaic virus (TMV) by freeze-fracture microscopy. The twist disclination core is small, of the order of a virus length, and the virus particles reorient abruptly by $\pi/2$ at the core. On the other hand, core of the wedge line is large (several virus lengths), the rod-like particles are seen oriented along the disclination line,[6] and they twist out to a direction perpendicular to the core alignment, in such a way that the overall configuration has a well-defined three-fold symmetry, as expected from the standard model of a $k = -\frac{1}{2}$. Similar conclusions follow from optical microscopy observations[7] of a main-chain polyester nematic phase: The $k = -\frac{1}{2}$ wedge lines show, even at that scale, a three-fold symmetry, and a mobility lower than that of the $k = +\frac{1}{2}$ wedge lines, whose optical contrast is different. It was concluded in that investigation that the $k = -\frac{1}{2}$ wedge lines obey the same model as for the TMV case, whereas the molecules in the core of the $k = +\frac{1}{2}$ wedge lines are perpendicular to the core direction. Because one expects the rotational diffusivity $D_{rot} \sim K/\eta$ to be much smaller than the translational diffusivity $D_{trans} \sim \eta/\rho$ (η being a viscosity and ρ a mass density), this could also explain the difference in mobility. Monte-Carlo simulations of $k = -\frac{1}{2}$ wedge lines in polymers clearly show a biaxial ring at the disclination core[8] and a three-fold symmetry; the details depend strongly on the ratio of the elastic constants, which is determined by the length-to-diameter ratio of the rod-like molecules.

11.1.5. Singular Points

Singular points form either in the bulk (Figs. 11.1 and 11.8c) or at the surfaces (Fig. 11.7). Point defects in the bulk are called "hedgehogs," whereas point defects at the surface are called "boojums"; the reason for this terminology will become clear in Chapter 13. A practical way to observe bulk and surface point defects is to disperse nematic droplets in an isotropic fluid (such as glycerol or water) (Fig.11.13). As will be discussed in Sect. 13.2.5, point defects occur in the equilibrium state of these droplets due to boundary conditions and topological constraints. Point defects can spread into disclination loops (Fig. 11.11).

The energy of singular points scales as $K''R$, where R is a macroscopic length (for example, the distance between successive singular points along a disclination line) and K'' is some combination of the Frank coefficients depending on the geometry. The Frank

---

[6]J.A.N. Zasadzinski, M.J. Sammon, R.B. Meyer, M. Cahoon, and D.L.D. Caspar, Mol. Cryst. Liq. Cryst. **138**, 211 (1986).

[7]G. Mazelet and M. Kleman, Polymer **27** (1986) 714–720.

[8]S.D. Hudson and R.G. Larson, Phys. Rev. Lett. **70**, 2916 (1993).

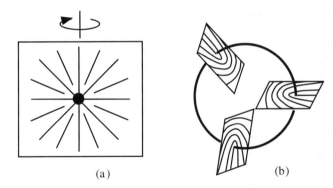

Figure 11.11. (a) A point defect–radial hedgehog in a uniaxial nematic spreads into (b) a disclination loop.

energy density, which scales like $\frac{K''}{r^2}$, does not diverge on the core:

$$\int_{r_c}^{R} \frac{K''}{r^2} r^2 \, dr \sim K''(R - r_c). \tag{11.13}$$

However, the core exists physically, because topologically there is now way to escape the singularity and preserve uniaxial order. Assume indeed that the core is a sphere of radius r_c (at least of the order of a molecular length) with a constant energy density $\varepsilon > 0$ (either isotropic or biaxial[9]). Minimizing the total energy, one gets $r_c^2 \sim K''/4\pi\varepsilon$, and a total energy for the singular point $W_p = K''(R - \frac{2r_c}{3})$, i.e., smaller than $K''R$.

So far, all calculations neglected the divergence elastic terms (see Chapter 5). These terms become important when the deformations are not planar. This is the case of integer disclinations with an escaped core, and this is the case of point defects. For example, the elastic energy of a "radial" $\mathbf{n} = (x, y, z)/\sqrt{x^2 + y^2 + z^2}$ and a "hyperbolic" $\mathbf{n} = (-x, -y, z)/\sqrt{x^2 + y^2 + z^2}$ hedgehogs reads as, respectively,

$$W_r = 8\pi R(K_1 - K_{24}) \quad \text{and} \quad W_h = 8\pi R \left(\frac{K_1}{5} + \frac{2K_3}{15} + \frac{K_{24}}{3} \right); \tag{11.14}$$

K_{13} is assumed to be zero. Note the difference in the sign of the K_{24} term. Difference in the elastic constants might be an important factor in the stability of hedgehogs vs disclination loops.[10]

---

[9]E. Penzenstadler and H.-R. Trebin, J. Phys. France **50**, 1027 (1989).

[10]H. Mori and H. Nakanishi, J. Phys. Soc. Japan **57**, 1281 (1988); O.D. Lavrentovich, T. Ishikawa, and E.M. Terentjev, Mol. Cryst. Liq. Cryst. **299**, 301 (1997).

11.1.6. Confinement-Induced Twists

In the next section, we will discuss cholesteric phase, which is, loosely speaking, a twisted version of the nematic phase. It is of interest to discuss twist deformations that occur in the confined nematic samples, even when there is no chemical chirality of the molecules. A pure twist can be produced by placing a nonchiral nematic liquid crystal between two parallel rubbed solid surfaces and then rotating one plate in its own plane. Such a structure is optically active; The twist is maintained by the surface "azimuthal" anchoring.

One would expect that when **n** is allowed to rotate in the plane of one of the plates (no azimuthal anchoring), the twist and optical activity would disappear. Surprisingly, this is not what one can observe by placing a nematic droplet on a rubbed plate and letting the upper nematic surface free: The sessile droplets demonstrate significant optical activity.[11] The phenomenon can be explained if one takes into account that the free surface of a sessile drop is usually curved. Thickness gradients produce an in-plane aligning torque, even when the sample is bounded by isotropic media which is called the "geometrical anchoring" effect.[12] For example, the wedge profile shown in Fig. 11.12b forces **n** to align normally to the thickness gradient (say, along y-axis in Fig. 11.12b) in order to reduce the elastic energy of splay. When one of the plates is rubbed along any direction different from y, competition between the geometrical and physicochemical easy axes can result in twist.

To show this, let us calculate the energy per unit area of the wedge, neglecting director distortions in the plane of the cell. We parameterize the director through the polar angle θ and the azimuthal angle φ as $(n_x, n_y, n_z) = [\sin\theta(z)\cos\varphi(z), \sin\theta(z)\sin\varphi(z), \cos\theta(z)]$. At the bottom plate, $\theta(z = 0) = \pi/2$ and $\varphi(z = 0) = 0$. The director is tangential to the upper surface. If the two bounding surfaces were parallel, then in equilibrium $\theta(z) = \pi/2$ and $\varphi(z) = 0$ (Fig. 11.12a). Suppose now that the upper surface is tilted around the y-axis by an angle γ (Fig. 11.12b). The polar angle $\theta(z = d)$ now depends on γ and on the azimuthal parameter φ_0, which is the angle between **n** and a fixed axis x' in the inclined upper plane: $\theta(z = d) = \arccos(\sin\gamma\cos\varphi_0)$. Small deviations from the uniform state, $\theta(z) \to \pi/2 + \theta_1(z)$ and $\varphi(z) \to 0 + \varphi_1(z)$, lead to the free energy density $f = \frac{1}{2}K_1\theta_{1,z}^2 + \frac{1}{2}K_2\varphi_{1,z}^2$. The bulk equilibrium equations, $\theta_{1,zz} = 0$ and $\varphi_{1,zz} = 0$, together with the boundary conditions above, lead to the energy per unit area[12]

$$F = \frac{K_1}{2d}[\arcsin(\sin\gamma\cos\varphi_0)]^2 + \frac{K_2}{2d}\left[\arctan\left(\frac{\tan\varphi_0}{\cos\gamma}\right)\right]^2.$$

According to the last equation, the equilibrium azimuthal angle at the upper surface can be nonzero (Fig. 11.12c). It implies twist and hence optical activity of the sessile droplet. The twist angle increases as the ratio K_2/K_1 decreases so that the effect may be strongly

[11]D. Meyerhofer, A. Sussman, and R. Williams, J. Appl. Phys. **43**, 3685 (1972).
[12]O.D. Lavrentovich, Phys. Rev. A **46**, R722 (1992).

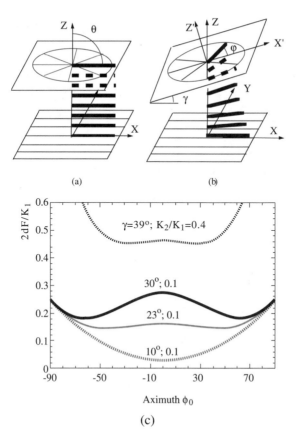

Figure 11.12. (a) Tangentially aligned nematic liquid crystal confined between two plates; the bottom plate is rubbed; the top plate is isotropic. (b) The tilt γ of the upper plate tends to reorient the director normally to the plane of the figure and results in twist; (c) Elastic energy vs azimuthal angle at the top surface for different γ and K_2/K_1.

pronounced for nematic polymers such as PBG, where the ratio K_2/K_1 can be as small as 0.1 or even smaller.

Twist relaxation of splay and bend is a general phenomenon in materials with small K_2. The well-known example is the periodic pattern of stripes that occur in the geometry of splay Frederiks transition in polymer nematics with a small (less than 0.33) ratio K_2/K_1. A field applied normally to the planar nematic cell causes stripe structures[13] composed mostly of twist rather than of a uniform splay response observed in regular materials.

Especially clear demonstration of twist relaxation is given by tangentially anchored spherical nematic droplets suspended in an isotropic matrix (glycerin) (Fig. 11.13). The

---

[13]F. Lonberg and R.B. Meyer, Phys. Rev. Lett. **55**, 718 (1985).

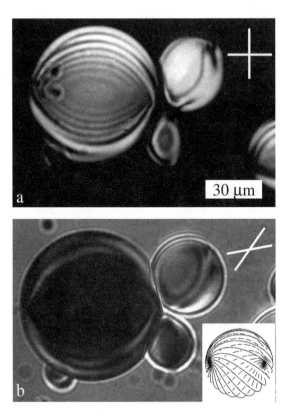

Figure 11.13. Double-twisted nematic droplets suspended in an isotropic matrix. (a) The central part of the largest droplet is bright when the axis is along one of the crossed polarizers (white bars) and (b) dark when the polarizer and analyzer are not crossed at right angle. The droplet is thus optically active due to the director twist. The insert shows the director configuration at the droplet's surface. Nematic n-butoxyphenyl ester of nonylhydrobenzoic acid dispersed in glycerin. Volovik and Lavrentovich, Zh. Eksp. Teor. Fiz. **85**, 1997 (1983) [Sov. Phys. JETP **58**, 1159 (1983)].

director lines join two point defects, boojums, at the poles of the droplet. However, instead of a naive picture of pure splay and bend, with lines being meridians that lie in the planes of constant azimuth, one observes a twisted structure. The director lines are tilted with respect to the meridional planes. This tilt decreases as one approaches the axis of the droplet. Each droplet is thus twisted and optically active despite the nonchiral nature of the molecules of both the nematic and the matrix. Of course, there is an equal number of "left"-handed and "right"-handed droplets in the dispersion.

The droplets in Fig. 11.13 present in fact a double-twist rather than a simple unidirectional twist. Double-twist is discussed in Section 11.2.2 in relation to the saddle-splay modulus K_{24}.

11.2. Cholesterics

The director field in the ground state of chiral phases is nonuniform because molecular interactions lack inversion symmetry. The cholesteric phase in which the director **n** is twisted into a helix is the simplest in the broad variety of spatially distorted chiral structures. Spatial scale of background deformations, e.g., the pitch p of the helix, is normally much larger ($p \geq 0.1\mu$m) than is the molecular size because the interactions that break the inversion symmetry are weak. The twisted ground state of chiral liquid crystals willingly accepts additional deformations imposed by external fields, by surface interactions, or by a tendency of molecules to form smectic layers, hexagonal order, or double-twist arrangements. Very often, such additional deformations result in topological defects. In this chapter, we discuss basic properties of defects such as disclinations and dislocations in pure cholesterics. More can be learned from Chapter 12 on topological properties. Even at the simplest level, the subject of defects in cholesterics is more complicated than in uniaxial nematics, as the properties of defects and deformations at scales smaller than p and larger than p are different.

11.2.1. Elastic Theory at Different Scales

There are two complementary approaches to describe distortions in the cholesteric phase, depending on the ratio L/p, where L is the characteristic scale of the deformations or the size of the liquid crystal sample. We distinguish weakly twisted cholesterics ($L/p \ll 1$) and strongly twisted ($L/p \gg 1$) cholesterics. The elasticity of weakly twisted cholesterics is described by (5.3). Any weak deformation $|\nabla\mathbf{n}| \ll q_o$ should show the same characters as in a nematic. In other words, there is little difference between a slightly twisted nematic and a cholesteric observed on a scale smaller than the pitch. And indeed optical observations of large pitch cholesterics in samples of size $L < p$ reveal thick and thin disclinations, as in nematics, Fig. 11.1.

The cholesteric characteristics prevail at larger scales $L \gg p$. These characteristics fall into two categories: the layered structure (of periodicity $p/2$), which entails properties analog to those of smectics; and the existence of a local trihedron of directors as in biaxial nematics. One therefore might expect at the same time dislocations, focal conic domains (see Chapter 10), and three types of disclinations. This happens to be true, but the actual situation is somewhat more subtle, as we will see in Section 11.2.3.

At $L \gg p$, the elastic properties of the cholesteric layered structure have been tentatively described within the framework used for the lamellar phases (Chapter 5). Here again, two different situations are possible. First, the cholesteric "layers" might be only slightly bent and preserve the topology of flat surfaces. The free energy density describing layers dilatation and small tilts writes in terms of the displacement field $u(x, y, z)$

$$f = \frac{1}{2}K\left(\frac{\partial^2 u}{\partial x^2} + \frac{\partial^2 u}{\partial y^2}\right)^2 + \frac{1}{2}B\left[\left(\frac{\partial u}{\partial z}\right) - \frac{1}{2}\left(\frac{\partial u}{\partial x}\right)^2 - \frac{1}{2}\left(\frac{\partial u}{\partial y}\right)^2\right]^2, \qquad (11.15)$$

where one introduces renormalized constants $B = K_2 q_0^2$ and $K = \frac{3}{8} K_3$; this coarse-grain model is called the de Gennes–Lubensky model.

When deviations from the flat geometry are substantial, it is more appropriate to use the principal curvatures σ_1 and σ_2 of the layers:

$$f = \tfrac{1}{2} K (\sigma_1 + \sigma_2)^2 + \tfrac{1}{2} B \gamma^2, \tag{11.16}$$

where $\gamma = |p - p_0|/p_0$ is a relative dilatation. In the coarse-grained picture, the curvature energy scales as $F_c \sim KL$, whereas the dilatation energy scales as $F_p \sim BL^3$; hence, $F_p/F_c \sim (L/p)^2 \gg 1$. In other words, at $L/p \gg 1$, the theory treats the cholesteric medium as a system of equidistant (and thus parallel) layers with predominantly curvature distortions. Generally, the boundary conditions can be satisfied only by the appearance of large-scale defects, such as *FCDs* and *oily streaks*.

11.2.2. Weak Twist Deformations: Double Twist

One may inquire about the meaning of the K_{24} term in a weakly twisted cholesteric; the solution is in the double-twist tendency of cholesterics.[14] Let n_0 be some director, e.g., along the axis Z (see Fig. 2.28). In the local state of the smallest energy, the chiral molecules in the vicinity of n_0 tend to rotate helically along *all* of the directions perpendicular to n_0. This double-twist is energetically preferable than is the one-dimensional twist, at least for some chiral materials.

In cylindrical coordinates, the elementary double-twist configuration writes as

$$n_r = 0, \quad n_\theta = -\sin\psi(r), \quad n_z = \cos\psi(r), \tag{11.17}$$

with $\psi(0) = 0$. The free energy is

$$f = \frac{1}{2} K_2 \left(q_o - \frac{\partial \psi}{\partial r} - \frac{1}{r} \sin\psi \cos\psi \right)^2 + \frac{1}{2} K_3 \frac{\sin^4 \psi}{r^2} - \frac{K_{24}}{r} \frac{d}{dr} (\sin^2 \psi). \tag{11.18}$$

There is no K_{13} term, because div $\mathbf{n} \equiv 0$. Integrating f, we see that K_{24} term contributes to the energy of a cylinder of matter of radius R by the quantity

$$F_{24} = -K_{24} \int\limits_0^R 2\pi \frac{d}{dr} (\sin^2 \psi) \, dr = -2\pi K_{24} \sin^2 \psi(R), \tag{11.19}$$

which is negative for any value of $\psi(R) \neq \pi n$, when K_{24} is positive. Nucleation of a double-twisted cholesteric geometry is favored in such a case, in particular when K_1 is

[14]M. Kleman, J. Phys. (Paris) **46**, 1193 (1985); J. Phys. Lett. **46**, L-723 (1985).

large compared with K_3. Double-twist can be found in nonsingular disclinations of strength $k = 2$ in cholesteric spherulites (often observed in biopolymers) that are discussed later.

Another example with double twist is met in the chromosome of a microscopic algae, Prorocentrum Micans (dinoflagellate chromosomes).[15] As proposed in Kleman,[16] the structure contains two $k = \frac{1}{2}$ disclinations that rotate helically about the chromosome axis. The double-twist geometry has a limited size, beyond which double twist decreases and frustrations in the system become too large. The layers have a negative Gaussian curvature. This geometry is favored over the spherulitic geometry, probably when K_1 is smaller than K_3, because the $k = \frac{1}{2}$ lines cause splay.

At last, if K_{24} is positive and very large, the cylindrical geometry can become stable versus the cholesteric phase: This is the origin of the *blue phases* (BPs). In Fig. 2.28, as the distance from the z-axis increases, the cholesteric cylindrical shells become flatter and the double twist smoothly disappears. The director far-field distribution becomes closer to the one-dimensional twist of the usual cholesteric phase; the energy gain is reduced. Thus, the double twist cannot be extended over the whole 3D space. A typical radius of the energy-gaining cylindrical region about the $\mathbf{n}_0$ axis is the half-pitch $p/2$. (This is the reason why we discuss the double twist as a weakly twisted structure; the situation should not be confused with the fact that the BPs usually occur in materials with a short submicron pitch). Now, these cylinders of finite radius cannot tile space continuously. According to the current models of BPs, this frustration is relieved by disclinations, either regularly distributed or in disorder. Figure 2.28 illustrates how three cylinders of double twist generate a singularity in the region where they merge. A word of caution should be said about the interpretations of planar disclination lines as a source of saddle splay. There is no K_{24} nor K_{13} contribution to the elastic energy of a straight planar disclination, as already indicated in Section 11.1.1. Both terms vanish when the energy density is integrated over the azimuthal angle around the disclination core. A nonvanishing saddle-splay energy may come from the regions where the disclinations cross or from point defects if such are present.

The blue phases of types BPI and BPII are modeled as regular networks of disclinations with periodicity of the order of p. Indeed, the 3D periodic structure of these phases is revealed in their nonzero shear moduli, ability to grow well-faceted monocrystals, and Bragg reflection in the visible part of the spectrum (which is natural because p is of the order of a few tenths of a micron). The third identified phase, BPIII, that normally occurs between the isotropic melt and BPII, is less understood. It might be a melted array of disclinations. Note that although most blue phases have been observed in thermotropic systems, double-twist geometries are relatively frequently met in textures of biological polymers, like DNA.

DNA, polypeptides (such as PBG mentioned above) and polysaccharides (such as xanthan), and many other biological and nonbiological polymers have a definite handedness due to the chiral centers. Rod-like long molecules of these materials in water solutions

[15]F. Livolant and Y. Bouligand, Chromosoma **80**, 97 (1980).
[16]M. Kleman, Physica Scripta **T19**, 565 (1987).

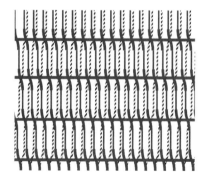

Figure 11.14. Coexistence of twist and close hexagonal packing in a system of chiral rods that form a twist grain boundary phase with lattices of screw dislocations; unidirectional twist perpendicular to the plane of the figure.

often crystallize into a hexagonal columnar phase so that the cross section normal to the rods reveals a triangular lattice. Because the polymers are chiral, close hexagonal packing competes with the tendency to twist. Macroscopic twist can proliferate by introducing screw dislocations into the system, in a way akin to twist grain boundary phases of chiral smectics. Two types of defect-stabilized phases that combine close packing and twist are possible.[17] One is a polymer tilt grain boundary phase, a direct analog of the twist-grain boundary phase and a usual cholesteric with a unidirectional twist (Fig. 11.14). Another is a Moiré grain boundary phase, similar to the blue phases with double twist. In the center of a cylindrical element, there is a polymer rod; the neighboring polymers twist around it, preserving the hexagonal close packing; the cylinders are packed together thanks to the honeycomb lattice of screw dislocations.

11.2.3. Disclinations λ τ, and χ

At large scales, the cholesteric order is specified by three mutually perpendicular directors: λ along the local direction defined by the molecules, χ along the helix axis, and $\tau = \chi \times \tau$. One can immediately envision that there would be three classes, which we denote C_λ, C_τ, and C_χ, of half-integer disclinations, that correspond to π rotations of two out of three directors around the disclination's core. For example, C_λ or λ lines relate to rotations of χ and τ, whereas λ remains nonsingular and oriented along the disclination line (Fig. 11.15). Furthermore, there should be a class, call it $\overline{C}_0$, of integer-strength disclinations with 2π-rotations. Unlike their uniaxial nematic counterparts, these lines cannot escape into the third dimension. Really, if one of the three directors is escaped, as in Fig. 11.8b, two others still remain singular. It is only when the disclinations carries 4π rotation, i.e., $|k| = 2$,

---

[17]R.D. Kamien and D.R. Nelson, Phys. Rev. **E53**, 650 (1996).

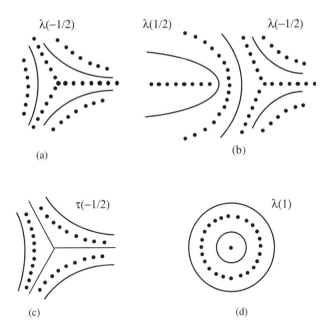

$\lambda(-1/2)$ $\qquad$ $\lambda(1/2)$ $\qquad$ $\lambda(-1/2)$

(a) $\qquad$ (b)

$\tau(-1/2)$ $\qquad$ $\lambda(1)$

(c) $\qquad$ (d)

Figure 11.15. Disclinations λ and τ in the cholesteric phase.

it becomes topologically unstable and can be continuously transformed into a uniform cholesteric. We denote this class C_0.

The striking difference between 2π (stable) and 4π (unstable) lines will become clear in Chapter 12, when we discuss disclinations in the biaxial nematics: Both phases have the same topological classification of defects. The difference between 2π and π (both stable) lines can be illustrated using χ-lines (no singularity in the χ-field) as an example. Suppose the χ-disclination is perpendicular to equidistant cholesteric layers. When one approaches the core, the elastic energy $\sim K(\nabla \mathbf{n})^2$ increases, until at distances $\sim p$, it becomes comparable to the energy difference $\sim K/p^2$ between the cholesteric and the nematic states. At scales smaller than $\sim p$, one deals with the nematic order. Therefore, the 2π-lines with integer k should have a thick core of typical diameter $\sim p$ that is nonsingular from the nematic point of view: The director is uniform (escaped in third dimension) inside the cylinder of diameter $\sim p$. In contrast, χ-lines with half-integer k are singular both for the uniaxial nematic and cholesteric order parameters.

The energy of π-disclinations strongly depends on how the trihedron $\boldsymbol{\lambda}, \boldsymbol{\chi}, \boldsymbol{\tau}$ is distorted. The three directors bear different physical meaning and different distortion energy. Only $\boldsymbol{\lambda}$ is a real director, whereas $\boldsymbol{\tau}$ the $\boldsymbol{\chi}$ are "immaterial" directors; singularities C_τ and C_χ would be generally more costly than would C_λ. The core size of λ-disclinations is of a radius p ("thick" lines) (Fig. 11.15a,b,d), whereas the core of τ-disclinations ("thin" lines) is of molecular size (or somehow larger, as discussed above) (Fig. 11.15c). The line tensions thus differ by an amount $\sim K \ln \frac{p}{a}$, where a is of the order of (1-10) molecular sizes.

Disclinations λ and τ are often observed in the *fingerprint textures*, in which the χ-axis is in the plane of the sample. On the ground of energy estimate above, one would expect that λ defects are more frequent. However, this analysis might be altered if the cholesteric phase is biaxial: Then all three directors might have the same energy weight. Cholesteric textures of biological polymers DNA, PBG, and xanthan show that the λ lines are frequent indeed.[18] On the other hand, the τ lines often appear in pairs with λ lines to replace the χ disclinations of semiinteger k.

11.2.4. Dislocations

The symmetry of rotations $n\pi$ around the χ-axis in cholesterics is equivalent to the symmetry of translations $n(p/2)\chi$. Therefore, the χ disclinations can be equivalently treated as dislocations,[19] with Burgers vector

$$b = -kp. \tag{11.20}$$

The values of the Burgers vector are included in Table 11.1, which summarizes the classification of all line defects in cholesterics. Figure 11.16a pictures a $\chi(1/2)$ wedge disclination (χ is continuous). It can be constructed by a Volterra process performed along the line, by opening the cut surface by an angle π: Each cholesteric layer yields a 2D $k = \frac{1}{2}$ configuration that rotates helically along the line with a pitch p.

The equivalence demonstrated for screw dislocations vs wedge χ disclinations in Fig. 11.16a can be extended to edge dislocations (Fig.11.17) vs twist χ disclinations and, even further, to mixed dislocations and disclinations, for the simple reason that the two corresponding Volterra processes are the same.

Table 11.1. Volterra classification of defect lines in cholesterics; n is integer.

C_0	$\overline{C}_0$	C_λ	C_τ	C_χ
$b = -2np$	$b = -(2n+1)p$			$b = -(n+\frac{1}{2})p$
$\lambda(2n)$	$\lambda(2n+1)$	$\lambda(n+\frac{1}{2})$		
$\tau(2n)$	$\tau(2n+1)$		$\tau(n+\frac{1}{2})$	
$\chi(2n)$	$\chi(2n+1)$			$\chi(n+\frac{1}{2})$

---

[18]F. Livolant, J. Phys. **47**, 1605 (1986).
[19]Y. Bouligand and M. Kléman, J. Phys. (Fr.) **31**, 1041 (1970).

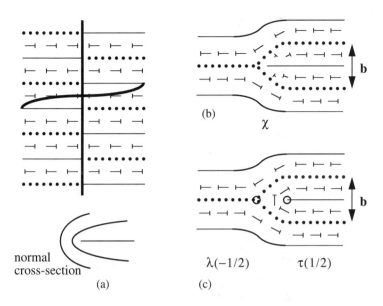

normal
cross-section

(a)

(b)

χ

$\lambda(-1/2)$ $\tau(1/2)$

(c)

Figure 11.16. Equivalence in presentation of χ lines: (a) wedge $\chi(1/2)$-disclination = screw dislocation; (b) χ-twist disclination = edge dislocation; (c) splitting of the core of a dislocation into a pair of disclinations.

An important property of χ dislocations is their ability to split into combinations of λ and τ disclinations. For example, a χ line from the class C_χ can split into a pair of λ and τ lines (classes C_λ and C_τ, respectively). An example is shown in Fig. 11.16b, c: The core splits into a $\lambda(-1/2)$ and a $\tau(1/2)$ separated by a distance $p/4$; the Burgers vector is $b = p/2$, i.e., twice the distance of pairing. In SMLC cholesterics a $b = p/2$ line usually splits into a $\tau(-1/2)\lambda(1/2)$ pair. Figure 11.17 shows a split dislocation with $b = p$, composed of a $\lambda(-1/2)\lambda(1/2)$ pair.

11.2.5. Other Effects of the Layer Structure

11.2.5.1. Focal Conic Domains and Polygonal Textures

The SMLC cholesterics most frequently present domains analog to the FCDs of the first species in smectics (Chapter 10). The layers have then a negative Gaussian curvature, but their thickness varies (contrarily to smectics), at the moderate expense of some twist energy K_2. These so-called *polygonal textures* are less frequent in biopolymers in solution (DNA, polypeptides, etc), because this twist adds a considerable bend contribution K_3, due to rigidity of the long molecules, and some (not too large) splay K_1 (see below).

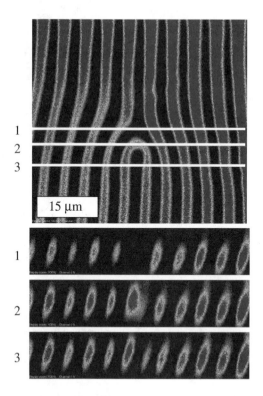

Figure 11.17. Confocal-microscope image of a dislocation $b = p$ in the fingerprint cholesteric texture (χ axis in the plane of the sample). Confocal microscope technique allows one to obtain the image of the director pattern not only in the plane of the sample (top), but also in the vertical cross section (1, 2, 3) as well (Photos courtesy: D. Voloschenko).

11.2.5.2. Robinson Spherulites

The most rigid cholesteric biopolymers have other types of layer textures, the so-called Robinson spherulites: The layers are approximatively along concentric spheres (positive Gaussian curvature). The molecular orientation is necessarily singular, because it is impossible to outline a continuous field of directors on a sphere. This statement can be easily visualized by considering the field of parallels or the field of meridians: Both have a $k = 1$ singularity at each pole, and the total strength of these singularities is $k = 2$ (see Fig. 11.13 and Section 13.2.5.1). It can be shown that the same total strength characterizes any field outlined on a sphere. Coming back to the spherulites, note that the surface singularities must necessarily continue in the bulk, and indeed the observations tell us that either two radial lines $k = 1$ or one radial line $k = 2$ connect the surface to the center of the sphere.

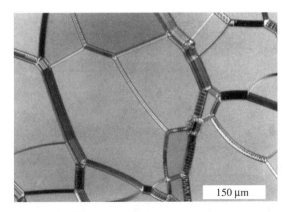

Figure 11.18. Cholesteric planar texture with a network of oily streaks. Most of the oily streaks are of zero total Burgers vector and divide regions of the same color; if the color of two adjacent domains is different, $b \neq 0$.

11.2.5.3. Oily Streaks

Oily streaks and liquid crystals were discovered simultaneously. In 1888, F. Reinitzer studied cholesteryl benzoate and noticed elongated "fluid" inclusions in the cholesteric sample.[20] Oily streaks, as FCDs, are common for many lamellar liquid crystals. In a flat cell with layers paralel to the bounding plates, oily streaks appear as long bands that divide ideal domains of flat layers (Fig. 11.18). Their inner structure is complicated. According to G. Friedel's model,[21] oily streaks are made of pairs of edge dislocations of (large) opposite Burgers vectors $nd, n'd$, making a total Burgers vector $b = (n - n') d$; here d is the characteristic interlamellae distance, such as the thickness of a smectic A layer or the half-pitch in cholesterics. A large Burgers vector dislocation $b = n d$ can further split into two disclinations of opposite signs, as explained in Chapter 9. The simplest variety of an oily streak is shown in Fig. 11.19: two parallel $k = 1/2$ disclinations with a wall between them. The total Burgers vector is zero. There is no transversal striation so that the Gaussian curvature is zero everywhere except at the end region (where it is negative). Normally, the oily streaks do not show free ends and terminate at nodes as in Fig. 11.18; the semicircular ends might be observed when the oily streaks grow, e.g., under an applied field.

11.3. Beyond the Classic Volterra Process, First Step

The above presentation of defects in nematics and cholesterics starts with the application of the Volterra process for solids to nematics, but soon forgets it for a more intuitive pre-

---

[20]F. Reinitzer, Monatsh. Chem. **9**, 421 (1888).
[21]G. Friedel, Ann. Phys. (Paris) **18**, 237 (1922).

Figure 11.19. An oily streak of zero Burgers vector composed of two disclinations.

sentation, just keeping in mind the Volterra classification by the elements of the symmetry group. This choice results from a number of drawbacks relating to the application of the Volterra process to media with continuous symmetries.

We have illustrated Fig. 11.3 the Volterra process for wedge disclinations $|k| = 1/2$, but it is difficult to understand how it can be extended to angles equal to or larger than 2π. Second, twist disclinations cannot be constructed with a Volterra process, except locally. This later point is well illustrated in the figures of Harris.[2] Third, the "escape in the third dimension" obviously does not come out of the Volterra analysis.

The question therefore arises of whether disclinations are really classified by the rotation vectors of the group of symmetry—this classification is classically related to the Volterra process—and if they are, how should the Volterra process be modified. The answer is twofold.

1. Keeping in the frame of the Volterra classification, the Volterra *process* has to be modified, and the key ingredients to do so are the concepts of continuous dislocation densities (see Section 10.2) and continuous disclination densities. Such considerations explain the mobility and flexibility of disclinations as a result of the viscous relaxation of these densities (J. Friedel), and they are the object of the sections that follow.

2. The Volterra *classification* of defects has to be revisited, if one wants for instance to include in a general picture the possible escape of the core, the existence of point defects, and so on. This extended analysis requires topological considerations, Chapter 12.

11.3.1. Dislocations and Disclinations Densities in Relation with Disclinations

The following approach was first introduced in Friedel and Kleman.[22]

[22]J. Friedel and M. Kleman, J. Phys. Paris, Colloq. **30**, C4, 43 (1969); NBS Special Publication **317**, 607–636 (1970).

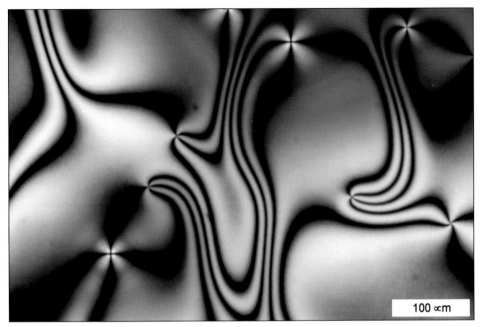

Fig. 3.14. Schlieren texture of a thin (~1μm) film of the nematic 5CB on a glycerin substrate. Note the interference colors; dark brushes mark the regions in which the director is parallel to either the polarizer or the analyzer. Nodes in which four brushes meet are cores of topological point defects.

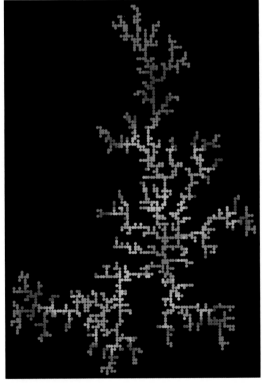

Fig. 7.14. Computer-simulated DLA aggregate of 1500 particles. The total number of released particles was 12421. Algorithm used: Gaylord and Wellin (1995).

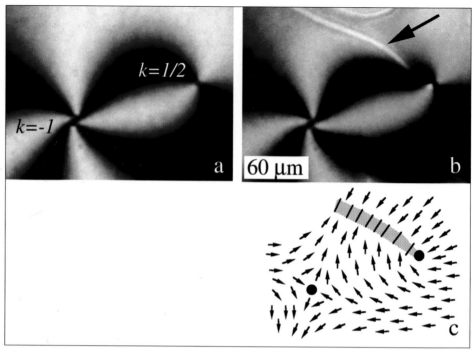

Fig. 11.6 Nematic Schlieren texture with $k=-1$ and $k=1/2$ defects. Crossed polarizers. Director surface orientation is tangential in (a) and slightly tilted in (b). Note the wall defect attached to the $k=1/2$ line, marked by an arrow. Scheme (c) shows surface director field for texture (b); small arrows indicate tilted director. Photos courtesy Yu. A. Nastyshyn.

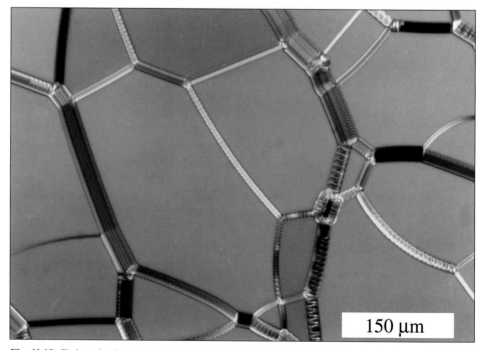

Fig. 11.18. Cholesteric planar texture with a network of oily streaks. Most of the oily streaks are of zero total Burgers vector and divide regions of the same color; if the color of two adjacent domains is different, $b \neq 0$.

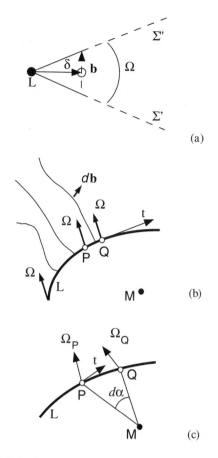

Figure 11.20. Densities of infinitesimal dislocations. (a) Displacement of a wedge disclination parallel to itself (this sketch assumes for the sake of simplicity that Ω is small; (b) densities of dislocations
attached to a curved disclination; M is a running point on the cut surface; (c) dislocations and disclinations densities.

11.3.1.1. Dislocations Densities

Consider first a straight disclination line L in a nematic, of wedge type, and displace it parallel to itself by a distance δ (Fig. 11.20a). Assume also that the rotation vector $\Omega = \Omega \nu$
does not move during this displacement. In terms of the Volterra process, this is equivalent
to displacing the lips of the cut surface with respect to each other by a supplementary quan-
tity $2 \sin \frac{\Omega}{2} (\nu \times \delta)$, i.e., introducing a dislocation of Burgers vector $\mathbf{b} = 2 \sin \frac{\Omega}{2} (\nu \times \delta)$
perpendicular to the disclination. In a solid, such a process would require a large energy,
and it is then forbidden. In a nematic, the *viscous* dispersion of the dislocation into infinites-

imal defects $d\mathbf{b}$ relaxes the elastic energy, and it is in fact the process by which $\mathbf{\Omega}$ follows the line to its new position. The movement of a disclination in a liquid is accompanied by a generic process of emission (absorption) of dislocations.

Consider now a curved disclination line (Fig. 11.20b). Curved lines exist in nematics at rest, and one also observes lines moving while changing shape. We show that the flexibility of a nematic disclination is compatible with a $\mathbf{\Omega}$ varying in position (but not in direction) from point to point along L. Consider two points $\mathbf{P}$ and $\mathbf{Q} = \mathbf{P} + \mathbf{t}\,ds$ infinitesimally close on L. The infinitesimal dislocation introduced by the variation of position of $\mathbf{\Omega}$ from $\mathbf{P}$ to $\mathbf{Q}$ is, by reasoning on the cut surface as above, equal to

$$\mathbf{d_Q}(\mathbf{M}) - \mathbf{d_P}(\mathbf{M}) = 2\sin\frac{\Omega}{2}(\boldsymbol{\nu} \times \mathbf{QM}) - 2\sin\frac{\Omega}{2}(\boldsymbol{\nu} \times \mathbf{PM})$$

$$= -2\sin\frac{\Omega}{2}(\boldsymbol{\nu} \times \mathbf{t})\,ds, \tag{11.21}$$

where $\mathbf{M}$ is any point on the cut surface. This dislocation can be thought of as attached to the line. Compared with the result of the usual Volterra process, the new shape of the disclination line is obtained through viscous relaxation of densities of infinitesimal dislocations $d\mathbf{b} = \mathbf{d_Q}(\mathbf{M}) - \mathbf{d_P}(\mathbf{M})$, which are allowed by the symmetry of the nematic phase. This relaxation optimizes the energy carried by the disclination. The Volterra process defined in Chapter 8 is not relevant to the present geometry, because it can be performed only if the rotation vector $\mathbf{\Omega}$ is fixed in space. The modified Volterra process, which respects the variation in position of $\mathbf{\Omega}$ along the line, consists in a relative rotation $\mathbf{\Omega}$ of the director from one lip to the other at each point of the cut surface, without a relative displacement of the two lips.

The same possibilities do not exist for all lines in a cholesteric. It is easy to show that disclination lines of the λ or τ type must belong to a cholesteric plane and must be rectilinear, either parallel or perpendicular to the director, as in Fig. 11.15. On the other hand, χ disclinations can take any shape, due to the existence of continuous translations in the cholesteric plane and, consequently, of infinitesimal dislocations that can curve χ lines.

Note that the whole set of dislocations densities attached to the curved disclination line has the same status as the dislocations densities alluded to in Section 10.2. Their knowledge is equivalent to the knowledge of the distortion (contortion) of the director field, which is not unique and depends on boundary conditions, material constants, and so on. Dislocations densities can indeed take any shape away from their attachment points on L (Fig. 11.20b).

11.3.1.2. Disclinations Densities

Assume (Fig. 11.20c) that $\mathbf{\Omega_P} \neq \mathbf{\Omega_Q}(\Omega\boldsymbol{\nu_P} \neq \Omega\boldsymbol{\nu_Q})$. $\mathbf{\Omega_P}$ and $\mathbf{\Omega_Q}$ are nearly equal for two infinitesimally close points P and Q, and one can write $|\boldsymbol{\nu_P} \times \boldsymbol{\nu_Q}| = \sin(d\alpha) \approx d\alpha$.

The director of a nematic is an axis of infinitesimal rotational symmetry $d\Omega = d\Omega\mathbf{n}$; the helical axis of a cholesteric is an axis of infinitesimal translation-rotation symmetry $d\mathbf{b}, d\Omega = -\frac{2\pi}{p}d\mathbf{b}$. There are no other continuous rotational symmetries in these phases. We generalize (11.21) to the case when $\Omega = \pi$, i.e., for disclinations of minimal strength in nematics and cholesterics. The product of two π rotations about two different axes is the sum of a translation [already taken into account in (11.21)] and of a rotation by an angle $2d\alpha$ about an axis that is perpendicular to $\Omega_{\mathbf{P}}$ and $\Omega_{\mathbf{Q}}$ and intersects the axes of these rotations. After some calculation, one obtains $d\Omega = \frac{2}{\pi^2}\Omega_{\mathbf{P}} \times \Omega_{\mathbf{Q}}$ (we have defined $\Omega = \pi\boldsymbol{\nu}$). To leading order, the axis of $d\Omega$ passes through P or Q. Eventually, the difference of displacement of the cut surface in P and Q can be written as

$$\mathbf{d_Q}(M) - \mathbf{d_P}(M) = \left[d\Omega \times \mathbf{PM} - 2\sin\frac{\Omega}{2}\boldsymbol{\nu} \times \mathbf{t}\,ds \right], \tag{11.22}$$

i.e., as the sum of an infinitesimal dislocation $d\mathbf{b} = -2\sin\frac{\Omega}{2}(\boldsymbol{\nu} \times \mathbf{t})\,ds$ (as above) and an infinitesimal disclination $d\Omega$. Again, in these expressions, $\Omega = \pi$.

Let L be a curved line in a nematic, and Ω everywhere be tangent to L. Because $d\Omega$ must be along $\mathbf{n}$, the director in the undeformed nematic is perpendicular to the line, which is therefore locally of a wedge type (Fig. 11.21). Note that the director rotates by an angle $\int d\alpha = 2\pi$ when traversing a loop around L (for example, the loop γ in Fig. 11.21). Therefore, one expects that the surface bounded by the loop L is pierced by a $|k| = 1$ line, which in the case of Fig. 11.21 is "escaped."

Infinitesimal disclinations in cholesterics allowed by symmetry are defects related to continuous rotations along the χ-axis, coupled to continuous translations along the same direction. These defects, if interacting with a χ disclination line, may transform it by viscous relaxation into a dislocation (Section 11.2.4).

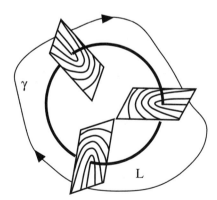

Figure 11.21. Closed wedge line in a nematic.

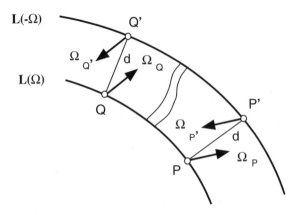

Figure 11.22. Splitting of a dislocation into two disclinations of opposite signs. The wavy lines stand for defect densities.

11.3.2. Extension to Finite Dislocations

A λ or τ disclination of strength $|k| = \frac{1}{2}$ has to be straight, but a pair of λ or τ discli-
nations of opposite signs Ω and $-\Omega$ coupled at a short distance **d** in the same cholesteric
plane and parallel one to the other can take more general shapes, at the expense, however,
of some densities of defects, attached to the lines and extending in the strip region between
the disclinations. These densities are not valid densities from the point of view of the symmetries, but they are present only in a limited core region between the lines. The detailed
analysis[22] shows that helical shapes are favored. Sections patterns of such pairs are already
illustrated in Fig. 11.16. The general case is schematized in Fig. 11.22. With $\boldsymbol{\nu} \times \mathbf{d}$ constant,
the pair is equivalent to a unique dislocation with a Burgers vector

$$\mathbf{b} = 2\boldsymbol{\nu} \times \mathbf{d}. \tag{11.23}$$

Because **b** is quantified, so is the distance **d** between the disclinations.

Splitting of χ disclination lines have been observed in Grandjean–Cano wedges (see
Fig. 8.22): One of the disclinations is always a λ; i.e. it does not carry any material singularity. The first dislocations near the center of the wedge have a small Burgers vector
($b = p/2$), whereas b increases for dislocations far from the center.

11.3.3. Core Structure and Physical Properties

Dislocations in solids are impeded in their movement by a (generally anisotropic) lattice
friction, so that under small or even vanishing applied stresses, a dislocation does not stabi-

lize along a shape of minimal energy. This lattice friction originates from the core structure (Section 8.5.2). The movement of a disclination is of a different nature. It is related to the emission-absorption of dislocations, and how this mechanism relates to the core structure of a mobile disclination. Very little is known on this question. Measurements[23] of the velocity U_0 of a nematic twist disclination line under the action of a magnetic field and observations of the core show that the core reaches a large size of the order of $0.04d$, d being the sample thickness (the line is parallel to the sample boundaries). Such a huge mobile core implies a considerable reorganization of the director. The large core radius seems to be in agreement with a model that states that the viscous dissipation, caused by local rotation of the director, takes place only in the bulk of the material, outside the core region. Other models, which take into account other dissipation processes (backflow, isothermal relaxation of the OP in the core, etc.), yield even larger cores.[24] In fact, experiments on moving disclinations display a number of characteristic lengths attached to the core.[25] A simple dimensionality argument tells us that the length over which any flow vorticity created by the line movement vanishes is of the order of $r_v = 2\nu/U_0$, where ν is the kinematic viscosity for fluid motion. We can visualize this production of vorticity as due to the change of configuration in the core related to the emission-absorption of the density of dislocations, with rate $db/dt \approx U_0$. The region of size r_v extends in the wake of the disclination, like a vortex sheet of effective thickness $\approx (\nu r/U_0)^{1/2}$, where $0 < r < r_v$, if the concept of emitted-absorbed dislocations is valid. The core of a moving disclination should therefore be very different from the static core. Furthermore, as suggested by Friedel,[26] the core may show a longitudinal instability. The ratio r_v/R_c, where $R_c \approx K/\gamma U_0$ is the length above which the director reorientation and the dissipation vanish (Section 11.4.1), is large; hence, the dislocation mechanism affects a large region. We leave aside the discussion of other characteristic lengths.

In cholesterics, the emitted-absorbed dislocations may be finite χ dislocations, as noticed by Frank (see Friedel and Kleman [22]). This process is then akin to what would occur in solids, were the disclinations present in them.

11.4. Dynamical Properties: General Features, Instabilities

How do topological defects such as disclinations and point singularities get into the liquid crystals samples? The causes can be of two categories. First, the defects might exist as a feature of the *equilibrium state* of the system, as illustrated by point defects in the freely suspended nematic droplets, Fig. 11.13. Equilibrium defects can also be generated by mechanical impurities, air bubbles, etc. For example, a spherical air bubble with per-

---

[23]J.A. Geurst, A.M. Spruijt and G.J. Gerritsma, J. Phys. (Paris) **36**, 653 (1975).

[24]O. Parodi, G. Durand, G. Malet, and J.J. Marignan, J. Physique lett. **43**, L727 (1982).

[25]M. Kleman, The relation between core structure and physical properties. In Dislocations 1984, P. Vayssière & al. eds. Editions du CNRS 1984,1.

[26]J. Friedel, J. Physique Coll., **40**, C3, 45 (1979).

pendicular director orientation acts as a radial point defect. To keep the total topological invariants unchanged (see Chapter 12), such a defect should be accompanied by yet another defect, for example, a hyperbolic defect (a radial-hyperbolic pair is shown in Fig. 11.8c). To induce the equilibrium topological defects, the liquid crystal droplet (or the mechanical particle inside the liquid crystal) should be sufficiently large, of the size $> K/w$, where w is a characteristic anchoring strength at the surface of the droplet or the particle (see Chapter 13). Second, the defects can be caused by *far-from-equilibrium* processes, such as thermal or pressure quenching of an isotropic fluid into the liquid crystalline state or by flow of the liquid crystal in the regimes of high Ericksen numbers. These types of defect production are little understood. Besides obvious fundamental interest, the question is of practical importance, e.g., for processing of high-strength LCPs that should be well oriented and defect-free. In what follows, we consider flow processes relevant to the problem of topological defects.

11.4.1. General Features

The Ericksen–Leslie and Harvard models of nematodynamics (Chapter 6) assume that there are no defects (disclination lines or points) of the order parameter. The models are therefore well fitted to the description of stationary instabilities close to the thresholds, when the OP amplitude is not perturbed by the flow, i.e., as long as the Deborah number $\mathrm{De} = \dot{\gamma}\tau_r \ll 1$ (6.73). Here, $\dot{\gamma}$ is a typical value of the shear rate acting on the sample and τ_r is a molecular relaxation time. Stationary instabilities of the director orientation are observable under the polarizing microscope and have thus been studied in thin samples steadily sheared between two parallel plates. This is the only geometry that will be alluded to in this chapter.

Most generally, the behavior of these instabilities is nonlinear above the threshold, which is often made visible by the appearance of disclinations. As a matter of fact, the order parameter is broken, in phase and amplitude, along these disclination lines, but this new situation can be treated theoretically, at least in principle, by the methods developed for disclinations at rest, extended to dynamical problems; i.e., the hydrodynamics equations summarized in Section 6.4.2 are still valid outside of the disclinations cores, as long as De is small enough.

Above $\mathrm{De} \approx 0.1$, say, whatever the nature of the appearing topological defects, the Leslie–Ericksen equations are no longer valid, and the amplitude of the OP couples to the deformation and rotation rates of the director. This regime is reached in LCPs, for which τ_r is large, at moderate values of the shear rate. For example, in a poly benzyl (right) glutamate (PBDG) 13% wt. solution with molecular weight $M = 298\,000$, one finds[27] $\tau_r \approx 0.1s$; hence, a shear rate as small as $\dot{\gamma} \approx 1s^{-1}$ suffices to invalidate the Leslie–Ericksen equations. They are replaced by the Doi molecular theory for polymers.[28]

---

[27]R.G. Larson and D.W. Mead, Liq. Cryst. **15**, 151 (1993) and **20**, 265 (1996).

[28]M. Doi and S.F. Edwards, The Theory of Polymer Dynamics, Oxford University Press, New York, 1986.

Although it is usual to describe large De situations in function of the value of De, other nondimensional quantities are in order when the Leslie–Ericksen equations are valid: The Reynolds number $\mathrm{Re} = \frac{\rho U d}{\eta}$ and the Ericksen number $\mathrm{Er} = \frac{\eta U d}{K}$ (6.73). Here U is a characteristic velocity, for example, the velocity of the upper plate with respect to the lower one, and d is a characteristic length such as the thickness of a sample; K and η are combinations of Frank moduli and Leslie viscosities. Because $\dot{\gamma} = U/d$ is the relevant shear rate, i.e., the relevant frequency, it is interesting to rewrite Re as $\mathrm{Re} = \tau_{\mathrm{lin}}\dot{\gamma}$, and Er as $\mathrm{Er} = \tau_{\mathrm{ang}}\dot{\gamma}$, where $\tau_{\mathrm{lin}} = \frac{\rho}{\eta} d^2$ is the characteristic time for the propagation of linear momentum, and $\tau_{\mathrm{ang}} = \frac{\eta}{K} d^2$ is the characteristic time for the propagation of angular momentum. Note that $\mathrm{Re} \ll 1$ for the thin samples under consideration. Because furthermore the ratio

$$\frac{\mathrm{Re}}{\mathrm{Er}} = \frac{\tau_{\mathrm{lin}}}{\tau_{\mathrm{ang}}} = \frac{\rho K}{\eta^2}$$

is extremely small, it is the director orientation that rules the dynamical processes, and Er is the relevant number. Remark also that (1) the Ericksen number Er is simultaneously the ratio of the viscous torque to the elastic Frank torque; (2) in the Doi regime, there is no Frank elasticity and the Deborah number does not depend on d, hence, no dependence on the sample thickness in the Doi regime.

In SMLCs, $\tau_r \approx 10^{-8}s$, and the Doi dynamical range is not reachable with usual shear rates. Contrarywise, the Doi dynamical range and the Leslie–Ericksen range are both visible in LCPs.

11.4.2. Instabilities of Initially Defect-Free Samples

Flows of nematic fluid practically always result in complex 3D director configurations that ultimately may produce topological defects, regardless of the initial geometry: The director may be uniformly aligned in the shear plane, or be perpendicular to it (Figs. 6.7 and 6.9 in Chapter 6). The crucial question is how exactly the defects nucleate and develop. There is no clear answer to this question. Theoretical models give an insight only into the initial stages of instabilities; most of the new results come from experiments. Below we give a brief account of data, restricting ourselves to the case when the unperturbed director is in the shear plane, formed by the velocity and the gradient of velocity. For this geometry, one distinguishes "flow-aligning" nematics and "tumbling" nematics, according to the value of the nondissipative coefficient λ, or the sign of the product $\alpha_2\alpha_3$ of two Leslie coefficients (Sections 6.4.2 and 6.5.2).

11.4.2.1. Flow-Aligning Regime, $\lambda^2 > 1, \alpha_2\alpha_3 > 0$

The director adopts steady zero-viscous orientation within the shearing plane, which is close to the flow direction (for rod-like molecules). This is the case of most SMLCs, such as 5CB. Flow aligns **n** along the axis

$$\theta_0 = + \arctan \sqrt{(\lambda + 1)/(\lambda - 1)}$$

[Fig. 11.23 and (6.75)]. When the nematic is confined between two plates with strong anchoring the director configuration is determined by the balance of elastic, surface anchoring and viscous forces, as sketched in Fig. 11.23b. At high Er, the viscous torques are predominant, and the director adopts constant orientation θ_0 almost everywhere, except in

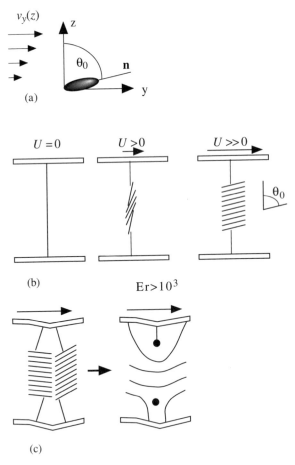

Figure 11.23. (a) Homogeneous alignment in the shear plane of a flow-aligning nematic along the direction $\theta_0 = + \arctan \sqrt{(\lambda + 1)/(\lambda - 1)}$; (b) director configurations in confined geometry with strong anchoring (at some easy axis θ_s) and increasing Ericksen number (increasing velocity of the upper plate); (c) a possible mechanism of disclination nucleation at surface inhomogeneities.

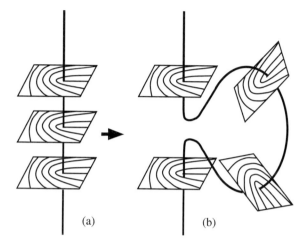

(a)   (b)

Figure 11.24. A disclination $k = 1/2$ emits a disclination loop. Redrawn from Y. Bouligand, Physics of Defects, eds. R. Balian, M. Kleman, J.-P. Poirier, North-Holland, Amsterdam, 1981, p. 665.

the thin boundary layers. Because for many flow-aligning nematics, θ_0 is close to $\pi/2$, there is a chance that in the surface region the director would wind around and form a disclination $k = \pm\frac{1}{2}$; the process can be facilitated by irregularities of the bounding plate that deviate the local easy axis from the vertical axis z (Fig. 11.23c). Experiments[29] indeed show that thin disclinations in the form of closed loops nucleate at bounding plates, starting with shear rates $\dot{\gamma} \approx 1s^{-1}$. Interestingly, one can also observe thick lines forming in the bulk of the sample.[30] Once a disclination nucleates, it can emit other disclination loops (Fig. 11.24). The loops elongate in the flow and tumble; at $\dot{\gamma} \approx 10s^{-1}$, the entire field of view is covered with a mesh-like network of interacting lines. When the shear ceases, the network relaxes by shortening and by collapsing into point defects.

11.4.2.2. Tumbling Regime, $\lambda^2 < 1, \alpha_2\alpha_3 < 0$

This is the case of some nematic SMLCs at temperatures close to the smectic phase (such as 8CB; Section 6.5.2), and this is the case of all LCPs, in accordance with Doi theory. If the director is in the shear plane, it has to rotate with the rate proportional to the sheart rate $\dot{\gamma}$ [see (6.74a)]:

$$d\theta/dt = \dot{\gamma}(\alpha_3 \sin^2\theta - \alpha_2 \cos^2\theta)/\gamma_1, \qquad (11.24)$$

[29]D.J. Graziano and M.R. Mackley, Mol. Cryst. Liq. Cryst. **106**, 103 (1984).
[30]P.T. Mather, D.S. Pearson, and R.G. Larson, Liq. Cryst. **20**, 527 (1996).

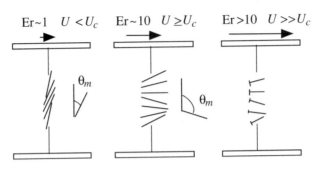

Figure 11.25. Initial stages of instability in a tumbling nematic; see text.

thus, no stationary state in the absence of elastic distortions. Of course, the idea that **n** remains in the shear plane is only an assumption. Its validity depends on elastic (especially, K_2) and viscous parameters. Direct comparison of flow-aligning 5CB and tumbling nematic 8CB shows that at a given Ericksen number, the steady-state density of disclinations is about an order of magnitude higher in 8CB.[31]

Qualitatively, development of director distortions under increasing velocity U of the upper plate (equivalently, the Ericksen number) is illustrated in Fig. 11.25. First, at Er $\sim$ 1, the elastic torques are capable of stabilizing the director in a steady regime with some maximum tilt angle $\theta_m < \pi/2$ in the midplane; θ_m increases with U. In the flow-aligning nematic, θ_m would reach a stationary value $\theta_0 < \pi/2$. In the tumbling nematic, when the viscous torques overcome the stabilizing elastic torques, there is no stable stationary value of θ_m. When U passes some critical velocity U_c, θ_m abruptly becomes larger than $\pi/2$. The transition occurs at Er $\sim$ 10 and is reminiscent of the first-order phase transition (Fig. 11.26). Further increase of the velocity results in out-of-plane director reorientation[32] (Fig. 11.25), and the director can become nearly parallel to the vorticity axis. The evolution of patterns does not end here, as at Er $> 10^3$, one observes roll-cell instabilities[33] involving director modulations along the vorticity axis; the rolls eventually produce disclinations through seemingly a bulk-nucleation process (see Mather et al. [30]). The flow becomes irregular in time; the regime is called director turbulence.[34]

The evolution of textures has been observed in many LCPs, such as PBG solutions in metacresol. High viscosities of LCPs make the Ericksen number large even at low

[31]P.T. Mather, D.S. Pearson, and R.G. Larson, Liq. Cryst. **20**, 539 (1996).

[32]P. Pieranski, E. Guyon, and S. Pikin, J. Phys. Paris, Colloq. **37**, C1, 3 (1976).

[33]P. Pieranski and E. Guyon, Phys. Rev. Lett. **32**, 924 (1974); P. Manneville and E. Dubois-Violette, J. Phys. (Paris) **37**, 285 (1976).

[34]P.E. Cladis and W. van Saarloos, in Solitons in Liquid Crystals, Edited by L. Lam and J. Prost, Springer Press, New York, 1992; G. Marrucci and F. Greco, Adv. in Chem. Phys. **86**, 331 (1993); M. Srinivasarao, in Liquid Crystals in the Nineties and Beyond, Edited by S. Kumar, World Scientific, Singapore, 1995.

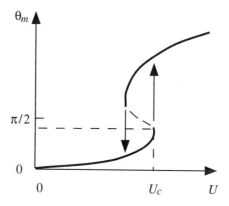

Figure 11.26. Director tilt angle vs velocity for a tumbling nematic. After T. Carlsson, Mol. Cryst. Liq. Cryst. 104, 307 (1984).

shear rates. As a result, disclinations are observed in abundance, both in lyotropic and thermotropic[35] LCPs. Formation of disclinations is normally heralded by development of nonsingular quasi-1D textures, "stripes" and "bands" (Table 11.2). The stripes are oriented along the flow direction, whereas the bands are perpendicular.

The evolution of defect textures, with at least a few initial defects at rest, has been observed in copolyesters and in solutions of biopolymers. It is not yet clear whether these observations are in the Doi range of shears that, as indicated above, is reached with moderate shear rates in high molecular weight polymers. After a stage at very low shear where the defects elongate in the direction of shear and do not interact (which is certainly the Leslie–Ericksen range), one first observes a so-called "worm texture," a kind of disordered polydomain texture at the scale of a few μm, with multiplication of the defects; in this

Table 11.2. Patterns in sheared PBG solutions (R.G. Larson, in Spatio-Temporal Patterns in Nonequilibrium Complex Systems, Eds. P.E. Cladis and P. Palffy-Muhoray, Addison-Wesley, MA, p. 219 (1995)).

Ericksen No. Er	Deborah No. De	Pattern	Flow condition
$10^2 - 10^3$	$\ll 1$	stripes	steady-state shear
$> 10^2$	$\ll 1$	bands	transient, during shear
$> 10^3$	≤ 1	bands	transient, after shear
$\gg 10^3$	≈ 2	stripes	steady-state shear

[35]T. DeNève, P. Navard, and M. Kleman, J. Rheol. 37, 515 (1993).

regime, the first normal stress difference N_1 is negative (Section 6.5.2.1); a negative N_1 might be indicative of the Doi regime. At higher shears, N_1 increases again sharply and reaches positive values; the corresponding so-called "ordered texture" is birefringent and becomes uniform for the highest shear rates. It is probable that the "worm texture" is akin to director turbulence.

11.5. Dynamics of Defects

11.5.1. Isolated Disclination, Drag Force

The first step in analyzing dynamics of an isolated disclination can be made by simplifying both dynamics and elasticity. Namely, one assumes that (1) the defect velocity is too low to cause flow of the nematic fluid; (2) the disclination is of the planar type (11.1), and all elastic constants are equal K. Assumption (1) simplifies the director equations (6.65) to

$$-\frac{\partial f}{\partial n_i} + \frac{\partial}{\partial x_j}\left[\frac{\partial f}{\partial(\partial n_i/\partial x_j)}\right] = \gamma_1 \frac{\partial n_i}{\partial t}, \tag{11.25a}$$

and (6.20), (6.25), (6.64b) for dissipation rate per unit volume to

$$T\sigma = \gamma_1 \sum_i \left(\frac{\partial n_i}{\partial t}\right)^2. \tag{11.26a}$$

Here, f is the Frank–Oseen elastic energy density, $\gamma_1 = \alpha_3 - \alpha_2$ is the twist viscosity, T is the absolute temperature, and σ is the entropy production per unit volume. According to (11.25a), the rate of change in the elastic free energy is exactly compensated by the viscous dissipation. Assumption (2) simplifies things even further:[36]

$$\frac{\partial^2\varphi}{\partial x^2} + \frac{\partial^2\varphi}{\partial y^2} = \frac{\gamma_1}{K}\frac{\partial\varphi}{\partial t}, \tag{11.25b}$$

$$T\sigma = \gamma_1 \left(\frac{\partial\varphi}{\partial t}\right)^2. \tag{11.26b}$$

If the disclination moves slowly in the direction x with the velocity $u = $ const, the solution of (11.25b) is approximately

$$\varphi \approx k \arctan\frac{y}{x - ut}; \tag{11.27}$$

---

[36] H. Imura and K. Okano, Phys. Lett. **42A**, 403 (1973); P.G. de Gennes, in Molecular Fluids, Proc. Les Houches Summer School, Edited by R. Balian and G. Weill, Gordon and Breach, London, 1976.

i.e., the line preserves the static geometry described by (11.6), in which $\theta = \arctan \frac{y}{x}$. The dissipation per unit length is the integral

$$\Sigma = \gamma_1 \int \left(\frac{\partial \varphi}{\partial t} \right)^2 dx \, dy \approx \pi \gamma_1 k^2 u^2 \ln \frac{R}{r_c}, \tag{11.28}$$

so that the friction coefficient $\eta = \Sigma/u^2$, and the drag force $f_{\text{drag}} = \Sigma/u$ experienced by the disclination,

$$\eta = \pi \gamma_1 k^2 \ln \frac{R}{r_c}, \quad f_{\text{drag}} = \pi \gamma_1 k^2 u \ln \frac{R}{r_c}, \tag{11.29}$$

are logarithmically diverging with the system's size R. At first sight, the logarithmic divergence is a natural feature of the problem: The same logarithmic term occurs in the static elastic energy of the disclination (11.7). However, more careful analysis[37] shows that the drag force does not diverge with R, because at large distances, $R > R_c \approx \frac{3.6K}{\gamma_1 u}$, the director reorientations and dissipation of energy practically vanish, so that

$$f_{\text{drag}} = \pi \gamma_1 k^2 u \ln \frac{3.6}{\text{Er}}; \tag{11.30}$$

here, $\text{Er} = \gamma_1 u r_c / K$ is the Ericksen number of the problem. Similar correction has been reported[38] for the line tension W of a moving disclination $W \approx \pi K k^2 \ln \frac{1.1}{\text{Er}}$. In essence, the result (11.29) corresponds to $\text{Er} = 0$. Note that both (11.29) and (11.30) neglect dissipation at the core; thus, they are valid when $R \gg r_c$ and $\text{Er} \ll 1$.

11.5.2. Interaction and Annihilation of Line and Point Defects

11.5.2.1. Lines

The model above can be immediately applied to the dynamics of annihilation of two disclinations with strengths of opposite signs. The potential of interaction (per unit length) of two planar disclinations of strengths k_1 and k_2, separated by a distance L in a system of lateral size R, is (Problem 11.4):

$$W_{12} = \pi K (k_1 + k_2)^2 \ln \frac{R}{r_c} - 2\pi K k_1 k_2 \ln \frac{L}{2r_c}. \tag{11.31}$$

Two disclinations of opposite strength $k_1 = -k_2 = k$ attract each other with the force $f_{12} = -\partial W / \partial L$

---

[37]G. Ryskin and M. Kremenetsky, Phys. Rev. Lett. **67**, 1574 (1991).
[38]C. Denniston, Phys. Rev. B**54**, 6272 (1996).

$$f_{12} = -2\pi K k^2 / L. \tag{11.32}$$

Balancing it with the drag force $f_{\text{drag}} = 2\pi \gamma_1 k^2 u \ln \frac{3.6}{\text{Er}}$ (11.30), one concludes that the two lines will approach each other according to the rule

$$L^2 \approx \text{const} \times \frac{K}{\gamma_1} (t_0 - t) \tag{11.33}$$

until annihilation at $t = t_0$.

The analogy between (11.32) and the Peach and Koehler force in solids has been discussed by Eshelby.[39] Although the Peach and Koehler force is a fictitious configurational force, f_{12} is a real force and, in equilibrium, must be balanced by an external force applied to the disclination.

11.5.2.2. Points

The result (11.32) also applies to point defects in 2D, but not to point defects in 3D. Dimensional analysis[40] suggests that point defects (with elastic energy $\sim K R$ when isolated) should interact via a linear potential $W_{12} \sim K L$ (as quarks). More careful analysis shows that two point defects are connected by a soliton "string" of nearly constant width; outside of the string, the director is practically uniform. For large separations, $L \gg r_c$, the energy of the pair of defects is determined by the elastic energy $W \sim L$ stored in the string. Thus, in first approximation, the attraction force between two elementary point defects in 3D bulk does not depend on L:[41]

$$f_{12} = -4\pi K \tag{11.34}$$

(note that the dimension of f_{12} is that of a true force). The drag force f_{drag} depends on the geometry of the region in which the director is distorted and the energy dissipates. Because the director configuration is strongly influenced by factors such as boundary effects, the situation is not clear; the result may be different, for example, when $L \gg R$ and when $L \sim R$; here, R is the lateral size, such as the radius of a cylindrical capillary.

11.5.2.3. Experiments

The dynamics of planar line defects, observed in nematic MBBA cells under a strong electric field (that ensures the planar director configuration everywhere in the bulk),[42] obey

---

[39] J. Eshelby, Phil. Mag. **A42**, 359 (1980).
[40] W.F. Brinkman and P.E. Cladis, Phys. Today **35**, 48 (1982).
[41] S. Ostlund, Phys. Rev. B **24**, 485 (1981).
[42] H. Orihara, T. Nagaya, and Y. Ishibashi, in Formation, Dynamics and Statistics of Patterns, Edited by K. Kawasaki and M. Suzuki, World Scientific, Singapore, vol. 2, 1993.

(11.33). The situation is more subtle with point defects, as most studies have been performed in capillary tubes, i.e., under the influence of the bounding walls. When $L \sim R$, dynamics follow the rule $L \propto (t - t_0)^{1/2}$; when $L \gg R$, there are indications of a linear dependence.[43]

The distinctive features of linear potential $W_{12} \sim KL$, namely, the linear rule $L \propto (t - t_0)$, is observed for pairs of surface point defects, boojums, connected by 2π "strings," in $\sim (1 - 10)\mu$m-thick hybrid-aligned films placed between two isotropic media[44] (Fig. 11.27). The strings in the experiment are of an approximately constant width D; thus, the dissipation should take place only within the region of a characteristic size D: $f_{\mathrm{drag}} \sim D\gamma_1 u$. With $f_{12} \sim K$, the string should shrink without acceleration:

$$ L \approx \mathrm{const} \times \frac{K}{D\gamma_1}(t_0 - t). \tag{11.35} $$

as observed experimentally, at least for $L > D$. The constant width of the string might be explained by the effect of geometrical anchoring, Ref. [12] and p. 402, associated with the gradients of the film thickness.[45]

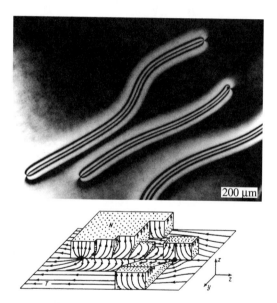

Figure 11.27. Strings connecting pairs of point defects, boojums, in a hybrid-aligned nematic film of 5CB placed onto a glycerin substrate [44].

[43] A. Pargellis, N. Turok, and B. Yurke, Phys. Rev. Lett. **67**, 1570 (1991).

[44] O.D. Lavrentovich and S.S. Rozhkov, Pis'ma Zh. Eksp. Teor. Fiz. **47**, 210 (1988) [Sov. Phys. JETP Lett. **47**, 254 (1988)].

[45] D.R. Link, M. Nakata, Y. Takanishi, K. Ishikawa, and H. Takezoe, Phys. Rev. Lett. **87**, 195507 (2001).

11.5.3. Coarsening of Disclination Networks

Dynamics of an isolated disclination allows one to consider, at least qualitatively, the coarsening dynamics of defects. As already indicated, dense networks of disclinations can be formed by strong flow or by sudden quench of the isotropic fluid into the nematic state. What would be the time dependence of the disclination density?

Let us first consider a much simpler problem of an isolated shrinking disclination loop. The time dependence of the radius R of the loop can be roughly estimated by equating the viscous drag force $\eta \, dR/dt$ to the elastic energy W/R, $W = \pi k^2 K \ln \frac{R}{r_c}$ per unit length of the curved disclination

$$\frac{W}{R} = -\eta \frac{dR}{dt}. \tag{11.36}$$

Integration shows that the loop shrinks according to the law[46]

$$R = \sqrt{\frac{2W}{\eta}(t_0 - t)} \tag{11.37}$$

($t_0 > t$ is the time at which the loop disappear), which seems to hold in experiments.[47]

In a similar fashion, one can consider[48] a mesh of disclinations in 3D with a characteristic segment length ξ. The disclination density per unit area is $\rho \sim \xi^{-2}$. The rate of loss of energy per unit volume is $Wu\rho/\xi = W^2\rho^2/\eta$. Equating this to the time derivative of the line tension per unit volume $W\rho$ results in

$$\frac{d\rho}{dt} = -\text{const} \times \frac{W}{\eta}\rho^2, \tag{11.38}$$

or, after integration,

$$\rho = \frac{\eta}{\text{const} \times W}t^{-\nu}, \tag{11.39}$$

with the exponent $\nu = 1$. The last result seems to describe the experimental situation, at least for the networks of extended singular disclinations formed during the quenching of the isotropic phase into the nematic state.[48] Also, it is the exact analogue of Friedel's result for the analysis of the growth of the mosaic structure of dislocations in crystals.[49] For a general theory of coarse-graining phenomena during phase transitions, see the review article by Bray.

[46] G.J. Gerritsma, A. Geurst, and A.M. Spruijt, J. Phys. Lett. **A43**, 356 (1973).
[47] W. Wang, T. Shiwaku, and T. Hashimoto, J. Chem. Phys. **108**, 1618 (1998).
[48] I. Chuang, R. Durrer, N. Turok, and B. Yurke, Science **251**, 1336 (1991).
[49] J. Friedel, Dislocations, Pergamon Press, Oxford, 1964, p. 239.

Problem 11.1. Consider a uniaxial nematic liquid crystal in a circular capillary (Fig. 11.8b). The walls are treated to keep the director normal to the cylindrical boundary. The director lines are allowed to escape along the z-axis, so that the director can be parameterized in cylindrical coordinates as $n_r = \cos \chi (r)$, $n_\varphi = 0$, $n_z = \sin \chi (r)$, where χ is the angle between the director and the (r, φ) plane; the boundary conditions write $\chi (r = R) = 0$ and $\chi (r = 0) = \pi/2$. Using (5.2) for the free energy density, find the equilibrium director distribution and the energy per unit length of the configuration when (a) $K_1 = K_3 = K$ and (b) $K_1 \neq K_3$ [P.E. Cladis and M. Kleman, J. Physique 33, 591 (1972); R. Meyer, Phil. Mag. **27**, 405 (1973)].

Answers:

(a) With

$$\operatorname{div} \mathbf{n} = \frac{1}{r} \frac{d(r n_r)}{dr} = -\sin \chi \frac{d\chi}{dr} + \frac{\cos \chi}{r}$$

and

$$\operatorname{curl}_\varphi \mathbf{n} = -\frac{d(n_z)}{dr} = -\cos \chi \frac{d\chi}{dr},$$

the elastic energy per unit length is

$$W = \frac{1}{2} K \int_0^R \left[\left(\frac{d\chi}{dr}\right)^2 + \frac{\cos^2 \chi}{r^2} - \frac{1}{r} \sin 2\chi \frac{d\chi}{dr} \right] r \, dr \int_0^{2\pi} d\varphi.$$

In problems with cylindrical symmetry, it is often useful to replace r with a new variable ξ: $r = e^\xi$. The energy integral transforms into

$$W = \pi K \int_{-\infty}^{\ln R} \left[(\chi')^2 + \cos^2 \chi - \chi' \sin 2\chi \right] d\xi,$$

where $\chi' = d\chi/d\xi$.

The Euler–Lagrange equation is then $\chi'' = -\cos \chi \sin \chi$. The first integration results in

$$(\chi')^2 = \cos^2 \chi + \text{const.} \tag{11.40}$$

According to the boundary condition, $\xi \to -\infty$ at the center of the capillary. It is possible when $\chi' = 0$ at $\chi \to \pi/2$; therefore, the constant of integration is zero. Specifying one of the two possible directions of the escape, e.g., $\chi' = -\cos \chi$, one finds the equilibrium solution from

$$\int_r^R \frac{1}{r} dr = -\int_\chi^0 \frac{dp}{\cos p}, \tag{11.41}$$

as

$$\frac{R}{r} = \frac{1 + \tan(\chi/2)}{1 - \tan(\chi/2)} \quad \text{or} \quad \chi = 2 \arctan\left(\frac{R - r}{R + r}\right). \tag{11.42}$$

The solution satisfies both boundary conditions and describes smooth reorientations of the director by $\pi/2$ between the axis and the wall of the cylinder. The energy per unit length is

$$W = 3\pi K. \tag{11.43}$$

(b) The analog of the first integral (11.40) is

$$(\chi')^2 = \frac{K_1 \cos^2 \chi}{K_1 \sin^2 \chi + K_3 \cos^2 \chi} + \text{const}, \tag{11.44}$$

which results in the solution

$$\int\limits_r^R \frac{1}{r}\, dr = -\int\limits_\chi^0 \frac{\sqrt{\sin^2 p + (K_3/K_1)\cos^2 p}}{\cos p}\, dp. \tag{11.45}$$

If $K_3 > K_1$, the solution writes

$$\ln \frac{R}{r} = \frac{\beta}{\sqrt{1-\beta^2}} \arcsin(\beta \sin \chi) - \tfrac{1}{2} \ln \frac{\sqrt{1-\beta^2 \sin^2 \chi} - \sqrt{1-\beta^2}\sin \chi}{\sqrt{1-\beta^2 \sin^2 \chi} + \sqrt{1-\beta^2}\sin \chi}, \tag{11.46}$$

where $\beta = \sqrt{1 - K_1/K_3}$. The energy per unit lenth is then

$$W = \pi K_1 \left(2 + \frac{\arcsin \beta}{\beta\sqrt{1-\beta^2}} \right). \tag{11.47}$$

Note that the last formula with a replacement $K_1 \rightarrow K_2$ and under a (reasonable) assumption $K_3 > K_2$, describes a circular disclination $k = 1$, $c = \pi/2$ (Fig. 11.4), that escapes via twist. Finally, the calculations above neglect the saddle-splay K_{24} contribution (5.4) that become nonzero for escaped disclinations. Inclusion of this term allows one to estimate K_{24} [D.W. Allender, G.P. Crawford, and J.W. Doane, Phys. Rev. Lett. **67**, 1442 (1991) and R.D. Polak, G.P. Crawford, B.C. Costival, J.W. Doane, and S. Zumer, Phys. Rev. E**49**, R978 (1994)]. Note also that (11.43) and (11.11) carry contributions from divergence term hidden in the bulk density (5.7).

Problem 11.2. Nucleation of disclination loops can relax the twist imposed onto a nematic slab between suitably oriented walls. Consider a loop L in a nematic slab located between two parallel plates with strong unidirectional surface anchoring in the plane of the plates. Let d be the thickness of the slab and $\alpha = qd$ the angle between the two anchoring directions; it is assumed that the elasticity is isotropic ($K = K_1 = K_2 = K_3$) and that the deformation takes place in a plane parallel to the anchoring directions. Show that:

(a) The angle of rotation of the director can be written as $\varphi(\mathbf{r}) = \tfrac{k}{2}\Omega(\mathbf{r}) + qz + cst$, where $\Omega(\mathbf{r})$ is the solid angle under which the loop is seen from point $\mathbf{r}$.

(b) The energy of the loop is $F_{\text{loop}} = K(\tfrac{\pi k}{4} L - \pi q \Sigma)$, where L is the self-inductance of the loop L considered as an electric circuit and Σ is the projected area of L on the plane of rotation.

Answers: J. Friedel and P.-G. de Gennes, C. R. Acad. Sci. Paris **B268**, 257 (1969). Let us first develop the electromagnetic analogy for the equation $K \Delta \varphi(\mathbf{r}) = 2\pi k \delta(L)$. Because the angle $\varphi(\mathbf{r})$ obeys the harmonic equation $\Delta \varphi(\mathbf{r}) = 0$, except on the line L, we introduce a divergenceless magnetic field vector $\mathbf{H} = -K \nabla \varphi(\mathbf{r})$. The solution of the harmonic equation is singular on L: $\varphi(\mathbf{r}) = \frac{j}{4\pi} \Omega(\mathbf{r}) + \psi(\mathbf{r})$, such that $\iint \mathrm{curl}\, \mathbf{H} \cdot d\mathbf{\Sigma} = \int -\nabla\varphi . d\mathbf{L} = -2\pi k$. Hence, $j = 2\pi k$.

Problem 11.3. Is it possible to estimate the polar anchoring coefficient W_a from Fig. 11.6b if it is known that the cell is filled with 5CB and its thickness h is about 5μm?

Answers: Yes. Let $\delta_0 \ll 1$ be the equilibrium angle between the director and substrate; at the center of the wall, $\delta_0 = 0$. For small tilt, the surface anchoring potential is $W_a(\delta - \delta_0)^2/2$. If the director distortions occur in the plane normal to the wall, as Fig. 11.6c suggests, then the elastic energy per unit length of the wall is $2Kh\,\delta_0^2/d$ (d is the walls' width) and the anchoring energy is $W_a\,\delta_0^2\,d/3$. Minimizing the total energy, one finds $W_a = \frac{6Kh}{d^2}$. Estimating $d \sim 10\mu$m from the photograph, one gets $W_a \sim 10^{-6}\,$J/m^2.

Problem 11.4. Show that the energy of two planar disclinations is given by (11.31), $W_{12} = \pi K(k_1 + k_2)^2 \ln \frac{R}{r_c} - 2\pi K k_1 k_2 \ln \frac{L}{2r_c}$.
Hint. Use the superposition of solutions

$$\varphi = k \arctan \frac{y}{x} + \mathrm{const}: \varphi = \sum_i k_i \arctan \frac{y - y_i}{x - x_i} + \mathrm{const}.$$

See C.F. Dafermos, Quart. J. Mech. Appl. Math. **23**, 549 (1970).

Further Reading

A. Bray, Adv. in Phys. **43**, 357 (1994).

Chirality in Liquid Crystals, Edited by C. Bahr and H.-S. Kitzerow, Springer Verlag, Series on "Partially Ordered Systems," New York, 2000.

T. De'Nève, M. Kleman, and P. Navard, C.R. Acad. Sci. Paris **316**, série II, 1037–1044 (1993).

M. Doi and S.F. Edwards, The theory of polymers dynamics, Clarendon Press, Oxford, 1986.

Formation and Interaction of Topological Defects, vol. **349** of NATO Advanced Study Institute, Series B: Physics, Edited by A.-C. Davis and R. Brandenberger, Plenum, New York, 1995.

M. Kleman, Points, Lines and Walls in Liquid Crystals, Magnetic Systems, and Various Ordered Media.

Defects in Liquid Crystals: Computer Simulations, Theory and Experiments, Editors: O.D. Lavrentovich, P. Pasini, C. Zannoni, and S. Žumer, Kluwer Academic Publishers, The Netherlands, 2001.

S. Chandrasekhar and G.S. Ranganath, Adv. Phys. **35**, 507 (1985).

P. Navard, J. Polym. Sci.: Polym. Phys. Ed. 24, 435 (1986).

R. Ribotta, and A. Joets, in *Cellular Instabilities*, Edited by J. E. Weisfreid and Z. Zaleski, Springer, New York, 1984.

CHAPTER 12

Topological Theory of Defects

In Chapter 3, we have considered the notion of order parameter, its amplitude and phase. The order parameter is a continuous field (scalar, vector, tensor, etc.) describing the state of the system at each point. Generally, it is a function of coordinates, $\psi(\mathbf{r})$. Distortions of $\psi(\mathbf{r})$ can be of two types: those containing singularities and those without singularities. At singularities, ψ is not defined. For a 3D medium, the singular regions might be either zero-dimensional (points), one-dimensional (lines), or two-dimensional (walls). These are the *defects*. Whenever a nonhomogeneous state cannot be eliminated by continuous variations of the order parameter (i.e., one cannot arrive at the homogeneous state), it is called *topologically stable*, or simply, a *topological defect*. If the inhomogeneous state does not contain singularities, but nevertheless is not deformable continuously into a homogeneous state, one says that the system contains a *topological configuration* (or soliton).

Very often the problems involving defects are too complex for analytical treatment within the framework of an elastic theory. The difficulties arise either from the complexity of the free energy functional (biaxial nematic, smectic C, anisotropic phases of superfluid 3He, etc.) or from the complexity of the defect configuration (e.g., crossing of disclinations). Even when the solutions are possible, they rely on certain assumptions and, thus, might be strongly model dependent.

An adequate description of defects in ordered condensed media requires introducing a new mathematical apparatus, viz. the theory of homotopy, which is part of algebraic topology. It is precisely in the language of topology that it is possible to associate the character of ordering of a medium and the types of defects arising in it, to find the laws of decay, merger and crossing of defects, to trace out their behavior during phase transitions, and so on. The key point is occupied by the concept of *topological invariant*, often also called a *topological charge*, which is inherent in every defect. The stability of the defect is guaranteed by the conservation of its topological invariant. The following simple example of twisted ribbon strips gives a flavor of the concept of topological invariant.

12.1. Basic Concepts of Topological Classification

12.1.1. Topological Charges Illustrated with Möbius Strips

Consider a set of elastic strips closed into rings. Each strip is characterized by a number k that counts how many times the ends of the strip are twisted by 2π before they are glued together to produce a ring (Fig. 12.1). The ring with $k = 1/2$ (Fig. 12.1.b) is the well-known Möbius strip. The deformation energy stored in any twisted strip is larger than the pure bend energy of the $k = 0$ ring. However, to transform a twisted strip into a state $k = 0$, one needs to cut the strip. There is no continuous deformation that transforms one strip into another if the two have different k. The energy needed to cut the ribbon, $F_{\text{cut}} \sim US/a^2$, is much higher than the stored twist energy $F_{\text{twist}} \sim k^2 KS/L$; here, L is the length of the strip, S is its cross-section area, and $K \sim U/a$ is some elastic constant of the order of the intermolecular energy; a is the molecular scale. The transitions between the states with different k's are prohibited by high-energy barriers.

The quantity k does not change under any continuous transformation and is a useful invariant to label topologically different states. Left and right twists can be distinguished by the sign of k. Obviously, one can create a pair of left and right twists without cutting the strip, what matters is the total sum of k's that should be preserved. Therefore, topological charges k's obey a conservation law.

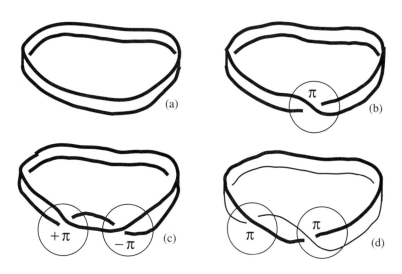

Figure 12.1. Topologically different rings of elastic strips: (a) nontwisted ring, $k = 0$; (b) Möbius strip, $k = 1/2$; (c) twisted strip with two identical edges, $k = 1$; (d) twisted strip with two distinctive edges, $k = 1$.

The allowed values of k are defined by the inner symmetry of the strip. For example, if the edges of the strip are different, e.g., marked by red and blue colors (Fig. 12.1.d), then only integer k's (2π-twists) are allowed.

Topological stability of twisted strips is similar to that of topological solitons; the issue of a singular core is not involved. Furthermore, one can draw a parallel between the twisted strips and singular defects. Imagine a circle around a π-disclination in a uniaxial nematic liquid crystal (Fig. 11.10). The set of molecules centered at this circle form a Möbius strip with $k = 1/2$. After going once around the circle, the director $\mathbf{n}$ flips into $-\mathbf{n}$, which is possible, because the nematic bulk is centrosymmetric, $\mathbf{n} \equiv -\mathbf{n}$. The number k would remain equal $1/2$ if the radius of the circle is taken larger or smaller (Fig. 11.10). Thus, the overall director configuration can be characterized by $k = 1/2$. At the disclination core, one faces the singularity: When the circle shrinks into a point, there is an infinity of director orientations at this point. If a disclination were created in a ferromagnet, a Möbius strip $k = 1/2$ around it would be impossible because the magnetization vector does not have the head-to-tail symmetry of the director.

To summarize, the examples above show that the topologically stable defects and configurations ("topological twists") obey the following general rules:

1. Defects types are related to the type of ordering of the system.

2. Defects are characterized by quantized invariants (topological charges) k.

3. The operations of merger and decay of the defects are described as certain operations (e.g., additions) applied to their charges k; conservation laws of topological charges control the results of merger and decay.

The topological invariants k's form groups. Because the concept of group is important for the homotopy classification of defects, we briefly consider it in the next sections. Before doing so, we briefly comment on, perhaps, the most intriguing twisted strips—the DNA molecules.

12.1.2. DNA and Twisted Strips, a Digression

Twisted strips with different k's are of relevance to the problem of configuration and replication of double-stranded DNA molecules. Two strands are arranged in a helicoid fashion in which a 2π-twist occurs per every 10.5 base pairs (Fig. 1.21). In many organisms ranging from viruses and prokaryotes to some eukaryotes, DNA molecules form closed loops. Topologically, these loops remind of a twisted strip with two distinctive edges and an integer Lk that is referred to as the *linking number* of the two strands (Fig. 12.2). Lk is preserved in any conformational change of DNA molecule that does not break the strands. If Lk is close to $Lk_0 = l/p$ (l is the total DNA length, and $p \approx 3.4$ is the helix pitch; Fig. 1.21), the DNA ring is relaxed and can lie flat on a planar surface without contortions. Often $Lk \neq Lk_0$: The ends of the relaxed linear DNA duplex may be additionally twisted (or untwisted) by some number of rotations $\pm 2\pi$ before forming the ring. There are two ways

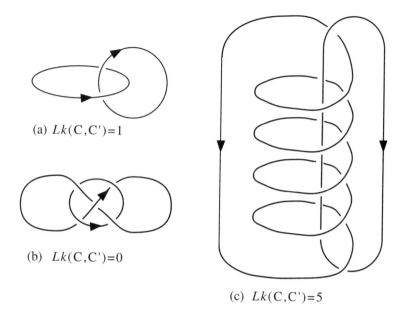

(a) $Lk(C,C')=1$

(b) $Lk(C,C')=0$

(c) $Lk(C,C')=5$

Figure 12.2. Linking numbers for pairs of oriented curves.

to deal with the induced strain. First, the number of base pairs per pitch can be changed; the ring remains *planar*, and the linking number is equal to the number of turns of one strand around another. In that case, $Lk = k$, the *topological* twist defined above. Second, the duplex axis can twist upon itself, leaving the number of pairs per pitch unaffected. Such a supertwisted DNA is no longer planar and coils in three dimensions, like a buckled twisted ribbon. Whatever the case, although k and Lk stay unaffected, and are still equal integral numbers of a *topological* nature, the global *geometry* (and consequently, the energy of the "twisted" ribbon and the way it relaxes) depends on the elasticity properties of the molecule and is better described by introducing two *geometrical* parameters: the *twist* Tw and the *writhe* Wr. The twist can be written as

$$Tw = \frac{1}{2\pi} \oint \Omega(s)\,ds,\tag{12.1}$$

where $\Omega(s)$ is the rate of wrapping of either strand about the duplex axis. This quantity can be defined equally for an open strip; Tw can take any value, and we can refer to it as the *geometrical* twist. However, if the duplex axis is planar, one gets $Lk = Tw = k$. The writhe Wr of a curve C is a much more subtle quantity. Introduced by Fuller, it is the number of averaged self-crossings (with sign) of the planar orthogonal projections of C (closed or not); in the DNA context, it describes the buckling of the duplex axis, so to speak. Like Tw,

Wr can take any value. We have the important relation:

$$Lk = Tw + Wr, \tag{12.2}$$

with *Lk* (for two oriented curves C and C′) and *Wr* (for an oriented curve C) given by double integrals:

$$Lk = \frac{1}{4\pi} \oint\!\!\!\oint_{C,C'} \frac{\mathbf{r}(s) - \mathbf{r}(s')}{|\mathbf{r}(s) - \mathbf{r}(s')|^3} \cdot [d\mathbf{s} \times d\mathbf{s'}];$$

$$Wr = \frac{1}{4\pi} \oint\!\!\!\oint_{C,C} \frac{\mathbf{r}(s) - \mathbf{r}(s^*)}{|\mathbf{r}(s) - \mathbf{r}(s^*)|^3} \cdot [d\mathbf{s} \times d\mathbf{s^*}]. \tag{12.3}$$

Here, C is the duplex axis, say, and C′ is anyone of the strands. *Wr* vanishes when C is planar. Note that for the example in Fig. 12.2 (b), $Lk(C, C') = 0$, because the two curves can be disentangled by crossing of a ∞-shaped line with itself; such crossings are not reflected in the integral *Lk* above.

To separate the DNA strands during replication, one needs to change the number *Lk* (see Problem 12.1). It can be done directly by topoisomerases that cut one or both strands. In other cases, the replication occurs through local binding of the DNA molecule to proteins that creates zones of negative and positive supertwisting.

12.1.3. Groups: Basic Definitions

Consider a set (finite or infinite, discrete or continuous) G of elements $a, b, c, \ldots$, for which there is an operation $\otimes$ that combines the elements in a prescribed way. The set G is a group if and only if the following requirements are satisfied:

1. Any two elements a, b in the set G can be combined by the operation $\otimes$ to produce a third element $a \otimes b$ in the set.
2. The operation is associative: $(a \otimes b) \otimes c = a \otimes (b \otimes c)$.
3. There is an identity element I of G, such that for any element a, $a \otimes I = I \otimes a = a$.
4. Every element a has an inverse element denoted a^{-1}, such that $a \otimes a^{-1} = a^{-1} \otimes a = I$.

A simple example of a group is a set Z of all integers with the operation of addition ($\otimes \rightarrow +$). Indeed, the axiom (2) is fulfilled when one adds integers; the identity element is 0; and the inverse to a is $-a$.

Groups are either commutative (also called Abelian) or noncommutative (or non-Abelian). For the Abelian groups, $a \otimes b = b \otimes a$ for any pair of elements. For non-Abelian groups, $a \otimes b \neq b \otimes a$. The group of integers is Abelian. Groups can contain a finite number of elements or infinitely many elements. Finite or denumerably infinite groups are called discrete groups. The additive group Z of integers is discrete.

A subgroup H of a group G is a subset of elements of G that is also a group. If h_i are the elements of the subgroup H and g is any element of G, then the set of elements-products $g \otimes h_i$ is called a *left coset* of H and the set $h_i \otimes g$ is called a *right coset* of H. It is easily proven that the cosets $g_1 \otimes h_i$ and $g_2 \otimes h_i$, formed by two elements g_1 and g_2 of G, are either identical or have no common elements whatever. In other words, a given subgroup H divides the group G into disjoint cosets that form a *coset space* or *orbit* denoted as G/H. The coset space is not necessarily a group. However, if the subgroup H is *normal* (also called *invariant*), meaning that the left and right cosets contain the same elements for each g of G, then the coset space G/H has a group structure and is called a *factor group*.

Two types of groups are important in the topological classification of defects of a given ordered medium, both related to the order parameter:

1. The (generally) continuous group G whose elements are in correspondence with all the permissible transformations of the order parameter. The group of symmetry H is a subgroup of this continuous group.

2. The discrete homotopy groups that are related to the topological structure of the order parameter space.

This will be detailed below. We first schematize how these groups are involved in the topological classification of defects.

12.1.4. General Scheme of the Topological Classification of Defects

Homotopy classification of defects in ordered media includes the following three steps:

First, one defines the order parameter (OP) ψ of the system. In a nonuniform state, the OP is a function of coordinates, $\psi(\mathbf{r})$.

Second, one determines the OP *(or degeneracy) space* $\mathcal{R}$, i.e., the manifold of all possible values of the OP that do not alter the thermodynamical potentials of the system. The function $\psi(\mathbf{r})$ maps the points of real space occupied by the medium, into $\mathcal{R}$.

The mappings of interest are those of i-dimensional spheres enclosing defects in real space. A point defect in a 2D system or a line defect in 3D can be enclosed by a linear contour, $i = 1$; a point defect in a 3D system can be enclosed by a sphere, $i = 2$; a wall defect can be "enclosed" by two points, $i = 0$, located at opposite sides of the wall.

Third, one defines the homotopy groups $\pi_i(\mathcal{R})$. The elements of these groups are mappings of i-dimensional spheres enclosing the defect in real space into the OP space. To classify the defects of dimensionality t' in a t-dimensional medium, one has to know the homotopy group $\pi_i(\mathcal{R})$ with $i = t - t' - 1$.

On the one hand, each element of the homotopy group corresponds to a class of topologically stable defects; all of these defects are equivalent to one another under continuous deformations. On the other hand, the elements of homotopy groups are topological invariants, or topological charges of the defects. The defect-free state (e.g., $\psi(\mathbf{r}) = $ const) corresponds to a unit element of the homotopy group and to zero topological charge.

12.1.5. Order Parameter Space. Groups That Describe Transformations of the Order Parameter

The Heisenberg isotropic ferromagnetic phase with a unit magnetization vector $\mathbf{d}$ as the order parameter is an example of a medium for which the OP space is easily found by a qualitative consideration. This phase is isotropic in the sense that the coupling between $\mathbf{d}$ and the crystallographic axes is neglected. Any rotation about a fixed $\mathbf{d}$ transforms the system into itself. The ends of vectors $\mathbf{d}$ with different orientations in space describe a sphere S^2. Thus it is obvious that the OP space is the sphere S^2. For many other media, the situation is not that clear. Below, we illustrate a general way to find the OP space that sheds some light on the relationship between the symmetry of the ordered medium and the OP space.

Consider a continuous group of 3D rotations. This group is the part of the full Euclidian group of translations and rotations, which leaves the thermodynamic state of the system invariant. A 3D rotation can be specified by a vector $\mathbf{k}$ that is parallel to the axis of rotation and has an absolute value equal to the angle of rotation φ. Rotations around all possible axes having one common point form a group called *the group of proper rotations in 3D Euclidian space*. This group is represented by a solid 3D sphere of radius π denoted $SO(3)$ and composed of points $\varphi\mathbf{k}/|\mathbf{k}|$, where $-\pi \leq \varphi \leq \pi$. Two diametrically opposite points at the surface of such a sphere are identical: π-rotations around axes directed in opposite directions give the same result.

In principle, $SO(3)$ can serve as the OP space of the Heisenberg ferromagnet. However, there are sets of points in $SO(3)$ that correspond to *indistinguishable* stable states of the ferromagnet. Because any rotation around $\mathbf{d}$ transforms the system into itself, all points $\varphi\mathbf{d}$ along any fixed radius of the solid sphere $SO(3)$ describe indistinguishable states. Because of this symmetry, the solid sphere $SO(3)$ is "reduced" to the sphere S^2 by rotations that leave the order parameter $\mathbf{d}$ unchanged. The process is called a *factorization* of the group $SO(3)$ by the group $SO(2)$ of 2D rotations around the fixed axis $\mathbf{d}$ (Fig. 12.3). $SO(2)$ is a *subgroup* of $SO(3)$. The OP space of the ferromagnet is represented as $S^2 = SO(3)/SO(2)$, where $SO(3)/SO(2)$ is the notation for the *coset space* of $SO(2)$ in $SO(3)$. Note that S^2 is an example of a manifold. General definitions are given below.

If the medium is a uniaxial nematic, then the directions $\mathbf{d}$ and $-\mathbf{d}$ are identical and there is an additional factorization of S^2 by a set of two diametrically opposite points on S^2. The OP space of the uniaxial nematic is thus $\mathcal{R} = SO(3)/SO(2) \times Z_2 = S^2/Z_2$, also called the projective plane RP^2 (see Chapter 3). Z_2 is the group of two numbers 0 and 1: $0 + 0 = 1, 0 + 1 = 1, 1 + 0 = 1, 1 + 1 = 0$. The symbol $\times$ denotes a *direct product* of two groups $SO(2)$ and Z_2. The direct product of two groups, say, G and T, is

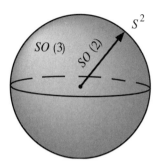

Figure 12.3. Factorization of the group $SO(3)$ of proper rotations by the group of rotations $SO(2)$ results in the sphere S^2.

the set $G \times T$ of pairs (g, t) that is a group under the combination law $(g_1, t_1) \otimes (g_2, t_2) = (g_1 \otimes g_2, t_1 \otimes t_2)$.

Examples above lead to the following generalization of the group-theoretical description of the OP space.

The order parameter of a continuous perfectly ordered medium can be associated with a "thermodynamic" group G (which is usually the Euclidian group) with elements g that transform a given value ψ_0 of the order parameter into another value $g\psi_0$ for which the thermodynamical potentials of the system remain the same. Rotations of a perfect ferromagnet as a whole are transformations of this kind. Among the elements g, there might be transformations that preserve not only the energy, but also the value of the order parameter, $g_H\psi_0 = \psi_0$. These elements form a subgroup H of G called the *isotropy group* of ψ_0 or the little group of ψ_0. The OP space is then the coset space $\mathcal{R}$, noted G/H:

$$\mathcal{R} = G/H. \tag{12.4}$$

Note that generally $\mathcal{R}$ is not a group. Furthermore, there is a certain arbitrariness in the choice of the group G (but not in $\mathcal{R}$!). The group G can be taken "larger" or "smaller," but the corresponding isotropy subgroup H must finally result in the same $\mathcal{R} = G/H$. If G is the full Euclidian group, then H is the group of symmetry of the ordered medium.

12.1.6. Homotopy Groups

Homotopy groups describe the topology of the OP space. Here, we briefly consider some abstract OP space $\mathcal{R}$ and the group of *oriented contours (loops)* in it that form the so-called *fundamental*, or *first homotopy group*.

Suppose that $\mathcal{R}$ is a connected surface: Any two points on $\mathcal{R}$ can be connected by a curve. Take an arbitrary point M that belongs to the surface and draw oriented continuous

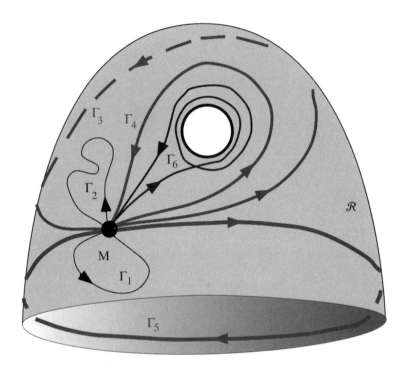

Figure 12.4. Oriented contours based at point M in the order parameter space $\mathcal{R}$. See text.

contours that start and end in M (Fig. 12.4). The contour is "oriented" when the direction of traversing is specified. Among the contours, there are some that can be continuously transformed into each other, such as contours Γ_1 and Γ_2 or Γ_3 and Γ_4 in Fig. 12.4. These contours are said to be *homotopic* or representing the same *homotopy class*. If $\mathcal{R}$ is connected (i.e., made of one piece only) and *simply connected* (no holes), then any contour can be contracted to a point M; thus, all the contours belong to the same homotopy class. If $\mathcal{R}$ is not simply connected (one example is a circle S^1, another is given by Fig. 12.4), then there are distinct classes of homotopic contours. For example, Γ_3 and Γ_5 in Fig. 12.4 that encircle different "holes," or Γ_4 and Γ_6 that encircle the same hole but a different number of times, belong to different homotopy classes. There is no continuous transformation between the contours from different homotopy classes.

One can introduce a product of two contours Γ_n and Γ_k as a contour Γ_{nk} obtained by first traversing Γ_n and then Γ_k: $\Gamma_{nk} = \Gamma_n \otimes \Gamma_k$. Figure 12.5 shows two homotopic representations of Γ_{nk}. Similarly to *the product of individual contours*, one can consider a *product of homotopy classes* as a set of products of representatives of these classes. This concept allows one to impose a group structure on the set of contours with the product of homotopy classes being the group operation. The elements of the group are the homotopic

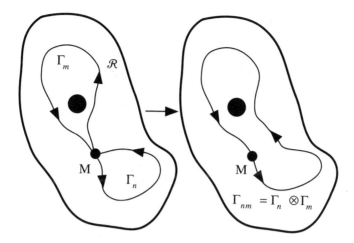

Figure 12.5. Two homotopic representations of Γ_{nk}, the contour product of the two contours Γ_n and Γ_k traversed in this order.

classes. The class of contours that are homotopic to zero (contractible to a single point) is the identity of the group. Each class has its inverse element that is the same set of contours but with opposite orientation. The product of classes satisfies the associative law. Thus, the set of homotopy classes of contours form a group called the *fundamental group of $\mathcal{R}$ at the base point* M and denoted $\pi_1(\mathcal{R}, M)$ or simply the *fundamental group of $\mathcal{R}$*, $\pi_1(\mathcal{R})$. This simplification in omitting the base point is possible when $\mathcal{R}$ is connected. If instead of M one chooses any other point M_1 in the connected space $\mathcal{R}$ as the base point, the resulting group $\pi_1(\mathcal{R}, M_1)$ is an isomorphic copy of $\pi_1(\mathcal{R}, M)$. The group isomorphism is a one-to-one mapping of one group onto another that preserves the group operation. Thus, a connected OP space $\mathcal{R}$ can be characterized by a single abstract group $\pi_1(\mathcal{R})$. $\pi_1(\mathcal{R})$ is also called the first homotopy group to distinguish it from the nth homotopy groups $\pi_n(\mathcal{R})$ that are discussed later.

The fundamental group $\pi_1(\mathcal{R})$ can be Abelian or non-Abelian, depending on the OP space $\mathcal{R}$. If $\mathcal{R}$ is a 2D plane with one punched hole (homeomorphic to a circle S^1), then $\pi_1(\mathcal{R})$ is Abelian (Fig. 12.5). If $\mathcal{R}$ is a 2D plane with two punched holes (homeomorphic to the figure "8"), then the fundamental group is non-Abelian, $\Gamma_k \otimes \Gamma_n \neq \Gamma_n \otimes \Gamma_k$. As shown in Fig. 12.6, there is no way to pass continuously from $\Gamma_3 = \Gamma_1 \otimes \Gamma_2$ to $\Gamma_3^* = \Gamma_2 \otimes \Gamma_1$ with A a fixed point on Γ_3 and Γ_3^*.

This concludes the discussion of the OP space and homotopy groups needed to understand the general scheme of classification of topological defects. To illustrate the scheme, we first consider point defects in 2D nematic, smectic, and crystalline phases classified by the fundamental group. These examples require simple topological considerations. In the

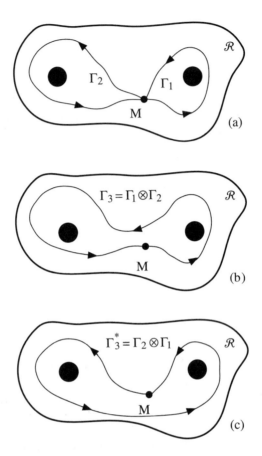

Figure 12.6. Contours in the order parameter space $\mathcal{R}$ (a plane with two punched holes). Base point M. $\Gamma_3(= \Gamma_1 \otimes \Gamma_2) \neq \Gamma_3^*(= \Gamma_2 \otimes \Gamma_1)$ yield a non-Abelian fundamental group $\pi_1(\mathcal{R})$.

general case, more sophisticated group-theoretical methods are required to calculate the homotopy groups from the structure of G/H.

12.1.7. Point Defects in a Two-Dimensional Nematic Phase

In our model of a 2D nematic, the centers of gravity of molecules lie in one plane, while the director **n** makes the angle $0 \leq |\psi_0| \leq \pi/2$ with the normal $\boldsymbol{\nu}$ to the plane. The order parameter can be chosen either as a unit vector $\boldsymbol{\tau} = \mathbf{n} - \boldsymbol{\nu}(\mathbf{n} \cdot \boldsymbol{\nu})$ (which is a projection of **n** on the plane) or as the wave function $\psi = |\psi_0| \exp(i\varphi)$, where $|\psi_0|$ is the polar and φ is the azimuthal angle of the tilt of the molecules, respectively. The free energy density of the uniform state does not depend on φ and can be represented as a certain function of the modulus $|\psi_0|$; e.g.,

$$F_{cond} = A|\psi|^2 + B|\psi|^4 \tag{12.5}$$

(cf. the description of superfluid helium; Section 3.1.1). The OP space $\mathcal{R}$ depends on the modulus $|\psi_0|$.

1. If $0 < |\psi_0| < \pi/2$, then $\mathcal{R} = S^1$. The phase φ can vary from 0 to 2π, and each point of S^1 corresponds to a certain value of φ. Fig. 12.7b shows the circle S^1 as the bottom of the free-energy density F_{cond} (see also Fig. 3.1).

2. If $|\psi_0| = \pi/2$, then $\mathcal{R} = S^1/Z_2$. Any two diametrically opposite points of the circle become identical, owing to the nonpolarity $\mathbf{n} \equiv -\mathbf{n}$ of the nematic. Topologically, S^1 and S^1/Z_2 are *identical*. However, there is an important physical difference between the defects at $0 < |\psi_0| < \pi/2$ and $|\psi_0| = \pi/2$, as we shall see below.

3. If $|\psi_0| = 0$, $\mathcal{R}$ is a single point.

Nonuniformity in the azimuthal orientation of molecules gives rise to an additional gradient energy term:

$$F_{cond} = A|\psi|^2 + B|\psi|^4 + \tfrac{1}{2}K|\nabla\psi|^2, \tag{12.6}$$

where K is the elastic constant of the in-plane splay and bend deformations. The energy density (12.6) determines a length scale that is called the coherence length $\xi = \sqrt{K/A}$. If the characteristic scale of distortions is much larger than ξ, the tilt angle is close to its equilibrium value $|\psi_{0,eq}| = \sqrt{-A/2B}$ and the inhomogeneity involves only the variations of the azimuthal angle $\varphi(x, y)$. This function $\varphi(x, y)$, or, equivalently, the function $\tau(x, y)$, maps the real 2D space (x, y) into $\mathcal{R}$. The study of mappings of closed contours around point defects in a 2D system enables one to determine whether the defects are stable.

As an example, let us elucidate the stability of three different points P_0, P_1, and P_2 in the τ field for the case when $0 < |\psi_0| < \pi/2$, and thus, $\mathcal{R} = S^1$ (Fig. 12.7).

Let us surround the "suspicious" point by an oriented loop, such as γ_0 around point P_0. The mapping $\tau(x, y)$ draws a corresponding oriented loop on S^1, such as Γ_0 in Fig. 12.7b. Γ_0 can be continuously contracted (without leaving the circle S^1) into a single point (Fig. 12.7d); accordingly, smooth rearrangement of the vector field $\tau(x, y)$ in the real space results in a uniform state $\tau(x, y) = $ const (Fig. 12.7c). The defect P_0 under test proved to be removable, or topologically unstable.

The situation differs for the radial-like configuration of $\tau(x, y)$ (Fig. 12.7e). The loop Γ_1 runs around the entire circle S^1 and cannot be continuously contracted. To eliminate the defect in Fig. 12.7e, one has to destroy the condensed state along the entire line starting at P_1 (to cut Γ_1) or to allow appreciable deviation of the tilt angle from $|\psi_{0,eq}|$ (to separate Γ_1 from the circle S^1). Both cases require overcoming a considerable energy bar-

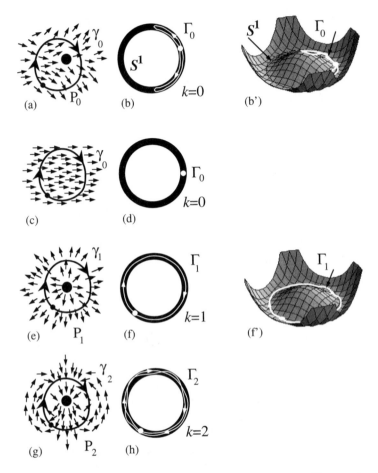

Figure 12.7. Point defects with different topological charges k in a polar 2D nematic with tilt angle $0 < |\psi_0| < \pi/2$; the arrows depict the projections of director on the plane: (a), (c) $k = 0$, defect-free state; (e) $k = 1$; g) $k = 2$; (b), (d), (f), (h) corresponding contours in the order parameter space S^1. (b') and (f') show S^1 as the circle of degenerate minima of the free energy potential; see text.

rier that greatly exceeds the energy of the in-plane distortions. In other words, the defect in Fig. 12.7e is topologically stable. The defect P_2 (Fig. 12.7g), whose contour Γ_2 runs twice around S^1, is also stable: It cannot be transformed into the uniform state nor into the radial-like defect.

The scheme above sets a correspondence between topologically stable defects and contours Γ_k that encircle S^1 k times in a given direction. All point singularities are divided into classes, each of which corresponds to its own class of homotopic contours Γ_k. The set of classes Γ_k forms the fundamental group $\pi_1(S^1)$. The definition of group is satisfied:

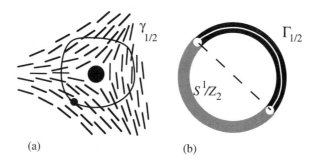

Figure 12.8. Defect $k = -1/2$ in a 2D, nonpolar nematic.

Any two contours Γ_k and Γ_n can be combined to give a third contour $\Gamma_{kn} = \Gamma_k \Gamma_n$ (we simplify the notations for product by dropping the symbol $\otimes$ from this point onward). The operation is associative; there is a special contour Γ_0 equivalent to a point, which is the identity element; contours Γ_k and Γ_{-k} that run around S^1 in opposite directions are the inverses of each other, $\Gamma_k \Gamma_{-k} = \Gamma_0$. Finally, $\pi_1(S^1)$ is Abelian.

Each element of $\pi_1(S^1)$ can be labeled by an integer k; thus, $\pi_1(S^1)$ is isomorphic to the group Z of integers. The number k is the topological charge of the defect. It cannot be changed by continuous deformations. Analytically, for point defects under consideration,

$$k = \frac{1}{2\pi} \oint_\gamma \nabla\varphi \, dl = 0, \pm 1, \pm 2, \ldots. \tag{12.7}$$

When $|\psi_0| = \pi/2$, the degeneracy space is $\mathcal{R} = S^1/Z_2$, i.e., topologically identical to $\mathcal{R} = S^1/Z_2$. Thus, there are as many (infinitely many) defects in the case $|\psi_0| = \pi/2$ as in the case $0 < |\psi_0| < \pi/2$. However, physically, the two sets of defects are different: With $|\psi_0| = \pi/2$, defects with an odd number of π-rotations are allowed and thus k can be integer or *half*-integer (Fig. 12.8a). In Fig. 12.8b, the contour $\Gamma_{1/2}$ that connects antipodal points of the circle is closed and cannot be contracted into a point; compare to Fig. 11.3d.

Finally, for $|\psi_0| = 0$, there are no defects at all, and the fundamental group is trivial, $\pi_1(0) = 0$.

12.1.8. Point Dislocations in a Two-Dimensional Crystal

As was established in Chapter 3, the OP space of a 2D crystal is the direct product of two circles, i.e., the torus, $\mathcal{R} = S^1 \times S^1$ (we neglect the symmetries of rotation) (Fig. 3.7 and Fig. 12.9). Any in-plane displacement of a 2D crystal lattice as a whole leads just to another presentation of the crystal but does not change its thermodynamic potentials. If the displacement vector coincides with the primitive lattice vector, the transformation leads

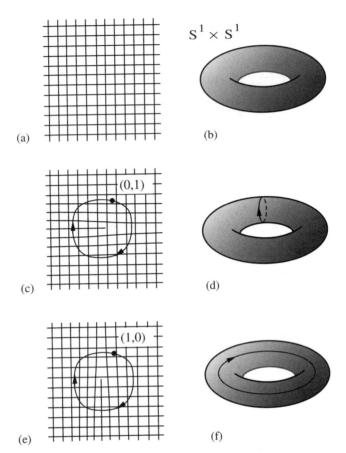

$$S^1 \times S^1$$

(a)   (b)

(c)   (d)

(e)   (f)

Figure 12.9. Point dislocations in a 2D crystal with the order parameter space $\mathcal{R} = S^1 \times S^1$.

to the identical state. The torus as the degeneracy space results from identification of the boundaries of a 2D Bravais cell.

The following basic types of loops cannot be contracted into a point on the torus: Those that run around the "small" circle (Fig. 12.9d), those that run around the "large" circle, (Fig. 12.9f), and their combinations. These loops correspond to point dislocations. Each dislocation is characterized by a pair of topological invariants (k_x, k_y) that shows how many times the loop runs about the "small" and the "large" circle, respectively:

$$\pi_1(S^1 \times S^1) = \pi_1(S^1) \times \pi_1(S^1) = Z \times Z = (k_x, k_y). \qquad (12.8)$$

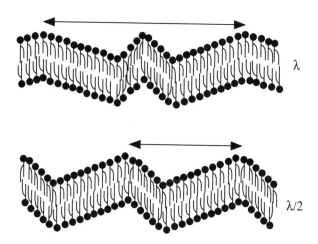

Figure 12.10. Molecular structure of the bilayers with λ- and $\lambda/2$-order in the $P_{\beta'}$ phase, shown in cross section by a plane normal to the bilayers. Arrows indicate the period of structures.

The numbers (k_x, k_y) in (12.8) are nothing else but the x and y-components of the Burgers vector **b** expressed in the units of the lattice repeat vectors $(\mathbf{a}_x, \mathbf{a}_y)$: $\mathbf{b} = k_x \mathbf{a}_x + k_y \mathbf{a}_y$.

A 2D smectic can be represented by a 2D medium with a 1D density wave. A close experimental model is the "rippled" lamellar lyotropic $P_{\beta'}$ phase composed of stacked lecithin bilayers with additional translational order along one of the two within-membrane directions. In-plane ordering makes the rippled membranes rigid; in principle, one can dilute the $P_{\beta'}$ phase and observe a behavior of a single bilayer in isolation. The bilayers of the $P_{\beta'}$ phase manifest two types of in-plane corrugated supermolecular structures— the λ-phase and the $\lambda/2$-phase (Fig. 12.10). The ridges and troughs of the membrane are equidistant and can be considered as a 2D smectic phase. The difference between the two is that the "layers" of the λ-phase are of equal width d, whereas in the $\lambda/2$-phase, the alternating odd and even "layers" are of different width; this asymmetric profile is observed for chiral surfactants.

To find the degeneracy space of the symmetric λ-phase, one has to consider both translation and rotation symmetries. A translation t by a multiple of d along the normal to the layers and a rotation r by an angle multiple of π bring the system into itself. The OP space is thus a rectangle $\{(t, r), 0 \leq t \leq d, 0 \leq r \leq \pi\}$ (Fig. 12.11a). A point (t, r) within the rectangle corresponds to a different representation of the 2D smectic; all of these representations have the same energy. The sides of the rectangle are identified by the rules $(0, r) \rightarrow (d, r)$ and $(t, 0) \rightarrow (d - t, \pi)$, as illustrated by the arrows in Fig. 12.11a. Note that after a rotation by π, a translation t transforms into $d - t$. This subtle point makes the resulting degeneracy space a Klein bottle (Fig. 12.11) rather than a torus. The fundamental group of the Klein bottle is non-Abelian and is isomorphic to a semidirect product of the

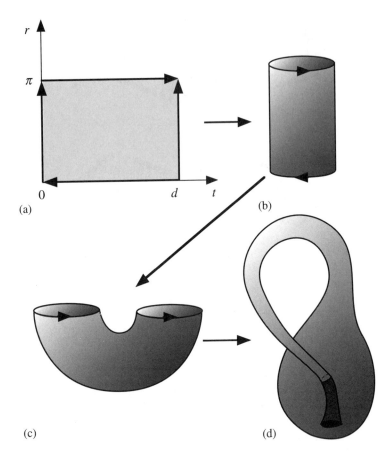

Figure 12.11. Equivalent presentations of the order parameter space of a 2D smectic: (a) a rectangle $\{(t, r), 0 \le t \le d, 0 \le r \le \pi\}$ with edges identified; (d) the same rectangle transformed into a Klein bottle through steps (b) and (c).

group Z_{tr} describing translations by the group Z_{rot} describing rotations, which are both isomorphic to Z:

$$\pi_1(\mathcal{R}_\lambda) = Z_{tr} \wedge Z_{rot}. \tag{12.9}$$

Consequently, every point defect in the λ-phase corresponds to a pair of numbers (n, m). Elements $(n, 0)$ correspond to point dislocations with Burgers vector $\mathbf{b} = nd$, and elements $(0, m)$ correspond to point disclinations of integer (even m's) and half-integer strength (odd m's). Here, the values of m are taken twice as large as those of the usual k's in order to better represent the combinations of defects. The combination law for the semidirect product is different from that of a direct product. If n_1 and n_2 are two translations, and m_1 and m_2 are two rotations, then the combination law is $(n_1, m_1)(n_2, m_2) = (n_1 + m_1 n_2, m_1 + m_2)$.

Noncommutativity of the fundamental group $\pi_1(\mathcal{R}_\lambda)$ results in subtle physical effects. For example, the result of merger of two defects in the presence of a third defect is ambiguous and depends on the path of merger. In a non-Abelian medium, a point defect no longer corresponds to one element of the fundamental group, but to an entire class of *conjugated* elements. By definition, two elements a and b of a group G are said to be conjugate to one another if there is an element q of G such that $b = q^{-1}aq$. Figure 12.12 shows that a $(1, 0)$ dislocation after passing around a $(0, 1)$ disclination should be characterized by the pair $(-1, 0)$. In other words, the same defect is described by different elements $(1, 0)$ and $(-1, 0)$ that belong to the same conjugacy class. In the OP space, the contour $\Gamma_{(1,0)}$ transforms into the contour $\Gamma_{(-1,0)} = \Gamma_{(0,1)}^{-1}\Gamma_{(1,0)}\Gamma_{(0,1)}$ after the corresponding defect $(1, 0)$ goes around the $(0, 1)$ disclination. In an Abelian medium, $\Gamma_{(1,0)}$ and $\Gamma_{(0,1)}$ commute, and thus $\Gamma_{(1,0)}$ and $\Gamma_{(-1,0)}$ would be homotopic; however, in the noncommutative case, $\Gamma_{(1,0)}$ and $\Gamma_{(-1,0)}$ are not homotopic.

Because the same defect can correspond to different elements of the conjugacy class in a non-Abelian π_1, the coalescence of two defects is not uniquely defined. For example, a $(1, 0)$ dislocation and a $(-1, 0)$ "antidislocation" upon merging either annihilate (Fig. 12.13b) or form a double dislocation $(2, 0)$, if the point $(-1, 0)$ passed around the $(0, 1/2)$ disclination on the path to the merger site (Fig. 12.13c). The result is determined not by multiplication of individual elements of the homotopy group, as for Abelian media, but by all results of multiplication of classes of conjugated elements. Thus, to predict the result of merging, one has to know the global configuration of the order parameter; local topology around the defects is not enough.

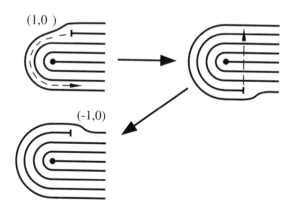

Figure 12.12. Conversion of a dislocation $(1, 0)$ into an antidislocation $(-1, 0)$ after circumnavigating around a disclination in the 2D smectic λ-phase.

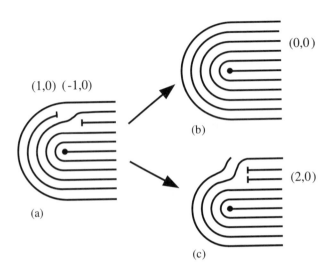

(1,0) (-1,0)

(0,0)

(b)

(2,0)

(a)

(c)

Figure 12.13. The result of merging of $(1, 0)$ and $(-1, 0)$ dislocations in the 2D smectic λ-phase depends on the path of merger around the disclination: (b) uniform state $(0, 0)$; (c) double dislocation $(2, 0)$.

The other variety of the $P_{\beta'}$ phase, the $\lambda/2$ structure, seemingly hardly differs from the λ structure (Fig. 12.10): Only the twofold symmetry axis C_2 has disappeared. However, now the order parameter space is a 2D torus T^2, and thus,

$$\pi_1(\mathcal{R}_{\lambda/2}) = Z_{\text{tr}} \times Z_{\text{rot}} \qquad (12.10)$$

is a direct product of groups, and is commutative: The result of the merger of two defects is always unambiguous. Moreover, in contrast to the λ-phase, the $\lambda/2$-phase lacks isolated disclinations with an odd m.

Simple examples of λ- and $\lambda/2$-phases also reveal some restrictions of the homotopy theory for classifying defects in media with broken translational symmetry. According to the homotopy theory, (12.9) and (12.10), there might be disclinations with any integer m in a layered system. In reality, disclinations with m not equal to 1 or 2 ($k \neq 1/2, 1$) break the equidistance of layers and thus are energetically unfavorable.

12.2. The Fundamental Group of the Order Parameter Space. Line Defects

The fundamental group classifies topologically stable point defects in 2D-ordered media and topologically stable line defects in 3D.

12.2.1. Unstable Disclinations in a Three-Dimensional Isotropic Ferromagnetic

The OP space of the Heisenberg isotropic ferromagnet is a sphere S^2. Consider a disclination in the magnetization vector field. An oriented loop surrounding the disclination core in real space is mapped by the vector field into the OP space and, thus, produces a contour on S^2. Obviously, any contour drawn on a sphere can be contracted into a point. Thus, $\pi_1(S^2) = 0$: Any disclination in the 3D isotropic ferromagnet can be continuously transformed into a uniform state (Fig. 12.14).

The *topological* simplicity of the OP space should not let one believe that the question of defects in ferromagnets is trivial. There are interesting similarities between defects (points and lines) in nematics and ferromagnets, because the order parameters look very much alike, a vector in the case of a ferromagnet, a director in a nematic. But again the knowledge of nematic defects does not cover the knowledge of magnetic defects. The *physical* differences are indeed considerable. The subject of defects in ferromagnets will not be

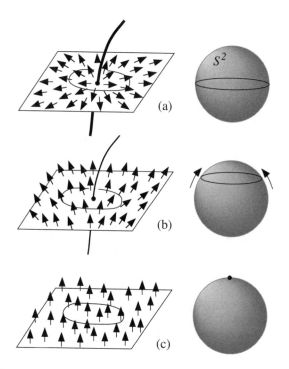

Figure 12.14. Continuous transformation of a disclination $k = 1$ (a) into a uniform state $k = 0$ (c) in a 3D ferromagnetic phase (left); corresponding contraction of a loop in the order parameter space (right).

pursued further. The interested reader might appreciate the specificity and the richness of this subject in the texts cited in "Further Reading."

12.2.2. Stable Disclinations in a Three-Dimensional Uniaxial Nematic Phase

In a 3D nonpolar uniaxial nematic phase, $\mathbf{n} \equiv -\mathbf{n}$, and the OP space is the sphere S^2/Z_2 or the projective plane RP_2.

There are two types of contours in S^2/Z_2: This is visible at once in the 2D representation of the projective plane (Fig. 3.6), where one notices contours that are actually closed (e.g., circles) and contours that are terminating at two diametrically opposite points. The first ones can shrink into a point; they correspond to disclinations of integer strength k that are topologically unstable. The second class of contours is not contractible to a point: under any continuous deformations, the ends of the contours remain fixed at the diametrically opposite points. These contours correspond to disclinations of half-integer k.

It is easy to see that all contours corresponding to half-integer k's can be transformed one into another (Fig. 12.15). Therefore, there is just one class of topologically stable

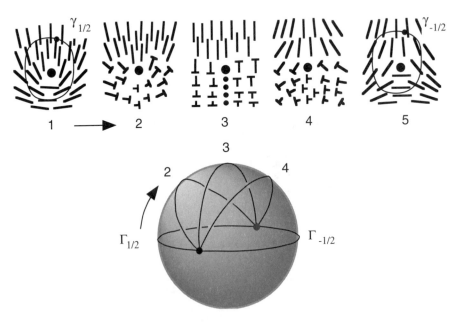

Figure 12.15. Continuous transformation of a disclination $k = 1/2$ into a disclination $k = -1/2$ in real space (director configurations above) and the corresponding transformation of contours in the OP space RP^2.

disclinations in the uniaxial nematic phase:

$$\pi_1(S^2/Z_2) = Z_2. \tag{12.11}$$

The laws of conservation are simply $1/2 + 0 = 1/2$ and $1/2 + 1/2 = 0$, if expressed in terms of k's.

12.2.3. Disclinations in Biaxial Nematic and Cholesteric Phases

Biaxial nematic order is specified by a tripod of mutually perpendicular directors $\mathbf{l} \equiv -\mathbf{l}$, $\mathbf{n} \equiv -\mathbf{n}$, and $[\mathbf{nl}] \equiv -[\mathbf{nl}]$. The OP space is the group $G = SO(3)$ of rotations of the triad $\mathbf{l}$, $\mathbf{n}$, $[\mathbf{nl}]$, factorized by the four-element point group D_2 of π-rotations about the directions $\mathbf{l}$, $\mathbf{n}$, and $[\mathbf{nl}]$:

$$R_{bx} = SO(3)/D_2. \tag{12.12}$$

Any two diametrically opposite points at the surface of $SO(3)$ are identical: π-rotations about axes oriented in opposite directions yield the same result, as discussed in Section 12.1.5. Thus, $SO(3)$ can be equivalently represented as the projective space $RP^3 = S^3/Z_2$ or as a 3D sphere in 4D space at which the antipodal points are identified, $SO(3) = S^3/Z_2$ (compare with the uniaxial nematic with OP space $RP^2 = S^2/Z_2$).

This is the place for some comments of general interest, which will be useful relating to media whose OP is a local triad (biaxial nematics, cholesterics). A standard method to calculate homotopy groups in the context of ordered media is *to lift* the topological space G to a *simply connected* space $\overline{G}$ (called the *universal cover* of space G), i.e., such that $\pi_1(\overline{G}) = 0$. In this process, any point $g \in G$ is lifted to a set of n (n independent of g) points g_i. Reciprocally, the *projection* on G of any $g_i \in \overline{G}$ and of a neighborhood V_i of g_i maps g_i and V_i on g and a neighborhood V of g. A path γ_{ij} on $\overline{G}$ from g_i to g_j maps on a closed loop Γ_{ij} in G, i.e., is in correspondence with one element of $\pi_1(G, g)$; i.e., n is the index of the fundamental group, or one of its subgroups (Massey). Note also that a subgroup $H \subset G$ lifts to a subgroup $\overline{H} \subset \overline{G}$, with relations of the type $g_i h_j = f_p$; $g_i, h_j, f_p \in \overline{G}$ if $gh = f$; $g, h, f \in G$. A simple ilustration of the properties above is with $G = S^1$, the 1D circle; its universal cover is a helix ($n = \infty$), which for simplicity, the reader can figure out as having the same radius as the circle, and staying "above" to visualize the projection process, although one must keep in mind that this "metrical" representation has no relation whatsoever with the topological properties that are considered here. Eventually, let us consider a subgroup $\overline{H} \subset \overline{G}$ that is the lift of a subgroup $H \subset G$. The fundamental group $\pi_1(G/H) = \pi_1(\overline{G}/\overline{H})$, because the coset spaces are the same, $R = G/H = \overline{G}/\overline{H}$. Now, consider the manifold obtained by identifying any point $\overline{g}$ in $\overline{G}$ with all points $\overline{h}^{-1}\overline{g}\overline{h}$; $\overline{h} \in \overline{H}$ belonging to the subgroup conjugated to $\overline{g} \in \overline{G}$. Clearly, this manifold represents R. Furthermore, $\pi_1(\overline{G}/\overline{H}) = \overline{H}$, because $\overline{G}$ is simply connected; a complete demonstration requires more sophisticated methods than those possible in the frame of this textbook.

The universal cover of the projective plane is the sphere S^2; similarly, the universal cover of $SO(3)$ is S^3. In both cases, $n = 2$. Consider $\overline{G} = S^3$; the symmetry group of the triad of directors is D_2, an Abelian subgroup of $SO(3)$ with four elements, which we shall denote I, i, j, and p, with the relations $i^2 = I, ij = ji = p$, where I is the identity and each other element is a π rotation about one of three different perpendicular axes. The lift of D_2 into S^3 is a group with eight elements. We call these elements, as the above, I, i, j, and p, adding the supplementary elements $J, -i, -j$, and $-p$, whose notations mean that they are, one by one, on the same "fibers" lifted over $SO(3)$ (therefore, their geometrical interpretation is the same). This is enough to obtain the new group relations, remembering the remark of the latter paragraph (viz. $g_i h_j = f_p$; $g_i, h_j, f_p \in \overline{G}$ if $gh = f$; $g, h, f \in G$). This eight-element group, noted Q, is non-Abelian and obeys the multiplication rules:

$$ij = -ji = p, \quad jp = -pj = i, \quad pi = -ip = j,$$
$$JJ = I, \quad ii = jj = pp = J, \quad ijp = J. \tag{12.13}$$

It is the group of quaternion units. Eventually, one finds

$$\pi_1(S^3/D_2) = Q \tag{12.14}$$

The elements of the quaternion group form five conjugacy classes $C_0 = \{I\}, \overline{C}_0 = \{J\}$, $C_x = \{i, -i\}, C_y = \{j, -j\}$, and $C_z = \{p, -p\}$.

Disclinations in biaxial nematics differ sharply from disclinations in uniaxial nematics. Among them, one should distinguish five, rather than one, classes of topologically stable lines, which correlate with the five classes of conjugated elements of the group Q. Correspondingly, the topological charge can acquire the values $I, J, (i, -i), (j, -j), (p, -p)$, with the multiplication rules (12.13). Different disclinations are shown in Fig. 12.16. The strength k can be half-integer (π rotation of a director around the core, classes C_x, C_y, and

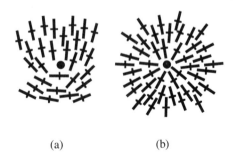

(a)   (b)

Figure 12.16. Director fields $\mathbf{n}$ (long rods) and $\mathbf{l}$ (short rods) for topologically stable disclinations in a biaxial nematic: (a) $k = 1/2$, class C_z; (b) $k = 1$, class $\overline{C}_0$. Cross sections in a plane perpendicular to the axes of the disclinations are shown.

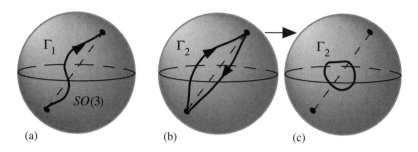

Figure 12.17. Closed contours Γ_1 (a) and Γ_2 (b), (c) corresponding to $|k| = 1$ and $|k| = 2$ disclinations in the OP space of the biaxial nematic. Both contours connect diametrically opposite and equivalent points at the surface of $SO(3)$. Γ_1 cannot continuously shrink into a point. Γ_2 runs between the two antipodal points twice (b) and can smoothly leave these points and shrink into a point (c).

C_z) or integer (2π-rotation, class $\overline{C}_0$). The singular core of the 2π disclination *cannot* be eliminated by the "escape in third dimension" as in the uniaxial nematic (Figs. 12.16 and 12.17). In contrast, 4π-disclinations, class C_0, $|k| = 2$, are topologically unstable. The striking difference between a 2π- and a 4π-lines is illustrated in Fig. 12.17.

The merger and decay of disclinations in the biaxial nematic obey the multiplication rules that are specific to the classes of elements, rather than the elements themselves. The results are given in Table 12.1.

If two disclinations belonging to two different classes merge, then a disclination is formed that belongs to the class of the product of the first two. The result of merger of disclinations of the same class from the set C_x, C_y, C_z is ambiguous: Either a nonsingular trivial configuration (class C_0) or a disclination from class $\overline{C}_0$ can be formed, depending on the path of merger with respect to other defect lines in the system.

The cited features of the disclinations merger stem from the noncommutativity of the group Q. Another spectacular consequence shows up in the entanglement of disclinations in biaxial nematics.

Table 12.1. Multiplication rules of five classes of elements of the quaternion group that control the merger and decay of disclinations in a biaxial nematic.

	C_0	$\overline{C}_0$	C_x	C_y	C_z
C_0	C_0	$\overline{C}_0$	C_x	C_y	C_z
$\overline{C}_0$	$\overline{C}_0$	C_0	C_x	C_y	C_z
C_x	C_x	C_x	C_0 or $\overline{C}_0$	C_z	C_y
C_y	C_y	C_y	C_z	C_0 or $\overline{C}_0$	C_x
C_z	C_z	C_z	C_y	C_x	C_0 or $\overline{C}_0$

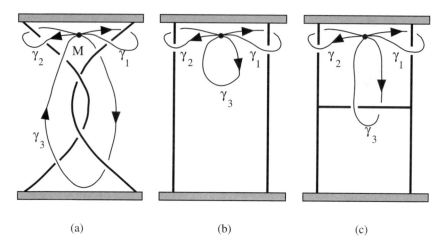

(a)   (b)   (c)

Figure 12.18. (a) Entanglement of disclinations in a biaxial nematic; (b) topologically trivial; (c) non-trivial.

Figure 12.18a shows two entangled disclinations. The question is whether they can be transformed by continuous variations of the directors into an unlinked configuration (Fig. 12.18b), if we require that the ends of the disclinations remain fixed.

To find the answer, let us draw three contours γ_1, γ_2, and γ_3 from a point M of real space: γ_1 and γ_2 encircle the defect lines, and γ_3 encircles the entangled region (Fig. 12.18). Their images in OP space will be some contours Γ_1, Γ_2, and Γ_3. Evidently, the defects can be unlinked only when Γ_3 is homotopic to zero. If this is not so, then separation of the disclinations will leave a topologically nontrivial trace in space, a third disclination (Fig. 12.18c). The result depends on the nature of the linked disclinations. One can show (Fig. 12.19) that the contour Γ_3 is homotopic to the product $\Gamma_1\Gamma_2\Gamma_1^{-1}\Gamma_2^{-1}$; an element of this form is called a *commutator* in the fundamental homotopy group. For Abelian groups, the commutator is the identity element, because $\Gamma_1\Gamma_2 = \Gamma_2\Gamma_1$. This is not true for non-Abelian groups; in particular, for the group Q, the contour Γ_3 can belong either to the class $C_0(\Gamma_1\Gamma_2\Gamma_1^{-1}\Gamma_2^{-1} = 1)$ or to the class $\overline{C}_0(\Gamma_1\Gamma_2\Gamma_1^{-1}\Gamma_2^{-1} = -1)$. The latter situation occurs when the two entangled disclinations belong to different classes from the set C_x, C_y, C_z. Therefore, after drawing two different disclinations $|k| = 1/2$ through one another, they prove to be connected by a disclination $|k| = 1$ belonging to $\overline{C}_0$.

The topological classification of defects in biaxial nematics can be applied to cholesterics, when the cholesteric pitch is much smaller than is the characteristic scale of deformations, as discussed in Chapter 11, where for convenience, we used the same notations for the classes of disclinations. However, the topological classification considered above does not apply, in its full generality, to the coarse-grained picture of cholesterics that takes into account the high energy cost of changing the cholesteric pitch.

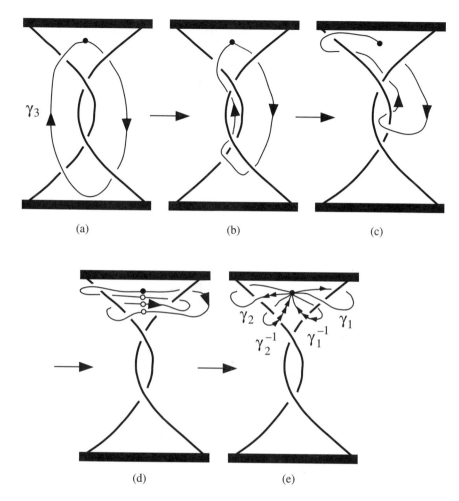

Figure 12.19. Continuous deformations of the contour γ_3 from Fig. 12.18 into the product contour $\gamma_1\gamma_2\gamma_1^{-1}\gamma_2^{-1}$, demonstrating that the image Γ_3 of γ_3 in OP space is homotopic to the product $\Gamma_1\Gamma_2\Gamma_1^{-1}\Gamma_2^{-1}$. At the step (d), one pinches together four points marked by circles.

12.3. The Second Homotopy Group of the Order Parameter Space and Point Defects

Point defects in 3D ordered phases are classified by the elements of the second homotopy group.

12.3.1. Point Defects in a Three-Dimensional Ferromagnet

In a 3D isotropic ferromagnet, the magnetization field $\mathbf{m(r)}$ may contain singular points at which the direction of $\mathbf{m}$ is not specified. To elucidate the topological stability of such a point, one encloses it with a closed surface σ (Fig. 12.20).

The radius of the sphere should be much larger than the core size of the point defect. The function $\mathbf{m(r)}$ produces a mapping of the surface σ into some surface in the OP space $\mathcal{R} = S^2$ (Fig. 12.20). If the resulting surface Σ can be contracted to a single point (Fig. 12.20a), the point defect is topologically unstable. If Σ is wrapped $N \neq 0$ times around the sphere S^2, the point singularity is a stable defect with topological charge

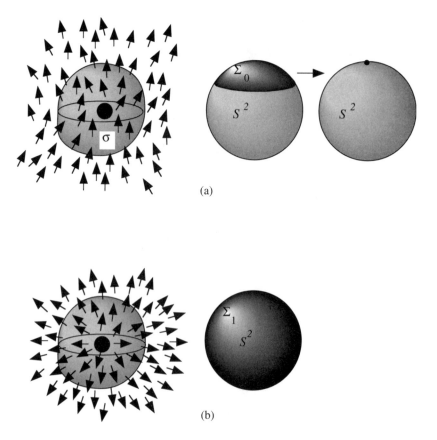

(a)

(b)

Figure 12.20. Topological stability of a point defect in a ferromagnet, core region (black ball) surrounded by a closed surface σ. (a) Unstable point defect; (b) topologically stable point defect.

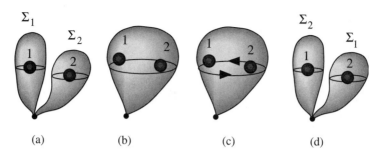

Figure 12.21. Multiplication of classes of surfaces, e.g., Σ_1 and Σ_2, is a commutative operation: $\Sigma_1\Sigma_2 = \Sigma_2\Sigma_1$. The arrow in (c) shows deformation of rotation. Redrawn from Mineev.

$N \neq 0$. For example, Fig. 12.20b illustrates a "hedgehog" radial defect, $\mathbf{m}(\mathbf{r}) = \mathbf{r}/|\mathbf{r}|$, for which Σ covers the entire sphere S^2 once, $N = 1$. The classes of all images Σ's (including the surfaces homotopic to a single point) form the *second homotopy group* $\pi_2(S^2)$, which in the case of the ferromagnet is isomorphic to the group of integers Z. The topological charge N of a point defect is also called the degree of mapping of σ on S^2. N shows how many times the vector $\mathbf{m}$ runs over the sphere S^2 in moving over the closed surface σ. In the example above, $\mathbf{m}(\mathbf{r}) = \mathbf{r}/|\mathbf{r}|$, and $N = 1$; if one reverses the orientation of $\mathbf{m}$, i.e., $\mathbf{m}(\mathbf{r}) = -\mathbf{r}/|\mathbf{r}|$, then the degree of mapping also changes the sign: $N = -1$. When the point defects coalesce, the charges N add up.

In contrast to $\pi_1(\mathcal{R})$, groups $\pi_2(\mathcal{R})$ are always Abelian; i.e., the multiplication of classes of surfaces Σ is commutative. Figure 12.21 shows a continuous deformation that establishes $\Sigma_1\Sigma_2 = \Sigma_2\Sigma_1$.

12.3.2. Topological Charges of Point Defects

Analytically, the topological charge of a point defect in a 3D unit vector field $\mathbf{m}$ is defined as an integral[1] over the sphere σ:

$$N^{(3)} = \frac{1}{4\pi} \int_{\sigma} \mathbf{m} \cdot \left[\frac{\partial \mathbf{m}}{\partial u_1} \times \frac{\partial \mathbf{m}}{\partial u_2} \right] du_1\, du_2. \tag{12.15a}$$

The integrand contains the Jacobian of the transformation from the coordinates u_1 and u_2 on the sphere σ to the vector components $\mathbf{m}$ parameterizing the surface Σ that covers the OP S^2-sphere N times. If the vector field is parameterized as $\{n_x; n_y; n_z\} = \{\sin\theta\cos\varphi; \sin\theta\sin\varphi; \cos\theta\}$, with both angles θ and φ being the functions of the two

---

[1]M. Kleman, Phil. Mag. **27**, 1057 (1973); N.D. Mermin and T.-L. Ho, Phys. Rev. Lett. **36**, 594 (1976).

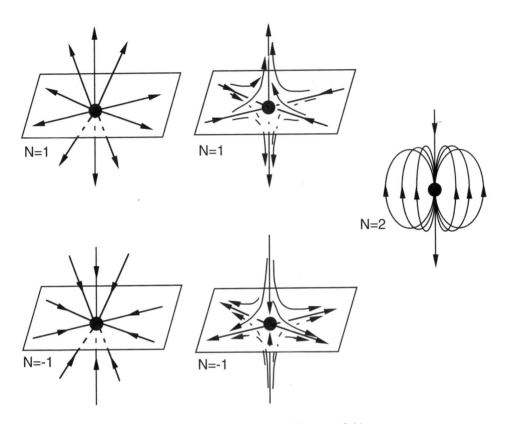

Figure 12.22. Point defects in a 3D vector field.

angular spherical coordinates u_1 and u_2, then

$$N^{(3)} = \frac{1}{4\pi} \int_0^{2\pi} du_2 \int_0^\pi \left(\frac{\partial\theta}{\partial u_1} \frac{\partial\varphi}{\partial u_2} - \frac{\partial\theta}{\partial u_2} \frac{\partial\varphi}{\partial u_1} \right) \sin\theta \, du_1. \qquad (12.15b)$$

For a radial hedgehog, $N^{(3)} = 1$, as expected. Different 3D point defects are shown in Fig. 12.22.

In a similar way, one can define the topological charge of a 2D unit vector field τ with components $\tau^1(l)$ and $\tau^2(l)$ as the integral over the coordinate l on a contour around the defect [compare with (11.4)]:

$$N^{(2)} = \frac{1}{2\pi} \oint \left(\tau^1 \frac{d\tau^2}{dl} - \tau^2 \frac{d\tau^1}{dl} \right) dl \quad (= 0, \pm 1, \pm 2, \ldots). \qquad (12.16)$$

Note that the two 2D radial configurations: "sink" (τ directed toward the core) and "source" (τ directed outward from the core), have the same invariant $N^{(2)} = 1$, in contrast to the 3D radial configurations, in which a reversal in the direction of **m** changes the sign of $N^{(3)}$.

12.3.3. Point Defects in a Three-Dimensional Nematic Phase

Classification of point defects in a 3D uniaxial nematic, $\mathcal{R} = S^2/Z_2$, is similar to that in a ferromagnet: $\pi_2(S^2/Z_2) = Z$. However, because $\mathbf{n} \equiv -\mathbf{n}$, each point can be equally assigned a charge N and a charge $-N$ (12.15). This ambiguity in the sign of the topological charge is the consequence of nontriviality of the fundamental group $\pi_1(S^2/Z_2)$ and its action on the group $\pi_2(S^2/Z_2)$.

Assume that the nematic volume contains a point defect and a π-disclination line. The director field provides a mapping of degree N of the sphere σ enclosing the point defect and part of the line defect, on S^2/Z_2. A point $\mathbf{r}_0$ of real space is mapped into an image point $\mathbf{n}(\mathbf{r}_0)$ of OP space. If one moves the point $\mathbf{r}_0$ over a closed contour γ around the disclination line, then the point $\mathbf{n}(\mathbf{r}_0)$ goes over a contour $\Gamma_{1/2}$ into an antipodal point $-\mathbf{n}(\mathbf{r}_0)$ and N reverses sign (Fig. 12.23). The contour $\Gamma_{1/2}$ connecting $\mathbf{n}(\mathbf{r}_0)$ and $-\mathbf{n}(\mathbf{r}_0)$ is a nontrivial element of the group $\pi_1(S^2/Z_2) = Z_2$. If the contour $\Gamma_{1/2}$ were the identity element, the degree of mapping would preserve the sign.

Thus, when $\pi_1(\mathcal{R})$ is nontrivial, each point singularity corresponds to a few elements (such as N *and* $-N$ above) of $\pi_2(\mathcal{R})$, rather than to a single element of $\pi_2(\mathcal{R})$ (such as N *or* $-N$). These elements are tranformed one into another by moving the image points along contours that are nontrivial elements of $\pi_1(\mathcal{R})$. The coalescence of two point defects N_1 and N_2 in the presence of disclination can result in a defect with a charge $|N_1 + N_2|$

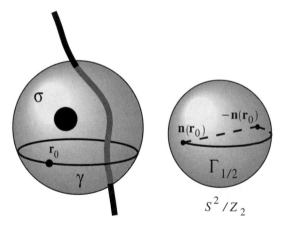

Figure 12.23. Action of the fundamental group π_1 on the second homotopy group π_2 in a uniaxial nematic: Moving point $\mathbf{r}_0$ of real space around a π-disclination transforms the image point $\mathbf{n}(\mathbf{r}_0)$ into an antipodal point $-\mathbf{n}(\mathbf{r}_0)$.

or $|N_1 - N_2|$, depending on the path of coalescence. Owing to this feature, all hedgehogs in a nematic system with a π-disclination can be annihilated, or at least all but one with $N = 1$ (if the total charge is odd).

12.4. Solitons

12.4.1. Planar Solitons

Let us study a uniaxial nematic placed in a plane capillary, both surfaces of which impose planar anchoring in one direction **h**. The director in the bulk is set to be oriented along **h**: $\mathbf{n} = \pm\mathbf{h}$. In other words, the interaction with the walls contracts the OP space of the nematic to a single point. Let a vertical disclination $k = \pm 1/2$ exist in the specimen. When the disclination is present, it is impossible to conserve a uniform configuration $\mathbf{n} = \pm\mathbf{h}$: at a certain surface supported by the disclination, the director will undergo a π-rotation (Fig. 12.24a). The width of the wall is fixed by the balance of elastic and surface anchoring

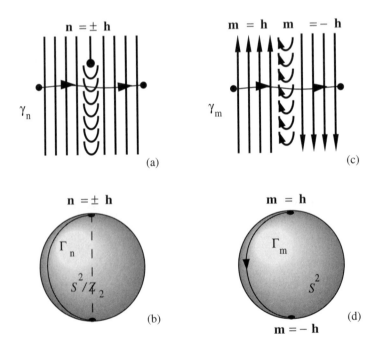

Figure 12.24. (a), (b) Topologically stable planar soliton in a uniaxial nematic; (c), (d) Bloch domain wall in an anisotropic ferromagnetic with an "easy magnetization" axis.

energies. If we write the energy of the wall as

$$F = \frac{Kd}{\rho} + w\rho, \tag{12.17}$$

then its equilibrium width ρ_0 is

$$\rho_0 = \left(\frac{Kd}{w}\right)^{1/2}. \tag{12.18}$$

Here, d is the thickness of the capillary, K is a nematic elastic constant, and w (dimension J/m^2) is the anchoring coefficient, calculated as the energy per unit surface area needed to deviate the director from the "easy axis" $\mathbf{n} = \pm\mathbf{h}$ by the angle $\pi/2$. Such walls with a nonsingular director configuration are called planar solitons of topological type.

Not only does the soliton under consideration preserve a constant width, but it also possesses a nontrivial topological charge. Indeed, let us study the mapping of the line γ_n threaded through the soliton into the OP space (Fig. 12.24a). The ends of the line are mapped into antipodal identical points $\mathbf{n} = \pm\mathbf{h}$, whereas the line γ_n is mapped onto the closed contour Γ_n, linking these points on S^2/Z_2 (Fig. 12.24). This contour cannot be contracted to a point by any continuous transformations, which determines the topological stability of the planar soliton.

In the general case, the classes of homotopic mappings of the line γ threaded through a planar soliton form the *relative homotopy* group $\pi_1(\mathcal{R}, \overline{\mathcal{R}})$, where $\overline{\mathcal{R}}$ is the region of possible values of the order parameter far from the core of the soliton, narrowed in comparison to $\mathcal{R}$ by additional interaction (external field, boundary conditions, etc.). If $\overline{\mathcal{R}}$ consists of a single point, as in the case being studied, the group $\pi_1(\mathcal{R}, \overline{\mathcal{R}})$ coincides with the fundamental group $\pi_1(\mathcal{R})$. Therefore, soliton walls in nematics exist in a mutually one-to-one correspondence with the disclinations that have produced them and are described by the same group $\pi_1(S^2/Z_2) = Z_2$. If $\overline{\mathcal{R}}$ is not a point, then to find $\pi_1(\mathcal{R}, \overline{\mathcal{R}})$, one first finds $\pi_1(\mathcal{R})$ and then excludes from $\pi_1(\mathcal{R})$ the elements that correspond to $\pi_1(\overline{\mathcal{R}})$. In other words, one must find the factor group of $\pi_1(\mathcal{R})$ by its subgroup, which is the image of the homomorphism $\pi(\overline{\mathcal{R}}) \to \pi(\mathcal{R})$:

$$\pi_1(\mathcal{R}, \overline{\mathcal{R}}) = \pi_1(\mathcal{R})/\mathrm{Im}\left[\pi_1(\overline{\mathcal{R}}) \to \pi_1(\mathcal{R})\right]. \tag{12.19}$$

There are other examples of walls, different from the planar solitons above, that occur in media with disconnected $\overline{\mathcal{R}}$ (Fig. 12.24c). For example, consider a ferromagnet in which additional interaction between the magnetization vector and the crystal lattice is anisotropic. In equilibrium, the vector $\mathbf{m}$ orients along a particular crystallographic axis, say, $\mathbf{m} = \pm\mathbf{h}$. The states $\mathbf{m} = \mathbf{h}$ and $\mathbf{m} = -\mathbf{h}$ are distinguishable, unlike in the nematic phase. Thus, $\mathcal{R}$ is reduced from the sphere S^2 into a set of two disconnected points $\mathbf{m} = \pm\mathbf{h}$ (Fig. 12.24d). The set of disconnected pieces of OP space $\mathcal{R}$ is denoted $\pi_0(\mathcal{R})$. In our case, $\pi_0(\mathbf{m} = \pm\mathbf{h}) = Z_2$ and any two domain walls can be merged to produce a uniform state. Alternatively, one can use the relative fundamental group $\pi_1(\mathcal{R}, \overline{\mathcal{R}}) = \pi_1(S^2, \mathbf{m} = \pm\mathbf{h})$; however, with a disconnected $\overline{\mathcal{R}}$, $\pi_1(\mathcal{R}, \overline{\mathcal{R}})$ is no longer a group. Domain walls of the type of the Bloch and Néel walls that provide a connection between different pieces of the disconnected OP space can be called "classical domain walls" to distinguish them from soliton walls ending at linear defects. This terminological distinction has a physical basis: To remove a wall associated with a linear singularity, it suffices to create a ring of disclinations in the plane of the wall. The latter, in expanding, "eats up" the wall; at the same time, to remove a classic wall requires overcoming a considerably larger energy barrier and transformation of the order parameter over the entire half-space on one side of the wall.

12.4.2. Linear Solitons

Just as a disclination in an external field can give rise to a planar soliton, a point defect can give rise to a linear soliton (Fig. 12.25). Linear solitons are described by the classes of mapping of the surface σ crossing the soliton into the OP spaces $\mathcal{R}$ and $\overline{\mathcal{R}}$, i.e., by the elements of the relative group $\pi_2(\mathcal{R}, \overline{\mathcal{R}})$. If a uniaxial nematic is oriented by a magnetic field $\mathbf{B}$ along the axis $\mathbf{h} \| \mathbf{B}$, then $\overline{\mathcal{R}}_N$ reduces to one point, and $\pi_2(\mathcal{R}_N, \overline{\mathcal{R}}_N) = \pi_2(\mathcal{R}_N) = \pi_2(S^2/Z_2) = Z$; i.e., the classification of linear solitons coincides with the classification of hedgehogs.

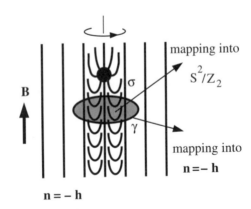

Figure 12.25. Linear soliton terminating at a point defect (a radial hedgehog in a nematic liquid crystal). The magnetic field $\mathbf{B}$ is along the axis $\pm\mathbf{h}$. See text.

12.4.3. Particle-Like Solitons

The distribution of the order parameter in particle-like solitons depends on all three coordinates. They are described by the group $\pi_3(\mathcal{R}, \overline{\mathcal{R}})$ of homotopy classes of the mappings of the 3D spherical volume D^3 containing the soliton into the space $\mathcal{R}$. Here, the boundary of the spherical volume, the sphere σ, is mapped into the narrowed space $\overline{\mathcal{R}}$. If $\overline{\mathcal{R}}$ consists of one point, then the particle-like soliton is described by the group $\pi_3(\mathcal{R})$. The spherical volume D^3 with all point of its surface σ being equivalent, is homotopic to a 3D sphere S^3 in a 4D space. Thus, the elements of $\pi_3(\mathcal{R})$ are mappings $S^3 \to \mathcal{R}$. Special cases $S^3 \to S^2$ and $S^3 \to S^2/Z_2$ are called Hopf mappings and correspond to $\pi_3(S^2) = \pi_3(S^2/Z_2) = Z$ (Fig. 12.26).

In a uniaxial nematic, the particle-like soliton amounts to a director configuration distorted in a region of finite size, outside of which the director field is uniform. As a rule, such solitons are unstable with respect to decrease in size and subsequent disappearance on scales smaller than the coherence length ξ. The decrease in size $L \to \mu L (\mu < 1)$ entails an increase in the elastic-energy density by a factor of $1/\mu^2$ and a decrease in the soliton's volume by a factor of μ^3, so that the total elastic energy decreases: $F \to F\mu$. Stabilization of particle-like solitons can be facilitated by an additional interaction, in particular, by helical twisting of the structure. In a weakly twisted cholesteric mixture, Bouligand observed two linked disclination rings $k_1 = k_2 = 1$, each of which by itself is topologically

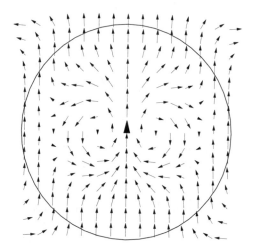

Figure 12.26. A nontrivial Hopf texture in a 3D vector field, as seen in the vertical cross section. The vector field is directed north everywhere outside of the sphere and at the origin. The vertical axis is the rotational symmetry axis. When going along any radius from the center to the surface of the sphere, the vector rotates by an angle $2\pi r/R$ around this radius. The length of the arrows is proportional to the length of vector projection in the XY plane.

unstable, whereby all points of the cores of the disclination are mapped into a single point of the degeneracy space S^2/Z_2. In going from one ring to the other, the director undergoes a 180° rotation, and one can represent the rings as inverse images of two diametrically opposite points on the sphere S^2. Evidently, one cannot convert the configuration into a homogeneous state because the rings are linked: Upon trying to unlink the rings, they must intersect one another and singularities would arise in the configuration. The degree of linking of the rings, equal in this case to unity, coincides with the Hopf invariant, which is an element of the group $\pi_3(S^2/Z_2) = Z$. The stability of the configuration as a whole is guaranteed by the conservation of the Hopf invariant.

Problem 12.1. Prepare a paper strip with a 2π-twist. Mark the edges of the strip in different colors. Imagine that the strip models a simple loop of a duplex DNA molecule. The DNA must be duplicated every time a cell divides. The two strands of the initial DNA must separate, and then each should synthesize its new partner to form a double-stranded DNA. Try to prepare the paper strip for a "replication" by cutting it along the central line; in a real DNA, it would correspond to the cutting of the hydrogen bonds between the base pairs. Repeat the cutting along the central line two times. Describe the topology of the result at each step. Compare the behavior of 2π-, π-, and untwisted strips.

Hint. The exercise should demonstrate the need in topoisomerases invented by Nature to change topology of the duplex DNA by cutting one or two of its strands.

Problem 12.2. The linking number Lk of two curves is an integer (12.3) that does not change when the curves are deformed without crossing each other. Calculate the linking number for the pair of circles $x^2 + y^2 = 1$ and $(y - 1)^2 + z^2 = 1$.

Answers: $Lk = 1$. Parameterize the circle $x^2 + y^2 = 1$ as

$$\mathbf{r}_{c'}(s') = (\cos s', \sin s', 0), 0 \le s' \le 2\pi;$$

enlarge the circle $(y - 1)^2 + z^2 = 1$ (without crossing the first circle) so that it can be parameterized as $\mathbf{r}_c(s) = (0, 0, s), -\infty < s < \infty$. According to (12.3),

$$Lk = \frac{1}{4\pi} \int_{-\infty}^{\infty} \int_0^{2\pi} \frac{ds\,ds'}{(1 + s^2)^{3/2}} = \frac{1}{2} \int_{-\infty}^{\infty} \frac{ds}{(1 + s^2)^{3/2}} = 1.$$

Problem 12.3. Let G be a group of transformations g that transform a given value of the order parameter ψ into another value $g\psi$ for which the thermodynamical potentials of the system remain the same. Let H be a set of all transformations g_H that leave the order parameter unchanged, $g_H\psi = \psi$. Prove that H is a subgroup of G.

Answers: If g_{H_1} and g_{H_2} leave the OP unchanged, so does $g_{H_1} g_{H_2}^{-1}$.

Problem 12.4. Find the OP spaces and fundamental groups of the 3D smectic A and smectic C [M. Kleman and L. Michel, J. Phys. Lett. **39**, L-29 (1978)].

Answers: In view of the complexity of the OP spaces of SmA and SmC phases, the direct algebraic approach to calculations is difficult. One can use an intuitive graphic scheme.

The density function of SmA and SmC is modulated along the normal to the layers with a period d_0 of the order of the molecular length (or larger, as in some diluted lyotropic lamellar phases). Within the layers, the density is constant, and the molecules are either normal to the layers (SmA) or tilted (SmC). The orientational order of both SmA and SmC, just like a nematic phase, is described by the director $\mathbf{n}$. Taking into account also the translational order, the OP space for the SmA phase can be written as a filled torus $\mathcal{R}_A = (S^2/Z_2) \times S^1$. The vertical cross sections of the torus in the form of two circles amount to hemispheres S^2/Z_2 stretched into disks whose points characterize the orientation of $\mathbf{n}$. The points along the large circles of the torus correspond to points along the segment $[0, d_0]$ closed into a circle S^1. At $(S^2/Z_2) \times S^1$, there are two types of elementary contours not homotopic to zero: the ones that join diametrically opposite points of the disks S^2/Z_2 (describing disclinations) and the ones that run around the hole of the torus (describing dislocations). The fundamental group $\pi_1(\mathcal{R}_A) = \pi_1(S^1) \times \pi_1(S^2/Z_2) = Z \times Z_2$ is composed of elements (b, k), where b is an integer and k is either 0 or $1/2$. Combinations of dislocations and disclinations with both b and k being nontrivial are called *disgyrations*.

For the SmC phase, the order parameter can be easily represented by using the relationship $\mathcal{R}_A = \mathcal{R}_C/S^1$, which implies that each point of $\mathcal{R}_A$ corresponds in $\mathcal{R}_C$ to an entire family S^1 of points that specify the orientation of the tilted molecules in the plane of the SmC layers. Direct calculations show that $\pi_1(\mathcal{R}_C) = Z \wedge Z_4$, where $Z_4 = (I, a, a^2, a^3)$ is the group of subtractions modulo 4 with the unit element I.

Problem 12.5. A nonbounded biaxial nematic contains a π-disclination C_x and a 2π-disclination $\overline{C}_0$. Find the way to eliminate the disclination $\overline{C}_0$.

Hint. Split the $\overline{C}_0$-disclination into two (which ones?) and bring them together (along which paths?).

Problem 12.6. Find the relative homotopy group $\pi_1(\mathcal{R}, \overline{\mathcal{R}})$ when $\mathcal{R}$ is a torus $S^1 \times S^1$ and $\overline{\mathcal{R}} = S^1$.

Answers: $\mathrm{Im}\,[\pi_1(\overline{\mathcal{R}}) \to \pi_1(\mathcal{R})] = \pi(\overline{\mathcal{R}}) = Z$, $\pi_1(\mathcal{R}, \overline{\mathcal{R}}) = Z \times Z/Z = Z$ [V.P. Mineyev and G.E. Volovik, Phys. Rev. B **18**, 3197 (1978)].

Problem 12.7. Calculate the topological charge of the following vector fields in 2D:

(a) $(x, -y)/\sqrt{x^2 + y^2}$;

(b) $(-x, -y)/\sqrt{x^2 + y^2}$;

(c) $(x^2 - y^2, 2xy)/(x^2 + y^2)$;

(d) $(x^2 - y^2, -2xy)/(x^2 + y^2)$;

(e) $(x^3 - 3xy^2, -y^3 + 3x^2y)/(x^2 + y^2)^{3/2}$.

Hints. Parameterize the vector field using the coordinate $0 \le l \le 2\pi$ at the circle $(\cos l, \sin l)$. For example, the field (e) adopts the form $(\cos 3l, \sin 3l)$.

Answers: (a) -1; (b) 1; (c) 2; (d) -2; (e) 3.

Problem 12.8. Calculate the topological charges and the elastic energy of the following two hedgehogs in the nematic bulk of radius R:

$$\mathbf{n}_1 = (x, y, z)/\sqrt{x^2 + y^2 + z^2} \quad \text{and} \quad \mathbf{n}_2 = (-x, -y, z)/\sqrt{x^2 + y^2 + z^2}.$$

Use the Frank–Oseen free energy density with elastic constants

$$K_1 \neq K_2 \neq K_3 \neq K_{24}, K_{13} = 0.$$

Answer: See (11.14) in Chapter 11.

Further Reading

General Courses

M. Kleman, Points, Lines, and Walls in Liquid Crystals, Magnetic Systems, and Various Ordered Media, John Wiley & Sons, Chichester, 1983.

T. Frankel, The Geometry of Physics: An Introduction, Cambridge University Press, 1997.

D.J. Thouless, Topological Quantum Numbers in Nonrelativistic Physics, World Scientific, Singapore, 1998.

Knots and Strips

Louis H. Kauffman, Knots and Physics, 3rd Edition, World Scientific, Singapore 2001.

W. F. Polh, J. Math. Mech. **17**, 975 (1968).

J. H. White, Am. J. Math. **91**, 693 (1969).

F. B. Fuller, Proc. Nat. Acad. Sci. USA **68**, 815 (1981).

DNA Topology and Its Biological Effects, Edited by N. Cozzarelli and J.C. Wang, Cold Spring Harbor Laboratory Press, 1990.

Geometry and Algebraic Topology

D. Hilbert and S. Cohn-Vossen, Geometry and the Imagination, Chelsea Publishing, New York, 1952.

N. Steenrod, Topology of Fiber Bundles, Princeton University Press, 1951.

W. S. Massey, Algebraic Topology: An Introduction, Harcourt, Brace & World, Inc., New York, 1967.

Reviews of Homotopy Theory Applied to Defects in Ordered Media

N.D. Mermin, Rev. Mod. Phys. **51**, 591 (1979).

L. Michel, Rev. Mod. Phys. **52**, 617 (1980).

V.P. Mineev, Sov. Sci. Rev. Sect. A, Vol. 2, Edited by I.M. Khalatnikov, Chur, London, Harwood Academic, New York, 1980.

H.-R. Trebin, Adv. Phys. **31**, 194 (1982).

Topological Stability

G. Toulouse and M. Kleman, J. Phys. Lett. (Paris) **37**, L-149 (1976); M. Kleman, J. Phys. Lett. (Paris) **38**, L-199 (1977).

G.E. Volovik and V.P. Mineev, Pis'ma Zh. Eksp. Teor. Fiz. **24**, 605 (1976) [JETP Lett. **24**, 595 (1976)]; Zh. Eksp. Teor. Fiz. **72**, 2256 (1977) [Sov. Phys. JEPT **46**, 1186 (1977)].

Y. Bouligand, Physics of Defects, Edited by R. Balian, M. Kleman, and J.-P. Poirier, North-Holland, Amsterdam, 1981, p. 665.

M. Kleman and L. Michel, Phys. Rev. Lett. **40**, 1387 (1978).

Y. Bouligand, B. Derrida, V. Poénaru, Y. Pomeau, and G. Toulouse, J. Phys. (Paris) **39**, 863 (1978).

Comparison with Experiments in Liquid Crystals

M.V. Kurik and O.D. Lavrentovich, Usp. Fiz. Nauk **154**, 381 (1988)/Sov. Phys. Usp. **31**, 196 (1988).

Biaxial Nematics

G. Toulouse, J. Phys. Lett. (Paris) **38**, L-67 (1977).

Defects in Anisotropic Superfluids

G.E. Volovik, Exotic Properties of Superfluid 3He, World Scientific, Singapore, 1992.

Defects in Ferromagnets

M. Kleman, Dislocations, disclinations, and magnetism, in "Dislocations in Solids," 5, Edited by F.R.N. Nabarro, North-Holland, Amsterdam,1980.

M. Kleman, Magnetization processes in ferromagnets, in "Magnetism of Metals and Alloys," Edited by M. Cyrot, North-Holland, Amstredam, 1980.

A. P. Malozemoff and J. C. Slonczewski, Magnetic Domain Walls in Bubble Materials, Academic Press, New York, 1979.

CHAPTER 13

Surface Phenomena

Real samples of condensed matter phases are necessarily bounded. The most obvious effect of the confinement is local: Molecular interactions in the bulk and at the boundaries are different. These differences lead to local changes that normally occur over length scales of a few atomic radii. Because the volume-to-surface ratio grows with the system size, the local surface effects are often neglected when a large system is considered. However, there can also be global consequences of surface interactions. For example, bounded volumes of liquid crystals in equilibrium contain topologically stable defects that are stable exclusively because of the surface conditions, as in Fig. 11.13. In this chapter, we consider surface effects both for isotropic and for anisotropic media. Monolayers and bilayers of surfactants are reviewed in Chapter 14.

13.1. Surface Phenomena in Isotropic Media

13.1.1. Surface Tension and Thermodynamics of Flat Interfaces

The surface tension is most easily illustrated for a liquid-gas interface. In the bulk of the fluid, the net force acting on a molecule averages to zero over time scales larger than the relaxation time of the molecular neighborhood. At the surface, the net force does not vanish (Fig. 13.1). To increase the surface area, one needs to perform work to transfer the molecules from the bulk toward the surface. The work needed to increase the surface area by a small quantity dA is proportional to dA: $dW = \sigma \, dA$. The positive-definite coefficient σ is called the surface tension.

In the thermodynamic description of interfaces, one often uses the Gibbs's concept of a "dividing surface" to eliminate the microscopic details. Most often, the dividing surface is considered as a mathematical plane of zero thickness located somewhere in the interfacial region (Fig. 13.1). The two bulk phases in contact, e.g., 1 and 2, are assumed to be homogeneous right up to the dividing surface. Because the actual properties of the two phases change across the interface, the actual value of an extensive thermodynamic quantity Φ (such as the internal energy or the entropy) of the system is not necessarily equal to the

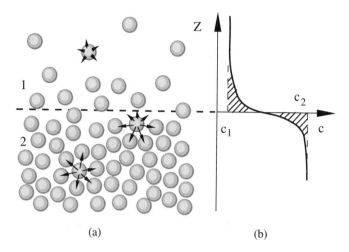

(a)   (b)

Figure 13.1. (a) Liquid-gas interface and (b) changes in the concentration of particles across the interface (b). Molecular forces in the bulk and at the interface are different. The position of the Gibbs dividing surface (dashed line) can be fixed by the convention that the dashed areas in (b) are equal; i.e., there are only particles of type 1 and type 2, but no surface particles.

sum of the bulk quantities $\Phi_1 + \Phi_2$ of the two homogeneous phases. An exception is made for the total volume,

$$V = V_1 + V_2, \qquad (13.1)$$

if the dividing surface is assumed to be of zero thickness. To make up the difference, one introduces a surface "excess" quantity ς_{surf}:

$$\Phi = \phi_1 V_1 + \phi_2 V_2 + \varsigma_{\text{surf}} A, \qquad (13.2)$$

where ϕ_1 and ϕ_2 are calculated per unit volume and A is the area of contact. Note that V_1, V_2, and thus ς_{surf}, depend on the actual position of the dividing surface. This ambiguity is of no importance for many practical purposes and can be avoided if one sets, beside the surface excess volume, some other excess quantity to be zero. For example, if the surface excess number of particles N_{surf} is zero, then V_1 and V_2 are uniquely defined from (13.1) and from the equation for the total number of particles:

$$N = c_1 V_1 + c_2 V_2. \qquad (13.3)$$

Here, c_1 and c_2 are the volume densities (usually called the concentrations) of the number of particles calculated for infinitely large phases 1 and 2, respectively.

Let us find the surface excesses of thermodynamic parameters for a one-component, two-phase system in contact with a reservoir at fixed temperature T and chemical potential μ. According to the first law of thermodynamics, the differential of the internal energy E of the system is

$$dE = -p\,dV + T\,dS + \mu\,dN + \sigma\,dA, \qquad (13.4)$$

where p is the pressure and S is the entropy. In (13.4), there are no terms explicitly assigned to the phase 1 or 2, because in equilibrium,

$$T_1 = T_2 = T_0; \quad p_1 = p_2 = p_0; \quad \mu_1 = \mu_2 = \mu_0 \qquad (13.5)$$

(the subscript "0" implies that the conditions are written for a flat interface of zero curvature; see below). Subtracting from both sides of (13.4) the term $d(TS + \mu N)$, we obtain the grand potential $\Omega = E - TS - \mu N$, with V, T, μ, and correlatively A being the natural parameters:

$$d\Omega = -p\,dV - S\,dT - N\,d\mu + \sigma\,dA, \qquad (13.6)$$

$$\Omega = -pV + \sigma A; \qquad (13.7)$$

the last equation is obtained after differentiating the internal energy $E = -pV + TS + \mu N + \sigma A$. For the homogeneous phases 1 and 2, $\Omega_1 = -pV_1$ and $\Omega_2 = -pV_2$. Therefore, the surface excess $\Omega_{\text{surf}} = \Omega - \Omega_1 - \Omega_2$ of the grand potential is

$$\Omega_{\text{surf}} = \sigma A, \qquad (13.8)$$

regardless of the position of the dividing surface.

The rest of the surface excesses can be found through standard thermodynamic relations, such as between the Helmholtz free energy F and Ω: $F = \Omega + N\mu$. For the entropy, $S = -\left(\frac{\partial\Omega}{\partial T}\right)_{\mu,A}$, the surface excess $S_{\text{surf}} = -\left(\frac{\partial\Omega_{\text{surf}}}{\partial T}\right)_{\mu,A}$ is

$$S_{\text{surf}} = -A\left(\frac{\partial\sigma}{\partial T}\right)_\mu. \qquad (13.9)$$

If $N_{\text{surf}} = \left(\frac{\partial\Omega_{\text{surf}}}{\partial\mu}\right)_T = 0$, then

$$S_{\text{surf}} = -A\frac{d\sigma}{dT},$$

because $\left(\frac{\partial\sigma}{\partial\mu}\right)_T = 0$ and $\frac{d\sigma}{dT} = \left(\frac{\partial\sigma}{\partial T}\right)_\mu$ in this case.

One can qualitatively connect the temperature dependence of σ to the structure of the liquid-gas interface. Because the separation between the particles near the surface increases, the interactions between them become weaker and one might expect a positive surface excess of entropy, $S_{surf} > 0$. Thus, σ decreases with T, which is observed for many systems, including organic and metallic (e.g., Sn) liquids. However, in some cases the very presence of the interface induces an additional order, such as density oscillations along the normal to the interface, enhancement of orientational order of anisometric molecules, or polar arrangement of molecular dipoles. Then, $S_{surf} < 0$ and σ increases with T. At sufficiently high T, close to the critical point, at which the boundary between the liquid and the gas disappears ($\sigma = 0$), spatial delocalization effects should take over and $\sigma(T)$ is expected to decrease again. Inversions of the slope $d\sigma/dT$ are often observed near phase transitions in liquid crystals. If the experiments exclude the possible role of elastic deformations, these data indicate that the surface region is more ordered than is the bulk.

13.1.2. Adsorption

The equilibrium distribution of chemical species becomes spatially nonuniform in the presence of surfaces. Foreign molecules can be expelled from the bulk to the interface to lower the surface tension. This phenomenon, called *adsorption*, plays an important role in colloidal systems.

Consider a two-phase system with k components. It might be, for example, a water solution of ethanol and glycerol in equilibrium with its gaseous phase. Equation (13.6) generalizes to

$$d\Omega = -p\,dV - S\,dT - \sum_{i=1}^{k} N_i\,d\mu_i + \sigma\,dA. \tag{13.10}$$

Differentiating (13.7) and subtracting the result from (13.10), one obtains the *Gibbs-Duhem equation*

$$S\,dT - V\,dp + \sum_{i=1}^{k} N_i\,d\mu_i + A\,d\sigma = 0. \tag{13.11}$$

Similar equations for the bulk 1 and the bulk 2 are

$$S_1\,dT - V_1\,dp + \sum_{i=1}^{k} N_{i,1}\,d\mu_i = 0, \quad S_2\,dT - V_2\,dp + \sum_{i=1}^{k} N_{i,2}\,d\mu_i = 0.$$

$$\tag{13.12}$$

Subtracting both equations (13.12) from (13.11), and employing the definitions $S_{surf} = S - S_1 - S_2$, $N_{i,surf} = N_i - N_{i,1} - N_{i,2}$, $V - V_1 - V_2 = 0$, one arrives at the *Gibbs adsorption equation*:

$$d\sigma = -\frac{S_{surf}}{A} dT - \sum_{i=1}^{k} \Gamma_{i,surf} d\mu_i, \tag{13.13}$$

where $\Gamma_{i,surf} = N_{i,surf}/A$ is the excess number of the i-species per unit area.

For a two-component solution at constant temperature, the Gibbs absorption equation reduces to

$$d\sigma = -\Gamma d\mu - \Gamma' d\mu', \tag{13.14}$$

where Γ and μ refer to the solute, whereas Γ' and μ' refer to the solvent. The relationship (13.14) between σ and Γ is ambiguous because both Γ and Γ' are unknown. However, it easily can be seen that the *relative* adsorption (Problem 13.2)

$$\Gamma_{relative} = \Gamma - \Gamma' \frac{c_1 - c_2}{c_1' - c_2'} \tag{13.15}$$

is expressed exclusively in experimentally measurable quantities such as V, A, the concentration c' of the solvent, and the concentration c of the solute in the bulk of phases 1 and 2. If one places the dividing surface so that the surface excess mass of the *solvent* is zero, then the Gibbs adsorption equation is simply

$$d\sigma = -\overline{\Gamma} d\mu \tag{13.16}$$

(the bar refers to the choice $\Gamma' = 0$). In equilibrium, the chemical potential μ of the solute should be uniform throughout the system. This circumstance can be used to express μ through the concentration of the solute in one of the phases, say, in the liquid, by employing the general thermodynamic relationship

$$d\mu = v \, dp, \tag{13.17}$$

valid at constant temperature. Here, v is the volume per one solute particle, equal to $v_1 = 1/c_1$ in phase 1. If the solution is very dilute and behaves as an ideal one, then $d\mu = k_B T \, d \ln c_1$ [see (1.4) in Chapter 1], and

$$\overline{\Gamma} = -\frac{c_1}{k_B T} \frac{d\sigma}{dc_1}. \tag{13.18}$$

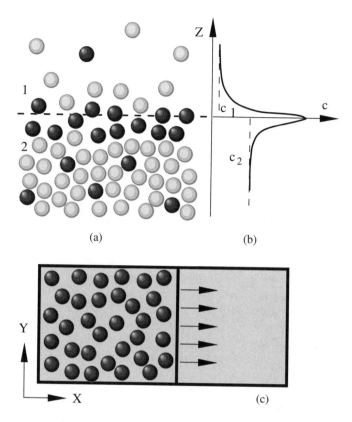

Figure 13.2. Adsorption at the liquid-gas interface. (a) Surface-active solute particles (dark spheres) accumulate in the interfacial region(redrawn from D.H. Everett. (b) Schematic concentration profile of the solute. (c) The film of the adsorbed solute pushes the floating barrier toward the pure solvent.

   According to (13.18), one can distinguish two types of solutes: (a) *surface-active* agents (or surfactants) that decrease the surface tension, $d\sigma/dc_1 < 0$, and accumulate at the interface (*positive adsorption*, $\overline{\Gamma} > 0$) (Fig. 13.2); (b) *surface-inactive* agents that increase the surface tension, $d\sigma/dc_1 > 0$, and are expelled from the interface (negative adsorption or *desorption* $\overline{\Gamma} < 0$). In water, surface-active agents are alcohols (e.g., ethanol), proteins, and especially amphiphilic molecules with polar and nonpolar parts, such as molecules of soaps and detergents. Strongly polar electrolyte materials (such as NaCl) that dissociate into ions are surface-inactive at the water-gas interface.
   If the surface tension does not change significantly with the concentration (weak adsorption), then one can expand the function $\sigma(c_1) \approx \sigma_0 + \sigma_1 c_1$ for small c_1 (σ_0 is the surface tension of the pure solvent; σ_1 is a constant), and (13.18) describes the adsorbed solute as a 2D ideal gas,

$$\pi = \overline{\Gamma} k_B T, \tag{13.19}$$

with a 2D "pressure"

$$\pi = \sigma_0 - \sigma. \tag{13.20}$$

A hemipermeable barrier floating at the surface of the solvent and separating the area of pure solvent from the area of adsorbed solute would experience a force tending to expand the area of the adsorbed solute (Fig. 13.2c). Experiments with floating barriers confirm the validity of (13.19) for low concentrations; for higher concentrations, (13.19) can be modified into a 2D analog of the van der Waals equation.

The experimental setup shown in Fig. 13.2c and known as the *Langmuir trough* is extensively used in modern technologies of molecular self-assemblies based on nonco-valent association of molecules into aggregates at the nanometer scales. A monolayer of adsorbed surfactant (the *Langmuir monolayer*) can be compressed from a gaseous state into different condensed phases by shifting the barrier and reducing the area of the film. By dipping and removing a smooth plate, one can transfer the monolayer on the plate. The procedure can be repeated to assemble a stack of monolayers. The films transferred from the trough onto the solid substrate are called Langmuir-Blodgett films. These films may be polar (the axes connecting heads to tails of the surfactant molecules belonging to neighboring monolayers point in the same directions) or nonpolar. An interesting and sim-ple variation of this technique is the so-called layer-by-layer electrostatic deposition.[1,2] A plate with an electrically charged (say, negatively) surface is dipped directly into a solution of a charged material such as a polyelectrolyte. A polyelectrolyte is a polymer composed of zwitterionic structural units. For example, a positively charged group is covalently bound to the macromolecule, whereas a negatively charged group dissociates when the polyelec-trolyte is dissolved in water. Electrostatic interactions of such a polymer with the negatively charged substrate lock the adsorbed layer. Then the excess of the polymer and counterions are washed out. The procedure is repeated by dipping the plate into a solution of a second polyelectrolyte with the electric polarity opposite to that of the first polymer. The technique allows one to assemble dyes, proteins, SiO_2 nanoparticles, and so on.

13.1.3. Curved Interfaces

13.1.3.1. Laplace–Young Equation

Suppose a liquid drop 1 is surrounded by a gas 2. The system is closed and kept at constant temperature and total volume. What is the equilibrium shape of the liquid drop? The answer

---

[1]R.K. Iler, J. Colloid Interface Sci. **21**, 569 (1966).
[2]G. Decher, J.D. Hong, and J. Schmitt, Thin Solid Films, **210/211**, 831 (1992).

follows immediately if one represents a thermodynamic potential in the spirit of the Gibbs approach as the sum of the bulk and surface parts. The equilibrium is achieved when the surface part σA is minimum (no evaporation). At fixed temperature, σ in an isotropic system (liquid/gas) is constant, and thus, the equilibrium requires minimum interfacial area A. The minimum area of a body with a fixed volume is that of a sphere.

Mechanical equilibrium of the interface poses certain conditions on the pressure of the coexisting phases. If the interface is flat, then $p_1 = p_2$. However, if the interface is curved, then a shift of the interface changes its surface area. The surface energy change is $\sigma \, dA$, or, if the volume 1 is a sphere of radius r, $8\pi\sigma r \, dr$. The equilibrium condition $-p_1 \, dV_1 - p_2 \, dV_2 + \sigma \, dA = 0$, where $dV_1 = -dV_2 = 4\pi r^2 \, dr$, leads to the *Laplace–Young equation*

$$p_1 - p_2 = \frac{2\sigma}{r}. \tag{13.21}$$

For an interface of arbitrary shape, characterized by two principal curvatures r_{max}^{-1} and r_{min}^{-1}, the *local* excess pressure depends on the mean curvature $(1/r_{max} + 1/r_{min})$,

$$p_1 - p_2 = \sigma \left(\frac{1}{r_{max}} + \frac{1}{r_{min}} \right) = \sigma \operatorname{div} \boldsymbol{\nu}, \tag{13.22}$$

where $\boldsymbol{\nu}$ is the unit normal to the interface. Here and henceforth, the convention is that $\boldsymbol{\nu}$ points from phase 1 toward phase 2, and that a radius of curvature is positive when the center of curvature is in the phase 1.

The Laplace–Young equation manifests itself in numerous phenomena such as a capillary rise of fluids and shaping of soap films spanning wire frames (Fig. 13.3).

13.1.3.2. Capillary Rise and Capillary Length

Consider a cylindrical capillary of a small inner radius r immersed in a dish of a large radius $R \gg r$ (Fig. 13.3a). The dish is filled with a fluid that wets the wall of the capillary. The fluid surface will thus be concave. The pressure jump across this surface, $\Delta p = p_1 - p_2$, also called the capillary pressure, is given by the Laplace–Young equation (13.21). The height h of the fluid rise in the capillary above the level of the (flat) fluid surface in the dish is such that Δp equals the hydrostatic pressure drop, $\Delta p = \Delta\rho gh$. Here, g is the acceleration due to gravity and $\Delta\rho$ is the density difference between the fluid and the air. In addition, the equation $2\sigma/r = \Delta\rho gh$ defines a characteristic *capillary length*:

$$\lambda_c = \sqrt{\frac{2\sigma}{\Delta\rho g}}. \tag{13.23}$$

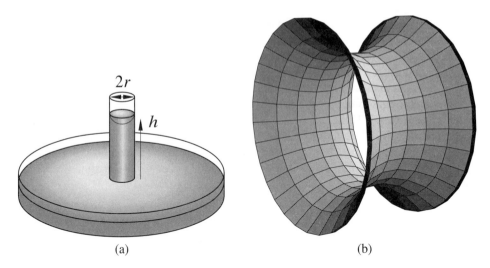

Figure 13.3. Illustrations of the Laplace–Young equation: (a) capillary rise; (b) a soap film between two coaxial rings adopts the shape of a catenoid with a zero mean curvature.

At length scales $x \ll \lambda_c$, gravity effects can be neglected as if the system were in the weightless state; at scales $x \gg \lambda_c$, gravity dominates.

13.1.3.3. Soap Films and Minimal Surfaces

If the soap film spanning a wire frame is not closed, there should be no pressure difference across the film. This is possible only when the mean curvature of the film is zero at every point (13.22); i.e., the film is a *minimal surface*. A trivial example is a flat surface $1/r_{max} = 1/r_{min} = 0$. If the film is drawn between two slightly separated coaxial rings, it forms a catenoid, $1/r_{max} = -1/r_{min} \neq 0$ (Fig. 13.3b).

13.1.3.4. Kelvin Equation

Let us return to the problem of two coexisting phases, e.g., a liquid drop 1 in a gas 2, kept at a constant temperature. Mechanical equilibrium is specified by (13.21), whereas the condition of chemical equilibrium is

$$\mu_1(p_1) = \mu_2(p_2). \tag{13.24a}$$

Note that the two chemical potentials refer to two different pressures. The conditions (13.21) and (13.24a) allow one to derive the *Kelvin equation* that describes how the equilibrium pressure over a curved interface changes with curvature.

First, we rewrite (13.24a) as

$$\mu_1(p_1) - \mu_0(p_0) = \mu_2(p_2) - \mu_0(p_0) \tag{13.24b}$$

(the subscript "0" refers to the flat interface), or, using (13.17), as

$$\int_{p_0}^{p_1} v_1\,dp = \int_{p_0}^{p_2} v_2\,dp. \tag{13.25}$$

In liquid 1, v_1 depends minimally on the pressure, and one can integrate the left-hand side of (13.25) with $v_1 = \mathrm{const}$:

$$\int_{p_0}^{p_1} v_1\,dp = v_1(p_1 - p_0). \tag{13.26}$$

The right-hand side of (13.25) is simplified when gas 2 is an ideal gas, see (1.4):

$$\int_{p_0}^{p_2} v_2\,dp = \int_{p_0}^{p_2} \frac{k_B T}{p}\,dp = k_B T \ln \frac{p_2}{p_0}. \tag{13.27}$$

Employing now the equivalent form of the Laplace–Young equation,

$$p_1 - p_0 = p_2 - p_0 + \frac{2\sigma}{r}, \tag{13.28}$$

and noticing that at not very high temperatures, $v_1 \ll v_2$ and thus $p_1 - p_0 \gg p_2 - p_0$, as (13.25) suggests, we eventually arrive at

$$k_B T \ln \frac{p_2}{p_0} = \frac{2\sigma v_1}{r}, \tag{13.29}$$

which is the Kelvin equation. More generally, the coefficient $2/r$ of spherical curvature in (13.29) should be replaced by the mean curvature of the interface, $2/r \rightarrow (1/r_{\max} + 1/r_{\min})$.

The Kelvin equation (13.29) predicts that the gas pressure over the liquid drop is larger than that over the flat interface (Fig. 13.4a). Conversely, for a gas bubble in a liquid, or for a gas over a liquid condensed in a capillary tube, as shown in Fig. 13.4b,c, the pressure is lowered. The pressure changes become noticeable for submicron scales. A droplet of water ($\sigma \approx 0.07\,\mathrm{J/m^2}$ and $v_1 \approx 3 \times 10^{-29}\,\mathrm{m^3}$, at room temperature) with radius $r = 1\,\mu\mathrm{m}$

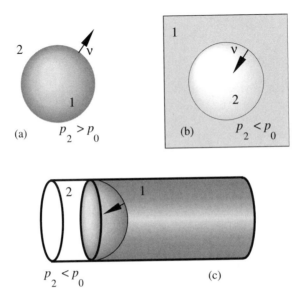

(a) $p_2 > p_0$

(b) $p_2 < p_0$

(c)

$p_2 < p_0$

Figure 13.4. The gas pressure over the liquid surface depends on the curvature of the interface: (a) $p_2 > p_0$ over a liquid droplet; (b) $p_2 < p_0$ for a gas bubble inside the liquid; (c) $p_2 < p_0$ over the liquid condensed in a capillary.

changes the gas pressure by only 0.1%, whereas for $r = 0.01 \ \mu$m, one finds $p_2/p_0 \approx 1.1$. Of course, at smaller scales, the validity of many assumptions (such as the independence of σ on r) becomes questionable.

13.1.4. Surface Tension and Nucleation of the New Phase

The concept of surface tension between two phases helps us to understand the phenomena of supercooling and overheating and the mechanism of nucleation during first-order phase transitions. Consider, as an example, condensation of a liquid 1 from its gas 2 in the range of temperatures T and pressures p below the critical point.

Figure 13.5 shows schematically how the chemical potentials of gas and liquid vary with p at constant T in the vicinity of the coexistence point (p_0, μ_0). The comparative phase stability is defined by the relative magnitudes of μ_1 and μ_2: The lower chemical potential corresponds to the more stable phase. At pressures $p < p_0$, the gas is stable. Heterophase fluctuations[3] appear as small liquid droplets (embryos); however, these droplets

---

[3]The term is used to distinguish the liquid embryos in vapor from regular fluctuations of the vapour density; see J. Frenkel, Kinetic Theory of Liquids, Dover Publications, Inc., New York, 1955, p. 375. When the embryos occur in the bulk of a pure system, the nucleation is called homogeneous, in contrast to heterogeneous nucleation at foreign inclusions or bounding walls.

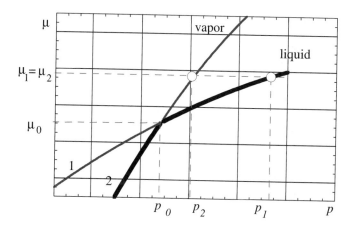

Figure 13.5. Chemical potentials of (nonideal) liquid 1 and gas 2 vs. pressure at constant temperature. p_0 and μ_0: coexistence across a flat interface. Circles: equilibrium of a liquid drop and of its gas.

decay because $\mu_1(p) > \mu_2(p)$ at low pressure. If the pressure is raised above p_0, the gas becomes *metastable*, $\mu_2(p) > \mu_1(p)$. In principle, the gas should transform into the stable liquid by allowing the liquid embryos to grow. However, because the liquid embryos appear inside the gas, there is surface area and surface energy associated with the embryo. The phase transformation is hindered by this surface energy.

Consider the work needed to create a spherical embryo as a function of its radius r. This work is defined by three terms that refer to (a) creation of the interface, (b) compensation of the pressure difference, and (c) compensation of the difference in the chemical potentials of the two phases:

$$W(r) = 4\pi r^2 \sigma - \frac{4}{3}\pi r^3 (p_1 - p_2) + \frac{4}{3v_1}\pi r^3 [\mu_1(p_1) - \mu_2(p_2)], \qquad (13.30)$$

where $4\pi r^3/3v_1$ is the number of particles in the phase 1. Equation (13.30) simplifies if the chemical potentials are brought to the same pressure p_2 of the phase 2. The Taylor expansion $\mu_1(p_1) \approx \mu_2(p_2) + v_1(p_1 - p_2)$ leads to

$$W(r) = 4\pi r^2 \sigma + \frac{4}{3v_1}\pi r^3 [\mu_1(p_2) - \mu_2(p_2)]. \qquad (13.31)$$

The behavior of the embryo depends on the balance between the surface term $\sim r^2$ and the bulk term $\sim r^3$ (Fig. 13.6). The surface term is positive definite. Thus when $\mu_1(p) > \mu_2(p)$, any embryo should decay. The bulk term becomes negative when $\mu_2(p_2) > \mu_1(p_2)$

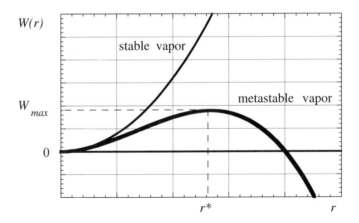

Figure 13.6. Work needed to create a liquid embryo from gas, as a function of the radius r of the embryo. Stable gas: The embryos decay. Metastable gas: The growth of the liquid phase is possible when the fluctuations are capable of creating an embryo of radius $r > r^*$.

and dominates the surface energy but only for the embryos larger than

$$r^* = \frac{2\sigma v_1}{\mu_2(p_2) - \mu_1(p_2)}. \tag{13.32}$$

The large embryos, $r > r^*$, expand until the whole system is transformed into the liquid. The amplitude of the potential barrier is $1/3$ of the embryo's surface energy:

$$W_{\max} = \frac{4}{3}\pi \sigma r^{*2}, \tag{13.33}$$

and it can be considered as the free energy of activation of the thermally activated nuclei. The probability of a fluctuative appearance of an embryo of size r^* is proportional to $\exp(-W_{\max}/k_B T)$. The rate at which the nuclei appear per unit time in a unit volume is $J = cb\exp(-W_{\max}/k_B T)$, where b is a kinetic factor that is difficult to calculate. Its value is estimated for many liquids as $b \sim 10^{10}\,\mathrm{s}^{-1}$. The time expected to form a new growing nucleus in volume V is $\tau = 1/JV$. For realistic values of the variables $V \sim 1\,\mathrm{cm}^3$, $\tau \sim 1\,\mathrm{s}$, and $c \sim 10^{21}\,\mathrm{cm}^{-3}$, one finds that the gas loses its metastability when $W_{\max} \leq 71k_B T$. One needs to apply a kinetic theory[4] rather than equilibrium thermodynamics in order to trace the growth of the nuclei beyond their critical size r^*.

---

[4]E.M. Lifshitz and L.P. Pitaevskii, Physical Kinetics, Pergamon, New York, 1979, Chapter 12.

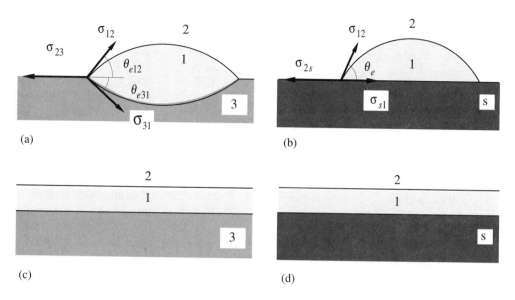

Figure 13.7. Three phases system in the regime of (a, b) partial and (c, d) complete wetting; the substrate is either (a, c) fluid or (b, d) solid.

13.1.5. Wetting

13.1.5.1. Neumann Triangle and Young Law

The geometrical manifestations of surface tension become much richer when the system contains three, rather than two phases. The three interfaces intersect along the so-called triple or contact line (Fig. 13.7a). Mechanical equilibrium of the triple line requires that the vector sum of the three interfacial tensions vanishes:

$$\sigma_{12} + \sigma_{23} + \sigma_{31} = 0, \tag{13.34}$$

which is called the *Neumann triangle construction*. If one of the phases is solid (Fig. 13.7b), then (13.34) simplifies. The vertical component of the vector sum is compensated by the elastic forces caused by the deformation of the solid; in most cases, these deformations are very small (a few Ångströms), and thus one can consider only the horizontal component of (13.34), known as the *Young law*:

$$\sigma_{s2} - \sigma_{s1} = \sigma_{12} \cos \theta_e. \tag{13.35}$$

Here, θ_e is the *equilibrium* contact angle, measured as shown in Fig. 13.7b.

The Young equation (13.35) predicts two distinct situations: (a) *partial wetting,* with a finite contact angle $0 < \theta_e < \pi$, defined from (13.34) or (13.35); the liquid 1 in Fig. 13.7a,b preserves a droplet appearance with a shape dictated by capillary and gravity forces; and (b) *complete wetting,* $\theta_e = 0$, when the drop spreads out (Fig. 13.7c,d). An important quantity that distinguishes the two scenarios is the spreading parameter S_{spr}, defined as the free energy difference between the bare substrate ('s' or 3), directly in contact with phase 2, and a substrate covered by a flat, thick (strictly speaking, infinitely thick, see below) layer of the fluid 1:

$$S_{spr} = \sigma_{s2} - (\sigma_{s1} + \sigma_{12}), \quad \text{or} \quad S_{spr} = \sigma_{23} - (\sigma_{31} + \sigma_{12}). \tag{13.36}$$

$S_{spr} > 0$ corresponds to the total wetting, $S_{spr} < 0$ leads to a partial wetting with $\cos \theta_e = 1 + S_{spr}/\sigma_{12}$. The sign of S_{spr} may be reversed by temperature changes, in which case a so-called *wetting transition* is observed.

A few remarks about (15.34) and (15.35) are in order. First, both formulae describe a small region near the triple line and are not affected by bulk forces such as gravity. Gravity influences the central part of a sessile droplet, when the characteristic size of the droplet is comparable or larger than the capillary length λ_c (Problem 13.5). The equilibrium shape of partially wetting droplets smaller than λ_c is that of spherical caps: The hydrodynamic pressure inside the droplet is equilibrated; thus, the curvature is constant to satisfy the Laplace–Young law (13.22). Lens-like fluid droplets floating at the surface of another fluid are bounded by two truncated spheres. The second remark is that the classic equations above are not applicable near the very "core" of the triple line, comparable to the molecular size. The "microscopic" contact angle may be different from the "macroscopic" one predicted by (13.36) or (13.35). Third, one should also bear in mind that the equilibrium contact angle is difficult to measure experimentally, especially at solid substrates. The observed contact angle θ usually corresponds to a metastable state, stabilized by surface roughness, chemical imputities, and so on. As a result, θ depends on the history of the sample: One distinguishes the advancing contact angle $\theta_A > \theta_e$ achieved by advancing the liquid droplet edge and receding contact angle $\theta_R < \theta_e$ obtained when the droplet retracts.

13.1.5.2. Disjoining Pressure

When the fluid film is thin, in the submicron and nanometer range, the energy of the system becomes thickness dependent, due to the long-range forces originating in van der Waals interactions (Chapter 1) and electric double layers (Chapter 14) as well as due to short-range (steric) molecular forces. For example, the excess of the Gibbs free energy G (related to the Helmholtz free energy as $G = F + PV$) can be represented as

$$\frac{G_{surf}(h)}{A} = \frac{G_{surf}(h \to \infty)}{A} + \int_h^\infty \Pi(z)\, dz = \sigma_{s1} + \sigma_{12} + \int_h^\infty \Pi(z)\, dz, \tag{13.37}$$

where A and h are the area and the thickness of the film. The quantity accesible to experimental measurements is the *disjoining pressure*, introduced by Derjaguin[5] as

$$\Pi(h) = -\frac{1}{A}\left(\frac{\partial G}{\partial h}\right)_{T,P,N_i}. \tag{13.38}$$

$\Pi(h)$ is the pressure that must be applied to prevent the tendency of the film 1 to thicken, i.e., to separate ("disjoin") the two adjacent phases 2 and 3. A positive $\Pi(h)$ tends to thicken the film 1, whereas a negative $\Pi(h)$ tends to thin the film; $\Pi(h) \to 0$ when $h \to \infty$ and $\Pi(h) \to S_{spr}$ when $h \to 0$. For example, the nonretarded long-range van der Waals forces contribute to the disjoining pressure as

$$\Pi_{vdW}(h) = -\frac{A_H}{6\pi h^3}, \tag{13.39}$$

where A_H is the Hamaker constant controlled by the dielectric permittivities of all three phases; $A_H \propto (\varepsilon_s - \varepsilon_1)(\varepsilon_2 - \varepsilon_1)$ in the first approximation (see Section 1.4). At molecular scales, short-range steric forces may cause oscillatory $\Pi(h)$. A nonmonotonic $\Pi(h)$ also can be observed at mesoscales, due to the balance of electrostatic and van der Waals forces. The concept of disjoining pressure can be applied to ordered fluids as well, in which case, there are additional sources of long-range interactions (Section 13.2.4).

The disjoining pressure is of crucial importance in the regime of complete wetting. In practice, the total volume of the spreading liquid is finite, and the wetting film adopts a "pancake" shape. Neglecting edge effects and gravity, the thickness of the film can be determined from the balance of the spreading parameter S_{spr} and $\Pi(h) > 0$. The excess energy of a pancake of thickness h and area A, calculated with reference to the solid-vapor interface, equals $A(-S + \int_h^\infty \Pi(z)\,dz)$. Minimizing this energy with the constraint of constant volume $Ah = V = \text{const}$, one obtains an equation that determines the thickness:

$$S = h_c\Pi(h_c) + \int_{h_c}^{\infty} \Pi(z)\,dz. \tag{13.40}$$

For example, when the disjoining pressure is determined by nonretarded van der Waals forces (13.39), then (13.40) yields

---

[5] See, e.g., B.V. Derjaguin, Theory of Stability of Colloids and Thin Films, Consultants Bureau, New York, 1989, 258 pp.

$$h_c = \frac{1}{2}\sqrt{\frac{-A_H}{\pi S_{\text{spr}}}} \tag{13.41}$$

(the Hamaker constant A_H should be negative for the two phases 2 and 3 to repel each other). This minimum thickness h_c may be significantly *larger* than the molecular size, despite the fact that $S_{\text{spr}} > 0$ (complete wetting).

We shall not discuss the dynamics of wetting. One of the most interesting things here is that for $S_{\text{spr}} > 0$, the spreading film develops a microscopic part of thickness h_c that advances in front of the macroscopic edge; this microscopic "tongue" is called a precursor film. We refer the reader to the reviews by de Gennes and Léger and Joanny for detailed analysis of both static and dynamics aspects of wetting.

13.2. Surface Phenomena in Anisotropic Media

For anisotropic media, σ depends on the orientation of the interface with respect to the symmetry axes, such as the crystallographic axes in a solid crystal or the director in a nematic liquid crystal. Furthermore, the creation of a new interface in liquid crystals and solid crystals implies distortions of the order parameter both at the surface and in the bulk. Quantities such as the Helmholtz free energy density f become functions of the order parameter deformations. Thus, the equilibrium shape and internal configuration of the order parameter of a bounded anisotropic system are determined by the minimum of the sum of the surface and bulk terms:

$$F = F_s + F_v = \oint \sigma \, dA + \int f \, dV, \tag{13.42}$$

where f is the bulk free energy density.

For *isotropic liquid* droplets, $\sigma = $ const and the minimization of the surface area yields spherical shapes. One can also imagine a *solid crystal* shaped as a sphere with atomic arrangements at the interface that correspond to a possible minimum σ. This configuration would inevitably be distorted in the bulk. The strain energy grows with the crystal size faster than does the surface energy, and a spherical crystal would become energetically too costly. Thus, the equilibrium shape of a solid crystal is considered under the assumption that there are no strains in the bulk, $f = 0$ in (13.42): The shape is defined by the dependence of σ on crystallographic directions. *Liquid crystals* present the most difficult case because both the surface and the bulk energies depend on molecular order (unlike in isotropic fluids, where there is no elasticity in the bulk), and unlike in solids, the assumption $f = 0$ does not apply because the spatially integrated energy of bulk deformations $\cong KR$ is smaller than the integrated surface energy $\cong \sigma R^2$ for sufficiently large size R of the system; thus, generally both integrals in (13.42) should be considered.

13.2.1. Equilibrium Shape (Wulff Shape) of Solid Crystals

Consider first the shape of solid crystals. Let $z = z(x, y)$ be the surface of a solid crystal in the Cartesian coordinates (x, y, z). The unit normal to this surface is

$$\nu = \frac{-z_x \mathbf{i} - z_y \mathbf{j} + \mathbf{k}}{\sqrt{z_x^2 + z_y^2 + 1}}, \tag{13.43}$$

and the element of the surface area is $dA = \sqrt{z_x^2 + z_y^2 + 1}\, dx\, dy$; here, $z_x = \partial z / \partial x$ and $z_y = \partial z / \partial y$. The equilibrium is achieved when the surface energy integral

$$F_s = \iint \sigma(\nu) \sqrt{z_x^2 + z_y^2 + 1}\, dx\, dy \tag{13.44}$$

is extremal under the condition that the crystal volume is constant

$$V = \iint z\, dx\, dy = \text{const}, \tag{13.45}$$

or, in other words, when there is an unconditional extremum of the integral

$$F_s - 2\lambda V = \iint \left[\sigma(\nu) \sqrt{z_x^2 + z_y^2 + 1} - 2z\lambda \right] dx\, dy; \tag{13.46}$$

λ is the Lagrangian multiplier. The Euler–Lagrange equation is

$$2\lambda + \frac{\partial}{\partial x} \frac{\partial (\sigma \sqrt{z_x^2 + z_y^2 + 1})}{\partial z_x} + \frac{\partial}{\partial y} \frac{\partial (\sigma \sqrt{z_x^2 + z_y^2 + 1})}{\partial z_y} = 0. \tag{13.47}$$

It can be verified that one of the solutions is

$$\sigma(\nu) \sqrt{z_x^2 + z_y^2 + 1} = \lambda(z - z_x x - z_y y). \tag{13.48}$$

Noticing that

$$\nu \cdot \mathbf{r} = \frac{z - z_x x - z_y y}{\sqrt{z_x^2 + z_y^2 + 1}},$$

where $\mathbf{r} = (x\mathbf{i} + y\mathbf{j} + z\mathbf{k})$ is the radius-vector drawn to the origin of the normal $\boldsymbol{\nu}$ (13.43) at the crystal surface, (13.48) can be rewritten as

$$\frac{\sigma(\boldsymbol{\nu})}{\lambda} = \boldsymbol{\nu} \cdot \mathbf{r}. \tag{13.49}$$

Equation (13.49) explains how to construct the equilibrium shape of a crystal if the polar plot $\sigma(\boldsymbol{\nu})$ (Wulff plot, Fig. 13.8) is known. First, connect the origin O and any point M on the curve $\sigma(\boldsymbol{\nu})$, obtained as follows: The direction of $\mathbf{OM}$ is $\boldsymbol{\nu}$; the length is proportional to $\sigma(\boldsymbol{\nu})$, say, $|\mathbf{OM}| = \sigma(\boldsymbol{\nu})/\lambda$. Through the point M, draw a plane perpendicular to $\mathbf{OM}$. The operation is repeated for all points of the Wulff plot $\sigma(\boldsymbol{\nu})$. The envelope of the resulting family of planes is the equilibrium shape of the crystal: (13.49) is satisfied for any radius-vector $\mathbf{r}$ drawn from the origin to the crystal surface, $\boldsymbol{\nu} \cdot \mathbf{r} = |r| \cos\alpha = |\mathbf{OM}|$, where α is the angle between $\mathbf{r}$ and $\mathbf{OM}$ (Fig. 13.8).

The equilibrium crystal shape exhibits extended planar regions, called "facets." The facets are caused by "*cusps*," or pointed minima of $\sigma(\boldsymbol{\nu})$. These pointed minima of $\sigma(\boldsymbol{\nu})$

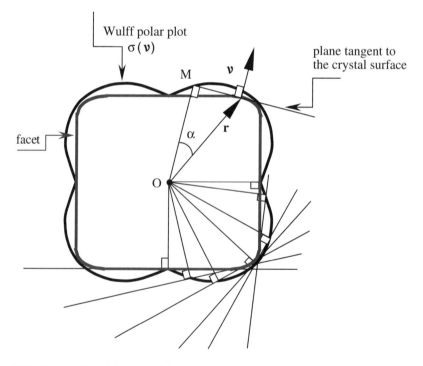

Figure 13.8. Construction of the 3D equilibrium crystal shape from the Wulff polar plot (the line with cusps).

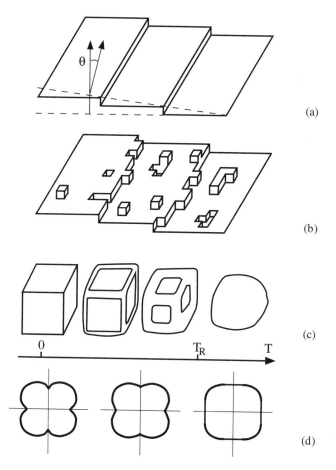

Figure 13.9. Steps at the crystal surface (a) at $T = 0$ and (b) at $T > 0$. (c) Evolution of the equilibrium crystal shape and (d) the cross-sections of the Wulff polar plot with temperature. The cusps and facets disappear at the roughening transition point T_R [redrawn from C. Rottman and M. Wortis].

occur along directions of symmetry. The surface tension of an interface slightly tilted away from the direction of symmetry (a vicinal surface) is increased by the formation of steps, separated by flat terraces (Fig. 13.9a). If the tilt angle θ is small, the steps are well separated and their interactions can be neglected. Thus, the surface tension in the vicinity of the cusp behaves as

$$\sigma(\boldsymbol{\nu}) \approx \sigma_0 + \frac{\varepsilon}{a}|\theta|, \tag{13.50}$$

where σ_0 is the surface tension of the flat reference surface, ε is the energy per unit length of the step, and a is the height of the step. The Wulff construction transforms the cusps of $\sigma(\nu)$ into the flat facets of the crystal shape. Note that there are parts of the plot $\sigma(\nu)$ that do not contribute to the crystal shape. The corresponding crystallographic orientations are forbidden at the crystal surface, and the surface tension for these orientations cannot be measured.

Faceting of crystals progressively disappears as the temperature increases. The reason is that thermal fluctuations disorder the edges (Fig. 13.9b). The entropy contribution decreases the free energy of the steps. The cusps become shallower, and the planar facets shrink. At some temperature, called the roughening transition temperature (different for different crystallographic orientations), the entropy of surface fluctuations wins over the increase of surface tension, the facet disappears, and the corresponding region of the crystal becomes smoothly rounded (Fig. 13.9c,d). The theoretical approach to the roughening transition[6] is much akin to the 2D Kosterlitz-Thouless transition (see Section 4.5). It happens at a temperature $T_R = 2\sigma a^2/\pi$, where σ is the (temperature dependent) surface tension, and a is the height of the steps, i.e., the periodicity of the lattice perpendicularly to the facet. The roughening transition is also of practical importance in understanding the process of crystal growth.

This section has concerned the equilibrium shapes, including fluctuations at $T > 0$. For these shapes, the growth theory discussed above should work, and it yields very large supersaturations. In fact, small supersaturations are the rule rather than the exception, experimentally. This is due to the presence of defects that pierce the surface; in particular, dislocations with a screw component perpendicular to the surface carry a (growth) step that is a region of easy growth, because the atoms along the step are even less bound to the crystal than are those that are elsewhere on the surface. These steps serve as centers of nucleation. For more details, see Burton et al.[7]

13.2.2. Surface Anchoring in Nematic Liquid Crystals

Anisotropic interactions between the molecules of liquid crystal and ambient media cause two basic effects: (1) modification of the liquid crystal structure in the vicinity of the interface and (2) lifting the degeneracy of the director orientation in the bulk. Examples of type (1) structural rearrangements are: (a) periodic modulations of density, which in some cases is smectic-like (similar modulations of density have been found also for ordinary liquids) (Fig. 13.10a); (b) polar ordering of molecular dipoles that eliminates degeneracy $\mathbf{n} = -\mathbf{n}$ near the interface (Fig. 13.10b); and (c) modification of the scalar order parameter (that may be either higher or lower than that in the bulk).

Phenomena of type (2) are commonly called *anchoring* phenomena and stem from the fact that the surface tension, defined as the surface excess of the grand thermodynamic

[6]S.T. Chui and J.D. Weeks, Phys. Rev. B**14**, 4978 (1976).

[7]W.K. Burton, N. Cabrera, and F.C. Frank, Phil. Trans. Roy. Soc. A**243**, 299 (1951).

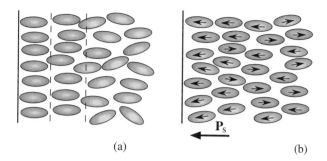

Figure 13.10. Surface modifications of the liquid crystalline order: (a) periodic modulation of the molecular density with a simultaneous increase of the scalar order parameter; (b) polar ordering of molecular dipoles that gives rise to the surface electric polarization $\mathbf{P}_s$. Spatial variation of the scalar order parameter (a) also leads to 'ordoelectric' polarization.

potential, depends on the director orientation.[8] For nonflat geometries of confinement, specific anchoring directions can be produced by the bulk elasticity of liquid crystal. Imagine, for example, a nematic slab between two flat plates that favor tangential director orientation. When the plates are parallel to each other, any director orientation parallel to the plates is an equilibrium orientation. However, when the plates are tilted with respect to each other, the equilibrium state is achieved only when the director is perpendicular to the thickness gradient (Section 11.1.6). This "geometrical anchoring" might contribute to the distortion of director patterns in samples with curved surfaces (freely suspended films, films at fluid or solid substrates). Another factor, important when the liquid crystal is aligned by a buffed polymer film, is a direct van der Waals interaction between the molecules of liquid crystal and those of the polymer.

Surface orientation of the director is characterized by a polar angle θ and an azimuthal angle φ. The particular direction (θ_0, φ_0) that minimizes the function $\sigma(\theta, \varphi)$ is called the "easy axis." Thus, $\sigma(\theta, \varphi)$ can be presented as a sum of the surface tension $\sigma(\theta_0, \varphi_0)$ at equilibrium orientation and the (generally unknown) anchoring energy function $w(\theta, \varphi)$ that depends on the details of molecular interactions at the interface

$$\sigma(\theta, \varphi) = \sigma_0(\theta_0, \varphi_0) + w(\theta - \theta_0, \varphi - \varphi_0). \tag{13.51}$$

When the liquid crystal is in contact with an isotropic fluid, w depends only on the polar angle θ. At the nematic-isotropic interface, there is no reason to expect any "cusps" in $w(\theta)$, because the nematic phase has no long-range translational order (Fig. 13.11).

---

[8]For a review of the Gibbs thermodynamics of a nematic interface, see H. Yokoyama, in Handbook of Liquid Crystal Research, Edited by P.J. Collings and J.S. Patel, Oxford University Press, New York, 1997.

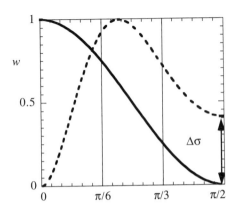

Figure 13.11. Possible dependencies of the anchoring energy on the polar angle θ between the director and the normal to the interface. Solid line: tangential anchoring, $\theta_0 = \pi/2$; dashed line: perpendicular anchoring, $\theta_0=0$.

The strength of anchoring in the vicinity of the equilibrium orientation (θ_0, φ_0) is often characterized by the so-called polar (w_θ) and azimuthal (w_φ) anchoring coefficients, which measure the work (per unit area) needed to deviate the director from the easy axis (θ_0, φ_0):

$$w = \tfrac{1}{2} w_\theta (\theta - \theta_0)^2 \quad \text{or} \quad w = \tfrac{1}{2} w_\varphi (\varphi - \varphi_0)^2. \tag{13.52}$$

The latter expansions can be justified for small deviations from the easy axis. Very often, especially for the polar part of the anchoring, it is postulated that the surface anchoring energy follows a specific functional form, for example,

$$w = \tfrac{1}{2} w_\theta \sin^2(\theta - \theta_0). \tag{13.53}$$

The latter form is called the *Rapini-Papoular anchoring potential*. To describe some effects, such as consecutive alignment treatments or dependence of the anchoring potential on the in-plane coordinates, it is more convenient to operate with a traceless coordinate-dependent symmetric tensor[9] $w_{\alpha\beta}(\mathbf{r})$:

$$w = -\frac{1}{2} \sum_{\alpha,\beta} w_{\alpha\beta}(\mathbf{r}) n_\alpha n_\beta. \tag{13.54}$$

---

[9]S.V. Shiyanovski, A. Glushchenko, Yu. Reznikov, O.D. Lavrentovich, and J.L. West, Phys. Rev. E **62**, R1477 (2000).

In the nematic phase, the surface anchoring coefficients are usually much smaller than is the energy (per unit area) needed to extend the area of the surface preserving the equilibrium director orientation. A typical surface tension coefficient for the liquid crystal cyanobiphenyls-glycerin surface is $\sigma_0 \sim (10^{-3} - 10^{-2})\,\text{J/m}^2$, whereas the polar anchoring coefficient for the same interface is $w_\theta \sim (10^{-5} - 10^{-6})\,\text{J/m}^2$. At the nematic-isotropic interface of 5CB, $\theta_0 = 64^0$, $\sigma_0 \approx 2 \times 10^{-5}\,\text{J/m}^2$, whereas $w_\theta \approx 5 \times 10^{-7}\,\text{J/m}^2$.[10] Both surface tension and anchoring coefficients are temperature dependent. The anchoring coefficients decrease when the temperature T increases, because the nematic scalar order parameter decreases. The temperature behavior of σ_0 may be more complex. In isotropic fluids, σ_0 usually monotonously decreases with T. In liquid crystals, $\sigma_0(T)$ may be non-monotonous, especially in the vicinity of phase transitions.

It is usual to distinguish three types of polar orientation at the isotropic fluid-nematic interface: (1) perpendicular, $\theta_0 = 0$; perpendicular anchoring is also called "homeotropic" when the interface is flat; (2) tilted conical, $0 < \theta_0 < \pi/2$; and (3) tangential, $\theta_0 = \pi/2$. The in-plane easy axes occur when the substrate is anisotropic, e.g., crystalline. Below, we discuss the role of surface anchoring in two field-induced effects, namely, the Frederiks transition and the combined flexoelectric-surface polarization effect.

13.2.3. Field Effects Under Finite Anchoring

13.2.3.1. Frederiks Transition

In Section 5.4.3, we mentioned field-induced reorientation of the director in the bulk of the nematic cell, known as the Frederiks transition. The Frederiks transition has a well-defined threshold when the director $\mathbf{n}$ is initially in the planar or homeotropic orientation and when the field $\mathbf{B}$ is either perpendicular to $\mathbf{n}$ or parallel to $\mathbf{n}$. The instability occurs in $\mathbf{B}\|\mathbf{n}$ geometry when the diamagnetic (dielectric) anisotropy is negative, $\chi_a < 0$, and in $\mathbf{B}\perp\mathbf{n}$ geometry when $\chi_a > 0$. The threshold (5.74) was calculated by assuming that the director is strongly anchored at the bounding plates $z = 0$ and $z = h$, where h is the thickness of the cell. Consider now how the finite anchoring would modify the result (5.74), for a particular geometry of $\mathbf{n} = (0, 0, 1)$ and $\mathbf{B} = (B, 0, 0)$; $\chi_a > 0$.

We parameterize the director as $\mathbf{n} = [\sin\theta(z), 0, \sin\theta(z)]$. Near the threshold, distortions are weak, $\theta \ll 1$, and the free energy (per unit area) writes in a simplified form:

$$F = \frac{1}{2}\int_0^h \left[K_3 \left(\frac{d\theta}{dz}\right)^2 - \mu_0^{-1}\chi_a B^2\theta^2 \right] dz + \frac{1}{2}w_\theta\theta^2\big|_{z=0} + \frac{1}{2}w_\theta\theta^2\big|_{z=h}. \qquad (13.55)$$

The Euler–Lagrange equation

---

[10]S. Faetti, Mol. Cryst. Liq. Cryst. **179**, 217 (1990).

$$\xi^2 \frac{d^2\theta}{dz^2} + \theta = 0,$$

where $\xi = \frac{1}{B}\sqrt{\frac{K_3}{\mu_0^{-1}\chi_a}}$ is the magnetic coherence length, has the solution

$$\theta(z) = C_1 \cos \frac{z}{\xi} + C_2 \sin \frac{z}{\xi}, \tag{13.56}$$

where the constants of integration C_1 and C_2 are determined from the torque boundary conditions

$$\begin{cases} -K_3 \dfrac{d\theta}{dz}\Big|_{z=0} + w_\theta \theta|_{z=0} = 0, \\[2mm] K_3 \dfrac{d\theta}{dz}\Big|_{z=h} + w_\theta \theta|_{z=h} = 0. \end{cases} \tag{13.57}$$

The system (13.57) yields nonzero C_1 and C_2 when

$$\cot \frac{h}{2\xi} = \frac{K_3}{w_\theta \xi}. \tag{13.58a}$$

Using the notation $\mathcal{B}_\infty$ for the field threshold (5.74) in a cell with infinitely strong anchoring, one may rewrite the last result as

$$\cot \frac{\pi \mathcal{B}_w}{2\mathcal{B}_\infty} = \frac{\pi K_3 \mathcal{B}_w}{h w_\theta \mathcal{B}_\infty}, \tag{13.58b}$$

where $\mathcal{B}_w$ is the threshold for a finite w_θ. Series expansion of (13.58a) in terms of a small parameter $(1 - \mathcal{B}_w/\mathcal{B}_\infty)$ yields an expression transparent for interpretation:

$$\frac{\mathcal{B}_w}{\mathcal{B}_\infty} \approx 1 - \frac{2K_3}{h w_\theta} = 1 - \frac{2L_w}{h}, \tag{13.58c}$$

where $L_w = K_3/w_\theta$ is the characteristic anchoring length we already met in Chapter 5. The polar anchoring coefficient w_θ can be measured by comparing the threshold of the Frederiks transition in thick and thin cells.[11] For an MBBA oriented homeotropically by a surfactant layer, w_θ decreases from 6×10^{-5} to 1×10^{-5} J/m^2 when the temperature

---

[11]C. Rosenblatt, J. Phys. (France) **45**, 1087 (1985).

approaches the nematic-to-isotropic transition point. Experiments also indicate that the scalar order parameter near the homeotropic alignment layer is somehow higher than in the bulk; the surface region might be smectic-like.[12]

13.2.3.2. Flexoelectric and Surface Polarization Effects

The Frederiks transition is bulk effect: The maximum director deviations for a symmetric cell occur in the midplane of the cell, at $z = h/2$. The diamagnetic effect considered above has a dielectric analog; the threshold electric field can be estimated by making use of the replacement $\mu_0^{-1}\chi_a B^2 \to \varepsilon_0 \varepsilon_a E^2$ (more careful calculations should take into account that the electric field in the cell with distorted director is nonuniform). Electric field causes not only dielectric, but also other types of effects, such as polar (i.e., depending on the polarity of the applied field) flexoelectric, and surface-polarization instabilities. These two are of especial interest for surface phenomena because the maximum director distortions develop at the bounding plates[13] $z = 0; h$.

We refer to the same homeotropic cell as in the previous section: $\mathbf{n} = (0, 0, 1)$. The electric field $\mathbf{E} = (0, 0, 1)$ is applied along the director. If the material is dielectrically positive, $\varepsilon_a > 0$, one would expect no response, because $\mathbf{E}||\mathbf{n}$. However, a simple experiment would show that a low DC voltage of about 1V, applied to a homeotropic cell of, say, 5CB, causes director distortions near one of the electrodes (Fig. 13.12). The possible mechanism is flexoelectric $f_{\text{flex}} = -\mathbf{P}_f \cdot \mathbf{E} = -[e_1 \mathbf{n} \operatorname{div} \mathbf{n} - e_3(\mathbf{n} \times \operatorname{curl} \mathbf{n})] \cdot \mathbf{E}$ and surface polarization $(-\mathbf{P}_s \cdot \mathbf{E})$ contributions to the free energy density. Both contributions are linear in $\mathbf{E}$ and average to zero when the applied field is of a high frequency. However, when the field is DC or a low-frequency AC, both flexoelectric and surface polarization can cause an instability of the homeotropic cell, namely, an increase in the surface polar angle from 0 to some value θ_s. For example, if the surface polarization is directed toward the bounding plates, a DC field $(E_{\text{DC}}, 0, 0)$ acting along the z-axis will cause director deviation at the plate $z = 0$ and enhance the homeotropic alignment at the plate $z = h$ (Fig. 13.12). The experimental verification of this polar effect is not simple because it can be masked by other field phenomena, such as injection of ions from the electrodes of the cell and a consequent motion of the nematic fluid, i.e., an electrohydrodynamic instability.

To test the origin of the electrooptical effects in the homeotropic cell, one can use a simultaneous action of the AC (E_{AC}) and DC (E_{DC}) fields.[14] The combined torque of the flexoelectric and surface polarization can be estimated as $(-e^* E_{\text{DC}})\theta_s$, where $e^* = e_1 + e_3 + |\mathbf{P}_s|$. The positive sign of $|\mathbf{P}_s|$ refers to the case when the surface polarization favors director deviations at the anode. The destabilizing torque $(-e^* E_{\text{DC}})\theta_s$ is opposed by (a) surface anchoring torque $w_\theta \theta_s = K\theta_s / L_w$ and (b) dielectric torques: $K\theta_s / \xi_{\text{DC}}$ of the DC field and $K\theta_s / \xi_{\text{AC}}$ of the AC field. Here,

---

[12]C. Rosenblatt, Phys. Rev. Lett. **53**, 791 (1984).

[13]W. Helfrich, Appl. Phys. Lett. **24**, 451 (1974).

[14]O.D. Lavrentovich, V.G. Nazarenko, V.V. Sergan, and G. Durand, Phys. Rev. A **45**, R6969 (1992).

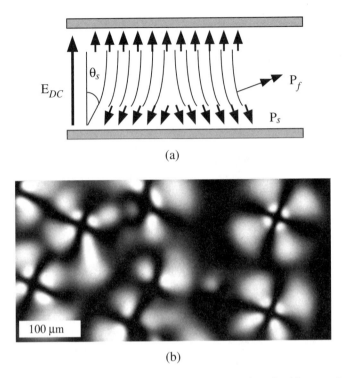

(a)

(b)

Figure 13.12. Polar surface instability in a homeotropic nematic cell with $\varepsilon_a > 0$: (a) mechanism of flexoelectric (double arrow) and surface polarizations (single arrows) coupling to the applied field $\mathbf{E}_{DC}$; (b) texture of a homeotropic layer of 5CB above the threshold of the polar instability.

$$\xi_{DC} = \sqrt{\frac{K}{\varepsilon_0 \varepsilon_a E_{DC}^2}} \quad \text{and} \quad \xi_{AC} = \sqrt{\frac{K}{\varepsilon_0 \varepsilon_a E_{AC}^2}}$$

are the corresponding coherence lengths. There is also an elastic torque that can be omitted when the cell is thick, $h \gg \xi_{AC}, \xi_{DC}$. The surface torque equation is

$$\left(\frac{K}{\xi_{AC}} + \frac{K}{\xi_{DC}} + \frac{K}{L_w} - e^* E_{DC} \right) \theta_s = 0,$$

and it yields the threshold of the surface polar instability:

$$E_{DC,\text{th}} = \frac{1}{e^* - \sqrt{\varepsilon_0 \varepsilon_a K}} \left[\sqrt{K \varepsilon_0 \varepsilon_a} \, | E_{AC} | + w_\theta \right]. \tag{13.59}$$

An important feature is that the threshold of polar instability $E_{DC,th}$ increases with $|E_{AC}|$; for electrohydrodynamic effects, the threshold should not increase with $|E_{AC}|$. Experiments indeed confirm the existence of the polar surface effect with $E_{DC,th} \cong |E_{AC}|$. Another important consequence of (13.59) is the possibility of measuring the anchoring length and e^*. To do this, one needs to compare the threshold $E^0_{DC,th}$ when there is no applied AC field, with the threshold $E_{DC,th}$, measured when the cell is subject to the stabilizing AC field:

$$\frac{L_w}{\xi_{AC}} = \frac{E_{DC,th}}{E^0_{DC,th}} - 1; \quad e^* = \sqrt{\varepsilon_0 \varepsilon_a K'} \left(1 + \frac{E_{AC}}{E_{DC,th} - E^0_{DC,th}} \right). \tag{13.60}$$

Experiments[14] yield $L_w \sim 1\mu m$ and $e^* \sim 5 \times 10^{-3}$ (cgs units) for 5CB at a silicone elastomer homeotropic layer. The total flexoelectric and surface polarization $e^* = e_1 + e_3 + |\mathbf{P}_s|$ is somewhat different from the pure flexoelectric quantity $(e_1 + e_3)$. Thus, the surface-polarization contribution can be significant. For further analysis and review of electric field effects, see the book by Pikin.[15]

13.2.4. Thin Liquid Crystal Films; Casimir Interactions

As already mentioned in Section 13.1.5, long-range interactions such as van der Waals forces play an important role in the behavior of systems with a mesoscopic characteristic lengths between 1 μm and a few nanometers. If the film is ordered, in the nematic phase, say, there are additional long-range interactions contributing to the disjoining pressure. The most obvious effect is caused by the balance of the anchoring and elastic forces. For example, if a nematic film has different easy axes at the two opposite surfaces, director distortions in the bulk tend to thicken such a hybrid-aligned film contributing to the disjoining pressure a positive term $\Pi_{elastic}(h) \sim K/h^2$ (Problem 13.6). In a similar way, two spherical particles in the nematic solvent may be prevented from close contact by spatial distortions of the director. Furthermore, even if the director is spatially uniform, there is yet another universal source of long-range interactions, of the Casimir type (Section 1.4.4).

The Casimir interactions originate in the entropy-costly restrictions that the geometry of the system imposes on the thermal fluctuations of the order parameter. Consider, for example, a flat nematic slab with an infinitely strong anchoring that keeps the director normal to the plates, $\mathbf{n} = (0, 0, 1) = $ const. There is no elastic energy stored in the slab. On the other hand, the surface anchoring suppresses a number of fluctuating director modes in the bulk. The system tends to decrease the volume in which these restrictions occur. The plates attract each other. The potential of this attraction should be proportional to $k_B T$ as the effect is caused by thermal fluctuations. Furthermore, the interaction should be extensive in the area A of the plates. To yield the correct dimension, one concludes that the potential is

---

[15]S.A. Pikin, Structural Transformations in Liquid Crystals, Gordon and Breach Science Publishers, New York, 1991, 424 pp.

roughly $(-k_B T)A/h^2$; i.e., it scales with the separating distance as $\sim 1/h^2$. The estimate turns out to be qualitatively correct, if one uses a one-constant approximation to the nematic elasticity. Generally, the elastic constants describing fluctuations with wave vectors perpendicular and parallel to the plates are not equal, and the amplitude of the Casimir interactions should be multiplied by a factor $(K_3/K_1 + K_2/K_1)$. Careful calculations[16] show that the attraction potential is

$$-\frac{\varsigma_R(3)}{16\pi}\frac{k_B T A}{h^2}\left(\frac{K_3}{K_1}+\frac{K_2}{K_1}\right).$$

Here, ς_R is the Riemann's zeta function, $\varsigma_R(3) \approx 1.202\ldots$. Using the definition (13.38), one finds the Casimir contribution to the disjoining pressure of the homeotropic nematic film:

$$\Pi_{C,N}(h) = -\frac{\varsigma_R(3)}{8\pi}\frac{k_B T}{h^3}\left(\frac{K_3}{K_1}+\frac{K_2}{K_1}\right), \tag{13.61}$$

which is obviously negative (hybrid alignment of the nematic film may cause a positive sign of the Casimir term). In a similarly oriented smectic A, with layers parallel to the plates, the dependence on separation distance is predicted to be different, because the smectic A elastic energy density (5.27) introduces the elastic length scale $\lambda = \sqrt{K_1/B}$. As follows from calculations by Ajdari et al.,[16]

$$\Pi_{C,\mathrm{SmA}}(h) = -\frac{\varsigma_R(2)}{8\pi}\frac{k_B T}{\lambda h^2}. \tag{13.62}$$

A rigorous experimental verification of Casimir interactions in liquid crystals (or other ordered media) is presently lacking.

13.2.5. Topological Defects in Large Liquid Crystal Droplets

The equilibrium of a bounded liquid crystal is defined by the minimum of the total energy that includes bulk elastic and surface (both surface "isotropic" and anchoring) terms. Because the general solution of the minimization problem is not known, one often uses different approximations based on a qualitative comparison of the different terms. For a nematic sample of a characteristic size R (say, a droplet confined in some matrix), representative estimates are K, $\sigma_0 R^2$, and $w_\theta R^2$; usually, $w_\theta \ll \sigma_0$. The bulk elastic energy scales *linearly* with R, which leads again to the anchoring scale $L_w = K/w_\theta$ as an important parameter. Nematic droplets of any reasonable supramolecular size, suspended in various fluids (water, glycerin), are practically spherical. The inner drop structure is different for $R \gg L_w$

---

[16]A. Ajdari, B. Duplantier, D. Hone, L. Peliti, and J. Prost, J. Phys. II France **2**, 487 (1992).

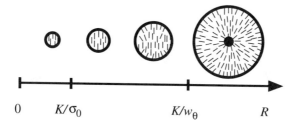

Figure 13.13. Large and small nematic droplets with perpendicular surface anchoring.

and $R \ll L_w$. Compare two extreme states of a drop with perpendicular anchoring $\mathbf{n}||\boldsymbol{\nu}$: (a) a spatially uniform director, $\mathbf{n} = \text{const}$, which violates the boundary conditions; and (b) a hedgehog structure, $\mathbf{n}||\boldsymbol{\nu}$ (Fig. 13.13). The energy of the uniform structure is that of the anchoring, $\frac{8}{3}\pi w_\theta R^2$ [in the Rapini-Papoular approximation (13.53)], whereas the hedgehog energy (11.14) $8\pi K R$ is pure elastic. Thus, a small-size droplet $R \ll L_w$ would prefer to be uniform, $\mathbf{n} \approx \text{const}$, whereas a large-size droplet $R \gg L_w$ would contain a hedgehog. Stable topological defects are a common feature of liquid crystal droplets as discussed in the next section.

13.2.5.1. Topological Conservation Laws

When the nematic drop is large, $R \gg K/w_\theta$, the director at the surface is fixed at the equilibrium polar angle θ_0. With a typical polar anchoring coefficient $w_\theta \sim (10^{-5} - 10^{-6}) \, \text{J/m}^2$ and $K \approx 10^{-11} \text{N}$, the supramicron droplets are "large." Even without solving the minimization problem, one can establish certain general topological properties of structures as long as θ_0 is fixed. This possibility stems from two theorems of differential geometry, namely, the Gauss–Bonnet theorem and the Euler-Poincaré theorem. These theorems allow one to connect the total charge of the point defects in the vector field to the Euler characteristic of the bounding surface, as we discuss below.[17]

There may be two types of topological defects in a bounded nematic: those in the bulk and those at the surface. Point defects in the bulk are hedgehogs (Chapters 11 and 12). Line defects in the bulk that are closed loops can be considered as hedgehogs with an extended core. At the surface, there may be another type of point defects: the so called "boojums" (Figs. 13.14 and 13.15). A boojum can be characterized by a "2D" topological charge k, defined as the index of the surface projection field $\boldsymbol{\tau} = \mathbf{n} - \boldsymbol{\nu}(\mathbf{n} \cdot \boldsymbol{\nu})$, see (12.4). However, boojums also cause distortions in the interior. For example, the defect shown in Fig. 13.14a is reminiscent of one-half of a radial hedgehog. One can assign to it a "bulk" characteristic $C = 1/2$. This characteristic is not, strictly speaking, a topological charge: It appears only

[17]G.E. Volovik and O.D. Lavrentovich, Zh. Eksper. Teor. Fiz. **85**, 1997 (1983) [Sov. Phys. JETP **58**, 1159 (1983)].

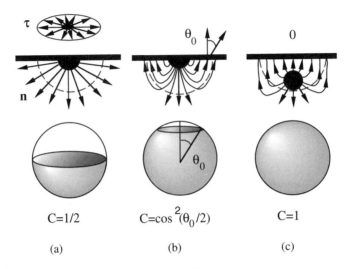

Figure 13.14. Point defects - boojums with $C = \cos^2(\theta_0/2)$, $k = 1$, and $N = 1$ for different easy anchoring angles: (a) $\theta_0 = \pi/2$, (b) $0 < \theta_0 < \pi/2$, and (c) $\theta_0 = 0$. Top parts (a): the projection field $\boldsymbol{\tau}$ with $k = 1$ on the nematic surface. Middle parts: the vector field $\mathbf{n}$ in the nematic interior. Bottom parts: an element of surface on the order parameter sphere S^2 whose area equals C in (13.63). Sketch by G.E. Volovik.

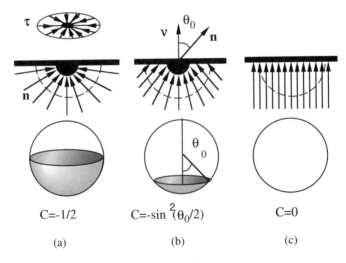

Figure 13.15. Point defects - boojums with $C = -\sin^2(\theta_0/2)$, $k = 1$, and $N = 0$ for the same easy angles as in Fig. 13.14: (a) $\theta_0 = \pi/2$, (b) $0 < \theta_0 < \pi/2$, and (c) $\theta_0 = 0$. Sketch by G.E. Volovik.

for anchoring conditions strong enough to confine the singular core, where the anchoring conditions cannot be satisfied, to a small region. To avoid the ambiguity with signs, let us regard **n** as a vector and leave disclinations out of the picture. When θ_0 varies from $\pi/2$ to 0, the boojum gradually transforms into a hedgehog $C = N = 1$ (Fig. 13.14). The opposite scenario, i.e., gradual disappearance, $C \to 0$, is shown in Fig. 13.15. In principle, any boojum can be characterized by a "bulk" charge N, in addition to the surface charge k. To determine N, it suffices to surround a boojum by a half-sphere from the nematic interior and to calculate the integral (12.15) in Chapter 12[18] over this half-sphere. The resultant quantity C connects N, k and the surface "easy angle" ($\mathbf{n} \cdot \boldsymbol{\nu} = \cos \theta_0$):

$$C = \frac{1}{4\pi} \int du_2 \int \mathbf{n} \left[\frac{\partial \mathbf{n}}{\partial u_1} \times \frac{\partial \mathbf{n}}{\partial u_2} \right] du_1 = \frac{k}{2} (\mathbf{n} \cdot \boldsymbol{\nu} - 1) + N. \tag{13.63}$$

From (13.63), one finds the topological charge N of the boojum because k is defined independently as the index of the projection field $\boldsymbol{\tau} = \mathbf{n} - \boldsymbol{\nu}(\mathbf{n} \cdot \boldsymbol{\nu})$. Note that the boojums with a noninteger C can never leave the surface and go off into the nematic bulk (in contrast, the hedgehogs can locate themselves in any place, including the surface).

Using continuously defined characteristics C's, one can find conservation laws for the charges k and N. Assume that there are p point defects on the surface of the drop and q hedgehogs in its interior. Surround all the defects in the bulk by a surface γ_1 and the entire nematic surface, together with the boojums, by a surface γ_2, as in Fig. 13.16. The total topological charge of the hedgehogs enclosed by γ_1 is $\sum_i N_i$. The total charge enclosed

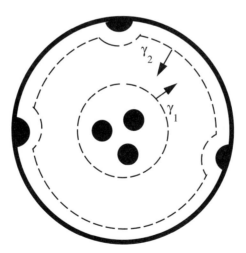

Figure 13.16. A bounded system with hedgehogs (solid circles) and boojums (half-circles).

[18]M. Kleman, Phil. Mag. **27**, 1057 (1973).

by γ_2 is equal to the sum of the boojums characteristics $\sum_j C_j$ and the characteristic of C_s of the droplet surface, which differs from zero because of the curvature of the surface. Taking the integral (12.15) over the surface of the drop with the boojums avoided, one finds

$$C_s = -\mathbf{n} \cdot \boldsymbol{\nu}. \tag{13.64}$$

C_s is equal to the integral, multiplied by $-\mathbf{n} \cdot \boldsymbol{\nu}/4\pi$, of the Gaussian curvature of the surface, which is equal to 4π in the case of a sphere of a unit radius. Now, because there are no defects in between γ_1 and γ_2, the two total topological charges $\sum_{a=p+1}^{p+q} N_a$ and $\sum_{b=1}^{p} C_b +$ C_s (taken with opposite signs) are equal to each other,

$$\sum_{a=p+1}^{p+q} N_a = -\sum_{b=1}^{p} C_b - C_s. \tag{13.65}$$

The definitions of C's and C_s above lead to the equality

$$\sum_{b=1}^{p} C_b + C_s + \sum_{a=p+1}^{p+q} N_a = \frac{1}{2}\left(\sum_{b=1}^{p} k_b - 2\right)(\mathbf{n} \cdot \boldsymbol{\nu} - 1) + \sum_{a=1}^{p+q} N_a - 1 = 0. \tag{13.66}$$

The last equation imposes restrictions on the surface 2D charges k's and the bulk 3D charges N's:

$$\sum_{b=1}^{p} k_b = 2 \tag{13.67}$$

(which is valid for any $\theta_0 \neq 0$) and

$$\sum_{a=1}^{p+q} N_a = 1. \tag{13.68}$$

Equation (13.67) is none other than the Euler-Poincaré theorem: The sum of the indices of a smooth (except a finite number of isolated singular points) vector field $\boldsymbol{\tau}$ on the closed surface is equal to the Euler characteristic χ of this surface:

$$\sum_{b=1}^{p} k_b = \chi. \tag{13.69}$$

As already discussed at the end of Chapter 5, χ of a spherical surface is equal to 2.

Equation (13.68) is a consequence of the Gauss–Bonnet theorem. Indeed, the total "bulk" charge does not depend on the "easy angle" and thus should be equal to the integral (13.63) taken over the bounding surface of the system with $\mathbf{n} = \boldsymbol{\nu}$ everywhere. The integrand is then the Gaussian curvature of the bounding surface, which makes the integral equal to half the Euler characteristic by virtue of the Gauss–Bonnet theorem (5.81); i.e.,

$$\sum_{a=1}^{p+q} N_a = \chi/2. \tag{13.70}$$

According to (13.67)–(13.70), large nematic droplets must contain topological defects in equilibrium. Trivial examples include a droplet with perpendicular anchoring that contains a hedgehog $N = 1$ and a tangentially anchored droplet with a pair of boojums $k_1 = k_2 = 1$ at the poles (Fig. 11.13). Intermediate surface orientations might induce disclinations as well.

13.2.5.2. Elastic Interactions

Topological constraints (13.67)–(13.70) and orientational elasticity of liquid crystals result in a specific type of colloidal interactions that arise when solid or liquid particles are dispersed in a liquid crystalline matrix.[19] If the dispersed particles are sufficiently large, $R \gg K/w_\theta$, they play a role of "seed" topological defects embedded in the nematic matrix. For example, each dispersed particle with radial anchoring is equivalent to a radial hedgehog $N = 1$ in the nematic matrix. Such a particle may be an air bubble or a water droplet with a dissolved surfactant that helps to orient $\mathbf{n}$ normally to the nematic-water interface. Suppose that there are many radially anchored particles dispersed in one very large nematic volume (a spherical drop). Because the global topological charge of the whole nematic should be equal to 1, each particle beyond the first that is added to the interior of the nematic drop must create an orientational defect in the matrix. One of the simplest ways to satisfy this global constraint is to create a hyperbolic hedgehog accompanying each particle in the nematic host (Fig. 13.17). These host defects prevent the dispersed droplets from approaching each other too closely and thus provide a repulsive barrier against coagulation.[20]

13.2.5.3. Growth of the Nematic Phase

The conservation laws given by (13.67)–(13.70) may influence the late stages of the first-order isotropic-to-nematic phase transition that occurs through nucleation of nematic droplets. The droplets grow by adding molecules from the surrounding isotropic matrix and by coalescence. At early stages, the droplets are small and thus practically uniform. However, the directors in different droplets are not correlated. When a few such domains

[19]P. Pieranski, P. Cladis, and M. Kleman, C.R. Acad. Sci. **273**, 275 (1971).
[20]P. Poulin, H. Stark, T.C. Lubensky, and D.A. Weitz, Science **275** 1770 (1997).

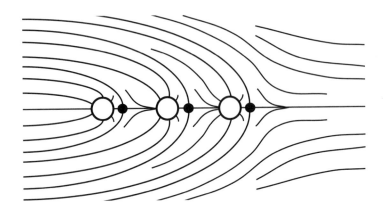

Figure 13.17. Normally anchored spheres (open circles) create companion hyperbolic hedgehogs (cores shown by closed circles) in the director field of the nematic host to preserve the total topological charge.

meet, there is a chance that their mutual disorientation is sufficient to form a defect at the point of merger (Fig. 13.18). This mechanism is valid for both the first- and second-order transitions. (Although the new phase appears approximately simultaneously in the second-order transition, different parts may have no time to communicate and to establish a globally uniform state). It has been proposed as a cosmological model for the structure of the universe.

Another mechanism takes place when the droplets grow beyond the size $\sim K/w_\theta$. The topological defects appear in each droplet because of surface anchoring and the topological

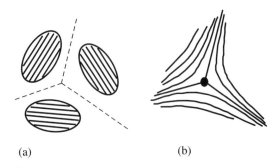

(a)   (b)

Figure 13.18. A possible (Kibble[21]) mechanism of defect formation: (a) differently oriented monodomains of an ordered phase can form (b) a topological defect upon merging.

[21]T.W.B. Kibble, J. Phys. A Gen. Phys. **9**, 1387 (1976).

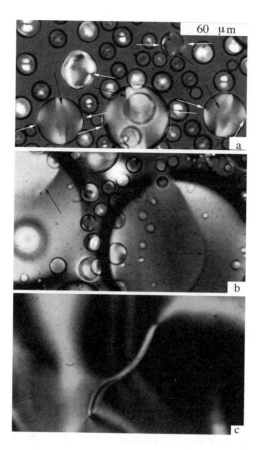

Figure 13.19. Biphasic region at the nematic-to isotropic phase transition. Nematic mixture E7 in a 200-micrometer thick glass container. White and black arrows point toward point and line defects, respectively.

constraints considered above. This is illustrated Fig. 13.19a,b, where one sees nematic droplets of supramicron size containing stable topological defects, because of the surface anchoring that favors a tilted conical director orientation. Most defects, except those with ends trapped at opposite walls (Fig. 13.19c), disappear when the droplets become large enough to fill the entire space between the glass bounding walls.

13.2.5.4. Monopoles: Hedgehogs with Attached Disclinations

Especially interesting consequences of the topological theorems (13.67)–(13.70) occur when the order parameter is characterized by more than one vector field. Consider, for ex-

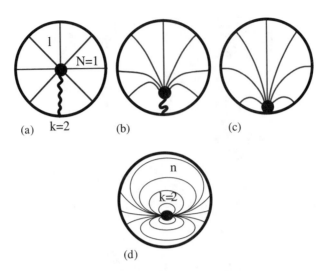

(a)   k=2   (b)   (c)

(d)

Figure 13.20. Defects in a droplet of a biaxial nematic: (a) monopole made of a hedgehog $N = 1$ in the field $\mathbf{l}$ and a disclination $k = 2$ (wavy line) in the field $\mathbf{n}$; (b) the line shrinks into a (c) boojum; (d) shows $\mathbf{n}$ around the $k = 2$ singularity at the spherical surface.

ample, a hypothetical uniaxial-biaxial nematic phase transition in a spherical droplet. Let $\mathbf{l}$ be the director that characterizes both phases and $\mathbf{n}$ the director that appears only in the biaxial phase; $\mathbf{l} \perp \mathbf{n}$. Suppose that the matrix sets the normal orientation of $\mathbf{l}$ at the surface so that a point defect $N = 1$ exists somewhere in the bulk, in accordance with (13.68). The position of the hedgehog is determined by the elastic constants. If the splay elastic constant is relatively small, the hedgehog will stay in the center (Fig. 13.20a). The director $\mathbf{n}$ in the biaxial phase is tangential to the spherical surface. Therefore, the $\mathbf{n}$-field should obey the Euler-Poincaré theorem (13.67) and, accordingly, must sustain either two singularities with $k = 1$ or one with $k = 2$ (Fig. 13.20d). The radial hedgehog with an attached defect line is reminiscent of a Dirac magnetic monopole,[22] in which the radial "hedgehog" of magnetic induction is accompanied with a string (disclination) in the vector potential (the two vector fields are perpendicular to each other). There is a way to reduce the energy of the system by moving the $\mathbf{l}$-hedgehog toward the surface and thus reducing the length of the linear discli-nations (Fig. 13.20b). The resulting surface singularity (Fig. 13.20c) is simultaneously a hedgehog in the $\mathbf{l}$-field and a boojum with $k = 2$ in the $\mathbf{n}$-field. The topological possibility that a line defect shrinks into a surface point singularity was first recognized by Mermin, who considered droplets of superfluid anisotropic A-phase of ^{3}He with an order parameter

[22]P.A.M. Dirac, Proc. Roy. Soc. (London) A **133**, 60 (1931).

Figure 13.21. Cholesteric monopole in a droplet (Robinson spherulite) with a disclination $k = 2$ in the director field. Droplet radius about 30μm.

similar to that of the biaxial nematic. Mermin called the singularity "boojum"[23] (a Lewis Carroll character in "The Hunting of the Snark").

A hedgehog with attached disclinations can be stable as against a boojum if there is some mechanism that prevents the disclinations from shrinking. Liquid crystals with layered structures (e.g., short pitch cholesterics and smectics C, SmC) offer such a stabilizing mechanism. Imagine, for example, a cholesteric droplet with concentric spherical packing of layers (Fig. 13.21). The normal **l** to the layers forms a radial hedgehog. Then the director field tangent to the layers should contain one disclination $k = 2$ or two disclinations $k_1 = k_2 = 1$. Any attempt to shorten these lines violates the layers equidistance.

13.2.6. Smectic A Droplets

A well-known feature of sessile (i.e., "sitting," as opposed to "pendant" or hanging drops) SmA droplets is that their free surface is faceted. The layers are parallel to the horizontal surface, and the curved shape of the droplet is accomodated by steps known as Grandjean terraces. As in solid crystals, the presence of facets indicates a cusp in the dependence $\sigma(\mathbf{n} \cdot \boldsymbol{\nu})$ at $\theta_0 = 0$: $w(\theta) \approx w_\theta|\theta|$ for $0 < |\theta| \ll 1$. The anchoring coefficient is large, because the tilted layers cannot fill space without creating dislocations or steps: $w_\theta \sim B d$ (or $w_\theta \sim \sigma_0$), where B is the SmA compressibility modulus and d is the layers spacing. Generally, $\sigma(\mathbf{n} \cdot \boldsymbol{\nu})$ in the SmA phase is expected to be nonmonotonic with two minima, at $\theta = 0$ and $\theta = \pi/2$, and a large maximum $w_{max} \sim B d$ in between.

---

[23]N.D. Mermin, Boojums all the way through, Cambridge Univ. Press, 1990.

For many materials, the equilibrium surface orientation at the SmA–isotropic interface is tangential, $\theta_0 = \pi/2$. The tangential anchoring on a closed surface is clearly incompatible with a uniform structure in the bulk. As we shall see below, one can again distinguish two limiting cases: "large" droplets, $R > K/\Delta\sigma$, and "small" droplets, $R < K/\Delta\sigma$, where $\Delta\sigma = \sigma_\perp(\theta = \pi/2) - \sigma_{||}(\theta = 0)$ is the surface energy anisotropy (smaller than w_{max}). Small droplets have a uniform structure and a Wulff shape. Large droplets contain focal conic domains (FCDs) that satisfy the surface orientation at the expense of low-cost bulk distortions (Figs. 13.22 and 13.23).

Imagine a large SmA droplet with spherical packing of layers and drill out from it a circular cone with its vertex at the center of the sphere (Fig. 13.22a). Now fill the hole with a circular cone belonging to an FCD (Fig. 13.22b, c). The base of the FCD turns the normal surface orientation into a tangential one, and thus reduces the surface energy by $\sim \pi a^2(\sigma_\perp - \sigma_{||})$. The smectic layers cross the conical surface between the spherically packed region and the FCD smoothly, because the normals to the layers are common normals for both the FCD and spherical layers. The FCD elastic energy, of order of Ka, is the only energy cost for relaxing the boundary conditions.

How many cones are to be removed from the sphere and replaced by the FCDs? One can start with the few largest possible domains, of radius $a_{max} < R$, then fill the gaps between them with smaller domains, and so on (Fig. 13.24). The resulting surface pattern is Apollonian packing of a sphere with circles. If the tangential anchoring is infinitely strong, the iterations should continue until the smallest domain's base is of molecular scale $\lambda = \sqrt{K/B}$ (see Chapters 5 and 10). However, if $(\sigma_\perp - \sigma_{||})$ is finite, the limit of iterations, as determined by the balance of the surface $\sim a^2(\sigma_\perp - \sigma_{||})$ and curvature $\sim Ka$ energies of the smallest FCD, is[24]

$$a^* \sim K/(\sigma_\perp - \sigma_{||}), \qquad (13.71)$$

which can be much larger than λ. Gaps of size smaller than a^* are filled with layers of spherical curvature.

The maximum size a_{max} of the FCDs in the iteration pattern is also restricted. The reason is that angular deviations $\sim a/2R$ from strictly tangential orientation at the base of FCDs increase as the size a of the domain increases. The largest domain base in Fig. 13.23 is visibly restructured to avoid these mismatches. The mismatching tends to decrease the size of the largest domains,[25] $a_{max} \ll R$.

Smectic patterns in flat samples with tangentially anchored plates are often similar to patterns in spherical droplets. Of course, the absolute minimum of the free energy in flat cells corresponds to the uniform structure with layers perpendicular to the plates. However, the history of the sample (phase transitions, flows) can result in metastable FCD textures, often called polygonal textures. For example, when the SmA phase appears di-

[24] O.D. Lavrentovich, Sov. Phys. JETP **64**, 984 (1986).

[25] J.B. Fournier and G. Durand, J. Phys. II France **1**, 845 (1991).

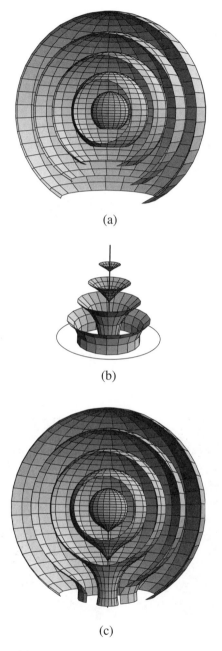

(a)

(b)

(c)

Figure 13.22. Filling a sphere with a focal conic domain. (a) The conical hole in the spherical packing of smectic layers is filled with (b) the focal conic domain (c) smoothly.

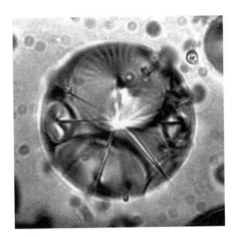

Figure 13.23. Smectic A droplet with tangential boundary conditions suspended in glycerin (diameter $\approx 90\,\mu$m). The droplet is filled with FCDs in geometry of Fig. 13.22. There are no domains with a base radius smaller than $a^* \sim 10\,\mu$m.

Figure 13.24. Apollonian filling of a surface with a set of focal conic domains. The bases of the domains (black) substitute unfavorable normal anchoring (white zones) with favorable tangential anchoring.

Figure 13.25. Polygonal texture in a SmA cell with degenerate tangential boundary conditions. Elliptical bases of the iterative sets of FCD's fill the polygons. The temperature decrease from (a) 40°C to (b) 21°C results in a decrease of the characteristic cutoff $a^* \sim K/\Delta\sigma$: Small FCDs disappear. Material: $CHF_2O(C_6H_4)CH = N(C_6H_4)C_4H_9$.

rectly from the isotropic melt, the seed FCDs are introduced through "bâtonnets" (SmA nuclei emerging from an isotropic melt) and then are trapped by the plates. The model of spherical droplet above is only the first step to understanding the features of FCD filling of the bâtonnets, first described by Friedel and Grandjean.[26] The reason is that the bâtonnets are not spherical due to the smallness of the surface tension at the smectic A-isotropic melt interface.

Figure 13.25 shows the typical SmA polygonal textures. By focusing a microscope on the upper and lower plate, one can distinguish two polygonal networks of defect lines. The sizes of the polygons are close to the sample thickness. Each polygon is the base of a pyramid whose apex lies on the opposite plate. The apices of the pyramids whose bases lie

---

[26]G. Friedel and F. Grandjean, Bull. Soc. Fr. Mineral. **33**, 409 (1910).

on the lower plane coincide with the apices of polygons lying on the upper plane, and vice versa.

To clarify the geometry of the filling, let us choose pyramids with quadrilateral bases for simplicity. The whole space can be divided into two sets of pyramids [of the type H(ABCD) and A(EFGH) in Fig. 13.26], with apices on opposite surfaces of the sample, and into a complementary family of tetrahedra of the types ABGH and ADEH (Fig. 13.26) that fill the gaps between the pyramids. Bold lines in Fig. 13.26a show defect lines that may be visible under the microscope. (Thin lines are not connected to singularities in **n** and serve

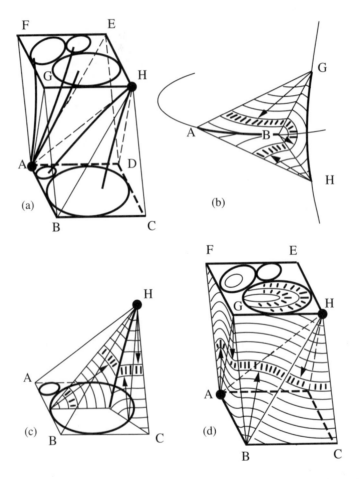

Figure 13.26. General scheme of space filling by FCDs; see text. Director orientation at the bounding plate (area FGHE) is tangential at the FCDs bases and normal (or slightly tilted) in the gaps between the bases.

only for the clarification of the structure). Visible defect lines are ellipses and hyperbolas inside pyramids, as well as edges of polygons. The edges of polygons are formed by the upper and lower edges of tetrahedra. These edges are parts of large hyperbola-ellipse pairs: Each tetrahedron is a part of one large FCD (Fig. 13.26b).

From the above account, it follows that the problem of space filling in a polygonal texture reduces to the problem of filling of pyramids. This filling takes place analogously to the filling of droplets considered above. The ellipses of an iterative family of FCDs are inscribed in the polygons. Inside the ellipses, the orientation is tangential (Fig. 13.26c, d). The hyperbolas of all of these FCDs converge at the apex of a pyramid (points A and H, Fig. 13.26a, c). The remaining gaps inside the pyramid are occupied by layers of spherical curvature with a center at the apex of the pyramid (Fig. 13.26c). At the sample surface, the gaps between the FCDs have normal (or slightly tilted) orientation of molecules (Fig. 13.26c). The iterative filling is truncated at some a^* defined as in (13.71), with modified numerical constants. Figure 13.25 shows that a^* might be a function of temperature. As the temperature decreases, the smallest FCDs disappear, presumably because of the increase of the splay elastic constant K.

Note that the SmA layers cross smoothly from pyramid to pyramid through tetrahedra: In the plane of crossing (ABH in Fig. 13.26d), the layers of both the pyramid H(ABCD) and the tetrahedron ABGH have a common center of curvature (point H). Thus, the curved smectic layers fill the entire space preserving their equidistance and satisfying boundary conditions at the expense of defect structures with dimensionality less than 2.

Problem 13.1. Find the surface excess of (a) the Helmholtz free energy and (b) the internal energy for a one-component system with two coexisting phases 1 and 2.

Answers: (a) $F_{surf} = \sigma A + \mu N_{surf}$; if the position of the dividing surface is fixed by choosing $N_{surf} = 0$, then the surface tension is equal the surface excess of the Helmholtz free energy per unit area $\sigma = \frac{F_{surf}}{A}$. (b) $E_{surf} = \sigma A + T S_{surf} + \mu N_{surf}$.

Problem 13.2. Express the relative adsorption (13.15) in experimentally measurable quantities such as V, A, the concentration c' of the solvent, and the concentration c of the solute in the bulk of the phases 1 and 2.

Answers: See, e.g., *Problems in Thermodynamics and Statistical Physics,* Edited by P.T. Landsberg, PION, London, 1971.

Using $N = N_1 + N_2 + A\Gamma$, one first obtains $\Gamma = (N - V_1 c_1 - V_2 c_2)/A$ and then, excluding V_1 and V_2,

$$\Gamma_{relative} = \frac{1}{A}\left[N - V c_1 - (N' - V c_1') \frac{c_1 - c_2}{c_1' - c_2'} \right].$$

Problem 13.3. Find the pressures p_1, p_2 and chemical potentials μ_1 and μ_2 of two phases in equilibrium when the phase 1 is a sphere of a large radius r.

Answers: For large radii of curvature, the parameters of system should be close to p_0 and μ_0. Expanding μ_1 and μ_2 in the vicinity of μ_0 (e.g., $\mu_1 = \mu_0 + (p_1 - p_0)v_1 + \cdots$), and using the Laplace–Young equation, one gets

$$p_1 = p_0 + \frac{2\sigma}{r} \frac{v_2}{v_2 - v_1}, \quad p_2 = p_0 + \frac{2\sigma}{r} \frac{v_1}{v_2 - v_1},$$

and

$$\mu_1 = \mu_2 = \mu_0 + \frac{2\sigma}{r} \frac{v_1 v_2}{v - v_1}.$$

Problem 13.4. Some parameters (which ones?) defining the critical radius of nucleation r^* in (13.32) are hard to measure experimentally. Reexpress r^* through the values that are readily measured: p_0, p_2, v_1, and v_2, assuming that the chemical potentials and pressures of the two phases do not differ much from μ_0 and p_0.

Answers: From the previous problem, it follows that

$$r^* = \frac{2\sigma}{p_2 - p_0} \frac{v_1}{v_2 - v_1}.$$

Problem 13.5. A large droplet is placed on a solid surface in vacuum. Gravity flattens the droplet in the central part, making it pancake-like. Balancing capillary and gravitation forces, estimate the thickness of the central part of the droplet as a function of the capillary length and the contact angle.

Answers: Combining the equation of equilibrium between capillary and gravitation forces, $S_{spr} + \frac{1}{2}\rho g h^2 = 0$, where $S_{spr} < 0$ is given by (13.36) and the Young law (13.35), one obtains $h = \sqrt{2}\lambda_c \sin(\theta_e/2)$, where λ_c is specified by (13.23).

Problem 13.6. Consider a hybrid-aligned nematic film with the director distorted in the plane normal to the film, due to the difference in polar anchoring directions $\bar{\theta}_0$ and $\bar{\theta}_d$ at the two bounding surfaces $z = (0, h)$ (see Section 5.4.2). Assuming that all Frank elastic constants are equal K and that the anchoring potentials are of the form $f_{si} = \frac{1}{2}w_i(\theta_i - \bar{\theta}_i)^2$, $i = 0, h$, find the disjoining pressure caused by director distortions in a film of thickness $h \gg K/w_i$.

Answers:

$$\Pi(h) = \frac{1}{2}K\left(\frac{\bar{\theta}_0 - \bar{\theta}_h}{h + K/w_0 + K/w_h}\right)^2 > 0.$$

The result may be different for films of small thickness $h < K/w_i$, as thin films undergo a variety of structural transitions [see the review by O.D. Lavrentovich and V.M. Pergamenshchik, Int. J. Mod. Phys. **B9**, 251 (1995)].

Problem 13.7. A semi-infinite nematic volume is bound by a surface with a sinusoidal profile $z = u \sin(qx)$, $u \ll q^{-1}$. Molecular interactions align the director parallel to the substrate. Calculate the elastic energy of distortions for director configurations (a) $\mathbf{n} = (n_x, 0, n_z)$ and (b) $\mathbf{n} = (0, 1, 0)$ in one-constant approximation.

Answers: D.W. Berreman, Phys. Rev. Lett. **28**, 1683 (1972). (a) Let ψ be the small angle between the director and the x-axis. The free energy density

$$\frac{K}{2} \left[\left(\frac{\partial \psi}{\partial x} \right)^2 + \left(\frac{\partial \psi}{\partial z} \right)^2 \right]$$

results in the Euler–Lagrange equation

$$\frac{\partial^2 \psi}{\partial x^2} + \frac{\partial^2 \psi}{\partial z^2} = 0$$

with a solution $\psi = uq \exp(-zq) \cos(qx)$. The energy per unit area is $F_{(a)} = Ku^2q^3/4$; (b) $F_{(b)} = 0$. If the vertical plane containing the director is turned by an angle φ away from the y-axis, the energy becomes $F = Ku^2q^3 \cos^2 \psi/4$.

Problem 13.8. Using (12.4) and (13.63), find the topological charges k, N and characteristics C for the boojums in Figs. 13.14 and 13.15.

Answers: $k = 1$ for all configurations, except those shown in parts (c) of the figures. To find N, parameterize the director field as

$$\mathbf{n} = [\sin \theta(u) \cos \varphi; \sin \theta(u) \sin \varphi; \cos \theta(u)].$$

For the boojum in Fig. 13.14, $\theta(u) = 2(\pi - \theta_0)u/\pi + 2\theta_0 - \pi$, for the boojum in Fig. 13.15, $\theta(u) = 2\theta_0 u/\pi - 2\theta_0$. Integration over $0 \le v < 2\pi$ and $\pi/2 \le u < \pi$ results in $C = \cos^2(\theta_0/2)$, $C = -\sin^2(\theta_0/2)$, $N = 1$, $N = 0$, respectively.

Further Reading

A.W. Adamson and A. P. Gast, Physical Chemistry of Surfaces, John Wiley & Sons, Inc., New York, 1997, 784 pp.

L.M. Blinov, E.I. Kats, and A.A. Sonin, Surface Physics of Thermotropic Liquid Crystals, Usp. Fiz. Nauk **152**, 449 (1987) [Sov. Phys. Usp. **30**, 604 (1987)].

P.S. Drzaic, Liquid Crystal Dispersions, Series on Liquid Crystals, vol. 1, Edited by H.L. Ong, World Scientific, Singapore, 1995, 430 pp.

D.H. Everett, Basic Principles of Colloid Science, Royal Society of Chemistry, London, 1988, 244 pp.

P.G. de Gennes, Rev. Mod. Phys. **57**, 827-863 (1985).

C. Isenberg, The Science of Soap Films and Soap Bubbles, Dover, 1992.

B. Jérôme, Rep. Prog. Phys. **54**, 391 (1991).

L.D. Landau and E.M. Lifshitz, Statistical Physics, 3rd edition, Pergamon Press, Oxford, 1978.

L. Léger and J.F. Joanny, Rep. Prog. Phys. 431–486 (1992).

P. Nozières, Shape and growth of crystals, in: Solids Far From Equilibrium, Edited by C. Godrèche, Cambridge University Press, Cambridge, 1992.

P.S. Pershan, Liquid Crystal Surfaces, J. Phys. (France) **50**, Coll. C7, C7-1 (1989).

J.S. Rowlinson and B. Widom, Molecular Theory of Capillarity, Oxford University Press, New York and London, 1982.

T.J. Sluckin and A. Poniewierski, in Fluid Interfacial Phenomena, Edited by C.A. Croxton, John Wiley & Sons, New York, 1986, Chapter 5.

J. D. Weeks, The roughening transition, in Ordering in Strongly Fluctuating Condensed Matter Systems, Edited by T. Riste, Plenum Press, New York, 1980, p. 293.

CHAPTER 14

Stability of Colloidal Systems

The microscopic stability of soft materials systems (liquid crystals, colloids, polymers) is a problem with many facets.

In the case of *thermotropic liquid crystals*, the nature of the interactions is relatively well known (repulsive steric forces, attractive van der Waals forces, dipolar interactions), although little has been worked out in detail. It is worth mentioning that, thanks to the power of modern computers, detailed Monte-Carlo calculations of thermodynamic functions and phase transitions—employing the Lebwohl-Lasher or Gay–Berne potentials alluded to in Section 1.3.5, as well as simulations of the liquid crystalline molecular structures of hard ellipsoids with steric repulsions—have already brought interesting results (for a review, see Pelcovits[1]).

Colloidal systems, a name that designates a wealth of different dispersed systems, are such that one of the components—the solute—is made of particles of a size r small enough, so that the surface effects ($\propto r^2$) dominate over the bulk effects ($\propto r^3$). The stability of colloidal suspensions depends, therefore, crucially on quantities such as surface tension, electrostatic interactions involving surface charges of the solute and the electrolyte content of the solvent, and van der Waals interactions. One should also add Brownian motion, because usually the particles are small enough to diffuse thermally.

The name of colloidal systems is also given to systems whose elements of the solute have only two small dimensions (e.g., PBLG rods, semiflexible polymers, glass fibers) or one small dimension (e.g., films of surfactants, clay platelets, mica platelets). For a certain range of concentrations, the elements of the solute may be self-assembled into various ordered phases, often called *lyotropic liquid crystals*, as opposed to thermotropic liquid crystals that do not need a solvent to show a mesomorphic state within an appropriate temperature region. For example, aggregates of surfactant molecules can form nematic (when the aggregates are ellipsoidal micelles), columnar (when the aggregates are cylindrical micelles), lamellar (when the aggregates are continuous bilayers), and other phases.

Colloidal systems *per se* are not given a detailed treatment here (Chapter 13 discusses some concepts relevant to colloids and of general use), and we shall focus more on lyotropic systems in the present chapter. There, the situation is richer theoretically (but no

---

[1]R.A. Pelcovits Theory and Computation, in Handbook of Liquid Crystals Research, Edited by R.J. Collins and J.S. Patel, Oxford University Press, New York, 1997.

significant computations yet) than in thermotropic systems. The intermolecular and inter-aggregate forces are of various origins: electrostatic, dipolar, van der Waals, entropic due to the presence of large fluctuations of the membranes (an effect not present in particulate colloidal solutions), and repulsive solvent-mediated complex forces, called solvation forces (or hydration forces when water is the solvent).

This chapter will present a short introduction to the question of electrostatic forces between rigid charged surfaces like plates and spheres (the Poisson–Boltzmann equation) and their competition with van der Waals forces (DLVO theory). The subject has long been a favorite of the colloid community, and a large, relevant literature exists. The question of nonrigid, fluctuating surfaces has been addressed more recently; here, two limit situations arise, which will be discussed in turn: a- In the presence of a *weak* electrolyte, the electrostatic forces extend over long distances and weaken the fluctuations to some extent, thus, contributing to the values of κ and $\bar{\kappa}$; b- in the presence of a *strong* electrolyte, the charges are screened, and large amplitude fluctuations develop that contribute to the repulsive forces between neighboring membranes; the Helfrich theory has been a key development in this perspective and has been met with a large experimental success.

Because of the broad scope of the subject, and the difficulties inherent in some approaches, we have skipped the full demonstration of some important formulae.

Finally, this chapter will also discuss, although to a smaller extent, systems made of zero-D building blocks—particles—some of them being colloidal crystals. This question is treated in fairly large detail in many excellent textbooks. Systems made of 1D building blocks (nematic packing of polymers, hexagonal phases, isotropic solutions of viruses like TMV, etc.) are briefly presented in the next chapter.

14.1. Interactions Between Rigid Surfaces

The models that are discussed below present a certain number of simplifications. They assume that apart from the electrostatic interactions, there are no molecular interactions between the charged species. Also, the discrete nature and finite size of the ions are not taken into account. These assumptions are well suited for small concentrations of charges. Therefore, one will adopt a continuous description of the medium, where the charges are smeared out through all space.

14.1.1. The Poisson–Boltzmann Equation

We start with the geometry of a rigid body (a sphere, a semi-infinite medium bounded by a plane, etc.) whose surface carries fixed charges, and that is in the presence of an electrolyte solvent whose dissociation rate is known. Typically, the solvent is water. When pure, it contains H^+ and OH^- ions in equal molar concentrations 10^{-7} M (M is the notation for moles/dm^3 units) far from the charged surface, i.e., the ions concentrations (number densities) $n^+ = n^- = 10^{-7} N_{Av}$ per dm³ (N_{Av} = Avogadro number) at room temperature. These are small concentrations compared to the usual concentration of counterions in water opposing the surface charges, or to the electrolytes concentrations (see Problem 14.1). In the sequel, n_i^+ and n_j^- will designate the concentration of positive (negative) ions of type $i(j)$ in the presence of the charged surface (surface density of charges σ), z_i^+ will be the (positive) valencies of cations, and z_j^- will be the (negative) valencies of anions. The total density of charges at some point M in the solution is, therefore,

$$\rho(M) = e \sum_{i,j} (z_i^+ n_i^+ + z_j^- n_j^-),$$ (14.1)

where e, the charge of the electron, is taken positive. The total charge of the sample vanishes; hence,

$$\int_V \rho(M)\, dV + \int_\Sigma \sigma\, d\Sigma = 0,$$ (14.2)

Surface charge Conservation

where σ/e is the concentration of surface ions per unit area.

As we shall see below, and as it is expected, ions of sign opposite to those carried by the surface gather in higher density near this surface, hence the name of *double layer* given to this situation. The theory goes as follows: The presence of the charges induces an electric field $\mathbf{E} = -\nabla\varphi$ (φ is the electric potential) that obeys the *Poisson equation*:

$$-\text{div}\,\mathbf{E} \equiv \nabla^2\varphi = \frac{-1}{\varepsilon\varepsilon_0}\rho(M),$$ (14.3)

where ε is the dielectric constant of the solvent (water 78.5; cyclodecane 2.0) and $\varepsilon_0 = 8.854 \times 10^{-12}\,C^2\,J^{-1}\,m^{-1}$ is the dielectric constant of vacuum. We shall use SI units; to return to electrostatic cgs units, replace ε_0 wherever it appears by $1/4\pi$.

Now, the distribution $\rho(M)$ is fixed by the electric potential: The chemical potential of a charged particle of valence z can be written as

$$\mu = \mu_0 + ez\varphi,$$ (14.4)

where μ_0 is the chemical potential in the absence of electric potential. In most real systems, the concentration of the charged species is small, so that they form a dilute ideal solution for which the chemical potential can be written as

$$\mu_0 = k_B T \ln n.$$ (14.5)

$\mu = k_B T \ln(n) + ez\phi \qquad \dfrac{\mu}{k_B T} = \ln(n) + \dfrac{ez\phi}{k_B T}$

In a system in equilibrium, the chemical potential μ per particle is a constant, and one obtains the *Boltzmann equation*:

$$n_i = n_{i,0} \exp -\frac{e z_i \varphi}{k_B T},$$ (14.6)

where $n_{i,0}$ is the particle concentration at points where $\varphi = 0$, conveniently taken at infinity (the "reservoir") in the case of a unique rigid boundary.

The *Poisson–Boltzmann* (PB) *equation* is obtained by substituting (14.6) into (14.3), viz.

$$\nabla^2 \varphi = -\frac{e}{\varepsilon \varepsilon_0} \sum_i z_i n_{i,0} \exp -\frac{e z_i \varphi}{k_B T}.$$ (14.7)

This differential equation must be supplemented by the *charge conservation law* (14.2) and by the *boundary conditions*, which are usually of one of the following types:

1. Fixed potential: $\varphi = $ constant (e.g., at infinity, $\varphi = 0$).

2. Fixed charges: $-\mathbf{E}_s.\mathbf{n} \equiv \frac{\partial \varphi}{\partial n} = \frac{\sigma}{\varepsilon \varepsilon_0}$. In this equation, the normal $\mathbf{n}$ to the surface is taken outward to the solvent.

The PB equation is generally difficult to solve, except in a few special cases, when the valencies of the charged particles are all equal, and the charged boundaries have a simple geometry (spheres, or an isolated plane, or two planes separated by the solvent and no added electrolyte). Simple solutions put into evidence two fundamental lengths, the Gouy–Chapman length b and the Debye–Hückel length λ_D, whose physical interpretation is of importance and makes easier a qualitative discussion of any system.[2]

14.1.2. Fundamental Lengths in the Poisson–Boltzmann Problem

14.1.2.1. The Gouy–Chapman Length for Weak Eectrolytes

Consider a planar, infinite, charged surface (e.g., the charged side of a rigid membrane, considered as the boundary of a nonconductive medium), in contact with a solvent with *no electrolyte* added, so that the only species of ions (apart H^+ and OH^-) present in the solvent are the counterions of the ionic species on the surface. We shall assume σ negative (because in most systems of practical interest, the fixed ionic species are negatively charged), and all the valencies equal to unity. The x-axis is perpendicular to the boundary. Then the PB equation can be written as

$$\frac{d^2 \varphi}{dx^2} = -\frac{e n_0}{\varepsilon \varepsilon_0} \exp -\frac{e \varphi}{k_B T},$$

---

[2]P. Pincus, J.-F. Joanny, and D. Andelman, Europhysics Lett. **11**, 763 (1990).

where n_0 is a reference concentration of the counterions (for which the potential is taken equal to zero), whose volume density reads as $n(x) = n_0 \exp(-e\varphi/k_B T)$. One can check that the solution is

$$\varphi(x) = \frac{2k_B T}{e} \ln\left(\frac{x+b}{\xi}\right).$$

(14.8)

Because $\frac{\partial \varphi}{\partial x}\big|_{x=0} = -\frac{\sigma}{\varepsilon\varepsilon_0}$ (boundary conditions for fixed charges), we have

$$b = -\frac{2\varepsilon\varepsilon_0}{\sigma} \frac{k_B T}{e},$$

(14.9)

and ξ is an arbitrary length that has no other effect than to add a constant to the potential. It must be chosen in agreement with the choice of the reference concentration. The calculation shows indeed that the product $n_0\xi^2 = -\sigma b/e$. This expression can also be written as

$$n_0\xi^2 = \frac{1}{2\pi\ell}; \text{ with } \ell = \frac{e^2}{4\pi\varepsilon_0\varepsilon k_B T},$$

(14.10)

where ℓ, the so-called Bjerrum length, is the distance beyond which the bare electrostatic interaction between two unit charges begin to be somewhat blurred by their Brownian movement: The charges cannot be considered as immobile, and entropy effects have to be taken into account; ℓ is of the order of 7 Å in water.

The Gouy–Chapman length b is the thickness of a solvent layer adjacent to the boundary and containing half the total amount of counterions. A typical value in water, for σ equal to one elementary charge per 0.16 nm^2 (i.e., 1 C.m^{-2} in SI units) and $\frac{k_B T}{e} = 25.69$ mV at room temperature (298K), is $b = 4 \times 10^{-2}$ nm, a pretty small region in the near vicinity of the plate. The density of counterions is parabolic:

$$n(x) = \frac{1}{2\pi\ell(x+b)^2}$$

(14.11)

(Problem 14.1). The logarithmic divergence (14.8) has little physical significance; it would disappear for a finite sample.

14.1.2.2. The Debye–Hückel Screening Length for Strong Electrolytes

The opposite situation is when the boundary is in the presence of a strong electrolyte, say, 1M NaCl in water. Let $n_0 = n_+(\infty) = n_-(\infty)$ be the density of ions of each sign introduced in the solvent. The calculation goes easily if these ions and the counterions have the same valency; we take it equal to unity. Therefore we have to solve the following PB equation:

$$\nabla^2 \varphi = \frac{2en_0}{\varepsilon_0 \varepsilon} \sinh \frac{e\varphi(\mathbf{r})}{k_B T},$$ (14.12)

where now n_0 is the electrolyte concentration in the reservoir. When the linearization is possible (Debye–Hückel approximation), i.e., when $\varphi < k_B T/e$ ($\approx 25\,\mathrm{mV}$ at room temperature, $\approx 0.9 \times 10^{-4}$ in cgs units), one gets

$$\nabla^2 \varphi = \frac{2e^2 n_0}{\varepsilon \varepsilon_0 k_B T} \varphi(\mathbf{r}) = k_D^2 \varphi(\mathbf{r}) \quad \text{with} \quad k_D = \lambda_D^{-1} = \left(\frac{2e^2 n_0}{\varepsilon \varepsilon_0 k_B T} \right)^{1/2},$$ (14.13)

i.e. an exponential decrease $\varphi(x) = \varphi_s \exp(-x/\lambda_D)$ with a characteristic length λ_D; φ_s is the potential at the charged boundary. Note that $k_D^2 = 8\pi \ell n_0$. The Debye–Hückel length λ_D (also called the Debye length) is the length beyond which the electrostatic effects due to the charged wall are screened; in the general case, it can be expressed as

$$\lambda_D = \left(\frac{e^2 \sum n_{i,0} z_i^2}{\varepsilon \varepsilon_0 k_B T} \right)^{-1/2}.$$ (14.14)

A typical value of λ_D for a strong electrolyte (1M NaCl) is 1 nm at room temperature. A more detailed discussion of (14.12) is given below in Section 14.1.5.

14.1.3. Free Energy and Maxwell Stress Tensor

14.1.3.1. Free Energy and "Disjoining" Pressure

The free energy density (per unit area of boundary) of the electric double layer has the following expression:

$$F_{\mathrm{el}} = \int_0^\sigma d\sigma' \varphi_s(\sigma'),$$ (14.15)

integrated over all the (rigid) charged surfaces. The quantity $d\sigma' \varphi_s(\sigma')$ is indeed the work necessary to bring a charge $d\sigma'$ from a potential $V = 0$ to a potential $V = \varphi_s(\sigma')$. This quantity is also obtained when the free energy is written as the sum of the excess electrostatic bulk energy over the homogeneous system

$$F_{\mathrm{bulk}} = \frac{1}{2} \varepsilon \varepsilon_0 \int E^2 \, dV = \frac{1}{2} \varepsilon \varepsilon_0 \int (\nabla \varphi)^2 \, dV,$$ (14.16)

and of the excess configurational entropy of the ions in the solvent over the ions in the reservoir, which in the case of two species of distinguishible ions and coions of same valency reads as

$$F_{entropy} = k_B T \int \left[\sum_{i=\pm} n_i \ln \frac{n_i}{n_0} - \sum_{i=\pm} n_i + 2n_0 \right] dV. \qquad (14.17)$$

The total energy is $F_{el} \equiv F_{bulk} + F_{entropy}$ (see Goldstein et al.[3] for a demonstration).

If the solvent is confined between two identical charged plates, the overall forces of electrostatic and osmotic origin are such that the two plates repel. Considering that one plate is kept fixed, the force exerted on the other reads as

$$p_d = -\frac{\partial F_{el}}{\partial d}\Big|_{T,n_i},$$

where d is the separation distance between the plates. Let $n(x = 0)$ be the total density of particles in the midplane $x = 0$. It can be shown[4] that the general expresion for the "disjoining" pressure p_d reads as

$$p_d = k_B T \left(n(x = 0) - \sum_i n_{i,0} \right). \qquad (14.18)$$

This quantity can be interpreted as the excess of the *osmotic* pressure of the ions in the midplane over the reservoir osmotic pressure $p_0 = k_B T n_0 = k_B T \sum_i n_{i,0}$, as it is made visible by its form reminiscent of the vant'Hoff equation for osmotic pressure. Note that because of the electrostatic potential, p_d is not equal to the pressure in the reservoir.

In the case when there is no electrolyte in the solvent, $n(x = 0)$ reduces to the density of counterions, and $\sum_i n_{i,0} = 0$. One can then also express p_d as a function of the counterions density $n_s = n(x = 0) + \frac{1}{2\varepsilon\varepsilon_0 k_B T}\sigma^2$ (Problem 14.2):

$$p_d = k_B T n(x = 0) = k_B T \left(n_s - \frac{1}{2\varepsilon\varepsilon_0 k_B T}\sigma^2 \right).$$

14.1.3.2. The Maxwell Stress Tensor; Longitudinal and Transversal Pressures

The total "osmotic" pressure p_d in (14.18) above contains two terms, one of a pure osmotic origin, and the other due to the electrostatic charges. The existence of these two types of

---

[3]R.E. Goldstein, A.I. Pesci, and V. Romero-Rochin, Phys. Rev. **A41**, 5504 (1990).
[4]J.Israelachvili, Intermolecular and Surface Forces, 2nd edition, Academic Press, New York, 1992.

terms can be made clearer by introducing the electrostatic forces acting on the ions in the solvent. These forces can be represented in the continuous limit by a stress tensor:

$$\sigma_{ij}^M(\mathbf{r}) = \varepsilon\varepsilon_0 \left(E_i(\mathbf{r})E_j(\mathbf{r}) - \tfrac{1}{2}\mathbf{E}^2(\mathbf{r})\delta_{ij} \right) - p(\mathbf{r})\delta_{ij} \qquad (14.19)$$

whose working out is rather delicate. This tensor originates in the forces per unit volume exerted by the electric field: (1) On the free charges $\rho(\mathbf{r})$, a force $\rho(\mathbf{r})\mathbf{E} = \varepsilon\varepsilon_0\mathbf{E}\,\text{div}\,\mathbf{E}$ [also noted $\varepsilon\varepsilon_0\mathbf{E}\nabla\cdot\mathbf{E}$, i.e. $(\varepsilon\varepsilon_0 E_k\partial_i E_i)$], (2) On the charges of polarization $\rho_p(\mathbf{r}) = -\nabla\cdot\mathbf{P}$, a force $\mathbf{P}\cdot\nabla\mathbf{E}$, i.e. $(P_i\partial_i E_k)$, and (3) Due to the local excess osmotic pressure $p(\mathbf{r})$ that would exist in the medium in the absence of an electric field, for the given concentration of ions in $\mathbf{r}$. The interested reader should consult Ref. [5] for details.

Longitudinal Stress: The *local* excess osmotic pressure $p(\mathbf{r})$ is such that $p(x) = k_B T(n(x) - n_0)$, where $n_0 = \sum_i n_{i,0}$. The σ_{xx}^M component reads $-p(x) + \tfrac{1}{2}\varepsilon\varepsilon_0\mathbf{E}^2$; this quantity is equal to $-p_d$ for $x = 0$; it is easy to see that σ_{xx}^M does not depend on x, because it is a first integral of the (general) PB equation. The longitudinal stress is nothing else than the disjoining pressure, changed sign.

Transversal Stress: Another force that will prove of interest in the sequel is the transverse force $p_t(x)$, i.e., with minus signs, the σ_{yy}^M and σ_{zz}^M components. Still considering planar charged plates, one gets

$$p_t(x) = p(x) + \tfrac{1}{2}\varepsilon\varepsilon_0 E^2, \qquad (14.20)$$

which also reads as

$$p_t(x) = p_d + \varepsilon\varepsilon_0 E^2 \qquad (14.21)$$

and is *not* independent of x.

14.1.4. Weak Electrolyte Solutions

We shall discuss in turn the case of weak electrolytes, when the screening effect of the fixed charges is small (large Debye length) and the (opposite) case of strong electrolytes. One can find a detailed treatment of the different types of solutions of the Poisson–Boltzman

---

[5]L.D. Landau and E.M. Lifshitz, Electrodynamics of Continuous Media, Pergamon, Oxford, 1960. W.B. Russel, D.A. Saville, and W.R. Scholwalter, Colloidal Dispersions, Cambridge Monographs on Mechanics and Applied Mathematics, 1991.

equation in function of the nondimensioned lengths $\xi_1 = \frac{\lambda_D}{d}$ and $\xi_2 = \frac{b}{d}$ (where d is the distance between two plates) in Andelman's review article.[6] We are interested in the variation of ξ_1, mostly. For large values of ξ_1, the fixed charges are screened little, and one expects that the electric potential decays algebraically with distance; in the opposite situation, one expects an exponential behavior.

In order to illustrate these general qualitative considerations, consider *two* parallel equally charged membranes, at a distance d, with *no electrolyte* in the solvent. The relevant PB equation and boundary conditions can be written as

$$\frac{d^2\varphi}{dx^2} = -\frac{e}{\varepsilon\varepsilon_0}n_0 \exp -\frac{e\varphi(x)}{k_BT}, \quad \frac{d\varphi}{dx}\bigg|_{x=d/2} = \frac{\sigma}{\varepsilon\varepsilon_0}, \tag{14.22}$$

where n_0 is the concentration of counterions at $x = 0$, in the midplane. Because of the symmetry of the problem, we take $\varphi(0) = 0$. It is easy to check that the solution is

$$\varphi = \frac{k_BT}{e} \ln \cos^2 k_Lx, \tag{14.23}$$

where $k_L^2 = 2\pi n_0\ell$ and $n_0 = -k_L\frac{\sigma}{e} \cot(k_Ld/2)$ is the counterions concentration in the midplane. The general expression for the disjoining pressure is $p(d) = k_BTn_0$. We also get [from the boundary condition and the definition of b, (14.9)] $k_Ld \tan(k_Ld/2) = d/b$. Notice that when $k_L d \gg 1$, i.e., for large membrane separation, this relation yields $k_L^{-1} \Rightarrow b$; for the opposite case $k_L d \ll 1$, one gets $k_L^{-1} \Rightarrow \sqrt{db/2}$. We consider these two extreme cases:

1. Gouy–Chapman regime: $k_L d \gg 1$ or, equivalently, $d/b \gg 1$ with b small, i.e., a large charge surface density on the membranes. For a completely ionized membrane, a typical value of σ could be 1 unit charge per 9 Å^2, i.e., ≈ 1.78 C m^{-2}; this yields $b < 1$ Å. Henceforth, most membranes in non electrolyte solvents are in this regime. The "disjoining pressure" is $p_d = k_BT\frac{\pi}{2\ell d^2}$; the repulsive potential $V_{el} = -\int_{\infty}^d dx' p(x')$ scales as d^{-1}.

2. Ideal-gas regime: $k_L d \ll 1$ or, equivalently $d/b \gg 1$; the surface charge density is small, and the counterions are uniformly distributed with a concentration $n \approx \frac{\sigma}{d}$. The force that keeps the membranes apart is $p_d \approx k_BT\frac{1}{\pi bd\ell}$, and the repulsive potential shows a logarithmic dependence on d; the electrostatic potential decays algebraically $\varphi \approx -k_BTk_L^2x^2/2e$.

All of these results are relatively easy to establish.

---

[6]D. Andelman, Electrostatic Theories of Membranes: the Poisson-Boltzmann theory, Handbook of Biological Systems **1**, 577 (1994).

14.1.5. Strong Electrolyte Solutions

We consider an isolated plate in an infinite electrolyte solution of monovalent ions. This is the first step to approach the double layer theory in the presence of van der Waals forces. At the time being, we assume that these forces are absent. Equation (14.7) takes the form (14.12); i.e.,

$$\frac{d^2\varphi}{dx^2} = \frac{2en_0}{\varepsilon_0\varepsilon} \sinh \frac{e\varphi(x)}{k_B T}. \tag{14.24}$$

This equation has a first integral

$$\frac{d\varphi}{dx} = -\left(\frac{8n_0 k_B T}{\varepsilon_0\varepsilon}\right)^{1/2} \sinh \frac{e\varphi(x)}{2k_B T};$$

the complete solution is illustrated in Fig. 14.1. The exact solution

$$\tanh \frac{e\varphi(x)}{4k_B T} = \tanh \frac{e\varphi_s}{4k_B T} \exp(-k_D x), \tag{14.25}$$

where φ_s is the potential at $x = 0$, shows the expected exponential behavior when using the *Debye–Hückel linear approximation*. The double layer is the region close to the charged surface, where there is a depletion of coions and accumulation of counterions, in a thickness of the order of λ_D. The concentrations of counterions $n_+(x)$ and of coions $n_-(x)$

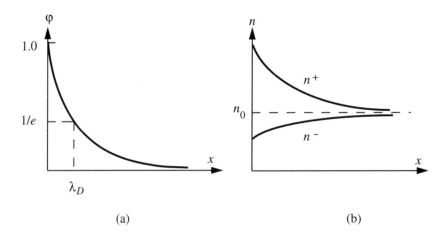

(a)   (b)

Figure 14.1. Electric potential and ions concentrations. Note the accumulation of counterions and the depletion of coions near the charged surface.

approach the electrolyte concentration n_0 at a distance from the plate. The plate charge density is easily obtained from the boundary condition on the gradient of the potential:

$$\sigma = 2(2\varepsilon\varepsilon_0 k_B T n_0)^{1/2} \sinh\left(\frac{e\varphi}{2k_B T}\right).$$
(14.26)

The total *local charge* is given by the Poisson relation:

$$e[n_+(x) - n_-(x)] = -\varepsilon\varepsilon_0 d^2\varphi/dx^2.$$
(14.27)

The concentrations of counterions and coions can be calculated if $\varphi(x)$ is known. One can also relate the total *concentration* of ions at the plate to the concentration of fixed charges (Grahame relation). We have, indeed,

$$\frac{dn_+(x)}{dx} = -n_0 \frac{e}{k_B T}\left(\frac{d\varphi}{dx}\right)\exp\left(-\frac{e\varphi}{k_B T}\right),$$

$$\frac{dn_-(x)}{dx} = +n_0 \frac{e}{k_B T}\left(\frac{d\varphi}{dx}\right)\exp\left(+\frac{e\varphi}{k_B T}\right),$$
(14.28)

i.e., adding these two derivatives, putting $n = n_+ + n_-$, and using (14.24),

$$\frac{dn(x)}{dx} = \frac{\varepsilon\varepsilon_0}{2k_B T}\frac{d}{dx}\left(\frac{d\varphi}{dx}\right)^2.$$
(14.29)

Equation(14.29) integrates easily, and yields for $x = 0$,

$$n_s = n_0 + \frac{\sigma^2}{2\varepsilon\varepsilon_0 kT}.$$
(14.30)

Equation (14.30) relates the concentration in the vicinity of the plate to the concentration in the reservoir. It has the same shape as that one obtained for a nonelectrolyte solvent between two plates (see Problem 14.2), but with a different meaning of n_0.

Let us now discuss some approximations from (14.25), which can also be written as

$$\frac{e\varphi}{2k_B T} = \ln\left(\frac{1 + \exp(-k_D x)\tanh\left(\frac{e\varphi_s}{4k_B T}\right)}{1 - \exp(-k_D x)\tanh\left(\frac{e\varphi_s}{4k_B T}\right)}\right).$$
(14.31)

1. The linear DH regime is obtained for $\frac{e\varphi_s}{4k_B T} \ll 1$.

2. For $\frac{e\varphi_s}{4k_B T} \gg 1$, the potential can be written $\frac{e\varphi}{2k_B T} \approx \ln(\frac{1+\exp(-k_D x)}{1-\exp(-k_D x)})$, i.e. it is independent of the surface potential.

3. For $x \gg k_D^{-1}$, one gets $\frac{e\varphi}{k_B T} = 4\tanh(\frac{e\varphi_s}{4k_B T})\exp(-k_D x)$; i.e., the decay is always exponential far from the boundary. Note that, if furthermore $\frac{e\varphi_s}{4k_B T} \gg 1$, the surface potential $\varphi_{s,\text{eff}}$ seen from a distance is a constant independent of the charge, $\frac{e\varphi_{s,\text{eff}}}{4k_B T} = 1$.

14.1.6. The DLVO (Derjaguin–Landau–Verwey–Overbeek)[7,8] Theory: van der Waals vs Electrostatic Interactions

The double layer interaction between two plates is globally repulsive. The van der Waals forces provide long-range attractive forces that counterbalance the electrostatic forces.

Figure 14.2 indicates the main characteristics of the DLVO potential $V_{\text{DLVO}} = V_{\text{elec}} + V_{\text{vdW}}$ as a function of the distance d between two plates. The van der Waals interaction is fairly independent of the electrostatic fixed charges and of the electrolyte strength. An $n = 6$ van der Waals interaction yields an attractive potential energy density (per unit area

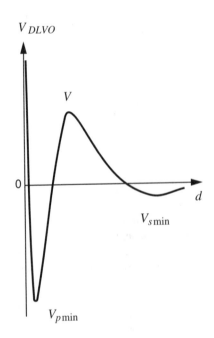

Figure 14.2. Total potential in the DLVO theory.

[7] B.V. Derjaguin and L. Landau, Acta PhysicoChemica USSR, **14**, 633 (1941)

[8] E.J.W. Vervey and J.Th.G. Overbeek, Theory of the Stability of Lyophobic Colloids, Elsevier, New York, 1948.

of plate) between two parallel plates of thickness δ, of molecular density ρ, which is of the form

$$V_{\text{vdW}} = -\frac{\pi C \delta^2 \rho^2}{2d^4}$$

($A = \pi^2 C \rho^2$ is the Hamaker constant) if $\delta \ll d$, and of the form

$$V_{\text{vdW}} = -\frac{\pi C \rho^2}{2d^2}$$

if $\delta > d$, see Table 1.5.

The key features of Fig. 14.2 are the presence of two minima, with a very deep *primary minimum* at very small distances ($< \lambda_D$), not infinitely deep, actually, because of short-range repulsive hydration or solvation forces, and an *intermediary maximum* that sets a repulsive barrier of energy.

For $d \gg \lambda_D$, i.e., very strong electrolytes, the electrostatic forces are negligible, and the colloidal objects (plate-like in these examples, but also spherical or cylindrical particles) should coalesce, except for the presence of a secondary minimum (Fig. 14.2). Another case, to be discussed in Section 14.2, is when the charged surfaces are flexible (not plates, but membranes), so that repulsive forces of entropic origin take over.

For $d \ll \lambda_D$, i.e., weak electrolytes, surfaces repel strongly, and one can expect a strong energy barrier between the charged particles. The deepest primary minimum is then not necessarily reached, and the system keeps in a state of kinetic stability, at the position of the more shallow minimum.

All intermediary situations exist. From this discussion, it appears that one of the most interesting features of the DLVO theory is the transition between a state of *kinetic stability*, when the barrier is high enough ($> k_B T$) to a state of *aggregation*, when, e.g., the concentration in electrolyte is increased. This transition has been the subject of numerous studies in the case of spherical charged particles (see Section 14.3), where the aggregates can form *flocs*, whose size is related to the range of the actual attractive forces between the particles. Contrarily, these effects of aggregation are not major effects in the case of membranes, because of short-range repulsive forces of another nature, provided by the entropic forces relating to the fluctuations of the membranes, and whose associated "Helfrich" potential V_{Hel} also varies steeply for small distances ($V_{\text{Hel}} = \beta (k_B T)^2 / \kappa d^2$) and can counterbalance V_{vdW} in this region (the Hamaker constant and $(k_B T)^2 / \kappa$ are comparable in magnitude).

14.2. Interactions in Lamellar Flexible Systems

As was already indicated in Chapters 1, 2, and 13, amphiphilic materials (Figs. 1.3, 1.5, and 2.17) readily adsorb at the interface between two media, e.g., water and oil or water

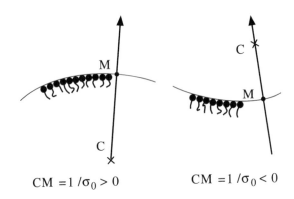

Figure 14.3. Conventions usually adopted for the sign of σ_0.

and air. Depending on the surface concentration of the molecules or, equivalently, the area available per molecule, these materials can form *monolayers* (Fig. 14.3) with different types of ordering, ranging from a 2D gaseous state to closely packed solid-like structures. The states differ by the average distance between the polar heads and by the conformations of the hydrocarbon tails. If the amphiphile is dissolved in a single-component solvent, a preferable type of assembling would be a *bilayer* formed by a pair of oppositely oriented monolayers (Fig. 14.4). When the amphiphile concentration is high enough, the bilayers can serve as building units of the lamellar (Fig. 2.23) and sponge (Fig. 2.29) phases. At low concentrations, the bilayers can be prepared in the form of closed *vesicles* of spherical, toroidal, or even higher genus[9] topology. Moreover, one can find different geometries of the vesicles within the same topology class. A well-known example is the biconcave discoidal shape of the bounding membrane of erythrocytes; it changes into ellipsoidal when the erythrocyte travels through the blood vessels or when the cell becomes abnormal. It appears

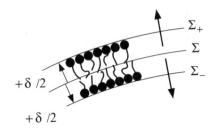

Figure 14.4. Partition of the bilayer to getting (14.34).

[9]B. Fourcade, M. Mutz, and D. Bensimon, Phys. Rev. Lett. **68**, 2551 (1992).

that the prime factor determining the shape of bilayers is their elasticity. The basic elements of this elasticity are discussed below. We consider only *fluid* membranes (monolayers, bilayers) that have no in-plane orientational order of the hydrocarbon tails (no cooperative tilt) and have vanishing in-plane shear modulus. The only deformations of interest are tangential compressions and expansions that change the area per molecule and curvatures of the membrane.

14.2.1. Elasticity of Neutral Membranes

14.2.1.1. Phenomenological Description

In this section, we discuss microscopic models that allow one to estimate material constants of a free membrane. The curvature free energy density *per unit area* of a *monolayer* has been proposed by Helfrich[10] as the function of the mean and the Gaussian curvatures:

$$f = \tfrac{1}{2}\kappa_m(\sigma_1 + \sigma_2 - \sigma_0)^2 + \bar{\kappa}_m\sigma_1\sigma_2 \qquad (14.32)$$

(the subscript stands for "monolayer"). The last expression is similar to those already introduced in Chapter 5, except for the presence of a *spontaneous curvature* σ_0, which takes into account the dissymmetry of the monolayer. If, by convention, the monolayer is oriented from the nonpolar aliphatic chains to the polar heads, σ_0 is taken to be positive when the "outer" side of the curved monolayer is the side of the polar heads (Fig. 14.3). Dimensionally, the moduli κ_m and $\bar{\kappa}_m$ are of the form $\kappa_m = K\,d_0, \bar{\kappa}_m = \bar{K}\,d_0$, where $K, \bar{K}$ are the elastic Frank moduli and d_0 is a length of the order of the monolayer's thickness.

Note that the curvature taken by a free membrane is not σ_0, but a value $\sigma_{eq} = \sigma_1 = \sigma_2$ such that f is minimized, that is:

$$\sigma_{eq} = \sigma_0 \frac{\kappa_m}{2\kappa_m + \bar{\kappa}_m}, \qquad (14.33)$$

to which has to be added the condition of stability $\partial^2 f/\partial\sigma_{eq}^2 > 0$, i.e. $2\kappa_m + \bar{\kappa}_m > 0$, because $\kappa_m > 0$.

A bilayer made of two chemically identical monolayers has evidently a vanishing spontaneous membrane curvature. The relationship between the materials constants $\kappa, \bar{\kappa}$ for the bilayer and $\kappa_m, \bar{\kappa}_m$ can be written as

$$\kappa = 2\kappa_m, \quad \bar{\kappa} = 2(\bar{\kappa}_m - \delta\sigma_0\kappa_m), \qquad (14.34)$$

where δ is an effective thickness of the bilayer represented in Fig. 14.4 with the other ingredients of the demonstration of (14.34) now detailed.

---

[10]W. Helfrich, Z. Naturforsch. **28**, 693 (1973).

We assume that the bilayer can be partitioned in a series of parallel physical surfaces, the midsurface being the continuous surface Σ that envelops the chains ends, and the outer surfaces being the surfaces Σ_+ and Σ_- that envelop the polar heads. The curvatures of the bilayer will be by convention those of Σ. Let $\delta/2$ be the (positive) distance measured from Σ at which (14.32) apply for each of the monolayers. We assume that there are no further interactions between the monolayers than those just defined by the geometry of steric hindrance. We have

$$f_\pm = \tfrac{1}{2}\kappa_m(\sigma_{\pm,1} + \sigma_{\pm,2} - \sigma_0)^2 + \bar{\kappa}_m\sigma_{\pm,1}\sigma_{\pm,2},\tag{14.35}$$

where the $+$ sign refers to the "upper" monolayer, and the $-$ sign to the "lower" one. Note that we have chosen the same signed spontaneous curvature, which implies that we have oriented the two monolayers in opposite directions. Now it can be shown (Problem 14.3) that

$$d\Sigma' = d\Sigma(1 + 2H\lambda + G\lambda^2),\tag{14.36a}$$

$$H' = H(1 - 2H\lambda) + G\lambda,\tag{14.36b}$$

$$G' = G(1 - 2H\lambda),\tag{14.36c}$$

where $2H = \sigma_1 + \sigma_2$, $G = \sigma_1\sigma_2$, and the primed quantities refer to the same variables for a surface that has been displaced by a signed length λ along the normal. Therefore,

$$f_+ = \frac{1}{2}\kappa_m\left(\sigma_1 + \sigma_2 - (\sigma_1 + \sigma_2)^2\frac{\delta}{2} + \sigma_1\sigma_2\delta - \sigma_0\right)^2$$
$$+ \bar{\kappa}_m\sigma_1\sigma_2\left(1 - (\sigma_1 + \sigma_2)\frac{\delta}{2}\right),$$

$$f_- = \frac{1}{2}\kappa_m\left(-\sigma_1 - \sigma_2 - (\sigma_1 + \sigma_2)^2\frac{\delta}{2} + \sigma_1\sigma_2\delta - \sigma_0\right)^2$$
$$+ \bar{\kappa}_m\sigma_1\sigma_2\left(1 + (\sigma_1 + \sigma_2)\frac{\delta}{2}\right),\tag{14.37}$$

and

$$f\,d\Sigma = f_+d\Sigma_+ + f_-\,d\Sigma_-.\tag{14.38}$$

We keep the first-order terms in δ only and obtain

$$f = \tfrac{1}{2}\kappa(\sigma_1 + \sigma_2)^2 + \bar{\kappa}\sigma_1\sigma_2,\tag{14.39}$$

with

$$\kappa = 2\kappa_m, \quad \overline{\kappa} = 2(\overline{\kappa}_m - \kappa_m\sigma_0\delta). \tag{14.40}$$

Consider the correction to the saddle-splay modulus; $\overline{\kappa}_m$ can be very small, as it has often been assumed in the L_α phase close to the L_3 sponge phase. In that case, the correction is dominant over the bare $\overline{\kappa}_m$ contribution; it yields a positive $\overline{\kappa}$ if the spontaneous curvature is negative, a situation that favours the appearance of the sponge phase.[11]

14.2.1.2. Microscopic Models; the Petrov–Derzhanski–Mitov Model

Various models have been built to relate the material constants of fluid membranes to microscopic parameters. We give the basic elements of the Petrov–Derzhanski–Mitov (PDM) model. This model and others like the Israelachvili–Mitchel–Ninham (IMN) model[12] are reviewed in some detail in Petrov.[13]

In the mechanical approach of the Bulgarian school, one starts from the free energy *per molecule* in a monolayer

$$f_m = \frac{k_H}{2}\left(\frac{A_H}{H} - 1\right)^2 + \frac{k_C}{2}\left(\frac{A_C}{C} - 1\right)^2, \tag{14.41}$$

where k_H (respectively k_C) are elastic constants characterizing the interaction between heads (resp. chains), A_H (respectively A_C) are the cross-section areas, and H (respectively C) are the cross-section areas of the molecules when free of interactions. The aim is to relate the elastic constants and the spontaneous curvature of the monolayer to the microscopic parameters, and particularly to their anisotropy. The result depends on the detailed model of the molecule that is used.

Consider first a single monolayer of a thickness $\delta/2$. We measure the head cross sections at the distance $\delta/2$ from the surface Σ formed by the ends of the chains and the chain cross sections at the distance $\delta/4$ from Σ. The relations between the curvatures (measured at Σ) and the cross-sectional areas are, according to (14.36a),

$$A_H = A\left[1 + \left(\frac{\delta}{2}\right)(\sigma_1 + \sigma_2) + \left(\frac{\delta}{2}\right)^2\sigma_1\sigma_2\right],$$

$$A_C = A\left[1 + \left(\frac{\delta}{4}\right)(\sigma_1 + \sigma_2) + \left(\frac{\delta}{4}\right)^2\sigma_1\sigma_2\right], \tag{14.42}$$

where A is the mean area per molecule measured at Σ.

[11]G. Porte, J. Appell, P. Bassereau, and J. Marignan, J. Physique France **50**, 285 (1989).
[12]J. Israelachvili, D.J. Mitchel and B.W. Ninham, J. Chem. Soc. Faraday Trans. II **72**, 1525 (1976).
[13]A.G. Petrov, The Lyotropic State of Matter, Gordon and Breach, 1999.

For a monolayer constrained to be flat ($A_H = A_C = A$), the equilibrium area minimizing f_m is

$$A_0 \approx \frac{k_H H^{-1} + k_C C^{-1}}{k_H H^{-2} + k_C C^{-2}}. \tag{14.43}$$

This is not the mean area for a monolayer. A monolayer, because of asymmetry between its two sides, gets a spontaneous curvature σ_0. Nevertheless, the result (14.43) can be directly used to consider a bilayer formed by two monolayers whose chains end at the same surface Σ. Equation (14.42) is applicable to both monolayers of a bilayer. The energy per unit area of a bilayer stretched isotropically by a quantity $\varepsilon = (A - A_0)/A_0$ can be written to second order $f_{bl,s} = 2\Delta f_m/A_0 = \kappa_s \varepsilon^2/2$. The stretching constant κ_s is easily obtained by using $\Delta f_m = f_m(A) - f_m(A_0)$. The corresponding tension is $\tau = \partial f_{bl,s}/\partial \varepsilon = \kappa_s \varepsilon$, and

$$\kappa_s = 2 \left(\frac{k_H}{H} + \frac{k_C}{C} \right). \tag{14.44}$$

Now, we have to substitute (14.42) in (14.41) if the bilayer is bent. In the sequel, we assume that there is no stretching, because the bilayers are liquid like. This condition can be written as $\tau = \tau^+ + \tau^- = 0$. Here, $\tau^i = \partial f_m^i/\partial \varepsilon^i$ is the tension that develops in the $i = +, -$ monolayer. We also assume that the two monolayers can exchange molecules (flip-flop); consequently, the chemical potential is the same in the two monolayers; i.e., $\mu(\tau^i, \sigma_1^i, \sigma_2^i) = f_m^+ - \tau^+ A^+ = f_m^- - \tau^- A^-$. Identifying the variation of $\sum_{i=+,-} f_m^i/A^i$ with the expression of the free energy of a bilayer (14.39), one gets[14]

$$\kappa = \frac{\delta^2 A_0}{8} \frac{k_H k_C}{H^2 C^2} \frac{1}{k_H H^{-2} + k_C C^{-2}} \equiv \frac{\delta^2}{16} \kappa_s \frac{(H - A_0)(A_0 - C)}{(H - C)^2};$$

$$\bar{\kappa} = -\frac{3\delta^2 A}{8} \frac{k_H k_C}{H^2 C^2} \frac{1}{k_H H^{-1} + k_C C^{-1}} (H - C) \equiv -\frac{3\delta^2}{16} \kappa_s \frac{(H - A_0)(A_0 - C)}{A_0(H - C)}. \tag{14.45}$$

It is interesting to notice that, although κ is always positive (which implies that A_0 is in the interval $[H, C]$), the sign of $\bar{\kappa}$ depends on the sign of the asymmetry of the molecule (H–C). Note also that the knowledge of κ, $\bar{\kappa}$, and κ_s and of the shape and sizes of the molecule (i.e., H and C) permits one to evaluate the local elastic constants.

Another quantity that can be expressed as a function of the microscopic elastic constants is the membrane edge energy for a pore of radius R (Fig. 14.5). One starts from the equality of the chemical potentials in the flat part of the layer and the bent part, which reads

---

[14]M.D. Mitov, PhD thesis, Bulgarian Acad. Sci, Sofia, 1980; A.G. Petrov and I. Bivas, Progress Surf. Sci., Edited by S. Davison, **16**, 389 (1984).

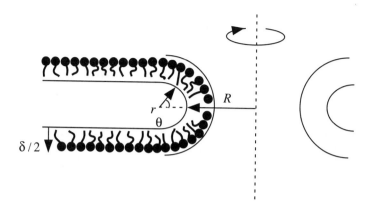

Figure 14.5. Model for the edge of a pore; the calculation is done for $r \to 0$.

formally as

$$\mu(\tau/2, 0, 0) = \mu(\tau^+, \sigma_1, \sigma_2), \quad \sigma_1 = \frac{\cos\theta}{R + r(1 - \cos\theta)}, \quad \sigma_2 = \frac{1}{r}, \tag{14.46}$$

i.e., yields τ^+ as a function of the applied stress τ, which tends to open the pore. The force per unit length of line exerted on the edge is given by the integral:

$$\gamma(\tau, R) \underset{r \to 0}{=} r \int\limits_{-\pi/2}^{+\pi/2} \tau^+(\tau, R, r, \theta)\, d\theta, \tag{14.47}$$

whose calculation yields an expansion of γ in powers of δ/R, the coefficients τ_i, γ_i depending on the microscopic elastic constants:

$$\gamma = \gamma_0 \left(1 + \frac{\tau}{\tau_0}\right) - \gamma_1 \left(1 + \frac{\tau}{\tau_1}\right)\frac{\delta}{R} + \gamma_2 \left(1 + \frac{\tau}{\tau_2}\right)\left(\frac{\delta}{R}\right)^2 + \cdots. \tag{14.48}$$

14.2.1.3. Microscopic Models; the Helfrich Model

The Helfrich model[15] (extended by Szleifer et al.[16]) relates the values of κ_m, $\bar{\kappa}_m$, and σ_0 to the distribution of stresses in the monolayer:

---

[15]W. Helfrich, Amphiphilic mesophases made of defects, in Physics of defects, Edited by R. Balian, M. Kleman, and J.P. Poirier, Les Houches, North-Holland, Amsterdam, 1980, 714.

[16]I. Szleifer, D. Kramer, A. Ben Shaul, W.H. Gelbart, and S.A. Safran, J. Chem. Phys., **92**, 6800 (1990).

$$\kappa_m \sigma_0 = \int x p_t(x)\, dx, \quad \overline{\kappa}_m = -\int x^2 p_t(x)\, dx,$$

$$\kappa_m = -\frac{1}{2} \int x\, dx \left. \frac{\partial p_t(x)}{\partial H} \right|_{H \to 0}. \tag{14.49}$$

The first and second formulae are valid for a planar layer; the third one applies to a bent surface with mean curvature H. The integrand must include all forces, in particular, those calculated in Section 14.1.3. The principle of the calculation is as follows (see also Safran[17]): The free energy variation of an *isotropic compressible liquid* changing volume from V_0 to V is $\Delta F = -\int p(V')dV'$, where $p(V')$ is the pressure in the state $V_0 \le V' \le V$; this expression generalizes for an anisotropic liquid (and for a solid) to the form $\Delta F = \int V' \sigma'_{ij} de'_{ij}$ (note that $dV' = V' de'_{ii}$; the e'_{ij} are the components of the elastic distorsion tensor). In the case of a 2D isotropic membrane, as already indicated, the relevant components of the stress tensor σ'_{ij} are expressed as functions of the longitudinal and the transversal "pressures" p_l and p_t. The longitudinal pressure opposes the work done by forces acting along the normal to the membrane, when the membrane changes thickness, and the transversal pressure opposes forces that bend the membrane and change the local area. The calculation assumes, furthermore, that the total volume is unchanged. Under such conditions, the quantity ΔF calculated for a unit membrane area is of the form of (14.39), with the moduli as in (14.49). The longitudinal pressure does not appear, because we expect it to vanish for an isolated membrane.

These formulae have been used successfully to calculate the contribution of electrolytes in solvent and charged layers on the elastic moduli for isolated membranes[18] and for lamellar systems[15] by taking for the transversal stress $p_t(x)$ the values given by (14.21). According to (14.49), these contributions to the moduli scale as the product of a pressure difference times a characteristic length cubed. For example,[15] in the case of an added electrolyte, one expects that the characteristic length is the Debye length λ_D, the pressure difference scales as $k_B T / b^2 \ell$ in the weak charge limit, and the moduli as $k_B T \lambda_D^3 / b^2 \ell$. The last quantity can be fairly negligible compared with other contributions, to be discussed later on (note that the repeat distance plays no role). On the other hand, in the case of no added electrolyte and strong surface charge, the pressure scales as $k_B T / d^2 \ell$, the characteristic distance is the repeat distance, and the moduli scale as $k_B T d / \ell$.

14.2.1.4. Experimental Values

κ_s: The stretching elastic constant that relates the 2D isotropic stress component τ (usually called a tension: It is a 2D pressure changed sign) to the relative variation of area ε, $\tau = \kappa_s \varepsilon$, has been measured in giant bilayer lecithin vesicles. The method consists in inflating the vesicle under a controlled overpressure, at constant molecular content of the

---

[17]S.A. Safran, Adv. in Phys. **48**, 395 (1999).
[18]H.N.W. Lekkerkerker, Physica **A159**, 319 (1989); Physica **A167**, 384 (1990).

vesicle.[19] Typical values for κ_s are 140 mN/m in lecithin, but 450 mN/m for the red blood cell membrane. For more details on the micromechanical (and also thermoelastic) properties of lecithin and biomembranes, see, e.g., Evans and Skalak[20] and Parsegian et al.[21]

κ: The bending modulus has been measured from the shape fluctuations of tubular and spherical vesicles. One typically gets $10k_BT$ for long, double-chain lipids,[14] and $0.1k_BT$ for short chains and mixtures. These low values of κ mean that thermal energy suffices to excite extensive shape fluctuations of amphiphilic films.[22,23] For a recent review see Safran.[17]

$\overline{\kappa}$: The PDM model predicts that the sign of the saddle-splay modulus $\overline{\kappa}$ depends on the shape anisotropy $(H - C)/C$. If the head cross section is larger than the chain cross section, one expects $\overline{\kappa} < 0$ for the bilayer. Note that this situation favors a positive spontaneous curvature for the monolayer. This is indeed what has been observed. This conclusion is also in agreement with (14.40), which predicts that the bilayer saddle-spay constant is smaller than twice the monolayer saddle-splay constant when $\sigma_0 > 0$. Reciprocally, one expects $\overline{\kappa} > 0$ when the spontaneous curvature is negative, viz. when the head cross section is smaller than the chain cross section. This is typically what happens in some swollen colloidal solutions of surfactants, namely, cetylpyridinium chloride plus hexanol as a cosurfactant, where it is believed that the chains are interdigitated (large C), whereas the heads show increasing area condensation when the concentration of alcohol increases. This process drives a transition between a lamellar L_α phase, at lower alcohol content, and a sponge L_3 phase, at higher alcohol content.[24] In this system, the value of $\overline{\kappa}$ controls the stability of large Burgers vectors dislocations (oily streaks) vs their splitting into focal conic domains: The topology of the core of the defect is not the same according to the case, which implies a role of $\overline{\kappa}$.[25] The value of $\overline{\kappa}_m$ can be obtained in oil-brine surfactant mixtures forming dilute droplet microemulsions (with a radius of a few nm), from the knowledge of the interfacial tension, the bending modulus, and the radius of the droplets (AOT–brine–oil system[26]).

Numerous lyotropic systems have been studied,[27] either with dominant electrostatic forces (SDS–water–pentanol, SDS-water-hexanol), or dominant undulation forces (SDS–brine –dodecane or pentanol). The experimental scattering profiles allow for a clear-cut distinction between these two categories of lamellar systems. Intermediate cases between undulation and electrostatic forces have been investigated.

The Caillé exponents (see Section 5.6.2), which appear in the shape of the diffraction peaks of a lamellar phase, permit a direct measurement of the quantity $\sqrt{KB}$. On the

---

[19]R. Kwok and E.A. Evans, Biophys. J. **35**, 637 (1981).

[20]E.A. Evans and R. Skalak, Crit. Rev. Bioeng., **3** and **4**, 181 (1979).

[21]V.A. Parsegian, N. Fuller and R.P. Rand, Proc. Nat. Acad. Sci. USA **76**, 2750 (1979).

[22]R.M. Servuss, W. Harbich, and W. Helfrich, Biochimica et Biophysica Acta **436**, 900 (1976).

[23]F. Brochard and J.F. Lennon, J. Physique (Paris) **36**, 1035 (1975).

[24]G. Porte, J. Appel, P. Bassereau, and J. Marignan, J. Phys. France **50**, 1335 (1989).

[25]P. Boltenhagen, O. Lavrentovich and M. Kleman. J. Phys. II, France **1**, 1233 (1991).

[26]H. Kellay, J. Meunier, and B.P. Binks, Phys. Rev. Lett. **70**, 1485 (1993).

[27]D. Roux, C.R. Safinya, and F. Nallet, Lyotropic lamellar L_α phases', in Micelles, Membranes, Microemulsions, and Monolayers, Edited by W.M. Gelbart, A. Ben-Shaul, and D. Roux, Springer-Verlag, New York, 1994.

other hand, in the case of L_α phases, the measurement of the periodicity in function of the dilution yield the value of $\kappa \approx K d_0$ [see (14.60)]. Therefore, K can be measured by two independent methods. Furthermore, the Caillé exponent exhibits in theory a different variation with d_0 whether the repulsive forces in the membranes are predominantly of electrostatic or of entropic (see below) origin.[27]

The following list of constants for lecithin is extracted from Mitov's thesis and Petrov and Derzhanski,[28] and has been collated by these authors from various measurements[16,20,29] ($\kappa = 0.26 \times 10^{-19}$ J, $\kappa_s = 0.14 \pm 0.2$ J/m^2, $\gamma_0 = 0.23 \times 10^{-11}$N)

$$k_H = 19.1 \times 10^{-21} \text{ J}, \qquad k_C = 32.3 \times 10^{-21} \text{ J},$$
$$H = 90.2 \times 10^{-20} \text{ m}^2, \qquad C = 66.2 \times 10^{-20} \text{ m}^2.$$

The value of $\bar{\kappa}$ can be deduced from these values of the microscopic constants, following (14.45): $\bar{\kappa} \approx 0.25 \times 10^{-19}$ J. A value of the same order of magnitude has also been inferred from light microscopy observation of saddle-splay instabilities[30] and passages.[31]

14.2.2. Flexible Layers and Excluded Volume

The stability of the L_1 (micellar), L_α, and L_3 (bilayer) phases can be discussed at a phenomenological level starting from the free energy surface density of an isolated membrane [see (14.39)]. The L_α lamellar phase is stable in the range of values $2\kappa + \bar{\kappa} > 0, \bar{\kappa} < 0$. We do not discuss the stability of the L_3 sponge and micellar phases. How do fluctuations enter in this picture?

Equation (14.39) is the same as the classic free energy surface density of an inextensional solid shell, with the following correspondence:

$$\kappa \Leftrightarrow \frac{Eh^3}{12(1-v^2)}, \qquad \bar{\kappa} \Leftrightarrow \frac{-Eh^3}{12(1+v)}, \tag{14.50}$$

where h is the thickness of the shell, E is the Young modulus, and v is the Poisson ratio.[32] The main differences between the two cases are as follows:

1. The inextensionality of the solid shell reduces its possible deformations to those that conserve lengths and angles, i.e., that conserve the Gaussian curvature $G = \sigma_1 \sigma_2$. The condition is less strict in a membrane; inextensionality is replaced by conservation of area div $\mathbf{u}_\perp = 0$, and Gaussian curvature can be modified by diffusion.

---

[28] A.G. Petrov and A. Derzhanski, Mendeleiev J. All Union Chem. Soc. (Moscow, in Russian) **28**, 197 (1983).

[29] W. Harbich and W. Helfrich, Z. Naturforsch. **34a**, 1063 (1979).

[30] A.G. Petrov, M.D. Mitov, and A. Derzhanski, Phys. Lett. **65A**, 374, (1978).

[31] W. Harbich, R.M. Servuss, and W. Helfrich, Z. Naturforsch. **33a** 1013 (1978).

[32] A.E.H. Love, A Treatise on the Mathematical Theory of Elasticity, Dover, 1944, article 298.

2. The fact that the bilayer suffers considerable thermal fluctuations (e.g. in systems in which electrostatic interactions are strongly screened by a strong electrolyte solution, κ is not much more than of the order of $k_B T$), which have to be averaged out in this large-scale description. Consequently, the bending modulus κ and the saddle-splay modulus $\bar{\kappa}$ are moduli renormalized over thermal fluctuations.[33] The integration $\int f \, dA_p$ for the total energy is to be carried over a *mean surface* A_p averaged over the thermal fluctuations.

14.2.2.1. The de Gennes Coherence Length

We first consider the case of an *isolated*, flat membrane, of area $A = L^2$, but crumpled by thermal fluctuations, without interaction with other membranes. de Gennes and Taupin[34] have shown that such a *free membrane* has a finite persistence length

$$\xi_\kappa = a \exp \frac{4\pi\kappa}{\alpha k_B T}, \tag{14.51}$$

i.e., that it makes sense to speak of a flat membrane as long as its size $L \leq \xi_\kappa$. In this expression, α is a coefficient of the order of unity, and a is the transversal size of a molecule of surfactant in the membrane. The demonstration goes as follows.

Let us assume, for the sake of simplicity of the calculations, that the amplitudes of the fluctuations are small. We can therefore use the small displacement approximation of the free energy density; that is, $f_{\text{bend}} = \frac{1}{2}\kappa(u_{xx} + u_{yy})^2$; in the same approximation the integration can be carried over the projected, flat, mean area, and not the true membrane area. There is no necessity to introduce the saddle-splay contribution, because the fluctuations do not change the topology of the surface. Let us now write the total energy in terms of the Fourier components of the displacement $u(\mathbf{r}) = A^{-1} \sum_\mathbf{q} u_\mathbf{q} \exp(i\mathbf{q} \cdot \mathbf{r})$ (with $u_{-\mathbf{q}} = u_\mathbf{q}^*$). One gets

$$\int f_{\text{bend}} d^2\mathbf{r} = \frac{1}{2}\kappa \int (u_{xx} + u_{yy})^2 d^2\mathbf{r}$$

$$= \frac{A^{-1}}{2}\kappa \sum_\mathbf{q} q^4 u_\mathbf{q} u_{-\mathbf{q}} = \frac{A^{-1}}{2}\kappa \sum_\mathbf{q} q^4 |u_\mathbf{q}|^2. \tag{14.52}$$

By the equipartition theorem, we have

$$k_B T = A^{-1}\kappa \mathbf{q}^4 |u_\mathbf{q}|^2, \tag{14.53}$$

---

[33] For a review see e.g., Statistical Mechanics of Membranes and Surfaces, Edited by D. Nelson et al., World Scientific, Singapore, (1989). The main results of the renormalization group approach in the calculation of κ and $\bar{\kappa}$ have recently been challenged by Helfrich, Eur. Phys. J. **B1**, 481 (1998).

[34] P.G. de Gennes and C. Taupin, J. Phys. Chem. **86**, 2294 (1982).

and by a simple integration over all the modes,

$$\langle u(\mathbf{r})^2 \rangle = \frac{1}{A} \int u(\mathbf{r})^2 d^3\mathbf{r} = \frac{1}{A^2} \sum_{\mathbf{q}} |u(\mathbf{q})|^2 = \frac{1}{A} \frac{k_B T}{\kappa} \sum_{\mathbf{q}} \frac{1}{q^4}. \tag{14.54}$$

These formulae of general use being established, let us calculate the director–director correlation function

$$g_\mathbf{n}(\mathbf{r}) = \left\langle (\mathbf{n}(\mathbf{r}+\rho) - \mathbf{n}(\rho))^2 \right\rangle = 2 \left(\left\langle |\mathbf{n}(\mathbf{r})|^2 \right\rangle - \langle \mathbf{n}(\mathbf{r}+\rho)\mathbf{n}(\rho) \rangle \right). \tag{14.55}$$

Note that the correlation function measures the angle ψ between two normals at a distance r. Indeed, $g_\mathbf{n}(r) = 2(1 - \langle \cos \psi(r) \rangle) \approx \langle \psi^2(r) \rangle$. Because, further, $\mathbf{n} = \{-\partial u/\partial x, -\partial u/\partial y, 1\}$, we have

$$g_\mathbf{n}(\mathbf{r}) = \frac{2}{A^2} \sum_{\mathbf{q}} q^2 \left\langle |u_\mathbf{q}|^2 \right\rangle (1 - \cos \mathbf{q} \cdot \mathbf{r}) \approx \alpha \frac{k_B T}{4\pi \kappa} \ln \frac{r}{a}. \tag{14.56}$$

The first relation in (14.56) is relatively easy to establish. The second one requires some care in the summation, which is replaced, as usual, by an integration. We have, straightforwardly, $2A^{-2}\Sigma q^2 \langle |u_\mathbf{q}|^2 \rangle \approx (kT/\pi\kappa) \ln q_{max}/q_{min}$. Writing $\int \cos \mathbf{q} \cdot \mathbf{r}\, d\theta = \int \cos(qr \cos \theta)\, d\theta = 2\pi J_0(qr)$, the second part of the integral reads as $-(kT/\pi\kappa) \cdot \int dq\, J_0(qr)/q$, where the integration is performed between $q_{min}r = 2\pi r/L$ and $q_{max}r = 2\pi r/a$. $J_0(qr)$ is the Bessel function of zeroth order. The upper limit is large, whereas the first one is smaller than 2π. Therefore, we integrate between $q_{min}r = 2\pi r/L$ and ∞. We get $-(kT/\pi\kappa) \int dq\, J_0(qr)/q = (kTt/\pi\kappa) \ln(q_{min}r/2)+$ terms that are negligible as long as $q_{min}r = 2\pi r/L$ is smaller than unity. All together we get the expression as in (14.56), with α being some coefficient of order unity or more. The important result is that the leading term in the correlation function does not depend on the size of the system. If we fix $r = \xi_\kappa$ such that $g_\mathbf{n}(r) \approx \langle \psi^2(r) \rangle$ is of the order of unity, i.e., the angle is large, it follows that the normals can be considered as decorrelated. This is the origin of the de Gennes–Taupin persistence length. The free bilayer is "rigid" at short scales ($L < \xi_\kappa$) and strongly wrinkled at larger scales ($L > \xi_\kappa$); of course, this does not preclude the existence of (small) thermal fluctuations at smaller scales.

The coefficient α varies according to the authors (Helfrich[35]: $\alpha = 1$; de Gennes and Taupin: $\alpha = 2$; calculations resulting from renormalization methods yield $\alpha = 3$—for a review, see David[36]).

---

[35] W. Helfrich, Z. Naturforsch. **33A**, 305 (1978).

[36] F. David, in '*Statistical Mechanics of Membranes and Surfaces*', edited by D. Nelson & al., World Scientific, Singapore, (1989) 157.

14.2.2.2. The Excess Area in a Fluctuating Membrane

Let A_p be the averaged area,[37] A the true area, and $\Delta A = A - A_p$ be the excess area of a membrane of typical size $L(A = L^2)$, Fig. 14.6. The (local) relative excess area is $\frac{\Delta A(\mathbf{r})}{A(\mathbf{r})} = \mathbf{n} \cdot \mathbf{z} - 1 \approx -\frac{1}{2}[(\frac{\partial u}{\partial x})^2 + (\frac{\partial u}{\partial y})^2]$, where $\mathbf{z}$ is a unit vector perpendicular to the membrane. The total relative excess area can therefore be written

$$\frac{\Delta A}{A} = -\frac{A^{-2}}{2} \sum_{\mathbf{q}} q^2 |u_q|^2;$$

i.e.,

$$\frac{\Delta A}{A} = \frac{k_B T}{4\pi\kappa} \ln \frac{q_{min}}{q_{max}}. \tag{14.57}$$

Note that this expression is valid for a *free* membrane as long as A_p is smaller than $A_p^* = \xi_\kappa^2$, because patches of sizes A_p^* are independent. The larger A_p is still smaller if the membrane is not free, for example, if it belongs to a L_α phase. Let us note $A_p^* = \xi_\perp^2$. According to (14.54), the mean squared displacement u_m^2 in the direction perpendicular to the membrane is of the order of $u_m^2 \sim (k_B T/\kappa)L^2$. Taking $u_m \sim d$, one gets[38] for $L = \xi_\perp$:

$$\xi_\perp = c\sqrt{\frac{\kappa}{k_B T}}d, \tag{14.58}$$

where c is a coefficient whose exact value is author-dependent. In Golubovic and Lubensky,[39] $c = (32/3\pi)^{1/2} \approx 1.84$.

Note φ_s the volume concentration of the membranes (the surfactant) in the lamellar phase, and δ the thickness of the bilayer. One has

$$\varphi_s = \frac{\delta A}{d A_p}, \tag{14.59}$$

Figure 14.6. Averaged A_p and chemical A areas to represent a membrane.

---

[37] Also called the projected area, because this theoretical approach is valid for quasi-planar surfaces.
[38] W. Helfrich and R.M. Servuss, Il Nuovo Cimento **3D**, 137 (1984).
[39] L. Golubovic and T.C. Lubensky, Phys. Rev. **B 39**, 12110 (1989).

where one has to make $A = A_p^*$. Using (14.57), with $q_{max} = 2\pi/a$, $q_{min} = 2\pi/\xi_\perp$, one gets

$$d = \frac{\delta}{\varphi}\left(1 + \frac{k_B T}{4\pi\kappa}\ln\left(c\frac{\delta}{a}\sqrt{\frac{\kappa}{k_B T}}\frac{1}{\varphi}\right)\right) \approx \frac{\delta}{\varphi}\left(1 + \frac{k_B T}{4\pi\kappa}\ln\left(c\frac{d}{a}\sqrt{\frac{\kappa}{k_B T}}\right)\right).$$

$$(14.60)$$

This logarithmic correction to the linear dilution law has been successfully used to measure the bending modulus.[40]

According to David,[41] the fluctuations renormalize the elastic moduli as follows:

$$\kappa(L) = \kappa\left(1 - \frac{3k_B T}{4\pi\kappa}\ln\frac{L}{a}\right), \quad \bar{\kappa}(L) = \bar{\kappa}\left(1 + \frac{5k_B T}{6\pi\bar{\kappa}}\ln\frac{L}{a}\right). \qquad (14.61)$$

Here, κ is the so-called *bare* elastic bending modulus, and it is related to the elastic properties at molecular scale.

14.2.2.3. An Interpretation as an Ideal Gas of Patches; the Helfrich Entropy Repulsion

In a free membrane, patches of area $A_p^* = \xi_\kappa^2$ are assumed to form a 2D ideal gas. In terms of energy density per unit area, this can be interpreted as yielding a surface tension $\gamma = k_B T/\xi_\kappa^2$.

In the case of a lamellar phase, the surface density of patches is $\xi_\perp^{-2}$. Let us adopt a mean field point of view: Consider that the fluctuations of a given membrane of the phase bump the two neighhboring membranes, supposedly immobile. The volume density of patches in the interval of width $2d$ is

$$n(d) = \frac{1}{2\,d\xi_\perp^2} = \frac{c^{-2}}{2\,d^3}\frac{k_B T}{\kappa}.$$

According to the Boyle–Mariotte law, the corresponding pressure is

$$p(d) = n(d)k_B T = \frac{c^{-2}}{2}\frac{(k_B T)^2}{\kappa\,d^3},$$

[40]D. Roux, F. Nallet, E. Freyssingeas, G. Porte, P. Bassereau, M. Skouri, and J. Marignan, Europhys. Lett. **17**, 575 (1992).

[41]F. David, in '*Statistical Mechanics of Membranes and Surfaces*', D. Nelson & al. eds, World Scientific, Singapore, (1989) 157.

and the energy of interaction of these fluctuations with either of the two neighboring membranes is

$$-\int_{\infty}^{d} p(d')\,dd' = \frac{1}{4c^2}\frac{(k_B T)^2}{\kappa\,d^2};$$

i.e. membranes at a separation d suffer a repulsive potential

$$V_{\text{rep}} = \frac{1}{4c^2}\frac{(k_B T)^2}{\kappa\,d^3}$$

(see Helfrich[42]). Another way to obtain this result is to notice that each patch has a thermal energy $k_B T$ (there are two degrees of freedom per patch; the patches are constrained on a 2D manifold), which is equally shared with the two neighboring surfaces, i.e., for each neighbor, an energy $\frac{1}{2}k_B T n(d)\,d$.

Assuming for a while that we regard the potential V_{rep} as a function of the distance $\Delta = d + u$ between the membranes, where d is the equilibrium distance and u is a small displacement,

$$V_{\text{rep}} = \frac{1}{4c^2}\frac{(k_B T)^2}{\kappa\,\Delta^3}.$$

By the definition of the modulus of compressibility, we have

$$\frac{1}{4c^2}\frac{(k_B T)^2}{\kappa\,\Delta^3} = \frac{1}{2}B\left(\frac{u}{d}\right)^2.$$

Hence,

$$B = d^2\frac{d^2 V_{\text{rep}}}{d\Delta^2}\bigg|_{\Delta=d} = \frac{3}{c^2}\frac{(k_B T)^2}{\kappa\,d^3}. \tag{14.62}$$

Many authors take for the coefficient $3/c^2$ the Helfrich value $9\pi^2/64$.

Another way of estimating the form of B is as follows. The penetration length perpendicular to the layers, which is of the order of $\xi_\perp^2/\lambda$ (because deformation extends laterally over a size $\xi_\perp$), should not be larger than the repeat distance d. On the other hand, it makes no sense to make it smaller. Let us therefore write, on the model of Chapter 5, $\xi_\perp^2/\lambda = d$, where $\lambda = \sqrt{K/B} = \sqrt{\kappa/dB}$. This equation yields

$$B \approx c^{-4}\frac{(k_B T)^2}{\kappa\,d^3}.$$

---

[42]W. Helfrich, Z. Naturforsch. **33a**, 305 (1977).

The full statistical–theoretical method for the question just discussed is given in ref. 41 and 42.

14.2.3. The Lamellar, Sponge, and Cubic Phases; Microemulsions

The respective stability of the different phases of *bilayers* has been alluded to in Problem 5.6. We recall the results, which are expressed in function of the signs of the curvature moduli κ and $\bar{\kappa}$, i.e. in function of macroscopic material parameters. There are three domains of stability:

1. for $\kappa + \frac{1}{2}\bar{\kappa} > 0, \bar{\kappa} < 0$, the Gaussian curvature vanishes, and the lamellar phase is stable.

2. for $\kappa + \frac{1}{2}\bar{\kappa} < 0, \bar{\kappa} < 0$, the Gaussian curvature has to be positive, and the micellar phase is stable.

3. for $\kappa + \frac{1}{2}\bar{\kappa} > 0, \bar{\kappa} > 0$ and the Gaussian curvature has to be negative; at the same time, one can make the mean curvature to vanish. The bilayer takes the shape of a minimal surface.

Case (1) also covers the case of a phase made of cylindrical bilayers, like the hexagonal phase. Such a phase is usually a phase of monolayers, like indeed the micellar phase [case (b)]. In both cases, one expects that $\bar{\kappa}_m < 0$. Note that $\bar{\kappa}$ is certainly also negative if the spontaneous monolayer curvature is positive; see (14.40).

The case in which the Gaussian curvature has to be negative is more interesting. It yields two subcases, cubic and sponge phases, which both are observed experimentally:

Cubic Phases: These phases, which appear in the classification of mesomorphic phases (Chapter 2), are known for moderately swollen surfactants.[43] It has been proven[44] that a periodic surface can be periodic in 3D space, without intersections. Experimentally, one encounters three cubic phases, called, respectively, G (space group Ia3d), D (space group Pn3m), and P (space group Im3m), but there are many more possibilities. The reader is referred to the literature for illustrations.[45] The stability is clearly related to the topological gain in energy due to the passages and the vanishing curvature energy (minimal surface $\sigma_1 + \sigma_2 = 0$). It has been proposed that the lamellar to cubic phase transition can be studied in the frame of a Landau theory.[46]

---

[43] Strontium soaps, Ia3d structure: V. Luzzati and P.A. Spegt, Nature **215**, 701 (1967); lipid-water systems, Ia3d structure: V. Luzzati, A. Tardieu, T. Gulik-Krzywicki, E. Rivas and F. Reiss-Husson, Nature **220**, 485 (1968).

[44] A.H. Schoen, NASA report TN D - 5541 (1970).

[45] See for example the special issue of the Journal de Physique on Geometry and Interfaces, colloquium C7 **51**, (1990).

[46] D.M. Anderson, S.M. Gruner and S. Leibler, Proc. Natl. Acad. Sci. USA **8**, 5364 (1988). W. Helfrich and H. Rennschuh, J. de Phys. **51**, C7-189 (1990).

Sponge Phase: The L_3 phase is documented in swollen surfactants, for which the curvature energy is known to be very small, as indicated above ($\kappa \approx k_B T$). The model of the sponge phase that is used is that of a bicontinuous random phase of solvent (two domains of solvent) separated by a random bilayer, similar to the model for microemulsions. Being random, it can be shown on mathematical grounds that the midsurface of the bilayer can be but only approximately minimal. The stability is ensured by the translational entropy of the configurations of passages, and the curvature energy is small.[47]

Finally, **microemulsions** are thermodynamically stable phases with at least three components: oil, water, and surfactant (most generally with a short alcohol acting as cosurfactant). The surfactant–cosurfactant form a *monolayer* that separates oil from water. One distinguishes globular (mostly spherical) microemulsions, whose existence is fairly well documented, and Winsor phases, whose model is that one of a random bicontinuous system, first advanced by Scriven[48] and Talmon and Prager.[49] The microemulsions can thus feature either a positive Gaussian curvature (microdroplets) or a negative Gaussian curvature (structurally similar to sponge phases). It is believed[50] that a discriminating factor between globular and random bicontinuous microemulsions is the spontaneous curvature σ_0, which is large in globular systems, and smaller and associated with a small bending modulus κ_m in random continuous systems. A number of theoretical models have been elaborated on to explain the stability of microemulsions. They are set out in great details in ref. [51]

14.3.   Solutions of Colloidal Particles; Stability Properties

The former sections have been devoted essentially to membranes and, by extension, to surfaces whose mean curvature is small compared with the inverse of the distance between two colloidal objects in solution. The core of the results is the DLVO theory for the competition between attractive dispersion forces (van der Waals forces, suitably modified for retardation effects at large distances, if necessary; see Chapter 1 and Problem 1.6) and repulsive electrostatic forces. At first sight, these results can be applied to moderately concentrated solutions of particles. But the two situations display important differences. The finite objects we now have in mind do not have the same topology as the infinite membranes of swollen surfactants. Because particles are of finite, small size (typically in the range 50 nm to 1μm), Brownian motion is an important factor in their stability and kinetic properties. Also, because the question of colloidal solutions has long been of such an in-

---

[47]M.E. Cates, D. Roux, D. Andelman, S.T. Milner, and S.A. Safran, Europhys. Lett. **5**, 733 (1988).
[48]L.E. Scriven, Nature **63**, 123 (1976)
[49]Y. Talmon and S. Prager, J. Chem. Phys. **69**, 2984 (1978)
[50]L. Auvray, The structure of microemulsions: experiments' in Micelles, in Membranes, Microemulsions, and Monolayers, Edited by W.M. Gelbart, A. Ben-Shaul, and D. Roux, Springer-Verlag, New York, 1994.
[51]Membranes, Microemulsions, and Monolayers, Edited by W.M. Gelbart, A. Ben-Shaul, and D. Roux, Springer-Verlag, New York, 1994.

dustrial importance (milk, inks, paints, etc., are colloidal solutions of particles), a wealth of experimental results are known, and significant theoretical efforts were already numerous in the beginning of the twentieth century, e.g., due to Smoluchowski and Einstein. The subject has therefore been treated in length in many excellent treatises and textbooks, and we shall therefore content ourselves with a somewhat simplified discussion, in which we shall put the accent on the question of the stability of the solutions of colloidal particles.

Colloidal solutions are of several types, according to the nature of the repulsive interactions between particles:

1. *Hard sphere interactions*, of great theoretical interest, can be experimentally modeled by silica spheres, dispersed in a nonpolar solvent whose refractive index matches the refractive index of silica (hence, no van der Waals nor electrostatic interactions), coated with a thin organophilic layer.[52] Brownian movement has to counterbalance sedimentation; hence, the particles must be small enough ($a^4 \Delta \rho g / k_B T < 1$; radius $a \leq 1 \mu$m; difference of mass density between the solvent and the particles $\Delta \rho$).

2. *Repulsive electrostatic interactions* (charged particles, e.g., polymeric, calibrated charged spheres in an electrolyte, often referred to as *latex* solutions, the word latex previously connoting the hevea sap from which rubber was manufactured). We recall that the DLVO potential (Fig. 14.2) displays a deep primary minimum, when the distance between the objects is very small, separated from a secondary minimum by a maximum that decreases when the concentration of electrolyte increases. The distance of approach of two *membranes* is in principle at the secondary minimum V_{smin}, either because when the electrostatic repulsive interactions are strong, the activation barrier opposed by the maximum is extremely high, or when the barrier is abolished by an electrolyte, the repulsive Helfrich fluctuations take over. On the other hand, in the case of *particles*, the primary minimum V_{pmin} is easier to maintain, if the activation barrier V_{max} has been passed, because the contact has to be effective in one point only. This aggregated state is irreversible if $|V_{\text{pmin}}|/k_B T$ is large enough, the time of escape being of the order of $a^2 D^{-1} \exp(|V_{\text{pmin}}|/k_B T)$ (where D is the diffusivity).

3. *Polymer-coated* particles. Repulsive interactions set in between polymer coatings of the particles (see Chapter 15).

For high enough concentrations, or in the presence of an agent that decreases the repulsive interactions (e.g., electrolyte, or polymer added to the solvent—see Section 14.3.2) the particles can form a *crystal*, a *glass*, or a solid-like *gel* phase. To summarize, the physics of colloidal particles is a physics of *aggregation* vs *solubilization*, and of *order* vs *disorder*. These are the topics that we discuss, briefly, in the sequel.

---

[52]C.G. de Kruif et al., in Complex and Supramolecular Fluids, Edited by S.A. Safran and N.A. Clark,Wiley-Interscience, New York, 1987, p. 673.

A problem in colloidal structure physics of a similar nature, besides the effects of anisotropy, is when the objects are rigid rods (like TMV viruses), or semiflexible polymers, like DNA, biopolymers, and so on. This situation will be briefly discussed in the next chapter.

14.3.1. Brownian Flocculation

Flocculation is a generic term often used to conote the various phenomena that yield the formation of loose, porous aggregates, or *flocs*. This happens, for example when the contacts between particles are provided by Brownian motion, and the primary minimum is deep enough compared with $k_B T$, so that the attractive force is large, and most of the contacts irreversible.

Rapid Brownian flocculation: It happens when at the same time the maximum of the pair potential V_{max} is negligible compared with $k_B T$. The only limiting process is diffusion. Consider one particle, supposed immobile, and count the number of contacts C it experiences per unit of time. This can be estimated by using Fick's first law of diffusion (see Chapter 8). The other particles move isotropically around the test particle, so that we have

$$C = 4\pi r_c^2 D \frac{dn}{dr},\qquad(14.63)$$

where $r_c \approx 2a$ is the distance at which the other particles (of the same radius a) touch the test particle, and D is the diffusivity. The gradient of concentration that is effective is $\frac{dn}{dr} \approx \frac{n}{a}$, where n is the density of particles. Therefore, the kinetic equation for the decrease of individual particles is

$$\frac{dn}{dt} = -16\pi a D n^2.\qquad(14.64)$$

The density of particles decreases to half its value in a time $t^* \approx (8\pi D a n_0)^{-1}$, which for $D = k_B T/(6\pi \eta a)$ (Stokes' law) is of the order of at most a few seconds, taking the initial concentration $n_0 \approx 10^{12}$ cm^{-3}. The kinetics of aggregation is very rapid.

Stability ratio: In most cases, such large rates of aggregation are not reached. This is due in part to the fact that the diffusion coefficient does not obey the Stokes law at short distances of separation (where the flow of solvent is more difficult), and, above all, to the repulsive forces at short distances. In order to describe quantitatively this latter effect, one introduces the stability ratio $W = J_0/J$, where $J_0 = 16\pi a D n_0^2$ is the (Brownian) rate of aggregation of individual particles, and J is the true rate, which is related to the details of the pair potential of interaction. Let $V(r)$ be the pair potential of interaction between two particles, $m = (6\pi a\eta)^{-1}$ the mobility, $D = mk_BT$ is the diffusivity according to Einstein relation. Equation (14.63) becomes

$$C = 4\pi r_c^2 \left(D\frac{dn}{dr} + mn\frac{dV}{dr} \right).$$ (14.65)

Assuming a steady state of aggregation (C constant), this equation can be integrated to

$$n(r) = n_0 \exp(-V(r)/k_BT)$$
$$+ C\frac{\exp(-V(r)/k_BT)}{4\pi D} \int_\infty^r \exp(-V(r')/k_BT)\frac{dr'}{r'^2},$$ (14.66)

where n_0 is the concentration far from the particle at the origin, and the boundary condition is $n(2a) = 0$. From (14.66), it is possible to deduce an expression for $W = J_0/J$. An approximate formula used in the case of electrostatic repulsion,

$$W = (2ak_D)^{-1} \exp(V_{max}/k_BT),$$ (14.67)

yields values as high as $W \sim 10^5$ if $V_{max}/k_BT \sim 15$, with $k_D \sim 10^8$ m^{-1}. Such values are experimentally observed.

Shape of the flocs: It is often fractal.[53] We refer the reader to the discussion of the structural properties of DLA (diffusion limited aggregation) in Chapter 7.

14.3.2. Depletion Flocculation

Two colloidal particles in a solution containing smaller particles suffer an attractive force, whose origin has first been recognized by Asakura and Oosawa[54], for the case when the subparticles are macromolecules. This phenomenon, called depletion flocculation, can be qualitatively understood as follows. Let n be the number density of subparticles, and consider (Fig. 14.7) a region of size h, area A, from which the subparticles are excluded, because h is smaller than or of the same order as the size of the subparticles. The Gibbs adsorption equation (13.13) applies to this case, with $2\Gamma A = -nhA$, where Γ is the excess number (here negative) of subparticles per unit area in the gap

$$\frac{d\sigma}{d\mu} = -\Gamma(h).$$ (14.68)

The force exerted by one particle on the other is $f = -2\frac{d\sigma}{dh}$. Assuming that the solution of subparticles is ideal, i.e., $\mu = k_BT \ln c + const$, μ being the chemical potential of the

---

[53] See e.g., S. Dimon et al. Phys. Rev. Lett. **57**, 595 (1986).
[54] S. Asakura and F. Oosawa, J. Chem. Phys. **22**, 1255 (1954).

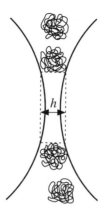

Figure 14.7. Mechanism for depletion flocculation. Redrawn from D.H. Everett, Basic Principles of Colloid Science, The Royal Society of Chemistry, 244 pp, 1988.

subparticle, we have

$$f = -ck_B T A. \tag{14.69}$$

The colloidal particles appear therefore as being submitted to an osmotic pressure $p_{\text{osm}} = ck_B T$. The gaps between them act as regions forbidden to the particles, as if they were separated from the bulk of the solution by hemipermeable membranes.

   These considerations extend to polymers that do not adsorb on the surface of the particles, with some modification in the values of the coefficients in the formulae above. It will be seen in the next chapter that an isolated, flexible, polymer chain takes the shape of a random coil, as a result from the competition between entropy and excluded volume interactions. The region in space occupied by the random coil has a typical radius R_G, say. If two particles are separated by a gap smaller than $\sim 2R_G$, the nonabsorbing macromolecules present in this region are squeezed, at the expense of some loss in entropy. They regain entropy by escaping from this region. This contribution of the internal degrees of freedom of the subparticle adds to the depletion effect described above for undeformable subparticles, and is taken into account in Joanny et al.[55]

14.3.3. Stability Under Shear; Rheological Properties

At low shear rates and low volume fraction ϕ, the effective viscosity η of a stabilized solution obeys Einstein relation[56]

[55]J.-F. Joanny, L. Leibler, and P.G. de Gennes, J. Poly. Sci.: Poly. Phys. Ed. **17**, 1073 (1979).

[56]A. Einstein, Ann. der Physik, **19**, 289 (1906); English translation in Investigations on the Theory of the Brownian Movement, Dover, 1956.

$$\frac{\eta}{\eta_s} = 1 + 2.5\phi + O(\phi^2), \tag{14.70}$$

where η_s is the viscosity of the solvent. This increase in viscosity originates in the extra dissipation due to the distorsion of the flow field $\dot{\gamma} = $ const. in the vicinity of the particles, along which the no-slip condition applies. Terms of higher order in (14.70) become important for higher concentrations: They are due to interactions involving pairs (ϕ^2), three particles (ϕ^3), and so on. The stability under shear and viscoelastic behavior of colloidal suspensions depends crucially on the nature of the interactions between particles and on their concentration (see Russel et al. and Larson textbooks for details). The coefficients of the ϕ^2, ϕ^3, and so on, terms in (14.70) depend on the nature of these interactions.

1. In the case of *hard spheres*, the effect of pair interactions at low-volume fraction ϕ has been calculated[57]

$$\frac{\eta}{\eta_s} = 1 + 2.5\phi + 6.2\phi^2 + O(\phi^3) \tag{14.71}$$

and is experimentally well documented. For all volume fractions, dimensional analysis anticipates the existence of a relation between the dimensionless ratio η/η_s and the dimensionless *Péclet number* $P_e = a^2\dot{\gamma}/D$ (which is the ratio of a time a^2/D characteristic of the Brownian motion over a time characteristic of the applied shear), at constant volume fraction, because there are no other interactions than those due to the excluded volume.[58] Indeed it has been shown experimentally that the variation of η/η_s can be represented along a unique master curve in function of the dimensionless quantity $\eta a^3\dot{\gamma}/k_BT$, which is equivalent to P_e (Fig. 14.8). One observes that the effective viscosity at constant volume fraction decreases (*shear thinning*) as the shear rate is increased. This dependence is experimentally visible for $\phi \geq 0.3$. Shear-thinning is due to the disappearance of the Brownian motion contribution at high shear stress (computer simulations[59]), and to the formation of strings of particles parallel to the flow (light scattering experiments[60]). However, these strings become unstable at still larger shear rates (or shear stresses), and *shear thickening* can take place.

2. In the presence of other interactions (*soft spheres*), the analysis of the results is more involved.

The effect of shear on the *stability* of the solution has received some attention on the theoretical side, but the comparisons with experimental results are scarce or elusive. At relatively small shear rates, i.e., $P_e \ll 1$, the number of collisions between particles due to

[57]G.K. Batchelor, J. Fluid Mech. **83**, 97 (1977).

[58]I.M. Krieger, Adv. Colloid Interface Sci. **3**,111 (1972).

[59]J.F. Brady, Curr. Opin. Collloid Interface Sci. **1**, 472 (1996).

[60]B.J. Ackerson and P.N. Pusey, Phys. Rev. Lett. **61**, 1033 (1988).

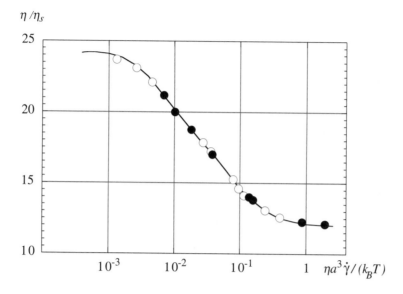

η/η_s

Figure 14.8. Variation of the relative viscosity η/η_s in function of the dimensionless number $\eta a^3 \dot{\gamma}/k_B T$ at constant volume fraction ϕ. Shear thinning. Adapted from Krieger[58]; $\phi = 0.40$.

Brownian motion exceeds the number of collisions due to the shear flow: The shear affects the results of the static case (Section 14.3.1) by a correction term to the stability ratio W, which is slightly reduced with respect to the static case (see Russel et al. textbook for elements of the theory). The increase in the number of pairs should result in an increase of the viscosity. At large shear rates, i.e., $P_e \gg 1$, larger aggregates can form, whose stability is related to the secondary minimum of the DLVO potential (smaller shear rates) or the primary minimum (larger shear rates).

In the case of *stabilized solutions* of charged particles, the theory[61] for small ϕ predicts that the viscosity can be described as in the hard sphere case, but with an effective particle diameter d_{eff} that is representative of the repulsive forces (see Larson textbook for details). One gets

$$\frac{\eta}{\eta_s} = 1 + 2.5\phi + \left[2.5 + \frac{3}{40} \left(\frac{d_{\text{eff}}}{a} \right)^5 \right] \phi^2 + O(\phi^3). \tag{14.72}$$

At low ionic strength, the theory predicts $d_{\text{eff}} \gg 2a$. The exponent 5 therefore yields a rapid increase of the effective viscosity with the volume fraction, due to the electrostatic interactions. This has indeed been observed. At large volume fractions ($\phi > 0.1$, say) and high shear rates, the hydrodynamic interactions become more important than do the

---

[61] W.B. Russel, J. Fluid Mech. **85**, 209 (1978).

electrostatic ones, d_{eff} decreases, and one gets a viscosity close to the viscosity anticipated for a system of hard spheres.

14.3.4. Order vs Disorder

Colloids offer remarkable examples of phase transitions that are of a fundamental interest. Contrarily to the case of solids, it is relatively easy to tune the interactions in the same system in order to cover a rich variety of bulk phases, from gas to 3D crystalline organizations.[62] The model systems that have been thoroughly studied are particles of gold, of silica, polystyrene in water, of polymethylmetacrylate (PMMA) in oil, binary colloidal mixtures like PMMA particles in polystyrene, and so on. In all of these cases, the size of the particles is small enough so that the Brownian motion is effective to prevent sedimentation. We discuss briefly only the effect of *repulsive* interactions on the stability of colloidal crystals. Recent experiments[63] have suggested the existence and role of attractive interactions between like charges in confined colloidal media, possibly yielding crystallization. That is to say that the subject of colloidal solutions and colloidal crystals is in full renewing.

14.3.4.1. Repulsive Forces; Hard Spheres

An essential element to understand the competition beween order and disorder is *entropy*. In a purely repulsive medium like a system of hard spheres, the reason for crystallization (the so-called Kirkwood–Alder transition) would be that the entropy of vibration of the particles in their cages, when ordered, can be larger than the entropy of translation of the liquid state. This is to be compared with the Onsager's model of liquid-nematic transitions in hard rods systems (Chapter 4), which also emphasizes repulsive interactions and the associated effects of the orientational-dependent excluded volume. The Kirkwood–Alder transition has been computed in molecular dynamics simulations[64] and observed[65] in a suspension of polystyrene spheres in a fluid so chosen that the van der Waals attractions were suppressed. At the transition, the volume fractions are for the liquid $\phi_{liq} = 0.494$ and for the crystal $\phi_{cr} = 0.545$. The structure is FCC, although this trasition takes place far below the densest packing $\phi_{cr} = 0.74$. However, densest packing can be reached continuously by increasing pressure. A glass transition[66] has been documented at $\phi_g = 0.58$. We refer the reader to the textbook of W.B. Russel et al. for details on the statistical theory.

---

[62] A.P. Gast and W.B. Russel, Physics Today, December 1998, p. 24.

[63] J.C. Crocker and D.G. Grier, MRS Bulletin **10**, 24 (1998).

[64] B.J. Alder, W.G. Hoover and D.A. Young, J. Chem. Phys. **49**, 3688 (1968).

[65] S. Hachisu and Y. Kobayashi, J. Colloid Interface Sci. **46**, 470 (1974).

[66] P.N. Pusey and W. van Megen, in Complex and Supramolecular Fluids, Edited by S.A. Safran and N.A. Clark, Wiley-Interscience, New York, 1987, p. 673.

14.3.4.2. Repulsive Forces; Soft Spheres

The repulsive interactions can be turned long range by permanent electric charges on the particles—the double-layer thickness is a distance of closest approach between two particles—or by grafting polymer chains on the particles—the two brushes do not interpenetrate (see Chapter 15). The transitions are still driven by entropy, the essential difference, compared with hard spheres, being that the polymer-grafted particles or the charged particles have a larger effective radius, so that the Kirkwood–Alder transition occurs at smaller volume fractions.

The case of charged particles was mentioned above (Section 14.3.3). The effective volume fraction is $\phi_{eff} = \phi (d_{eff}/2a)^3$, which can be much larger than ϕ for weak electrolytes. So charged polystyrene latices at very low ionic strength form BCC iridescent phases at $\phi \leq 0.01$, whereas FCC phases form at higher ϕ. Similarly, proteins can crystallize at very small volume fractions.

Being crystalline, colloidal crystals have a yield stress and plastic properties that are related to the movement of dislocations.[67] But the crystals so obtained are very soft and melt easily under shear.[68]

14.4. Measurements of Interactions in Colloidal Systems

The forces under which colloidal particles are acting are fairly small; for example, the force between two spheres of radius $a \approx 10^{-5}$ m, with one elementary charge per 0.16 nm^2 (i.e., $q \approx 1$ C.m^{-2} in SI units), $\frac{k_B T}{e} = 25.69$ mV, $k_D \approx 10^{-9}$ m^{-1}, and $h \approx 10a$, which is given by the expression[69]

$$-2\pi \varepsilon \varepsilon_0 (k_B T/ze)^2 aq^2 \ln(1 - \exp -hk_D),$$

is of the order of one femtoN (10^{-15}N). In complement is the difficulty to manipulate such small objects. Considerable achievements have been made in these directions (e.g., atomic force microscope to measure forces, optical tweezers to manipulate particles or biopolymers). By all means, specific ways of making accurate measurements have been developed in colloid science. We review briefly some that have acquired a great importance in the last twenty years. We have already mentioned the pipette aspiration technique to measure mechanical properties of unilamellar vesicles.

Osmotic Stress (OS).   The sample under study is embedded in a solution of dextrane, polyethylene glycol, and so on, in any case, a large molecule that cannot penetrate the sam-

---

[67]M. Jorand, F. Rothen, and P. Pieranski, J. Physique, **46**, C3-245 (1985).

[68]A.J. Ackerson and N.A. Clark, Phys. Rev. Lett. **46**, 123 (1981).

[69]See Russel et al. textbook, p. 559.

ple. The solvent is the same and circulates freely in the two media. The osmotic pressure of the surrounding medium is then transmitted mechanically to the sample under study. This method, which for example measures the forces acting between the lamellae in a lamellar phase in function of the repeat distance, has provided the direct measurement of the DLVO potential between surfactant bilayers.[70] The integrated pressure over the distances between layers yields the potential.

Surface Force Apparatus (SFA).   OS permits one to investigate the repulsive part of the potential only. On the other hand, the surface force apparatus (see Israelachvili textbook, p. 41) allows for the study of the whole interaction potential curve. The sample is introduced between a pair of atomically smooth mica cylinders, crossed at right angles. The apparatus measures the ratio of the force between the mica cylinders over their radius. This ratio depends on the smallest distance between the cylinders and has to be interpreted in function of the number of layers, if the sample is lamellar, say. SFA has also been used to study structural forces in micellar systems.[71] The accuracy is of the order of 10 picoN, comparable to the OS accuracy. For a comparison of SFA and OS, see Tao et al.[70] SFA is specially designed to measure attractive forces (e.g., the Hamaker constant[72]), but the fluctuations of the bilayers are strongly damped by confinement and anchoring. On the other hand, at very small separations, SFA can detect the molecular structure of the solvent and the breakdown at small distance of the standard models; the sponge-lamellar confinement-induced transition has been studied[73] in this way.

Atomic Force Microscopy (AFM).   This method permits at the same time imaging of a nonconducting surface at the atomic level—by scanning over the surface a non-conducting tip mounted on a cantilever with a resolution of a few Ångström—and measurements of the force between the tip and the explored atom or molecule of the surface, in the picoN (10^{-12}N) range.[74]

Optical Tweezers.[75]   Microscopic particles can be trapped inside an electromagnetic potential well created by laser light and manipulated in 3D. This technique has permitted one to follow the interaction between two independently trapped particles, or to follow the motion of particles. This second method yields the trajectories of particles, from which it is possible to measure the pair-correlation function

[70]Y.-H. Tao, D.F. Evans, R.P. Rand, and V.A. Parsegian, Langmuir **9**, 223 (1993).

[71]P. Richetti and P. Kékicheff, Phys. Rev. Lett. **68**, 1951 (1992).

[72]J.N. Israelachvili and P.M. McGuiggan, Science **241**, 795 (1988).

[73]D.A. Antelmi, P. Kékicheff, and P. Richetti, J. Phys. II France **5**, 103 (1995).

[74]D. Sarid, Scanning Force Microscopy with Applications to Electric, Magnetic, and Atomic Forces, Oxford University Press, 1991.

[75]A. Ashkin, J.M. Diedzic, J.E. Bjorkholm, and S. Chu, Opt. Lett. **11**, 288 (1986).

$$g(\mathbf{r}) \sim \frac{1}{\tau} \int_0^\tau dt \rho^{-2} \int d\mathbf{r}' \rho(\mathbf{r}' - \mathbf{r}, t)\rho(\mathbf{r}', t)$$

(where $\rho(\mathbf{r}, t) = \sum_{i=1}^N \delta(\mathbf{r} - \mathbf{r}_i(t))$ is the trajectory function of the N particles which are followed). From the pair-correlation function one gets, employing the relation $g(\mathbf{r}) = \exp[-V(\mathbf{r})/k_B T]$, the pair-interaction potential V (theory of stochastic processes for liquids[76]). This method of measurement of the potential was first used to put into evidence an attractive potential between polystyrene spheres of micron size confined between two plates, just by videomicroscopy observations.[77] The optical tweezer technique has permitted to follow the interaction of isolated pairs of particles,[78] and to obtain consistency with the DLVO theory, provided some parameters like the fixed charge of the spheres are adjusted.[63]

Problem 14.0.

(a) Calculate $n_0\xi^2$ (14.10) by using the charge conservation condition (14.2).

(b) Establish to which extent the presence of free H^+ and OH^- ions in water can be neglected in the calculation that yields (14.11).

Answers: (b) The concentration of counterions is large close to the charged plate and decreases with the distance x to the plate. The counterions concentration is of the order of the H^+ and OH^- ions concentration (i.e., $10^{-7} N_{AV}/dm^3$) for $x \approx (10^{-7} N_{AV} \times 1000/2\pi \ell)^{1/2}$ (in SI units) $\approx 2 \times 10^{-6}$m $= 2\mu$m.

Problem 14.1. Express the counterions density at the boundary as a function of the counterions density at the midplane between two equally charged plates, when the solvent contains no electrolyte.

Answers: The PB equation $\varphi'' = -(en_0/\varepsilon\varepsilon_0) \exp(-e\varphi/k_B T)$, which can be written as $\varphi'' = -en(x)/\varepsilon\varepsilon_0$ (where $n(x)$ is the density of counterions), transforms easily to $d(\varphi')^2/dx = \frac{2k_B T}{\varepsilon\varepsilon_0} \cdot dn(x)/dx$ after multiplying both sides by φ'. Let the origin of coordinates be at the midplane. By symmetry, $\varphi'(x = 0) = 0$, and $n(x) = n(0) + (\varepsilon\varepsilon_0/2k_B T)(\varphi')^2$. Employing the boundary conditions on the charged plates, one eventually gets $n_s = n(0) + (\varepsilon\varepsilon_0/2k_B T)\sigma^2$. Consequently, one can also express $p(d)$ as follows:

$$p(d) = k_B T n(x = 0) = k_B T (n_s - (\varepsilon\varepsilon_0/2k_B T)\sigma^2).$$

Problem 14.2. Solve the PB equation for a charged sphere of radius a in the Debye-Hückel approximation. Calculate the relation between the surface charge density σ and the potential $\varphi(a) = \varphi_s$ on the surface.

Answers: We assume an electrolyte with ions of valencies z_i and particle concentrations $n_{i,0}$. In the DH approximation, the nonlinear PB equation

---

[76] J.-P. Hansen and I.R. McDonald, Theory of Simple Liquids, 2nd edition, Academic Press, New York, 1986.
[77] G.M. Kepler and S. Fraden, Phys. Rev. Lett. **73**, 356 (1994).
[78] J.C. Crocker and D.G. Grier, Phys. Rev. Lett. **73**, 352 (1994).

$$\frac{1}{r}\frac{d}{dr}r^2\frac{d\varphi}{dr} = -\frac{e}{\varepsilon\varepsilon_0}\sum_i n_{i,0}z_i \exp\left(-\frac{ez_i\varphi}{k_BT}\right)$$

reduces to the simple form:

$$\frac{1}{r}\frac{d}{dr}r^2\frac{d\varphi}{dr} = k_D^2\varphi(r),$$

where k_D is the reciprocal Debye length (14.14). We have used the neutrality condition $\sum_i n_{i,0}z_i = 0$.

The most general solution of the DH equation is $r\varphi(r) = A\exp(-k_Dr) + B\exp(+k_Dr)$, with the condition that the potential vanishes at infinity and takes a given value on the surface of the sphere. One gets:

$$\varphi(r) = \varphi_s\frac{a}{r}\exp[-k_D(r-a)].$$

The boundary condition for the charges can be written $\sigma = -\varepsilon\varepsilon_0\left.\frac{d\varphi}{dr}\right|_{r=a}$; hence, $\sigma = \varepsilon\varepsilon_0\frac{\varphi_s}{a}(1 + k_Da)$. The total charge $q = 4\pi a^2\sigma$ on the particle opposes the total charge in the solvent $-q = \int_a^\infty 4\pi\rho r^2\,dr$. Eventually, one gets

$$\varphi_s = \frac{q}{4\pi\varepsilon\varepsilon_0 a} + \frac{-q}{4\pi\varepsilon\varepsilon_0(a + \lambda_D)}.$$

Problem 14.3. Prove (14.36)

Answers: When the area $d\Sigma = dx_1\,dx_2$ is transported by an infinitesimal quantity λ, $\sigma_i \to \sigma_i' \approx \sigma_i/(1 + \lambda\sigma_i)$ and $dx_i \to dx_i' \approx dx_i(1 + \lambda\sigma_i)$. These expressions are signed expressions.

Problem 14.4. Consider two membranes cohering along some element of area, according to the scheme of Fig. 14.9. They separate along some line L (perpendicular to the figure), and form an angle $2\vartheta_0$ at asymptotic distance from the line L. Let γ_{coh} be the energy per unit area gained when the contact between the two membranes is achieved, and γ the surface tension of each membrane.

(a) Show that the set of the two cohering membranes is submitted to a surface tension $\tau = 2\gamma - \gamma_{coh}$. We shall assume that this quantity is positive.

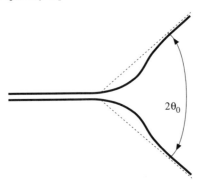

Figure 14.9. See Problem 14.4.

(b) Express ϑ_0 in function of γ and γ_{coh}.

(c) Justify the following expression of the free energy per unit area of the membrane away from the region of contact:

$$f_s = (1/2)\kappa \, d^2\vartheta/ds^2 + \gamma(\cos\vartheta - \cos\vartheta_0),$$

where s is the abscissa measured along the free membrane in a plane perpendicular to L. Find the first integral of the Euler–Lagrange equation of this energy density, and show that the membrane tends exponentially to its asymptotic direction ϑ_0 on a length of the order of $s_0 = (\kappa/\gamma)^{1/2}$.

Answers: (b) $2\gamma(1 - \cos\vartheta_0) = \gamma_{coh}$; (c) note that

$$\gamma(\cos\vartheta - \cos\vartheta_0) = \gamma\Delta A_p/A,$$

where ΔA_p is the opposite of the variation of the projected area of the membrane onto the direction $\vartheta = 0$; this expression, which is the 2D equivalent of the classical 3D expression $-p\,dV/V$, assumes that ϑ and ϑ_0 are small; $(1/2)\kappa(d\vartheta/ds)^2 + \gamma(\cos\vartheta_0 - \cos\vartheta) = 0$; for details, see W. Helfrich and R.-M. Servuss, Il Nuovo Cimento, **3D**, 137 (1984).

Further Reading

Membranes

A. W. Adamson and A. P. Gast, Physical Chemistry of Surfaces, John Wiley & Sons, Inc., New York, 1997, 784 pp.

Micelles, Microemulsions, Membranes and Monolayers, Edited by W. Gelbart, A. Ben-Shaul, and D. Roux, Springer-Verlag, New York, 1994.

Physics of Amphiphilic Layers, Edited by J. Meunier, D. Langevin, and N. Boccara, Springer-Verlag, New York, 1987.

Samuel A. Safran, Statistical Thermodynamics of Surfaces, Interfaces, and Membranes, Frontiers in Physics, Addison-Wesley Publishing, Reading, MA, 1994.

Colloid Science

D. F. Evans and H. Wennerström, The Colloidal Domain: where Physics, Biology, and Technology meet, Wiley-VCH, New York, 1994.

R. J. Hunter, Foundations of Colloid Science, Oxford University Press, Oxford, 1986.

J. N. Israelachvili, Intermolecular and Surface Forces, 2nd edition, Academic Press, New York, 1992.

R. G. Larson, The Structure and Rheology of Complex Fluids, Oxford University Press, New York, Oxford, 1999.

D. Myers, Surfaces, Interfaces, and Colloids, 2nd edition, Wiley-VCH, New York, 1999.

W. B. Russel, D.A. Saville, and W. R. Schowalter, Colloidal Dispersions, Cambridge University Press, Cambridge, 1995 (reprinted).

E. J. W. Vervey and J. Th. G. Overbeek, Theory of the Stability of Lyophobic Colloids, Elsevier, New York, 1948.

CHAPTER 15

Polymers: Structural Properties

Polymers are met under a large variety of chemical achievements, from biopolymers like DNA, xanthan, cellulose, proteins, and actin filaments, to synthetized polymers like polyethylene (PE), polybutadiene (PB), polystyrene (PS), polymethylmetacrylate (PMMA), and so on, all of great industrial interest. They are employed under numerous forms, in solutions, as gels, rubbers, synthetic fabrics, moulded pieces, and so on.

The first part of this chapter is inspired by the classic presentation of the subject elaborated by de Gennes. We believe indeed that the simplest way to introduce the fundamentals of the behavior of a polymer in solution or a molten polymer is through the presentation of the scaling properties, which are the main characteristics of chains with a large degree of polymerization (DP) N. Although the mathematics is easy, the physics is deep: It is not possible to get realistic scaling laws without a thorough analysis of the physical properties; One does not have even the usual help one can get from the free energy density, often used as a black box from which the results pop up as by magic. On the other hand, the nonlinear effects or the renormalization phenomena that are calculated at such painstaking efforts in the usual theories, appear there in their true light. This is not to underrate the usual presentations, which are absolutely necessary in order to take the material constants into account. In this first part, we limit our interest to the remarkable physical properties of flexible polymers, which are akin to critical phenomena arising at phase transitions. The physical properties are described in terms of conformation of flexible chains in function of the polymer molecular mass (often expressed as a number of monomers N per chain), of the volume concentration $c = n/V$ of monomers (n monomers in a volume V), possibly of temperature, and so on. In the phase transition picture, the analog of the temperature is $1/N$; we shall also find analogs of critical exponents, some of which can be interpreted as fractal dimensions.

Chains in melts are close to ideal chains, defined in the first part, and their physics can be approached in the frame of mean field theory. We therefore present the classic Flory–Huggins picture of melts and blends, and in particular the phenomena of blends phase separation, which are typical of polymers. This is the place where we come back to the structural point of view, which is one of the essential topics of this book, with a description of the ordered phases of diblock copolymers.

560

Finally, this chapter discusses, in the same vein, rigid and semiflexible polymers, which include all biopolymers, certain anisotropic viruses, and synthetic polymers used as textiles or building elements in high-strength materials used at room temperature, whose mechanical properties result from their trend to form liquid crystalline phases at high temperature or to align under shear.

We refer the reader to the rich available literature for a detailed description of the *microscopic* dynamical properties (relaxation times, diffusivities, etc.) of polymers (e.g., M. Doi and S.F. Edwards, 1988). The *macroscopic* rheological properties of polymers, either in solution or in the form of ordered or disordered melts, constitute such a vast body of research that it has seemed preferable to leave them completely aside. One will find an up-to-date presentation of these topics in Larson's book (1999). Neither have we touched on the question of *crystallization* in this book. This topic is covered in Chapter 4 of Strobl's book (1996). The stimulating review paper of Keller on chain-folding is also recommended.

15.1. Ideal and Flory Chains

We shall limit ourselves to *main chain flexible polymers*, whose prototype is polyethylene $[-CH_2-]_n$. Other standard examples are

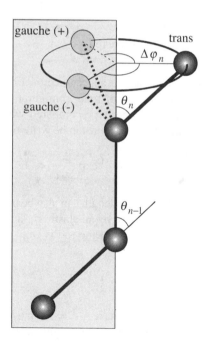

Figure 15.1. Configuration parameters along a chain of ethylene groups.

$$\text{polyoxyethylene } \left[-\text{O}-\text{CH}_2-\text{CH}_2-\right]_n, \text{ polydimethylsiloxane } \left[\begin{array}{c} \text{CH}_3 \\ | \\ \text{Si} \quad - \quad \text{O} \quad - \\ | \\ \text{CH}_3 \end{array}\right]_n,$$

$$\text{polystyrene } \left[\begin{array}{ccc} - & \text{CH}_2 & - & \text{CH} & - \\ & & & | \\ & & & \Phi \end{array}\right]_n, (\Phi = \text{phenyl ring}), \text{ and so on.}$$

Each of these polymers can be symbolized by a sequence of bonds linking successive elementary monomers; in the simple case of polyethylene, to which it will be referred as the model case, the angular variation between neighboring carbons is defined as $\Delta\theta_n = \theta_n - \theta_{n-1} = 0, \Delta\varphi_n = 0, \pm120°$ (Fig. 15.1). $\Delta\varphi_n = 0$ is called a *trans* configuration; $\Delta\varphi_n = \pm120$ is a *gauche* configuration. They do not have the same energy. Such a geometrical picture can be extended to any homopolymer, *mutatis mutandis*.

Taking the successive values of $\Delta\varphi_n$ at random, the chain appears, when observed at some scale much larger than a, the elementary length of a bond, as a *statistical coil*. The following ingredients are of interest:

1. *The Persistence Length* ℓ_p. Let $\Delta\varepsilon = \varepsilon_g - \varepsilon_t > 0$ be the difference in energy between the gauche configuration and the trans configuration. Then,

$$\ell_p = a \exp\left(\frac{\Delta\varepsilon}{k_BT}\right) \tag{15.1}$$

is the length on which the chain has the appearance of a rigid segment, because the segments are colinear for a succession of trans configurations. Hence, $L = Na$, the chemical length of the chain, has to be large compared with ℓ_p, for the chain to be considered as a statistical coil. This condition can be written as

$$x = \frac{\ell_p}{L} \sim N^{-1} \exp\frac{\Delta\varepsilon}{k_BT} \ll 1. \tag{15.2}$$

Phenomenologically, the persistence length can also be presented as a characteristic length of the thermal fluctuations carried by an elastic rigid rod. Let us assume, for the sake of simplicity, that the cross-section of the rod is isotropic. Then, the free energy per unit length of the rod reads as

$$F = \tfrac{1}{2}\kappa r_c^{-2},$$

where r_c is the radius of curvature.[1] We have $r_c^{-1} = \pm|\partial^2\mathbf{u}/\partial s^2| = \partial\omega/\partial s$, where $\mathbf{u}$ is the (small) displacement of the rod with respect to the unstrained straight rod, s is the

[1]L. Landau and L. Lifshitz, Statistical Physics, Pergamon, Oxford, 1986.

length measured along the rod, and ω is the angle of deflection of the rod. It is easy to show, by the Fourier transforming the free energy $F = \frac{1}{2} \int_0^L \kappa r_c^2 \, ds$ of a rod of length L, then applying the theorem of equipartition of energy, and summing over all modes, that

$$\langle \omega^2 \rangle = 2 \frac{k_B T}{\kappa} L. \tag{15.3}$$

Taking $\langle \cos \omega \rangle \approx 1 - L k_B T / \kappa$ equal to zero, which is the condition at which the elements of the rod at $s = 0$ and $s = L$ are decorrelated (see Landau and Lifshitz[1]), one gets $\ell_p = \frac{\kappa}{k_B T}$. Eventually, the free energy density can be written as

$$F = \frac{1}{2} k_B T \ell_p r_c^{-2}. \tag{15.4}$$

Of course, this expression of the free energy is valid for $L < \ell_p$. For lengths $L > \ell_p$, the chain is flexible, and its free energy is described in a completely different way (see below).

2. *The Persistence Time τ_p*, i.e., a time characteristic of a change of conformation. Let ΔE be the activation energy necessary for the trans-to-gauche isomerization. We have

$$\tau_p = \tau_0 \exp \frac{\Delta E}{k_B T}. \qquad \Delta E \quad \text{same as } \Delta \mathcal{E} \text{ ?} \tag{15.5}$$

The chain appears as flexible in any observation on a duration time $\tau > \tau_p$. The characteristic time τ_0 is related to the thermal vibrations of the chain, and it is of the order of 10^{-12}s.

15.1.1. Single-Chain Conformations

There are three rotational isomeric states per bond: t, g^+, and g^-. There are therefore 3^N different configurations for a polyethylene chain with DP equal to N. Among those, some are special, as follows.

1. *Helix Configurations:* These are configurations of the chain in which all monomeric units *repeat* along a direction of elongation of the chain (most often, this repeat pattern has a helical shape, hence, the denomination). Helical configurations have minimal internal energy and are met in *crystals of polymers*. For PE, this is the *all-trans* state: The chain shows a *zigzag* structure along its elongation, rather than a true 3D helical structure. For polytetrafluoroethylene (PTFE), it is a true helix, not far from the all-trans state, but differing from it by a superimposed twist along the elongation, caused by the repulsive interactions between neighboring fluorines. For polyoxymethylene (POM),

there are several minimal helical energy states, obtained by twisting the *all-gauche* configurations in slightly different ways.

2. *Coil Configurations:* We shall be interested in the sequel only in molecules that are *dynamically flexible*, i.e., whose chemical length aN is large compared with ℓ_p, and whose observation is made on durations much larger than τ_p. These molecules run through the 3^N configurational states. In such a case, a number of properties are not dependent on the local scales, and the chemical features can be ignored. On the other hand, all "mesoscopic" properties depend on N according to scaling laws, of the type

$$R_G \propto aN^\nu, \tag{15.6}$$

where, in this instance, R_G is the so-called gyration radius (to be defined rigourously later on), i.e., the typical half-size of the statistical coil in which the polymeric chain is folded, when isolated in some solvent. The next paragraphs are devoted to the calculation of the exponent ν in various theoretical cases.

15.1.2. The Ideal (or Gaussian) Chain

15.1.2.1. The Average Size of an Ideal Chain

Assume that the N elementary segments (sometimes loosely called bonds in the sequel) do not interact and are independent one from the other. Such a chain is called ideal. The vector that joins the origin of the chain to its end can be written as

$$\mathbf{r} = \sum_{i=1}^{N} \mathbf{a}_i. \tag{15.7}$$

Squaring $\mathbf{r}$, and taking averages on all possible configurations, one gets

$$\langle \mathbf{r}^2 \rangle = \sum_i \mathbf{a}_i^2 + 2 \sum_{i \neq j} \langle \mathbf{a}_i . \mathbf{a}_j \rangle = Na^2, \tag{15.8}$$

because two segments i and j are independent. Hence, the average separation of the ends, which can be also called the average size of the ideal chain, is

$$R_0 = \langle \mathbf{r}^2 \rangle^{1/2} = aN^{1/2}, \tag{15.9}$$

and $\nu = 1/2$ for the ideal chain. Note that this exponent does not depend on the dimensionality of the embedding space. The scaling law (15.9) is unchanged if one introduces interactions between different bonds along the same chain (called short-range interactions, in contrast to long-range interactions, between bonds belonging to different chains).

One can show, as a consequence of the central limit theorem (see Appendix 15.A), that the probability $P_N(\mathbf{r})$ for the N-link chain to terminate at $\mathbf{r}$, has the Gaussian distribution:

$$P_N(\mathbf{r}) = \left(\frac{3}{2\pi R_0^2}\right)^{3/2} \exp{-\frac{3}{2}\frac{r^2}{R_0^2}}. \tag{15.10}$$

Another demonstration is proposed as an exercise (Problem 15.1). $P_N(\mathbf{r})$ is the ratio between the number of configurations $\Gamma_N(\mathbf{r})$ susceptible to bring the chain from a fixed origin 0 to a point $\mathbf{r}$, over the total number of configurations $\Gamma_N = \sum_{\mathbf{r}} \Gamma_N(\mathbf{r})$. The entropy $\mathcal{S}$ of the ideal chain being $\mathcal{S} = k_B \ln \Gamma_N(\mathbf{r})$, we have

$$\mathcal{S}(\mathbf{r}) = k_B \ln P_N + k_B \ln \Gamma_N, \tag{15.11}$$

from which expression we get, but to an additive constant,

what happens to this term?

$$\mathcal{S}(\mathbf{r}) = -\frac{3}{2}k_B \frac{r^2}{R_0^2}, \tag{15.12}$$

which depends only on $r = |\mathbf{r}|$. The entropy is maximum when $r = 0$, from which we conclude that an ideal chain is a compact object with many crossings.

15.1.2.2. Free Energy

The free energy of the chain is entirely of entropic origin and can be written as

$$F(r) = F_0 + \frac{3}{2}k_B T \frac{r^2}{R_0^2}. \tag{15.13}$$

$G = U + PV - TS$

F_0

Assume that an isolated chain is submitted to the action of an applied external force $\mathbf{f}$. If the applied force is zero, the elongation r vanishes and the entropy is maximal. If not, a decrease in entropy creates an inner force $-\frac{\partial F}{\partial \mathbf{r}}$ that opposes the applied force

$$\mathbf{f} - \frac{\partial F}{\partial \mathbf{r}} = 0; \tag{15.14}$$

i.e., the elongation $\mathbf{r}$ is along $\mathbf{f}$ and takes the value

$$r = f\frac{R_0^2}{3k_B T}. \tag{15.15}$$

$f = k \cdot r$

This is the elongation of a spring of spring constant $3k_B T/R_0^2$.

de Gennes has shown that (15.15) can be derived directly from a scaling law reasoning, as follows. The elongation depends only on f, $k_B T$, and R_0. Assume that the dependence obeys a scaling law, i.e., that there is some exponent m such that

(15.9) $R_0 = a N^{1/2}$

$$r \cong R_0 \left(\frac{f R_0}{k_B T} \right)^m . = a N^{1/2} \left(\frac{f a N^{1/2}}{k_B T} \right) = \frac{a^2 N}{k_B T} \qquad (15.16)$$

A reasonable physical assumption is that r is proportional to the number of bonds N. According to (15.9), this implies $m = 1$. Hence, the result, but to a constant factor that cannot be deduced from this type of approach.

15.1.2.3. Rubber Elasticity, the Rouse Model (15.13) $F(r) = F_0 + \frac{3}{2} k_B T \frac{r^2}{R_0^2}$

Equation (15.13) is the simplest model of energy used to explain the elastic properties of rubbers[2] made of cross-linked polymer chains. We just give the result[3]: The stress (expressed by a symmetric tensor σ) carried by a deformation $\mathbf{E}$ can be written as $\sigma = G \mathbf{E}^T . \mathbf{E}$, where the elastic modulus $G = N_{cl} k_B T$ (N_{cl} is the volume density of cross links or, equivalently, of strands). By definition, $E_{ij} = \partial x_j / \partial x_i'$, where x_j (respectively, x_i') is the position of a material point after (respectively, before) the deformation. This expression of the deformation $\mathbf{E}$ is valid for large deformations; $\mathbf{B} = \mathbf{E}^T . \mathbf{E}$, the so-called Finger tensor, reduces to the usual strain tensor for small deformations. The Finger tensor takes a simple form in the case of a pure extensional deformation ($B_{ij} = \lambda_i^2 \delta_{ij}$), where the λ_i's are the elongations in three orthogonal directions, with $\lambda_1 \lambda_2 \lambda_3 = 1$ because of the conservation of volume.

A few comments about dynamic regimes. It is illuminating to compare (15.13) with (15.4), which refers to the elasticity at a scale smaller than the persistence length, or which applies to a rodlike polymer (Section 15.4).

In the rigid rod case, the dynamical equation of the macromolecule can be written as $\eta \partial u / \partial t + k_B T \ell_p \partial^4 u / \partial s^4 = 0$, which yield relaxation frequencies $\omega(q) \approx k_B T \ell_p q^4 / \eta$, where $2\pi q^{-1} > \ell_p$. This equation expresses that the sum of the elastic force opposes a force of friction proportional to the velocity. A typical value of the frequency is 10^8 Hz.

In the case of an ideal isolated chain, the (Rouse) dynamical equation is

$$\eta \partial r / \partial t - (3 k_B T / R_0) \partial^2 r / \partial s^2 = 0.$$

Again, this equation expresses that the elastic force $(3 k_B T / R_0^2) \partial^2 r / \partial s^2$ opposes a force of friction $m^{-1} \partial r / \partial t$, where $m \approx 1 / \eta R_0$ is the mobility of a sphere of size R_0 and η is the viscosity of the solvent. The characteristic frequencies scale as $\omega(q_n) \approx k_B T q_n^2 / R_0 \eta$, with $q_n = n / (R_0 N^{1/2})$, $n = 1, 2, \ldots N$. The smallest frequencies, of the order of 10^{-7} Hz (for

---

[2]L. Treloar, The Physics of Rubber Elasticity, Clarendon Press, Oxford, 1975.
[3]R. Larson, Constitutive Equations for Polymer Melts and Solutions, Butterworths, New York, 1988.

$R_0 = 10\,\text{nm}$ and $N = 100$), describe the motion of the chain as a whole. This is the limit where the Rouse model is more or less obeyed, provided the solution is dilute. The Rouse model has been improved by taking into account the hydrodynamic interactions, i.e., the effect of the motion of an element of the chain on the others. It no longer describes the physical reality if the polymer chains are entangled.

15.1.3. Pair Correlation Function and Radius of Gyration

An important quantity relating to a polymeric chain is the *pair correlation function* $g(\vec{r})$ that measures the density of bonds in $\mathbf{r}$, assuming one monomer at the origin. The pair correlation function is obviously normalized:

$$N = \int g(\mathbf{r})\, d^d\mathbf{r}. \tag{15.17}$$

There are on the average $n(r) \propto r^2/a^2$ bonds in a sphere of radius r, for the ideal chain, according to (15.9). Hence,

$$g(\mathbf{r}) = g(r) \sim \frac{1}{4\pi r^2}\frac{dn}{dr} \sim \frac{1}{a^2 r} \tag{15.18}$$

in 3D; in d dimensions, the volume of a spherical shell is proportional to r^{d-1}, so that

$$g(r) \sim \frac{1}{a^2 r^{d-2}}. \tag{15.19}$$

The above estimates are valid as long as $r \leq R_0$. For larger values of r, one expects an exponential decrease of $g(r)$.

The Fourier transform $S(\mathbf{q})$ of $g(\mathbf{r})$ can be measured by X-ray or (more accurately) by neutron scattering at incident wave vector $\mathbf{k} = \mathbf{q} + \mathbf{k}'$, where $\mathbf{k}'$ is the final scattered wave ($|\mathbf{k}'| = |\mathbf{k}|$):

$$S(\mathbf{q}) \sim \int g(\mathbf{r}) \exp i\mathbf{q} \cdot \mathbf{r}\, d^d\mathbf{r} \sim \frac{1}{q^{d-1}a^2}, \tag{15.20}$$

which is valid as long as $q R_0 \gg 1$. Note that the limit $q = 0$ is known from (15.17):

$$S(\mathbf{q} \to 0) \to N. \tag{15.21}$$

To be more precise, consider a chain of N segments, a particular segment n at $\mathbf{r}_n$. The density of bonds $g_n(\mathbf{r})$ at $\mathbf{r}_n + \mathbf{r}$ is $g_n(\mathbf{r}) = \sum_{m(\neq n)} \langle \delta(\mathbf{r}_n + \mathbf{r} - \mathbf{r}_m) \rangle$, where $\langle X \rangle$ is the thermal average of X. Also, $g(\mathbf{r}) = \frac{1}{N} \sum_n g_n(\mathbf{r})$. One eventually gets

$$g(\mathbf{r}) = \frac{1}{N} \sum_{n \neq m} \sum \langle \delta(\mathbf{r} + \mathbf{r}_n - \mathbf{r}_m) \rangle , \tag{15.22a}$$

$$S(\mathbf{q}) = \frac{1}{N} \frac{a^3}{V} \sum_{n \neq m} \sum \langle \exp[i\mathbf{q} \cdot (\mathbf{r}_n - \mathbf{r}_m)] \rangle . \tag{15.22b}$$

This last expression can be expanded for small $| \mathbf{q} \cdot (\mathbf{r}_n - \mathbf{r}_m) |$. One has $\langle \mathbf{r}_n - \mathbf{r}_m \rangle = 0$. Define

$$R_G^2 = \frac{1}{2N^2} \sum_n \sum_m \left\langle (\mathbf{r}_n - \mathbf{r}_m)^2 \right\rangle, \tag{15.23}$$

one gets

$$S(\mathbf{q}) = S(0) \left(1 - \tfrac{1}{3} q^2 R_G^2 + \cdots \right) \tag{15.24}$$

in dimension $d = 3$. It can be shown that the radius of gyration R_G can also be expressed as follows:

$$R_G^2 = \frac{1}{N} \sum_{n=1}^{N} \left\langle (\mathbf{r}_n - \mathbf{r}_{cm})^2 \right\rangle, \tag{15.25}$$

where $\mathbf{r}_{cm} = N^{-1} \sum_n \mathbf{r}_n$ is the center of mass.

Equations (15.22)–(15.24) are fairly general. Let us specialize them to the ideal chain. For $|n - m|$ large enough, one has $\langle (\mathbf{r}_n - \mathbf{r}_m)^2 \rangle = |n - m| a^2$, i.e. $R_G^2 = \frac{1}{2N^2} \sum_{m,n} |n - m| a^2$. Replacing the sum by an integral, one gets

$$R_G^2 = \tfrac{1}{6} N a^2. \tag{15.26}$$

15.1.4. The Flory Chain

The above model does not take into account the simplest possible, athermal interactions, due to the excluded volume. Different segments of an ideal chain can occupy the same location in space, and in fact they do, in this model, because the entropy is maximum for $\mathbf{r} = 0$. Radii of gyration have been measured in dilute solutions of polymers (dilute enough such that individual coils are independent) by light scattering, viscosimetry, or diffusivity measurement, and simulated numerically. All of these measurements yield a

Table 15.1. Critical exponents for R_F and Γ_F

d	4	3	2	1
$\gamma - 1$	0	$\sim 1/6$	$\sim 1/3$	0
ν	$1/2$	$\sim 3/5$	$\sim 3/4$	1

critical exponent ν close to 0.6, and not 0.5. An actual isolated polymeric chain is more swollen than is an ideal chain, as expected. In $d = 2$, numerical simulations yield $\nu \approx 0.75$. The Flory theory,[4] which accounts for the excluded volume, enables us to find results very close to those experimental values.

More precisely, numerical simulations of lattice models give critical exponents for R_F (the Flory radius of the swollen chain) and Γ_F (see Table 15.1):

$$R_F \propto aN^\nu, \quad \Gamma_F \propto N^{\gamma-1}, \tag{15.27}$$

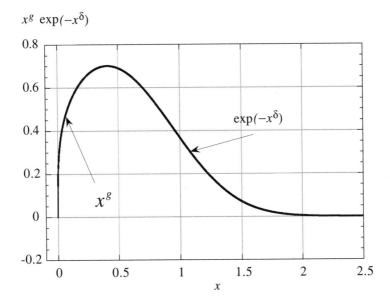

Figure 15.2. The probability function $x^g \exp(-x^\delta)$ for a swollen chain, calculated for typical exponents g and δ shown in Table 15.1 for $d = 3$.

---

[4]P.J. Flory, Principles of Polymer Chemistry, Cornell University Press, Ithaca, New York, 1971.

where γ is the analog of a susceptibility exponent in the phases transitions theory of polymer chains (see de Gennes, 1979).

Another useful exponent is δ, which describes precisely the excluded volume effect that moves the chain end away from the origin, as well as the "cage" effect, which traps the chain end inside the coil. Let the probability $P_N(\mathbf{r})$ that a Flory chain of fixed origin terminates at $\mathbf{r}$ takes now the form:

$$P_N(\mathbf{r}) \cong \frac{1}{R_F^d} f(x) \exp -x^{\delta}, \qquad (15.28)$$

where $x = r/R_F$. The probability must vanish for $r = 0$ and decrease exponentially for x large; hence, $f(x) \cong x^g$. These exponents are simply related to ν and γ, Fig. 15.2,

$$\delta = \frac{1}{1 - \nu}, \qquad g = \frac{\gamma - 1}{\nu}. \qquad (15.29)$$

Equation (15.28) is nothing else than an extension of (15.10), for which $g = 0$ (which expresses the fact that the ideal chain returns to the origin with a large probability) and $\delta = 2$.

1. *Flory Theory of the Scaling Law Exponent:* Let $c \cong NR^{-d}$ be the mean volume concentration of bonds in a coil occupying a spherical volume of radius R. The free energy density of the coil contains two contributions:

   (a) *an excluded volume term:*

   $$f_{ex} \cong \tfrac{1}{2} k_B T \langle c^2 \rangle v, \qquad (15.30)$$

   where v is the d-dimensional volume of a monomer. The $\langle c^2 \rangle$ term can be understood as measuring the number of pair interactions between bonds. More precisely, let $k_B T v \, \delta(\mathbf{r}_i - \mathbf{r}_j)$ be the energy of (steric) interaction between bonds i and j. The total free energy is $U = \tfrac{1}{2} k_B T v \sum_{i,j} \delta(\mathbf{r}_i - \mathbf{r}_j)$. We introduce the local bond density $c(\mathbf{r}) = \sum_j \delta(\mathbf{r} - \mathbf{r}_j)$ and write $\delta(\mathbf{r}_i - \mathbf{r}_j) = \int \delta(\mathbf{r}_i - \mathbf{r}) \delta(\mathbf{r} - \mathbf{r}_j) d^d\mathbf{r}$. This yields

   $$U = \tfrac{1}{2} k_B T v \int c^2(\mathbf{r}) \, d^d\mathbf{r}.$$

   We take $c^2(\mathbf{r})$ constant ($c = N/R^d$) and equal to its mean value in (15.30). This assumption neglects the effect of correlations and overestimates the interaction energy.

(b) *an elastic term of entropic origin,* for which we assume that the chain is ideal:

$$f_{el} \cong F_{el}/R^d \cong kT \frac{R^{2-d}}{Na^2}.$$
(15.31)

Again, this assumption neglects correlations between bonds and overestimates the total energy.

Minimizing the total energy of the coil $F = F_{ex} + F_{el} \approx R^d f_{ex} + F_{el}$, one gets

$$\nu = \frac{3}{d+2};$$
(15.32)

i.e., $\nu_3 = \frac{3}{5}$, $\nu_2 = \frac{3}{4}$. These values are extremely close to the experimental values.

More recent theories fully justify the use of scaling laws. de Gennes's "$n = 0$" theorem[5] establishes that the problem of the conformation of a chain with excluded volume (the self-avoiding walk or SAW problem) is formally similar to the problem of the correlations of n-dimensional spins in a ferromagnet, when $n = 0$! This remarkable analogy yields formal relationships between the SAW exponents and the critical exponents of the ferromagnetic case. The analog of the reduced temperature $(T - T_c)/T_c$ is N^{-1}. The analog of the correlation length is the radius of gyration $R \propto aN^{\nu}$. Numerical estimations made in the framework of the renormalization group theory have made possible the estimation of ν in various dimensions. The results are close to Flory's results, albeit a little smaller.

2. *The Concept of Blob:* Coming back to (15.28), and calculating the entropy contribution in the free energy of a coil, one gets

$$F = k_B T \left(\frac{r}{R_F} \right)^{\delta} + F_0;$$
(15.33)

i.e., an elongation

$$r \cong R_F \left(\frac{f R_F}{k_B T} \right)^{\frac{1}{\delta - 1}}$$
(15.34)

under the applied force f. As above, we assume that the elongation is proportional to N, which yields

$$\delta = \frac{1}{1 - \nu}.$$
(15.35)

This expression gives $\delta = 2$ for an ideal chain, as expected.

---

[5]P.G. de Gennes, Phys. Lett. **A38**, 339 (1972).

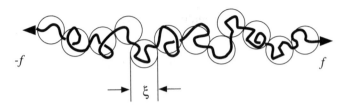

Figure 15.3. Splitting of a swollen chain into blobs under an applied force.

Note that (15.34) provides us with a new length, $\xi = k_B T / f$, such that

$$r \cong \frac{N}{n} \xi, \tag{15.36}$$

where

$$\xi \cong a n^\nu. \tag{15.37}$$

n is therefore the typical number of bonds belonging to a *blob* of size ξ. Equation (15.36) tells us that the elongation is the number of blobs N/n, times the typical size of a blob (Fig. 15.3).

3. *The $d \geq 4$ Case (Mean Field):* Flory's results predict mean field values (ideal chain) for the exponents ($\nu = 1/2$) in dimension 4. The same results apply for the whole range $d \geq 4$ in the $n = 0$ theory. Note that, from (15.30) and (15.31), one gets

$$\frac{F_{ex}}{F_{el}} \cong \frac{\nu}{a^d} N^{2-d/2}, \tag{15.38}$$

if F_{ex} is given its mean field value. This ratio vanishes when $d > 4$: The excluded volume terms become negligible, a feature typical of an ideal chain.

15.2. Chains in Interaction

Let us distinguish from the start two limit cases. In a very dilute solution, each chain has an individual behavior independent of the other chains. If the chain is "swollen," one refers to the solvent as a good solvent; the radius of gyration of the chain scales as the Flory radius ($R_G \propto N^{3/5}$). In the absence of solvent, on the other hand, each monomer of a "molten" polymer is submitted to attractive van der Waals interactions (due to the bonds of other neighbouring chains) and to repulsive interactions (due to the bonds of the same chain) otherwise responsible of the swelling behavior. These interactions compensate exactly at the scale of a few monomers, because the density fluctuations of the melt are so small. Therefore, each chain is expected to behave in a melt as an ideal chain ($R_G \propto N^{1/2}$), a fact that has been experimentally checked.

The object of this Chapter is to introduce a few methods relevant to interacting chains in cases different from the above limit cases. One shall first review the Flory–Huggins "mean field" approach and afterward the scaling laws approach.

15.2.1. The Mean Field Approach

15.2.1.1. The Flory–Huggins[4,6] Free Energy

Let us imagine that the monomers of the polymer chains are located on the sites of a lattice (edge length $= a$) and assume that each link occupies a volume a^3. Let ϕ be the fraction of sites occupied by the monomers, and let $(1 - \phi)$ be the fraction occupied by the solvent. We assume that each molecule of solvent also has volume a^3. The volume density of monomers is $c = \phi a^{-3}$. According to Flory–Huggins, the entropy of mixing $\mathcal{S}_{\text{mix}}$ can be written as

$$\mathcal{S}_{\text{mix}} = -k_B N_{\text{tot}} \left[\frac{\phi}{N} \ln \phi + (1 - \phi) \ln(1 - \phi) \right], \tag{15.39}$$

where N_{tot} is the total number of sites and N is the number of monomers per chain.

1. *Establishing (15.39):* The above expression is justified as follows. Assume that the two chemical species behave like ideal gases, whose thermodynamical potentials add linearly. The chemical potential of species i is

$$\mu_i = -k_B T \ln \frac{e V_i}{N_i} + f_i(T), \tag{15.40}$$

where V_i is the total volume occupied by species i; and N_i is the number of i-molecules. Note that one retrieves the Boyle–Mariotte law from (15.40): The free energy is indeed $F_i = \mu_i N_i$, and the partial pressure of species i is $p_i = -\partial F_i / \partial V_i = k_B T (N_i / V_i)$. Now the partial entropy of species i is $\mathcal{S}_i = -\partial F_i / \partial T$; i.e.,

$$\mathcal{S}_i = k_B N_i \ln \frac{e V_i}{N_i} - N_i f_i'(T). \tag{15.41}$$

When two species mix, they both occupy the total volume $V = V_1 + V_2$. Hence, the total entropy is

$$\mathcal{S} = k_B \left[N_1 \ln \frac{e V}{N_1} + N_2 \ln \frac{e V}{N_2} \right] + s(T),$$

[6]P.M. Huggins, J. Phys. Chem. **46**, 1712 (1942).

$s(T)$ being some additive function proportional to $\sum N_i$. The entropy of mixing is $S_{\text{mix}} = S - S_1 - S_2$, where S_i, $i = 1, 2$ are given by (15.41); i.e.,

$$S_{\text{mix}} = -k_B \sum_j N_j \ln v_j, \tag{15.42}$$

where $v_i = V_i/V$ is the volume fraction of species i. Equation (15.39) follows directly from (15.42) with substitutions $N_1 = \frac{\phi}{N} N_{\text{tot}}$ and $N_2 = (1 - \phi) N_{\text{tot}}$. The entropy of mixing per site is

$$S_{\text{mix/site}} = -k_B \left[\frac{\phi}{N} \ln \phi + (1 - \phi) \ln(1 - \phi) \right]. \tag{15.43}$$

2. *The Terms of Interaction:* To the entropy term (15.43), one adds a term of pair interactions, including monomer-monomer pairs, monomer-solvent pairs, and solvent-solvent pairs, viz:

$$U_{\text{/site}} = \tfrac{1}{2} k_B T \left[\chi_{\text{MM}} \phi^2 + 2\chi_{\text{MS}} \phi(1 - \phi) + \chi_{\text{SS}}(1 - \phi)^2 \right]. \tag{15.44}$$

Note the similarity of (15.44) with (15.30). The χ_{ij} coefficients are adimensional. The energy of mixing is $U_{\text{/site}} - \phi U_{\text{/site}}(\phi = 1) - (1 - \phi) U_{\text{/site}}(\phi = 0)$; i.e.,

$$U_{\text{mix/site}} = k_B T \chi \phi(1 - \phi), \tag{15.45}$$

where $\chi = \chi_{\text{MS}} - \tfrac{1}{2}(\chi_{\text{SS}} + \chi_{\text{MM}})$ is called the Flory interaction parameter. The pair interactions χ_{ij} are mostly due to van der Waals attractions and are proportional to the polarizabilities; i.e., $\chi_{\text{MM}} \propto -\alpha_{\text{M}}^2$, $\chi_{\text{MS}} \propto -\alpha_{\text{M}}\alpha_{\text{S}}$, $\chi_{\text{SS}} \propto -\alpha_{\text{S}}^2$. Therefore, χ is generally positive. Adding (15.43) and (15.45), one gets

$$\frac{F_{\text{mix/site}}}{k_B T} = \frac{\phi}{N} \ln \phi + (1 - \phi) \ln(1 - \phi) + \chi \phi(1 - \phi). \tag{15.46}$$

3. *Good Solvent and Poor Solvent:* When the attractive interactions are small, i.e., $\chi \to 0$, one has the so-called regime of good solvent. Strong atttractive interactions correspond to a regime of poor solvent. This conclusion can be made more precise by looking at the small ϕ behavior of (15.46) (note that linear terms in ϕ have been dropped from (15.47), because they have no physical meaning):

$$\frac{F_{\text{mix/site}}}{k_B T} = \frac{\phi}{N} \ln \phi + \frac{1}{2}\phi^2(1 - 2\chi) + \frac{1}{6}\phi^3 + \cdots, \tag{15.47}$$

which can be compared with the expression of the free energy one obtained in the discussion of the excluded volume, (Chapter 4). Let indeed $c = \phi a^{-3}$; then (15.47) can be written as

$$\frac{f_{\text{mix/site}}}{k_B T} = \frac{c}{N} \ln c + \frac{1}{2} v c^2 + \cdots, \tag{15.48}$$

where $f_{\text{mix/site}} = F_{\text{mix/site}}/a^3$ is now a free energy density and $v = (1 - 2\chi)a^3$ appears as an excluded volume. The regime of good solvent is satisfied when $v > 0$, $\chi < 1/2$. The regime of poor solvent occurs for $\chi > 1/2$.

4. Θ-*Temperature:* The temperature at which $1 - 2\chi = 0$ is called the Θ *(theta)-temperature*. It can be estimated from the expression of χ. Indeed, we expect that the van der Waals interactions depend little on temperature, so that

$$\chi \propto \frac{(\alpha_M - \alpha_S)^2}{k_B T} \quad \text{and} \quad \Theta \propto \frac{(\alpha_M - \alpha_S)^2}{k_B}.$$

One expects that because the interactions vanish, the chain is ideal at the Θ-temperature. This result can be understood by employing the Doi approach, which is the subject of Problem 15.2 (in this problem, in order to obtain the ideality of the chain at the Θ-temperature, introduce, instead of the volume $v = a^3$, the Flory–Huggins volume $v = (1 - 2\chi)a^3$, which vanishes at the Θ-temperature). For $T < \Theta$, the system is a poor solvent system, and the size of the polymer chain is smaller than the ideal chain radius of gyration. The polymer chain condenses upon itself, forming a dense *globule*. The transition from the regime of good solvent to the regime of poor solvent, across the Θ-temperature (i.e., decreasing T), is abrupt. It is called the coil-globule transition.

5. *The Athermal Regime:* $\chi = 0$ yields the largest possible value of v, i.e., $v = a^3$, and corresponds to a case in which the solvent and the monomers are chemically similar.

15.2.1.2. Osmotic Pressure

The osmotic pressure due to the polymer is

$$p_{\text{osm}} = -\left(\frac{\partial F_{\text{tot}}}{\partial V}\right)_{N_{\text{mono}}}, \tag{15.49}$$

where $N_{\text{mono}} = V\phi a^{-3}$. The differentiation in (15.49) is done under the condition $dN_{\text{mono}}/N_{\text{mono}} \equiv dV/V + d\phi/\phi = 0$; also, $F_{\text{tot}} = F_{\text{mix/site}} V/a^3 = f_{\text{mix/site}} V$. Hence

$$p_{\text{osm}} = \phi^2 \frac{\partial}{\partial \phi} \left\{\frac{f_{\text{mix/site}}}{\phi}\right\}, \tag{15.50}$$

$$\frac{a^3}{k_B T} p_{\text{osm}} = \frac{\phi}{N} + \ln \frac{1}{1-\phi} - \phi - \chi \phi^2. \tag{15.51}$$

(a) Dilute Regime, $\phi \to 0$, $\phi \ll 2N^{-1}$. A series expansion yields

$$p_{\text{osm}} \approx k_B T \left(\frac{c}{N} + \frac{vc^2}{2} \right). \tag{15.52}$$

The osmotic pressure depends on the molecular weight (and is proportional to the number of coils, which act as an ideal gas of coils). The excluded volume term is negligible as long as $2N^{-1} > \phi$, i.e., for small coils or very small concentrations. Now, it is of interest to consider the case of what is called the semi-dilute regime.

(b) Semi-dilute Regime, $2N^{-1} < \phi \ll 1$. Equation (15.51) now reduces to

$$p_{\text{osm}} \approx \frac{k_B T}{2} vc^2; \tag{15.53}$$

i.e., for large N, the osmotic pressure becomes independent of the DP. The prediction of a cross-over between two regimes of small concentrations (dilute vs semi-dilute) will be confirmed by scaling arguments. In fact, it does not happen at $\phi \sim N^{-1}$, as the above argument would let one think.

(c) Dense Limit, $\phi \sim 1$. The relevant variable is now the site fraction $\phi_s = 1 - \phi$ occupied by the solvent. Let us introduce the osmotic pressure of the solvent p_s, a concept that is unsatisfactory from a physical point of view, but will be useful anyway:

$$p_s = - \left(\frac{\partial F_{\text{tot}}}{\partial V} \right)_{N_{\text{solv}}} = \frac{k_B T}{a^3} \left\{ \phi_s + \phi_s^2 \left(\frac{1}{2N} - \chi \right) + \cdots \right\}. \tag{15.54}$$

The excluded volume term vanishes for the infinitely long polymer case, when the solvent is athermal. As already indicated above, the athermal solvent molecules are similar to the monomers of the polymeric chains. Therefore, $a^3/(2N)$, which appears in (15.54) as the excluded volume of the solvent, is also the excluded volume of the monomers. One can therefore expect that in the limit $\phi \to 1$, which is the polymer melt, the excluded volume interactions vanish, and one recovers the ideal chain behavior, as explained in the introduction of this section. We therefore expect that the Flory–Huggins model is valid for high concentrations of polymers of high DP. In effect, the Flory–Huggins model is mostly used with some success to describe dense polymer mixtures (with little or no solvent), their segregation or compatibility characters, stable or metastable states and the spinodal curve, and so on.

15.2.1.3. Compressibility; Measurement of the Excluded Volume

An important theorem in statistical physics relates the Fourier transform of the correlation function of the density, i.e., the scattering function, to the isothermal compressibility

$$\chi_T = -\frac{1}{V}\frac{\partial V}{\partial p_{osm}}\Big|_{T,N_1}.$$

See Appendix 15.B for more details and the establishing of (15.55a). We have

$$a^3 c^2 k_B T \chi_T = S(\mathbf{q} = 0), \tag{15.55a}$$

where $S(\mathbf{q}) = a^3 \int d\mathbf{r}\exp(i\mathbf{q}\cdot\mathbf{r})\langle c(0)c(\mathbf{r}) - c^2\rangle$ is the scattering function and c is the volume density of particles. Equation (15.55a) can also be written under the very convenient form as

$$S(\mathbf{q} = 0) = \left(\frac{\partial^2 F_{\mathrm{mix/site}}}{k_B T \partial \phi^2}\right)^{-1}, \tag{15.55b}$$

where the free energy per particle is expressed in function of the volume fraction (15.46). Here, we specialize the formulae to the case of a dilute solution ($\phi = ca^3 \rightarrow 0$). Employing (15.52) for the osmotic pressure, which we write in a standard form as $\frac{1}{k_B T}p_{osm} = \frac{c}{N} + A_2 c^2 + \cdots$, one gets

$$S(\mathbf{q} = 0) = \frac{1}{a^3 c}\frac{1}{N^{-1} + 2cA_2}, \tag{15.56}$$

which provides an experimental method to measure A_2, the second coefficient of the virial expansion, when N is large, i.e., the excluded volume.

15.2.2. Scaling Laws for Athermal Solutions

The Flory–Huggins approach treats the polymer species and the solvent as ideal gases to calculate their entropy. It does not take into account the excluded volume effect. This section discusses the case of athermal solutions ($\chi = 0$) from another perspective, the scaling laws. The athermal case is important because it is, experimentally, a case with no other interactions but the excluded volume, a "pure" case. Therefore, the Flory–Huggins approach should be particularly bad. We shall compare the two approaches.

15.2.2.1. Critical Concentration for the Semidilute Solution

We shall say that a solution is *dilute* when the polymeric coils are not in contact. Let $c^* \sim N/R_F^d$ be the mean concentration of monomers (segments) in the coil. We take for the radius of the coil the Flory radius, because $\chi = 0$ means that the polymer is in a good solvent. It is clear that the coils are in contact when $c > c^*$. Let us estimate the volume fraction for c^*. We have

$$c^* \cong \frac{N^{1-dv}}{a^d}; \quad \text{i.e.,} \quad \phi^* \cong N^{1-dv}. \tag{15.57}$$

In dimension $d = 3$, with $N = 10^4$, and $v = 3/5$, one gets $\phi^* \sim 10^{-3}$. The coils are in contact in a good solvent for small concentrations.

$c > c^*$, the solution is qualified *semidilute*. Notice the difference with the definition of a semidilute regime we were led to state in the frame of the Flory–Huggins theory; i.e., $\phi^* \cong N^{-1}$. The Flory–Huggins definition of the crossover between dilute and semidilute coincides with the present definition for $dv = 2$, i.e, precisely for the dimension $d = 4$ (where $v = 1/2$, see Table 15.1), for which mean field theory applies. de Gennes has noticed that, because in a semidilute solution the chains are interpenetrating, there is no way to tell to which coil belongs a monomer chosen at random. Therefore, the properties of conformation should not depend on N in this *universal regime*. Let us consider the consequences of this new situation on the osmotic pressure.

1. $c < c^*$. One might expect to improve the expression (15.52) of the osmotic pressure by adding a term of excluded volume proportional to R_F^d, viz.:

$$\frac{1}{k_B T} P_{\text{osm}} = \frac{c}{N} + A_2 c^2 + \cdots, \tag{15.58}$$

where $A_2 \propto N^{1-dv}$. This scaling law has been verified with great accuracy.

2. $c > c^*$. The osmotic pressure should not depend on N, on the one hand (following de Gennes's argument), and tends toward c/N when $c \to c^*$, on the other hand. Let us then write

$$\frac{1}{k_B T} P_{\text{osm}} \propto \frac{c}{N} \left(\frac{c}{c^*} \right)^x, \tag{15.59}$$

where x is obtained from the condition of independence on N. Using (15.57), one gets

$$x = \frac{1}{dv - 1} \quad \text{and} \quad \frac{1}{k_B T} P_{\text{osm}} \propto a^{-d} \phi^{\frac{dv}{dv-1}}. \tag{15.60}$$

Again, the Flory–Huggins mean-field prediction $P_{\text{osm}} \propto \phi^2$ is recovered for $d = 4$. For $d = 3$, the exponent $y = \frac{dv}{dv-1} = 9/4$ is well confirmed by the experiments; it is different from the mean field exponent 2, and the effect is very conspicuous if ϕ is very small.

15.2.2.2. Internal Energy for Semidilute Solutions

Let us consider a bond of the lattice model on which sits a monomer, with a probability ϕ. The conditional probability for a neighboring bond to be occupied is not ϕ, but ϕ^{y-1}. This conditional probability comes from a straightforward interpretation of (15.60): The osmotic pressure is due to a gas with pair interactions, yielding an exponent equal to 2 only in mean field theory (the ϕ^2 term; each element of the pair plays a mirror role with respect to the other). In $d \neq 4$ dimensions, the number of *effective* pairs of monomers is not $\sim \phi^2$, but $\sim \phi^y$. Consequently, the *effective* number of monomer-solvent molecule pairs is $\phi(1 - \phi^{y-1})$, and the internal energy can be written as

$$U_{\text{mix/site}} = k_B T \chi \phi (1 - \phi^{y-1}). \tag{15.61}$$

This expression is valid in the regime of good solvent.

15.2.2.3. Correlation Length in Semidilute Solutions

The fluctuations of density have a characteristic length ξ equal to the Flory radius R_F when $c = c^*$. Employing the same method as above, i.e., finding an exponent m

$$\xi \cong R_F \left(\frac{\phi^*}{\phi} \right)^m, \tag{15.62}$$

that makes ξ independent of N, we obtain from (15.27), (15.57), and (15.62),

$$m = \frac{\nu}{d\nu - 1}, \quad \xi \cong a\phi^{-m}. \tag{15.63}$$

The osmotic pressure can easily be expressed as a function of the correlation length: $p_{\text{osm}} \cong k_B T \xi^{-d}$, which suggests that the "blobs" of size ξ behave as independent objects. Let us for a while assume that all monomers of a blob belong to the same chain. This assumption is reasonable because, according to (15.63), the size ξ of a blob is smaller than is the Flory radius. The number of monomers per "blob" can then be estimated as $g \sim (\xi/a)^{1/\nu}$, which is also equal to $c\xi^d$. Therefore, the density of monomers in a blob is equal to the density in the solution, and one can think of the blobs as objects tiling space densely. One can then consider blobs belonging to the same chain as "suprasegments" of a molten polymer, in which the "suprachains" (i.e., the chains with their conformation of blobs) behave ideally. Hence, the radius of the suprachain scales as

$$R \cong \xi \left(\frac{N}{g} \right)^{1/2} = a N^{1/2} \phi^{(\nu - 1/2)/(d\nu - 1)}, \tag{15.64}$$

which is smaller than is the Flory radius, as expected.

Because the fluctuations of density are of size ξ, one expects that the semidilute polymer acts as a sieve of mesh ξ for particles that cross it through. This property has been employed to measure ξ and the scaling exponents, with good success. See also Daoud et al.[7] for neutron scattering experiments (measurements of the radius of gyration, of the range of the correlations, and of osmotic compressibility), a fundamental paper.

15.3. Phase Separation in Polymer Solutions and Polymer Blends

15.3.1. Liquid Equilibrium States vs. Nonequilibrium States

Polymers in poor solvents and *polymer blends* tend to phase separate. Consider a mixture of two types of polymers A and B, DP's N_A and N_B. In the Flory–Huggins approach, the free energy (15.46) can be generalized to

$$f_m \equiv \frac{F_{\text{mix/site}}}{k_B T} = \frac{\phi_A}{N_A} \ln \phi_A + \frac{\phi_B}{N_B} \ln \phi_B + \chi \phi_A \phi_B, \qquad (15.65)$$

where f_m is the dimensionless energy density and $\phi_A + \phi_B = 1$. If χ is positive and large enough, one expects that the energy decreases when ϕ_A (or ϕ_B) is small and ϕ_B (or ϕ_A) is large. This would happen in the case of a polymer in a poor solvent ($N_B = 1$, χ large). Alternatively, unless χ is extremely small, it is practically impossible to obtain a homogeneous mixture of two polymers present in comparable volume fractions. As we shall see, the incompatibility of two polymers can be used to yield structural features similar to those of surfactants in water (lyotropic phases), i.e., phase separation at a microscopic scale.

These phase separations are first order. First-order phase transitions display strong discontinuities of the state variables, latent heat, and hysteresis phenomena, like *supercooling* (retarded solidification) and *overheating* (retarded melting). In particular, it is well known that if supercooling is large, i.e., if the system keeps the disorder that is typical of the molten state down to very low temperatures, the viscosity can increase to such large values ($\eta > 10^{13}$P) that the system cannot crystallize over macroscopic times. One gets a glass or an amorphous system.

Such hysteretic, nonlinear, phenomena are all the more present in polymeric systems when the kinetics of phase transitions is controlled by slow processes of diffusion (all characteristic times are enlarged compared with atomic alloys and increase more quickly than linearly with the increase of the molecular weight M; diffusivity, for example, scales approximately as $M^{3.4}$). Polymers display further complications due to the actual poly-

---

[7]M. Daoud, J.P. Cotton, B. Farnoux, G. Jannink, G. Sarma, H. Benoit, R. Duplessix, C. Picot, and P.G. de Gennes, Macromolecules **8**, 804 (1975).

dispersity of the chains. We shall not discuss these out-of-equilibrium effects or transport properties in the sequel. In particular, we shall not discuss the complex kinetic and structural phenomena related to *polymer crystallization*, nor the formation of *gels*, i.e., networks of flexible chains, whose frozen liquid-like structures are due to chain reticulation (either irreversible—chemical reticulation, like in *rubber*, or reversible—physical reticulation); see Problem 15.6 for an illustration.

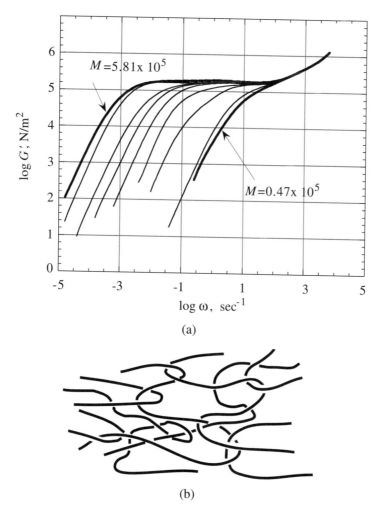

Figure 15.4. Rheology of entangled polymer melts: (a) storage modulus as a function of frequency for nearly monodisperse polysterenes with molecular weight in the range from 580,000 to 47,000. Replotted from data obtained by S. Onogi, T. Masuda, K. Kitawaga, Macromolecules **3**, 109 (1970); (b) entanglements in the plateau region.

As a matter of fact, the dynamics of melts and concentrated solutions display three regions in function of the characteristic time of observation. These regions are schematically visualized Fig. 15.4a, which is a plot of the storage modulus G' ($G'(\omega) = ReG^*(\omega)$, $G^*(\omega) = \sigma/e$, $\sigma = \sigma_0 \exp i\omega t$ the shear stress, $e = e_0 \exp i(\omega t - \varphi)$ the shear strain) in function of frequency. A distinctive feature of this plot is the presence of a *plateau*. The plateau modulus $G'_0 \sim \rho_{ek} k_B T$ can be related to the density of *entanglements* ρ_e or *effective cross-links*, i. e., Fig. 15.4b, the density of obstacles that each chain meets in its movement, due to the other chains. In the range of frequency of the plateau, the viscoelastic properties are those that are typical of a rubber. For higher frequencies, one observes a polymeric glass behavior, related to the relaxation of internal motions of the chains (trans ↔ gauche transitions, etc.). Finally, at low frequencies, the polymer melt behaves like a liquid, $G' \sim \omega^2$.

The reader can refer to Larson's book for a recent treatment of the large-scale (rheological) properties of polymers, and to Doi, Edwards, and de Gennes's books for the dynamical theory. We shall restrict ourselves here to the case of blends that are liquid-like under experimental conditions. Therefore, the states that we discuss are equilibrium states. They are particularly interesting from a conceptual point of view: They can be treated in the frame of the Flory–Huggins mean-field theory, even near critical points. We first need a short reminder of the elements of the theory of first-order phase transitions.

15.3.2. First-Order Phase Transitions, an Overview

15.3.2.1. Conserved and Nonconserved Order Parameter

It is useful to distinguish two types of order parameters, those that do not obey a *conservation law*, like the nematic order parameter (OP) that can take only one value at equilibrium at a given temperature (except at the phase transition temperature if it is a first-order transition), and those that obey a law of conservation, like the density at the liquid-gas transition, or the density of particles of each species A or B in a binary system. In such cases, the OP can take two values, and the system presents two phases in *coexistence*, at any temperature. Each phase takes the volume required by the condition of conservation of the total number of particles. For example (binary system),

$$\phi_{A,1}(T)V_1 + \phi_{A,2}(T)V_2 = \phi_A V, \quad V_1 + V_2 = V, \quad (15.66)$$

i.e., two equations for two unknowns V_1 and V_2. The transition is necessarily first order, except at some *critical temperature* where $\phi_{A,1} = \phi_{A,2}$.

15.3.2.2. Free Energy of a Binary Mixture

We discuss the case of a binary mixture, and we assume that the components A and B form ideal mixtures in each phase p1 and p2. In the absence of interaction, the free energy per particle is restricted to the entropic part, which can be written as

$$F = k_B T \left[\frac{\phi_A}{N_A} \ln \phi_A + \frac{(1 - \phi_A)}{N_B} \ln(1 - \phi_A) \right].$$ (15.67)

The minimum of this expression is unique and equal to

$$\phi_A / (1 - \phi_A) = \exp(N_B^{-1} - N_A^{-1})$$

at any temperature. One expects that the existence of this solution is favored at high temperatures, when the entropy contributions are predominant. At lower temperatures, the term of interaction may deform the curve $F = F(\phi_A)$ from a purely convex shape (Fig. 15.5a) to a shape with two minima (Fig. 15.5b).

15.3.2.3. Phase Separation

Let us now bring our atttention to the compositions ϕ_1 and ϕ_2, whose representative points are the contacts of the bitangent to the free energy curve. These two compositions can co-exist, because the chemical potentials $\mu(\phi, T) = \partial F / \partial \phi$ are equal in p1 and p2. Fig. 15.5c represents the chemical potential as a function of ϕ. Note that at a given temperature, it is not $F = F(\phi)$, but $G(\phi) = F(\phi) - \mu(T)\phi$ that is minimized; here, $\mu(T)$ is the chemical potential at the coexistence. Figure 15.5c is a plot of the slope $\mu(\phi)$ of the energy $F(\phi)$ in function of the volume fraction ϕ at a given constant temperature, i.e., of the chemical potential of all compositions at temperature T. The chemical potential $\mu(T)$ at the coexistence is represented by a horizontal line that cuts the $\mu = \mu(\phi)$ plot into two equal areas (Problem 15.4).

A solution whose global composition ϕ is between ϕ_1 and ϕ_2 will split into the phases p1 and p2, in a ratio x such that $\phi = x\phi_1 + (1 - x)\phi_2$, $F = x F_1 + (1 - x) F_2$. It is easy to show that the minimization of F under the condition that the chemical potentials are equal for the splitting compositions ϕ_1 and ϕ_2 yield $G(\phi_1) = G(\phi_2)$, i.e., precisely the condition of the Maxwell plateau, independent of x. The representative point of the set of coexisting phases is in p (Fig. 15.5b). One observes that this point is *below* the point on the free energy curve at the same global composition: The phase separation is favored. On the other hand, a solution whose composition ϕ' is outside the $\phi_1 - \phi_2$ interval does not split into two different phases.

Plotting all together the pairs of coexisting phases at all temperatures, one gets a *phase diagram*. A typical shape of the *coexistence curve* is represented in Fig. 15.5d, with critical temperature T_c and critical composition ϕ_c. We also plot the locus of the points of inflexion p*1 and p*2 of the free energy, the so-called *spinodal curve*.

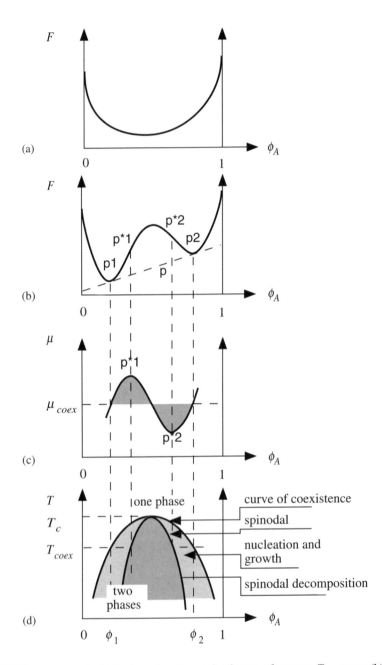

Figure 15.5. Free energy of a biphasic system: (a) predominance of entropy, $T = \mathrm{const}$; (b) total free energy, $T = \mathrm{const}$; (c) chemical potential, the Maxwell plateau; (d) a typical phase diagram.

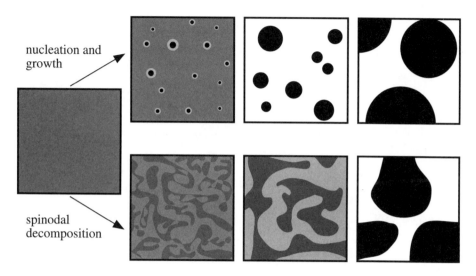

Figure 15.6. Structural evolution in phase-separating binary mixture of "black" and "white" species in the nucleation and growth (top) and spinodal decomposition (bottom) scenarios. The shades of gray roughly correspond to concentration of species. Coarsening at the later stages is driven by positive interfacial energy.

15.3.2.4. Spinodal Decomposition vs. Nucleation and Growth

The transition from a high-temperature homogeneous mixture ϕ, T_h to a low-temperature two-phase system ϕ_1, ϕ_2, T_l occurs differently, according to the position p of the mean composition ϕ_0 with respect to the spinodal curve, at least during the first stage of the phase separation process (Fig. 15.6). If p is inside the spinodal curve, where $\partial^2 F/\partial\phi^2|_{\phi=\phi_0} < 0$, the system is unstable and phase-separate according to the so-called spinodal decomposition process; if p is between the spinodal and the curve of coexistence, $\partial^2 F/\partial\phi^2|_{\phi=\phi_0} > 0$, the system is metastable and it does not phase-separate before the appearance of a nucleus of concentration ϕ_1 or ϕ_2. The related process of nucleation and growth has been discussed in Section 13.1.4 in some detail. The next paragraph presents a simplified theory of the spinodal decomposition.[8]

Let us expand the free energy in the vicinity of ϕ_0:

$$F = F(\phi_0) + (\phi - \phi_0)F_1 + (\phi - \phi_0)^2 F_2 + \cdots . \tag{15.68}$$

The chemical potential can be written as

$$\mu = F_1 + 2(\phi - \phi_0)F_2 + \cdots . \tag{15.69}$$

[8] J.W. Cahn, J. Chem. Phys. **42**, 93 (1965).

Consider the effect of fluctuations of the composition: The particles move under the effect of the chemical gradient $f = -\partial\mu/\partial x = -2F_2\partial\phi/\partial x$. Now, according to Einstein relation, the velocity of the particles can be written as

$$u = \frac{fD}{k_BT} = 2F_2\frac{D}{k_BT}\frac{\partial\phi}{\partial x}, \tag{15.70}$$

where D is the diffusivity. On the other hand, the law of conservation of the number of particles can be written as

$$\frac{\partial\phi}{\partial t} = -\frac{\partial(\phi u)}{\partial x}. \tag{15.71}$$

Therefore, we have

$$\frac{\partial\phi}{\partial t} = -2\frac{\phi_0 F_2 D}{k_BT}\frac{\partial^2\phi}{\partial x^2}, \tag{15.72}$$

i.e., with $\phi = \phi_0 + \delta\phi_0 \exp i(\omega t - qx)$,

$$i\omega = -2\frac{\phi_0 F_2 D}{k_BT}q^2. \tag{15.73}$$

Hence, if $F_2 < 0$ (inside the spinodal curve), any infinitesimal fluctuation exponentially increases with time. The fluctuations of smaller wavelengths grow quicker. The opposite is true in the "metastable" region, where the fluctuations decay with time. In this later case, only nuclei larger than a critical size can grow, as explained in Section 13.1.4. Consequently, their growth requires one to pass an energy barrier (nucleation energy).

The intensity and wavevector of fluctuations in the spinodal region can be directly probed by scattering experiments; it is observed that the peak position moves in the course of time to smaller scattering angles, i.e., larger wavelengths, as expected.

Many direct observations of the two types of growth process have been reported in polymers. In the growth and nucleation case, one observes the formation of spherical droplets, located at random in the matrix, of the composition "opposite" to that one in the matrix. In the spinodal case, a typical pattern would consist of interpenetrating continuous domains. As the decomposition proceeds, the domains enlarge and their boundaries become stiffer. For observations and a description of the spinodal decomposition in polymers, see Strobl's book.

15.3.3. Polymers Blends

The coexistence and spinodal curves obtain, respectively, for $\partial f_m/\partial\phi = 0$ and $\partial^2 f_m/\partial\phi^2 = 0$. The critical point also obeys $\partial^3 f_m/\partial\phi^3 = 0$. Let us first, for the sake of simplicity, consider the case in which the two polymers in the blend have the same DP $N = N_A = N_B$. The temperature is included in the Flory interaction parameter: $\chi \propto T^{-1}$ if the van der Waals interactions are predominant. One gets, starting from (15.65), for the coexistence curve,

$$\partial f_m / \partial \phi \equiv \frac{1}{N} \ln \left(\frac{\phi}{1 - \phi} \right) + (1 - 2\phi)\chi = 0, \tag{15.74a}$$

for the spinodal curve,

$$\partial^3 f_m / \partial \phi^3 \equiv \frac{1}{N} \left(\frac{1}{\phi} + \frac{1}{1 - \phi} \right) - 2\chi = 0, \tag{15.74b}$$

for the critical point,

$$\phi_c = 1/2, \; \chi_c = 2/N. \tag{15.74c}$$

The solutions of (15.74b) are real if $N\chi > 2$. This is also the condition of existence of a spinodal curve. Phase separation occurs for $\chi > \chi_c$, a condition that is physically easily achieved, because χ_c is so small when N is large. It is difficult to get χ very small, and segregation is unavoidable in practice.

The general case $N_A \neq N_B$ easily yields

$$\phi_{A,c} = \frac{N_B^{1/2}}{N_A^{1/2} + N_B^{1/2}}, \quad 2\chi_c = \frac{(N_A^{1/2} + N_B^{1/2})^2}{N_A N_B}, \tag{15.75}$$

which indicates that if $N_A \neq N_B$ are very different, χ_c can be large enough, and segregation avoided (e.g., for N_A small, N_B large, $\chi_c \sim 1/N_A$).

The critical fluctuations can be treated in the so-called random phase approximation (RPA), a method that was first developed for nearly free electrons in metals,[9] and then extended to (nearly ideal) polymer chains in melts.[10] The principle is given in Appendix 15.B: The Fourier transform of the pair correlations of the fluctuations of density, which is proportional to the response function, is the Fourier transform of the scattered intensity. One gets

$$\frac{1}{S(q)} = \frac{1}{\phi_A g(N_A, q)} + \frac{1}{\phi_B g(N_B, q)} - 2\chi, \tag{15.76}$$

where $g(N_A, q)$ (respectively, $g(N_A, q)$) is the pair correlation Fourier transform for an isolated A (respectively, B) chain. Equation (15.76) is the generalization of (15B.22) to a

---

[9]D. Pines, Elemementary Excitations in Solids, W.A. Benjamin, New York, 1963.
[10]P.G. de Gennes, Rep. Progr. Phys. **32**, 187 (1969).

binary system. Because $g(N_A, 0) = N_A$, $g(N_B, 0) = N_B$, one gets

$$\frac{1}{S(0)} = \frac{1}{N_A \phi_A} + \frac{1}{N_B \phi_B} - 2\chi = \frac{\partial^2 f_m}{\partial \phi^2}. \tag{15.77}$$

Therefore, the small angle scattering intensity diverges when $q \to 0$, not only at the critical point, but on the whole spinodal curve.

15.3.4. Microscopic Phase Separation into Block Copolymers

A way to force incompatible polymer chains to live together is to join them chemically end to end. These are the so-called *block* copolymers. Phase segregation is limited to the microscopic scale, and the resulting macroscopic phase shows structural similarity with surfactants in presence of a solvent (Fig. 15.7). The microphase separation from the melt is a fluctuation-induced, first-order phase transition.[11]

The morphology is dependent on the ratio N_A/N_B. One expects that the resulting phase is *lamellar* if $N_A/N_B \approx 1$ (assuming for the sake of simplicity equal monomeric volumes $v_A = v_B$ for the two polymers), and *micellar* (spheres of A in a continuous matrix of B, say) if $N_B \gg N_A$. Body-centered cubic micellar phases have indeed been observed, as well as *columnar* hexagonal and lamellar phases (see Fig. 15.7).[12] This sequence of phases in the range $0 \leq \phi_A \leq 0.5$ correlates to a decrease of the constant mean curvature of the so-called intermaterial dividing surface (IMDS) between the A and B chains ($\sigma_1 + \sigma_2 \cong 2R_A^{-1}$ for the micellar phase, $\sigma_1 + \sigma_2 \cong R_A^{-1}$ for the columnar phase, $\sigma_1 + \sigma_2 = 0$ for the lamellar phase).[13] Here, R_A stands for the radius of gyration of the ideal chain A (e.g., $R_A = aN_A^{1/2}$), but the true repeat distances present in the phases depend on a balance between the IMDS surface energy and the elastic energy of the chains, which is of the rubber type (see Problem 15.7). One has also observed *cubic bicontinuous* phases between the lamellar and the columnar phases, with double diamond symmetry of the rods of A chains tetracoordinated at points of junction (ordered bicontinuous double diamond, or OBDD).[14] It is expected that the mean curvatures of those cubic phases stand between those of the lamellar and columnar phases. But their stability necessarily involves, apart from the balance alluded to above, contributions from κ and $\bar{\kappa}$ curvature terms.

There is a wealth of other phases associated with more complex block (triblock, etc....) or side-grafted copolymers. For a review of the theoretical and experimental work on phases of block copolymers, see Bates and Fredrickson.[15]

[11]G.H. Fredrickson and K. Binder, J. Chem. Phys. **91**, 7265 (1989).

[12]G.E. Molau, in Block Polymers, Edited by S.L. Aggarwal, Plenum Press, New York, 1970, p. 79.

[13]E.L. Thomas, D.M. Anderson, C.S. Henkee, and D. Hoffman, Nature **334**, 598 (1988).

[14]E.L. Thomas, D.B. Alward, D.J. Kinning, D.C. Martin, D.L. Handlin, and L.J. Fetters, Macromolecules **19**, 2197 (1986).

[15]F.S. Bates and G.H. Fredrickson, Ann. Rev. Phys. Chem. **41**, 525 (1990).

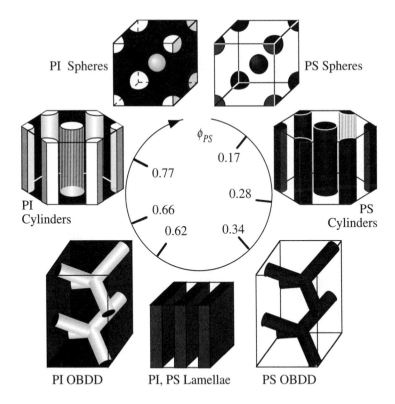

Figure 15.7. Various morphologies of polystyrene (PS) -polyisoprene (PI) diblock copolymers in function of the volume fraction ϕ_{PS} of PS (pictured in black). Data taken from Ref. [15].

A number of studies, experimental (mostly transmission electron microscope observations accompanied with simulations of the contrast) and theoretical, have been devoted to *grain boundaries* in lamellar block copolymers (polystyrene-polybutadiene[16]; polystyrene-poly ethylene, propylene[17]). Gido et al. [16] recognize two types of *twist* grain boundaries and a number of symmetric and asymmetric *tilt* boundaries. In the twist boundaries, the IMDS is continuous through the wall region and tends to take the shape of a minimal surface; the two morphologies approximate either the Scherk doubly periodic surface (over the whole twist range, from 0° to 90°) or a section of a right helicoid (in the small angle range). The *chevron* tilt boundary is the equivalent of the symmetric curvature wall—low angle tilt boundary— described in Chapter 10. In the *omega* symmetric tilt boundary, which occurs

---

[16](a) S.P. Gido, J. Gunther, E.L. Thomas and D. Hoffman, Macromolecules, **26**, 4506 (1993); (b) S.P. Gido and E.L. Thomas, Macromolecules **27**, 849 (1994); (c) S.P. Gido and E.L. Thomas, Macromolecules **27**, 6137 (1994); (d) B.L. Carvalho, R.L. Lescanec and E.L. Thomas, Macromol. Symp.,**98**, 1131 (1995).

[17]Y. Nishikawa, H. Kawada, H. Hasegawa, and T. Hashimoto, Acta Polymer. **44**, 192 (1993).

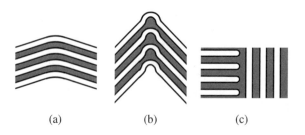

(a)   (b)   (c)

Figure 15.8. Tilt walls in the diblock copolymers lamellar phase. (a) chevron or curvature wall; (b) omega wall; (c) T-wall (adapted from Gido and Thomas [16c])

for larger angles, the wall distortions are relaxed by periodic protrusions of the IMDS; the simplest type of asymmetric wall is the T-wall. The tilt walls are illustrated in Fig. 15.8.

15.4. Rigid and Semiflexible Polymers

We are now considering a family of main chain polymers that differ considerably from the former ones by the ratio ℓ_p/d, where d is now the diameter of the cross section of the molecule. For *flexible* polymers, $\ell_p/d \approx 3$ to 5, and the persistence length is comparable to a monomer length. On the other hand, viruses with helical symmetry,[18] polypeptides, DNA, cellulose esters, or aromatic polyesters have $\ell_p/d \approx 20$ to 200. Actin filaments constitute an extreme case, but of considerable importance, because their presence is ubiquitous in cells of all living bodies. Actin is a globular protein ($\approx 42\,000$ Da) that assembles in broad ($d \approx 7$ nm) and long (chemical length up to $L = 50\,\mu$m) semiflexible filaments (ℓ_p/d is as large as 1500 to 3000). As a crude classification, we shall differentiate *stiff rod* polymers, for which $L < \ell_p$, and *semiflexible* polymers, $L > \ell_p$.

15.4.1. Rigid Rods

Rigid, rod-like polymers differ from semiflexible polymers by a number of features, as follows:

- The radius of gyration scales linearly with the molecular weight.

- Dynamical microscopic properties are characterized by a longer relaxation time and a smaller diffusivity.

- The high molecular anisotropy allows for an easier alignment of the isotropic phase under shear, magnetic, or electric fields.

---

[18]D.L.D. Caspar and A. Klug, Cold Spring Harbor Symp. Quant. Biol. **27**, 1 (1962).

- As the concentration increases, the interactions between rods become more and more important. Eventually, a transition to a liquid crystalline phase can occur, whose features are generally dominated by excluded volume effects (entropic forces), as already discussed in Chapter 4. Historically, the first observed phase transition of this type is by Bernal and Fankuchen in solutions of plant viruses.[19] The study of phase diagrams of viruses is still a very promising area of research (e.g., TMV,[20,21] fd, pf1;[22] even mixtures of TMV and spherical colloids[23]).

Dilute, semidilute, and concentrated solutions of rigid, rod-like polymers have been the subject of a number of theoretical and experimental studies (structural and dynamical properties), which are reviewed in Doi and Edwards's book and article.[24]

15.4.2. Semiflexible Polymers

With semiflexible polymers, a new characteristic length noted ℓ_e, comes into play; ℓ_e is the *deflection length*, which is the distance between contacts of a chain with neighboring chains[25,26] (Fig. 15.9).

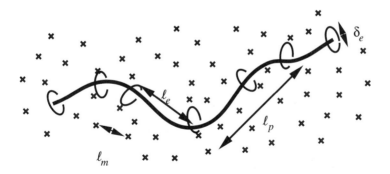

Figure 15.9. Characteristic lengths in a semidilute solution of semiflexible polymers (tube model). Surrounding polymer chains (shown as crosses) confine the given chain to a tube-like region.

[19]J.D. Bernal and I. Fankuchen, J. Gen. Physiol. **25**, 111 (1941).

[20]A. Klug, Fed. Proceed. **31**, 30 (1972).

[21]R.B. Meyer, in Dynamics and Patterns in Complex Fluids, Edited by A. Onuki and K. Kawasaki, Springer Proceed. Phys. **52**, 62 (1990).

[22]S. Fraden, in Observation, Prediction, and Simulation of Phase Transitions in Complex Fluids, Edited by M. Baus, L.F. Rull,and J.P. Ryckaert, NATO-ASI- Course CXXIX, Kluwer Acad. Pub., 113 (1995).

[23]M. Adams, Z. Dogic, S.L. Keller, and S. Fraden, Nature **393**, 349 (1998).

[24]M. Doi and S.F. Edwards, J.Chem. Soc., Faraday Trans. **74**, 1802, 1818 (1978).

[25]T. Odijk, Macromolecules **19**, 2313 (1986).

[26]A.N. Semenov, J. Chem. Soc., Faraday Trans. **82**, 317 (1986).

Let $c_s = \frac{n_L}{V}$ be the concentration of chains in a medium of volume V with n_L chains of length L. Let us also introduce the concentration of "monomers"' which here would be the concentration of segments of length ℓ_p, viz. $c = \frac{n_L}{V} \frac{L}{\ell_p}$, in accord with the notations used until now. The volume density of chain length is $\rho = c_s L = c \ell_p$, and the mean distance between chains is $\ell_m = \rho^{-1/2}$. But this is not the transversal size of the "tubes" in which the chain fluctuates.

Consider a segment of chain of length ℓ. The average angular span of the segment and the average transversal span are (see Section 15.1)

$$\theta(\ell) \approx 2 \left(\frac{\ell}{\ell_p} \right)^{1/2}, \quad \delta(\ell) \sim \ell\theta(\ell) \sim \ell^{3/2}\ell_p^{-1/2}. \tag{15.78}$$

The number of chains that pierce a tube of length ℓ and width $\delta(\ell)$ is $N_s = \delta(\ell) \times \ell \times \rho$. Making $N_s = 1$, one can estimate the length of a free segment as

$$\ell_e \sim \ell_p (\rho\ell_p^2)^{-2/5}, \tag{15.79}$$

and the tube size as

$$\delta_e \sim \ell_p (\rho\ell_p^2)^{-3/5}. \tag{15.80}$$

Our discussion assumes implicitly that the deflection length and the tube size are smaller than is the persistence length, $\ell_e < \ell_p$, $\delta_e \ll \ell_p$. This implies $\rho\ell_p^2 > 1$; i.e., $1/(L\ell_p^2) \ll c_s$. The last condition on the concentration has to be verified in the semidilute regime ($c_s > 1/R_G^3$, i.e. $L\ell_p^2 < R_G^3$), in order to get interactions between chains. The notion of deflection length extends to polymers the notion of Helfrich patches characteristic length (Section 14.2.2), which was developed for membranes.

The dynamical properties of this semidilute regime have been studied recently for actin filaments, experimentally[27,28] as well as theoretically.[29,30] But on the whole, this subject is far from being explored in detail.

At higher concentrations, a transition to a liquid crystalline phase is possible due to the excluded volume interactions. We have met such a transition for rigid rods (Chapter 4). In the case of semiflexible polymers, the first-order isotropic-nematic transition is obtained[31] for $\rho_{\text{nem}} \approx 6.7/(\ell_p d)$ (to be compared with Onsager's result: $\rho_{\text{nem}} \approx 4.3/(Ld)$).

---

[27] A. Ott, M. Magnasco, A. Simon, and A. Libchaber, Phys. Rev. **E48**, R1642 (1993).

[28] B. Schnurr, F. Gittes, F.C. MacKintosh, and C.F. Schmidt, Macromolecules **30**, 7781 (1997).

[29] H. Isambert and A.C. Maggs, Macromolecules **29**, 1036 (1996).

[30] D.C. Morse, Macromolecules **31**, 7030 (1998).

[31] A.R. Khokhlov and A.N. Semenov, Physica **108A**, 546 (1981); Sov. Phys. Uspekhi **31**, 988 (1988).

15.4.3. Chirality

Many semiflexible polymers—biopolymers—are chiral and get from that circumstance a number of interesting structural properties. Among those must be cited their hierarchichal structure and the nature of the long-range order. We shall content ourselves with a few remarks and citations, on a subject that has progressed, since Biot's discovery of rotary power, "with reckless abandon."[32]

Hierarchical order has been recognized in the conformation of protein macromolecules (Section 1.5), and in fibrous biological tissues[33] like collagen,[34] α-keratin,[35] and cellulose. The fibrillar organization yields outstanding mechanical properties to these structures, and it is of utmost importance to understand the nature of these arrangement, altogether for practical biological purposes and for the possible design of similar synthetic materials. Clearly, these structures depend on the chemical nature of the macromolecules, but the frequency of hexagonal fibrillar assemblies in biological tissues indicate the importance of the contribution of steric interactions. As indicated in Kleman,[36] the close packing of helically shaped molecules raises interesting questions of frustration, bearing similarities with the *double-twist* conformation discussed in Section 11.2. It may play a role at the origin of the hierarchy of the fibrous structures.

The same arguments of frustration are to be taken into account in a full description of the long-range order observed in phases of solutions of biopolymers,[37,38] like DNA, xanthan, PBLG, and so on. Finally, recent experimental studies of hexagonal phases of DNA under pressure, which revealed its hexatic structure, and theoretical analyses of hexagonal phases of polymers,[39] have pointed out the importance of the contribution of chirality.[40]

15.A. Appendix A: The Central Limit Theorem

This theorem states that the sum of N random independent variables x_i $(i = 1, \ldots, N)$ is a *Gaussian* random variable, when N goes to infinity, if the moments $\langle x_i^n \rangle$ exist. A Gaussian

---

[32]T.C. Lubensky, R.D. Kamien, and H. Stark, preprint, University of Pennsylvania, 1995.

[33]D.A. D. Parry and P.M. Steinert, Intermediate Filament Structure, Springer-Verlag, New York, 1995.

[34]L.J. Gathercole and A. Keller, in Structure of Fibrous biopolymers, Edited by E.D.T. Atkins and A. Keller, Butterworth, London, 1975; in The Periodontal Ligament in Health and Disease, Edited by B.K.B. Berkovitz, Pergamon, Oxford, 1982.

[35]F. Briki, B. Busson, and J. Doucet, Biochim. Biophys. Acta **1429**, 57 (1998).

[36]M. Kleman, J. Phys. Lett. **46**, L723 (1985), Physica Scripta **T19**, 565 (1987).

[37]F. Livolant, J. Physique **48**, 1051 (1987).

[38]F. Livolant and Y. Bouligand, J. Physique **47**, 1813 (1987).

[39]R. Podgornik, H.H. Strey, K. Gawrish, D.C. Rau, A. Rupprecht, and V.A. Parsegian, Proc. Nat. Acad. Sci. U.S.A. **93**, 4261 (1996).

[40]R.D. Kamien and D.R. Nelson, Phys. Rev. **E53**, 650 (1996).

random variable ξ is a variable that follows a probability law of the exponential type of (15A.7). We have $\sigma^2 = \int \xi^2 p(\xi) d\xi$; σ is called the variance.

Let $X = \sum_1^N x_i$. The mean square value of X is $\langle X^2 \rangle = \langle \sum_{i,j} x_i x_j \rangle$. Because the variables are independent, one has $\langle x_i x_j \rangle_{i \neq j} = 0$. We assume now that all mean square values $\langle x_i^2 \rangle = \sigma^2$. Define $\xi = N^{-1/2} X$. We get

$$\langle \xi^2 \rangle = \sigma^2, \tag{15A.1}$$

which does not depend on N.

We call *characteristic function* $\varphi_\xi(k)$ of a random variable ξ the Fourier transform of its distribution function, i.e.,

$$\varphi_\xi(k) = \langle \exp(ik\xi) \rangle \equiv \int \exp(ik\xi) p(\xi) \, d\xi. \tag{15A.2}$$

Reciprocally,

$$p(\xi) = \frac{1}{2\pi} \int \exp(-ik\xi) \varphi_\xi(k) \, dk. \tag{15A.3}$$

With the above expression of ξ, one gets

$$\varphi_\xi(k) = \left\langle \exp\left(ikN^{-1/2} \sum x_i\right) \right\rangle = \left[\varphi\left(kN^{-1/2}\right) \right]^N, \tag{15A.4}$$

because all the x_i have the same probability distribution.

Let N go to infinity, and one can expand the characteristic function $\varphi(kN^{-1/2})$:

$$\varphi(kN^{-1/2}) = 1 - \frac{k^2}{2N} \langle x_i^2 \rangle + \cdots = 1 - \frac{k^2}{2N} \sigma^2 + O(k^3 N^{-3/2}); \tag{15A.5}$$

hence,

$$\varphi_\xi(k) = \lim_{N \to \infty} \left[1 - \frac{k^2}{2N} \sigma^2 \right]^N = \exp -\frac{k^2 \sigma^2}{2}, \tag{15A.6}$$

whose Fourier transform is

$$p(\xi) = \frac{1}{(2\pi\sigma^2)^{1/2}} \exp -\frac{\xi^2}{2\sigma^2}. \tag{15A.7}$$

15.B. Appendix B: Isothermal Compressibility and Density Fluctuations; Static Linear Response

1. The *isothermal compressibility* is defined as

$$\chi_T = -\frac{1}{V}\left(\frac{\partial V}{\partial p}\right)_{N,T}.$$ (15B.1)

We prove that

$$\frac{1}{c\chi_T} = N\left(\frac{\partial \mu}{\partial N}\right)_{V,T},$$ (15B.2)

where μ is the chemical potential and $c = N/V$ is the number density of objects (particles, bonds, etc.). The interest of the expression (15B.2) of the compressibility will appear below.

We first use the Helmholtz free energy $F = F(T, V, N)$. We have

$$p = -(\partial F/\partial V)_{T,N}, \quad \mu = (\partial F/\partial N)_{T,V}.$$

Hence, by the equality of the second derivatives,

$$\left(\frac{\partial \mu}{\partial V}\right)_{T,N} = -\left(\frac{\partial p}{\partial N}\right)_{T,V}.$$ (15B.3)

The Gibbs–Duhem relation, written at constant temperature, is $V\,dp = N\,d\mu$, i.e.,

$$V\left(\frac{\partial p}{\partial N}\right)_{T,V} = N\left(\frac{\partial \mu}{\partial N}\right)_{T,V}.$$ (15B.4)

Now, using the Gibbs free energy $G = G(T, p, N) = N\mu$, we have

$$N(\partial \mu/\partial V)_{T,N} = (\partial G/\partial V)_{T,N} \equiv (\partial G/\partial p)_{T,N}(\partial p/\partial V)_{T,N}$$
$$= V(\partial p/\partial V)_{T,N},$$

which yields

$$N\left(\frac{\partial \mu}{\partial V}\right)_{T,N} = -\frac{1}{\chi_T}.$$ (15B.5)

Reporting in (15B.3) and (15B.4), one gets (15B.2).

2. In this section, we use the *grand canonical partition function* and (15B.2) to prove the following:

$$\frac{\langle N^2 \rangle - \langle N \rangle^2}{\langle N \rangle} = c k_B T \chi_T. \tag{15B.6}$$

The fluctuations of the number of particles appear in the grand canonical partition function, which is defined as

$$\Xi = \sum_{N=0}^{\infty} \frac{a_c^N}{N!} Z_N(V, T), \tag{15B.7}$$

where $a_c = \lambda_{dB}^{-3} \exp(\beta \mu)$ is the activity ($\lambda_{dB} = (2\pi\beta\hbar^2/m)^{1/2}$ is the de Broglie wavelength, $\beta = 1/k_B T$) and $Z_N(V, T)$ is the configuration integral (proportional to the canonical partition function,

$$Z_N(H_N, T) = \int \exp(-\beta H_N(\mathbf{r}_1 \ldots, \mathbf{r}_N, \mathbf{p}_1 \ldots, \mathbf{p}_N)) \, d\mathbf{r}^N \, d\mathbf{p}^N).$$

The grand canonical ensemble thermal average of any state variable

$$A(\mathbf{r}_1, \mathbf{r}_2, \ldots, \mathbf{r}_N; \mathbf{p}_1, \mathbf{p}_2, \ldots, \mathbf{p}_N)$$

(noted $A(\mathbf{r}^N; \mathbf{p}^N)$) can be written as

$$\begin{aligned}
\langle A \rangle &= \frac{1}{\Xi} \sum_{N=0}^{\infty} \frac{a_c^N}{N!} \int A(\mathbf{r}^N; \mathbf{p}^N) \exp(-\beta H_N) \, d\mathbf{r}^N \, d\mathbf{p}^N \\
&= \frac{\sum_{N=0}^{\infty} \mathrm{Tr}\,(A \exp(-\beta H_N + \beta \mu N))}{\sum_{N=0}^{\infty} \mathrm{Tr}\,(\exp(-\beta H_N + \beta \mu N))}.
\end{aligned} \tag{15B.8}$$

The grand canonical partition function is related to the grand potential $\Omega(T, V, \mu) = -pV$ by the relation $\Xi = \exp(-\beta\Omega)$. One gets the important relation:

$$\frac{\langle N^2 \rangle - \langle N \rangle^2}{\langle N \rangle} = \frac{k_B T}{\langle N \rangle} \frac{\partial \langle N \rangle}{\partial \mu}. \tag{15B.9}$$

In the thermodynamic limit, for an infinite system, we have $N = \langle N \rangle$, hence, (15B.6).

3. *A general relation for response functions.*[41,42] We want to characterize the variations of a set of state variables $A_\bullet$ ($\bullet = m, n, \ldots$) when the system is submitted to small applied fields $F_\bullet$ conjugate to $A_\bullet$. The perturbed Hamiltonian reads as $H = H_N + \sum A_\bullet F_\bullet$. We prove the following:

$$\chi_{mn} = \beta \langle (A_m - \langle A_m \rangle)(A_n - \langle A_n \rangle) \rangle \equiv \langle A_m A_n \rangle - \langle A_m \rangle \langle A_n \rangle, \qquad (15B.10)$$

where $\chi_{mn} = -\partial \overline{A}_m / \partial F_n \big|_{F_\bullet = 0}$ is a susceptiblity characterizing the effect of the field, $\overline{A}_\bullet$ is a thermal average under the perturbed Hamiltonian, and $\langle A_\bullet \rangle$ is a thermal average under the unperturbed Hamiltonian. The susceptibility does not depend on the perturbing field.

By definition, we have

$$\overline{A}_\bullet = \frac{\mathrm{Tr}\,(A_\bullet \exp(-\beta H))}{\mathrm{Tr}(\exp(-\beta H))};$$

hence,

$$\chi_{mn} = -\partial \overline{A}_m / \partial F_n \Big|_{F_\bullet = 0} = \beta \left\{ \frac{\mathrm{Tr}\, A_m A_n \exp(-\beta H_N)}{\mathrm{Tr}\,\exp(-\beta H_N)} - \langle A_m \rangle \langle A_n \rangle \right\}. \qquad (15B.11)$$

As an application of this process, let the total number of particles N be considered as a state variable, $A = N$. We have $AF = -N\mu$. One gets

$$\chi_{NN} = \partial \langle N \rangle / \partial \mu = \beta \left(\langle N^2 \rangle - \langle N \rangle^2 \right), \qquad (15B.12)$$

from which expression one recovers (15B.9) and (15B.6).

4. *Density-density correlation functions.* The state variable pair correlation can be replaced by a density pair correlation. By definition, the density $a_\bullet$ attached to the state variable $A_\bullet$ is such that $A_\bullet = \int a_\bullet d\mathbf{r}$. We also have $\langle a_\bullet \rangle = A_\bullet / V$. Consequently, one can write (15B.10) as follows:

$$\chi_{mn} = \beta \int \langle (a_m(\mathbf{r}) - \langle a_m \rangle)(a_n(\mathbf{r}') - \langle a_n \rangle) \rangle\, d\mathbf{r}\, d\mathbf{r}'$$

$$= \beta V \int \langle \delta a_m(\mathbf{r})\, \delta a_n(0) \rangle\, d\mathbf{r}, \qquad (15B.13)$$

[41]H.E. Stanley, Introduction to Phase Transitions and Critical Phenomena, Oxford University Press, Oxford, 1971.

[42]L.E. Reichl, A Modern Course in Statistical Physics, Arnold, London, 1991.

where $\delta a_\bullet(\mathbf{r}) = a_\bullet(\mathbf{r}) - \langle a_\bullet \rangle$. The above simplification of the double integral into a simple integral results from the fact that in a homogneneous system, the pair correlations between two state variables valued at $\mathbf{r}$ and $\mathbf{r}'$ depend only on $\mathbf{r} - \mathbf{r}'$. We, therefore, introduce the correlation function

$$S_{ab}(\mathbf{r}) = \langle \delta a_a(\mathbf{r}) \, \delta a_b(0) \rangle \equiv \langle a_a(\mathbf{r}) a_b(0) \rangle - \langle a_a \rangle \langle a_b \rangle \tag{15B.14}$$

and its Fourier transform

$$S_{ab}(\mathbf{q}) = \int \exp(-i\mathbf{q} \cdot \mathbf{r}) S_{ab}(\mathbf{r}) \, d\mathbf{r}. \tag{15B.15}$$

We can, therefore, write the susceptibility as

$$\lim \frac{\chi_{ab}}{V} = \beta S_{ab}(\mathbf{q} = 0). \tag{15B.16}$$

5. Let us now specialize to *density fluctuations*. The quantity

$$c(\mathbf{r}) = \sum_{m=1}^{N} \delta(\mathbf{r} - \mathbf{r}_m)$$

is the number density at some location $\mathbf{r}$ of a set of N identical particles (or bonds, when applying the present development to a polymer chain) in positions $\mathbf{r}_m$ occupying a volume V. The thermal average of the density is $c = \langle \sum_{m=1}^{N} \delta(\mathbf{r} - \mathbf{r}_m) \rangle = N/V$. We are interested (15B.14) in the correlation function

$$S(\mathbf{r}) = \langle \delta c(0) \, \delta c(\mathbf{r}) \rangle \equiv \langle c(0)c(\mathbf{r}) \rangle - c^2, \tag{15B.17}$$

where $\delta c(r) = c(r) - c$. We define the Fourier transform as

$$S(\mathbf{q}) = a^3 \int \exp(-i\mathbf{q} \cdot \mathbf{r}) S(\mathbf{r}) \, d\mathbf{r}, \tag{15B.18}$$

where the coefficient a^3 (the volume of a particle) has been introduced to make $S(\mathbf{q})$ dimensionless. Equation (15B.16), applied to density fluctuations, reads [see (15.55)] as

$$k_B T c a^3 \chi_T = c^{-1} S(\mathbf{q} = 0). \tag{15B.19}$$

Let us introduce the Fourier transform of the density:

$$c(\mathbf{q}) = \int c(\mathbf{r}) \exp(-i\mathbf{q} \cdot \mathbf{r}) d\mathbf{r} = \sum_m \exp(-i\mathbf{q} \cdot \mathbf{r}_m).$$

The scattered intensity in reciprocal space (the structure factor of the sample) can be written as a correlation function of this quantity:

$$I(\mathbf{q}) = \int \exp[-i\mathbf{q} \cdot (\mathbf{r}_\alpha - \mathbf{r}_\beta)]\langle c(\mathbf{r}_\alpha)c(\mathbf{r}_\beta)\rangle \, d\mathbf{r}_\alpha \, d\mathbf{r}_\beta$$

$$= \langle c(\mathbf{q})c(-\mathbf{q})\rangle = V(a^{-3}S(\mathbf{q}) + (2\pi)^3 c^2 \, \delta(\mathbf{q})). \qquad (15B.20)$$

We can safely ignore the delta function term, which corresponds to scattering in the forward direction and is not experimentally relevant.[43] Therefore, scattering experiments reveal the effects of the long-range fluctuations (at $\mathbf{q} = 0$) and in particular the compressibility.

The pair correlation function introduced in Section 15.1.3

$$g(\mathbf{r}) = \frac{1}{N} \sum_{m \neq n} \sum \langle \delta(\mathbf{r} + \mathbf{r}_n - \mathbf{r}_m)\rangle \qquad (15B.21)$$

is related to $S(\mathbf{r})$. We start from the following relationship (where $\mathbf{r} = \mathbf{r}_\alpha - \mathbf{r}_\beta$):

$$\sum_m \sum_n \langle \delta(\mathbf{r}_\alpha - \mathbf{r}_m) \, \delta(\mathbf{r}_\beta - \mathbf{r}_n)\rangle = \sum_{m \neq n} \sum \langle \delta(\mathbf{r}_\beta + \mathbf{r} - \mathbf{r}_m) \, \delta(\mathbf{r}_\beta - \mathbf{r}_n)\rangle$$

$$+ \sum_m \langle \delta(\mathbf{r}_\alpha - \mathbf{r}_m) \, \delta(\mathbf{r}_\beta - \mathbf{r}_m)\rangle$$

and transform the right-hand part. The first term in the right-hand part can be written as

$$\frac{1}{V} \int d\mathbf{r}_\beta \sum_{m \neq n} \sum \langle \delta(\mathbf{r}_\beta + \mathbf{r} - \mathbf{r}_m) \, \delta(\mathbf{r}_\beta - \mathbf{r}_n)\rangle = \frac{1}{V} \sum_{m \neq n} \sum \langle \delta(\mathbf{r}_n + \mathbf{r} - \mathbf{r}_m)\rangle,$$

and the second one as c^2 (see Section 15.1.4 for a demonstration of the last equality). Eventually, one gets the useful expressions:

$$cg(\mathbf{r}) = S(\mathbf{r}) \quad \text{and} \quad \phi g(\mathbf{q}) = S(\mathbf{q}), \qquad (15B.22)$$

where $\phi \equiv ca^3$ is the volume fraction.

Problem 15.1. Show that the probability $P_N(\mathbf{r})$ of an ideal chain of N bonds, fixed at the origin and whose extremity is at $\mathbf{r}$, obeys the differential equation

---

[43] J.-P. Hansen and I.R. McDonald, Theory of Simple Liquids, 2nd edition, Academic Press, New York, 1991, p. 99.

$$\frac{\partial P_N}{\partial N}(\mathbf{r}) = \frac{a^2}{6}\nabla^2 P_N(\mathbf{r}),$$

assuming N large. Use a lattice model of the polymer chain.

Answers: Assume that the polymer bonds $\mathbf{a}_i$ ($i = 1, 2, \ldots, z$) are along the edges of a regular lattice with coordination z. Obviously,

$$P(\mathbf{r}, N) = \frac{1}{z}\sum_i^z P(\mathbf{r} - \mathbf{a}_i, N - 1),$$

whose expansion with respect to the $\mathbf{a}_i$'s, which are small compared with $\mathbf{r}$, can be written as

$$P(\mathbf{r} - \mathbf{b}_i, N - 1) = P(\mathbf{r}, N) - \frac{\partial P}{\partial N} - \frac{\partial P}{\partial r_\alpha}a_{i\alpha} + \frac{1}{2}\frac{\partial^2 P}{\partial r_\alpha \partial r_\beta}a_{i\alpha}a_{i\beta}.$$

Also, because $\sum_i a_{i\alpha} = 0$, and $\frac{1}{z}\sum_i^z a_{i\alpha}a_{i\beta} = a^2\frac{\delta_{\alpha\beta}}{3}$, z disappears from the sum, and one gets the desired result (adapted from Doi, Introduction to Polymer Physics).

Problem 15.2.

(a) Show that in a lattice model with coordination number z, the number of configurations of an ideal chain of N segments with one extremity fixed at the origin and the other between $|\mathbf{r}|$ and $|\mathbf{r} + d\mathbf{r}|$ is $W_{id}\,dr$, with $W_{id} = z^N 4\pi r^2 P_N(r)$.

(b) If the probability of nonoverlapping of two segments of volume $v = a^3$ is $1 - a^3/R^3$ for a chain of radius of gyration R, show that the probability of nonoverlapping two by two for the N segments of a (nonideal) chain (N large) is $p(R) = \exp(-\frac{N^2 a^3}{2R^3})$.

(c) Deduce that the maximum value of the number of configurations of an nonideal chain of N segments is obtained for a value of $r = R \propto N^\nu$ ($\nu \approx 3/5$) (adapted from Doi, Introduction to Polymer Physics).

Answers: (b) $p(R) = (1 - a^3/R^3)^{N(N-1)/2} \approx \exp(-N^2 a^3/2R^3)$; (c) the quantity that has to be maximized is $W = p(r)W_{id}$. One gets $(R/R_{id})^5 - (R/R_{id})^3 = \frac{9\sqrt{6}}{16}N^{1/2}$, where $R_{id} = (2Na^2/3)^{1/2}$.

Problem 15.3. Calculate the three exponents of the Flory radius $R_F = v^\mu a^\rho N^\nu$ within the frame of the Flory theory of the scaling law exponent. In the same spirit, calculate the overlap concentration c^* and the correlation length ξ in the semidilute regime.

Answers: Start from the expressions (15.30) and (15.31) of the energy of a coil. The minimization with respect to R yields

$$\mu = \frac{1}{d+2}, \quad \rho = \frac{2}{d+2}, \quad \nu = \frac{3}{d+2}.$$

Also, $c^* = N^{1-\nu d}a^{2d/(d+2)}v^{-d/(d+2)}$, i.e., $\phi^* = N^{1-\nu d}(a/v)^{d/(d+2)}$, and so on.

Problem 15.4. Show that the line $\mu^* = (\partial F/\partial\phi)_{\phi=\phi_1} = (\partial F/\partial\phi)_{\phi=\phi_2}$ (the so-called Maxwell plateau) determines two equal areas with the curve $\mu = \mu(\phi)$ (see Fig. 15.5c).

Answers: The total signed area between the curve $\mu = \mu(\phi)$ and the line $\mu = \mu^*$ is given by the integral $\int_{\phi_1}^{\phi_2} [\mu(\phi) - \mu^*]d\phi = F(\phi_2) - F(\phi_1) - \mu^*(\phi_2 - \phi_1)$. This quantity vanishes because it is an equation satisfied by the points belonging to the bitangent to the curve $F = F(\phi)$.

Problem 15.5. Consider (see Fig. 15.10) a binary system with volume fraction ϕ_0 for one of the components, positioned at temperature T in the region of metastability. The first nucleus ϕ_1 to appear has the same chemical potential μ_0 as ϕ_0. Define the thickness and energy of the region of transition between the matrix and the nucleus, assuming that the wall is planar.

Answers: The free energy in the inhomogeneous region between the matrix and the bulk must contain a gradient term. We write the new energy to be minimized as $E(\phi, \nabla\phi) = F(\phi) - \mu_0\phi + \frac{1}{2}K(\nabla\phi)^2$, whose variational derivative can be written as $K\nabla^2\phi = \frac{d}{d\phi}[F(\phi) - \mu_0\phi]$.
   In the planar case, corresponding physically to nuclei whose radius of curvature is large compared with the thickness of the wall, there is a first integral

$$\frac{1}{2}K\left(\frac{d\phi}{dx}\right)^2 = F(\phi) - \mu_0\phi + \text{const} = h(\phi). \tag{15P.1}$$

The derivative must vanish for $x = \pm\infty$. Hence, in addition to the condition $\mu_0 = \mu(\phi_0) = \mu(\phi_1)$, we also have

$$F(\phi_0) - \mu_0\phi_0 = F(\phi_1) - \mu_0\phi_1, \tag{15P.2}$$

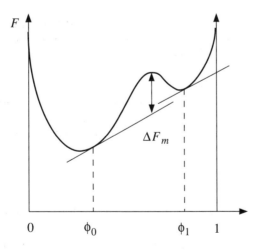

Figure 15.10. See Problem 15.5.

which means that if the wall is planar, the system is on the curve of coexistence, and $\mu_0 = \mu(T)$; (15P.2) is the condition for the Maxwell plateau. Integration of (15P.1) yields $x = \int d\phi \sqrt{K/2h(\phi)}$. The energy per unit area is $\gamma = \int_{-\infty}^{+\infty} dx\, E(\phi, \nabla\phi) = \int_{-\infty}^{+\infty} dx\, K(\nabla\phi)^2 = 2\int_{\phi_1}^{\phi_2} d\phi \sqrt{2Kh(\phi)}$.

These two quantities can be estimated by taking for $h(\phi)$ its value at the maximum of the free energy between the compositions ϕ_0 and ϕ_1, noted $\Delta h_m = \Delta F_m - \mu_0 \phi_m$,

$$\xi \approx (\phi_1 - \phi_0)\sqrt{K/\Delta h_m}, \quad \gamma \approx \xi\, \Delta h_m. \tag{15P.3}$$

Problem 15.6. Let a gel be prepared by reticulation of a semidilute solution of polymers in a good solvent. N is the mean number of monomers between nodes. No entanglements. One immerses the gel in a good solvent in excess, with a Flory interaction parameter χ. Show that the free energy per site per $k_B T$ reads as

$$f = (1 - \phi)\ln(1 - \phi) + \chi\phi(1 - \phi) + \frac{3}{2}\frac{\phi}{N}\frac{R^2}{R_0^2},$$

ϕ being the volume fraction of the monomers in the gel, $R_0^2 = Na^2$.

At swelling equilibrium, the gel is characterized by close-packed noninterpenetrating coils. Express f in function of the only variable ϕ. Calculate ϕ_{eq} and the size R_{eq} of the coils at swelling equilibrium. Note that $\phi_{eq} < 1$, and introduce the excluded volume parameter $v = a^3(1 - 2\chi)$. Show that the result corresponds to the crossover concentration c^* in a solution in a good solvent.

How is the free energy modified if the solvent is a polymer melt with $DP = P$ monomeres, of the same chemical nature as the coils of the gel?

Let P be small compared with N. Express ϕ_{eq} and R_{eq} in function of N and P. Show that for P of the order of $N^{1/2}$ the coils of the gel are ideal. What about the situation when $P \sim N$?

Problem 15.7. Calculate the repeat distance d of a lamellar phase of diblock copolymers, assuming $N = N_A = N_B$, $v = v_A = v_B$. The surface energy γ at the IMDS is supposed to be known.

Relate the surface energy γ to the Flory–Huggins parameter χ and the thickness δ of the region of transition between the A and the B chains.

Answers: Elastic energy per unit volume $f_{el} \approx k_B T d^2/(N^2 a^2 v)$, using (15.31); surface energy per unit volume: $f_{surf} \approx 2\gamma/d$. The sum is minimized for $d \approx [\gamma N^2 a^2 v/(k_B T)]^{1/3}$. The surface energy is proportional to the number of contacts between monomers of different species in the region of transition; i.e., $\gamma \approx \chi k_B T\, \delta/v$. Order of magnitude: Take δ of the order of a few monomer lengths.

Further Reading

J. des Cloizeaux and G. Janninck, Polymers in Solution, Their Modelling and Structure, Oxford University Press, 1990.

M. Doi, Introduction to Polymer Physics, Oxford Science Publications, 1995.

M. Doi and S. F. Edwards, The Theory of Polymer Dynamics, Oxford Science Publications, 1988.

P. J. Flory, Principles of Polymer Chemistry, Cornell University Press, Ithaca, New York, 1953.

P.-G. de Gennes, Scaling Concepts in Polymer Physics, Cornell University Press, Ithaca, New York, 1979.

N. G. van Kampen, Stochastic Processes in Physics and Chemistry, North-Holland, Amsterdam, 1981.

A. Keller, Chain-Folded Crystallization of Polymers from Discovery to Present Day: A Personalized Journey, in Sir Charles Frank, an Eightieth Birthday Tribute, Edited by R. G. Chambers et al., Adam Hilger, 1991.

R. G. Larson, The Structure and Rheology of Complex Fluids, Oxford University Press, New York and Oxford, 1999.

G. Strobl, The Physics of Polymers, Concepts for Understanding Their Structures and Behavior, Springer, New York, 1996.

Table of Constants
Conversion SI vs. cgs units

Avogadro
Number:
$$\mathcal{N}_A = 6.022 \times 10^{23}/\text{mole}$$

Boltzmann
Constant:
$$k_B = 1.381 \times 10^{-23}\,\text{J/K} = 1.381 \times 10^{-16}\,\text{erg/deg}$$

Modulus of the
Electron Charge:
$$e = 1.602 \times 10^{-19}\,\text{C} = 4.803 \times 10^{-10}\,\text{esu}$$

Planck Constant:
$$h = 6.626 \times 10^{-34}\,\text{J} \times \text{s} = 6.626 \times 10^{-27}\,\text{erg} \times \text{s}$$

Permittivity in Vacuum:
$$\varepsilon_0 = 8.854 \times 10^{-12}\,\text{C}^2/\text{J} \times \text{m} = 1\,\text{cgs}$$

$1\text{eV} = 1.602 \times 10^{-12}\text{erg} = 1.602 \times 10^{-19}\,\text{J}$

$1k_B T = 4.12 \times 10^{-14}\text{erg} = 4.12 \times 10^{-21}\text{J} \approx \frac{1}{40}\text{eV at} T = 298\,\text{K}$

$1\text{cal} = 4.184\,\text{J}$

$1\text{eV/molecule} = 23.057\text{kcal/mole} = 96.472\text{kJ/mole}$

$1\,\text{J} = 10^7\text{erg} = 0.239\text{cal} = 6.242 \times 10^{18}\,\text{eV}$

electric charge:
$1\text{esu} = 3.336 \times 10^{-10}\,\text{C}$

$1\text{dyne} = 10^{-5}\,\text{N}$

$1\text{dyne/cm} = 1\text{erg/cm}^2 = 1\text{mN/m} = 1\text{mJ/m}^2$

Electric Polarization: $D = \text{debye}$
$1\,D = 10^{-18}\text{esu} \times \text{cm} = 3{,}336 \times 10^{-20}\text{C} \times \text{cm}$

Pressure, Stress: $\text{Pa} = \text{pascal}$
$1\,\text{Pa} = 1\text{N/m}^2 = 10\text{dynes/cm}^2$

Dynamical Viscosity: $\text{Pl} = \text{poiseuille} = \text{Pa} \times \text{s}; \quad \text{P} = \text{poise} = \text{dyne} \times \text{s/cm}^2$
$1\,\text{Pa} \times \text{s} = 10\text{P}$

Note on relation symbols.
We use the following relation symbols:

$\sim$ "the same order of magnitude," numerical or algebraic dimensioned
$\cong$ dimensioned scaling laws
$\propto$ "proportional to," non-dimensioned scaling laws
$\approx$ "approximately equal" in numerical or algebraic evaluations; group
      isomorphism
$\equiv$ "identical"

Name Index

Subject Index

617